Handbook of Carbon Se

In an era dominated by environmental challenges and technological advancements, the need for precise and efficient monitoring tools has become paramount. Among these tools, carbon sensors stand as vanguards, revolutionizing our approach to sensing and detection. *Handbook of Carbon Sensors: Understanding and Applications* provides readers with a comprehensive and accessible guide to the world of carbon sensing. The book begins by exploring the basics of carbon sensing, detailing the underlying principles and their foundations before detailing their applications on real-world challenges, including monitoring air quality in urban environments, fine-tuning manufacturing processes in industries, or revolutionizing medical diagnostics. As the book develops, it moves from theoretical foundations to the impact of carbon sensors on our daily lives: from optimizing production lines to ensuring the purity of the air we breathe. It is a valuable reference for graduate students and researchers in environmental science, materials science, and engineering, in addition to scientists working in industry.

Key features

- Provides practical insights by incorporating real-world case studies that demonstrate how carbon sensors are actively solving challenges in industries, environmental monitoring, and healthcare.
- Caters to a wide range of readers, including students, researchers, and professionals.
- Offers a forward-looking perspective on carbon sensing technology, with a dedicated section exploring emerging technologies and future trends.

JG Manjunatha is Assistant Professor at the Department of Chemistry, Field Marshal KM Cariappa College, Madikeri, Mangalore University.

Series in Sensors

Series Editors: Barry Jones and Haiying Huang

Other recent books in the series:

Semiconductor X-Ray Detectors
B. G. Lowe, R. A. Sareen

Resistive, Capacitive, Inductive, and Magnetic Sensor Technologies
Winncy Y. Du

Sensors for Safety and Process Control in Hydrogen Technologies
Thomas Hübert, Lois Boon-Brett, William Buttner

Biosensors and Molecular Technologies for Cancer Diagnostics
Keith E. Herold, Avraham Rasooly

Compound Semiconductor Radiation Detectors
Alan Owens

A Hands-On Course in Sensors Using the Arduino and Raspberry Pi
Volker Ziemann

Radiation Sensors with 3D Electrodes
Cinzia Da Vià, Gian-Franco Dalla Betta, Sherwood Parker

Semiconductor Radiation Detectors
Alan Owens

Advanced Chromatic Monitoring
Gordon R. Jones, Joseph W. Spencer

Nanosensors: Physical, Chemical, and Biological, 2nd Edition
Vinod Kumar Khana

Multi-Component Torque Sensing Systems
Qiaokang Liang

Ultra Fast Silicon Detectors: Design, Tests, and Performances
Marco Ferrero, Roberta Arcidiacono, Marco Mandurrino, Valentina Sola, Nicolò Cartiglia

A Hands-On Course in Sensors Using the Arduino and Raspberry Pi, Second Edition
Volker Ziemann

Multifunctional Sensors: Design, Construction, Methodology and Uses
Bansari Deb Majumder and Joyanta Kumar Roy

IoT Sensors: An Exploration of Sensors for Internet of Things
Vinod Kumar Khanna

Handbook of Carbon Sensors: Understanding and Applications
JG Manjunatha

Handbook of Carbon Sensors

Understanding and Applications

Edited by
JG Manjunatha

CRC Press
Taylor & Francis Group
Boca Raton London New York

CRC Press is an imprint of the
Taylor & Francis Group, an **informa** business

Designed cover image: © Shutterstock

First edition published 2025
by CRC Press
2385 NW Executive Center Drive, Suite 320, Boca Raton FL 33431

and by CRC Press
4 Park Square, Milton Park, Abingdon, Oxon, OX14 4RN

CRC Press is an imprint of Taylor & Francis Group, LLC

ISBN: 9781032812083
ISBN: 9781032815473
ISBN: 9781003500322

DOI: 10.1201/9781003500322

Typeset in Times
by Newgen Publishing UK

Contents

Preface

Carbon, the cornerstone of life and the foundation of modern materials science, has captivated scientists and engineers for centuries. Its versatility, abundance, and unique properties have made it an indispensable component in countless applications, from everyday objects to cutting-edge technologies. This book delves into the fascinating world of carbon sensors, exploring their principles, design, and applications across various fields. Carbon sensors harness the remarkable properties of carbon materials, such as graphene, carbon nanotubes, and activated carbon, to detect and measure a wide range of analytes with unprecedented sensitivity and selectivity. These sensors offer a multitude of advantages, including their biocompatibility, flexibility, and potential for miniaturization. They have revolutionized industries such as healthcare, environmental monitoring, and industrial process control, enabling real-time analysis and early detection of critical parameters. In this book, we embark on a journey through the evolution of carbon sensors, from their humble beginnings to their current state-of-the-art capabilities. We explore the underlying principles of carbon-based sensing mechanisms, including chemisorption, physisorption, and electrochemical reactions. We delve into the design and fabrication techniques used to create these sensors, highlighting the importance of materials selection, surface modification, and device architecture. Furthermore, we showcase the diverse applications of carbon sensors, ranging from the detection of gases and volatile organic compounds to the monitoring of biological markers and environmental pollutants. We discuss the challenges and opportunities associated with the commercialization of carbon sensors, emphasizing the need for standardization, cost-effective manufacturing, and long-term reliability.

This book is intended for a broad range of audiences, including researchers, engineers, students, and professionals working in fields such as materials science, chemistry, physics, and environmental science. It provides a comprehensive overview of carbon sensors, offering valuable insights into their principles, design, and applications. By understanding the potential of carbon sensors, we can unlock new possibilities for innovation and address pressing global challenges.

Editor
Dr JG Manjunatha

Contributors

Vinayak Adimule
Angadi Institute of Technology and Management
Belagavi, India

Melike Akan
Ankara Hacı Bayram Veli University
Ankara, Türkiye

Nazife Aslan
Ankara Hacı Bayram Veli University
Ankara, Türkiye

Gangadhar Bagihalli
Visvesvaraya Technological University
Karnataka, India

Mohamed Bendany
Mohammed V University in Rabat
Rabat, Morocco

Hafssa Bensemlali
University Chouaib Doukkali
El Jadida,Morocco

Meryem Bensemlali
University Chouai Doukkali
El Jadida, Morocco

Khalid Ait Ben Brahim
Mohammed V University in Rabat
Rabat, Morocco

Murat Çelik
Ankara University
Ankara, Turkey

Seda Nur Topkaya Cetin
Izmir Katip Celebi University
İzmir, Türkiye

Meltem Ceylan-Ünlüsoy
Ankara University
Ankara, Türkiye

Ersin Demir
Afyonkarahisar Saglik Bilimleri University
Turkey

Çiğdem Kanbeş Dindar
Ankara University
Ankara, Turkey

Cem Erkmen
Istanbul Aydin University
Istanbul, Türkiye

Pattan Siddappa Ganesh
Korea University of Technology and Education
Cheonan-si, Chungcheongnam-do, Republic of Korea

Bojappa Ginni
Visvesvaraya Technological University
Karnataka, India

Sümeyye Güney-Kalkan
Ankara University
Ankara, Türkiye

Souad El Hajjaji
Mohammed V University in Rabat
Rabat, Morocco

Youssra El Hamdouni
Mohammed V University in Rabat
Rabat, Morocco

Cigdem Kanbes-Dindar
Department of Analytical Chemistry
Ankara University
Ankara, Turkey

Ipek Kucuk
Başkent University
Etimesgut, Ankara, Türkiye

Sevinc Kurbanoglu
Ankara University
Ankara, Turkey

Najoua Labjar
Mohammed V University in Rabat
Rabat, Morocco

Dipak Maity
Indiana University Indianapolis
IN, USA

Halima Mortadi
University Chouaib Doukkali
El Jadida, Morocco

H.P. Nagaswarupa
Davangere University
Davangere, India

Ramachandra Naik
New Horizon College of Engineering
Bangalore, India

Hamid Nasrellah
Chouaib Doukkali University
El Jadida, Morocco

Meltem Okan
METU MEMS Center
Ankara, Türkiye

Nida Aydogdu Ozdogani
Afyonkarahisar Health Sciences University
Afyonkarahisar, Türkiye

Nutan Rani
University of Delhi
New Delhi, India

Swati Rani
University of Delhi
New Delhi, India

C. R. Ravikumar
East West Institute of Technology
Bangalore, India

Selenay Sadak
Ankara University
Ankara, Türkiye

Kalawati Saini
University of Delhi
New Delhi, India

Seyda Nur Samanci
Ankara University
Ankara, Türkiye

Vildan Sanko
METU MEMS Center
Ankara, Türkiye

Hülya Silah
Bilecik Seyh Edebali University
Istanbul, Turkey

Nazlı Şimşek
Ankara University
Ankara, Turkey

D.M. Tejashwini
Davangere University
Davangere, India

Gözde Aydoğdu Tığ
Ankara University
Ankara, Turkey

Kübra Turan
Ankara University
Ankara, Turkey

Esra Ülker
Ankara University
Ankara, Turkey

Bengi Uslu
Ankara University
Ankara, Turkey

Sapna Yadav
University of Delhi
New Delhi, India

1 An Introduction to Carbon Sensors

Gangadhar Bagihalli, Vinayak Adimule, and Bojappa Ginni

1.1 INTRODUCTION

Carbon sensors play a pivotal role in several fields, ranging from environmental monitoring to industrial processes, health care, and beyond. Their ability to detect and quantify carbon-based compounds makes them an indispensable tool for understanding and mitigating the impact of human activities on the environment and for ensuring the safety and efficiency of numerous applications. In this chapter, we will delve into the fundamental principles, types, applications, and advancements and also have higher sensitivity, and reliability, quick response, low cost, reduced size, and some of the properties [1, 2]. Carbon-based materials have superior performance and are regarded as environmentally beneficial materials, making them a viable substitute for the presently costly electronic compound [3, 4] and also for the development of some carbon-based nanoparticles (Figure 1.1) such as graphene oxide materials and fullerene. Graphene oxide sensors and physical sensors are developed in the last few decades. Graphene sensors are widely used because of their specific properties such as high surface area, high electron mobility, low electrical noise, and quick response [5]. Carbon-based nanosensors hold immense significance due to their ability to detect analytes especially with sensitivity and selectivity at the nanoscale [6, 7]. These sensors offer advantages like high surface area, tunable surface chemistry, and excellent biocompatibility, making them well suited for a diverse range of applications in biomedical diagnostics, environmental monitoring, food safety, and industrial process control. By providing real-time, accurate, and reliable measurements, carbon-based nanosensors contribute to advancements in health care, environmental protection, and industrial optimization [8]. The properties of carbon-based nanomaterials, including high electric conductivity, mechanical strength, chemical stability, and biocompatibility, make them suitable candidates for sensor development. Carbon nanotubes (CNTs) exhibit extraordinary electrical conductivity and mechanical strength, enabling sensitive detection of analytes through various sensing mechanisms. Graphene, with its two-dimensional structure and high surface area, offers unparalleled electron mobility and sensitivity, making it an ideal candidate for ultrasensitive detection and biosensing applications. Carbon dots, then again, are fluorescent carbon nanoparticles with tunable optical properties, enabling sensitive and selective optical sensing and imaging [9, 10]. Fabrication is essential for carbon-based sensors because it enhances properties like electrical conductivity and thermal properties and also enhances the signal conversion [11, 12]. Figure 1.1 presents some of the carbon-based structures that are used in sensor applications, and it was opted and modified from Vinícius *et al.* [13] who discussed the performances of carbon-based sensors and their chemical and physical aspects.

DOI: 10.1201/9781003500322-1

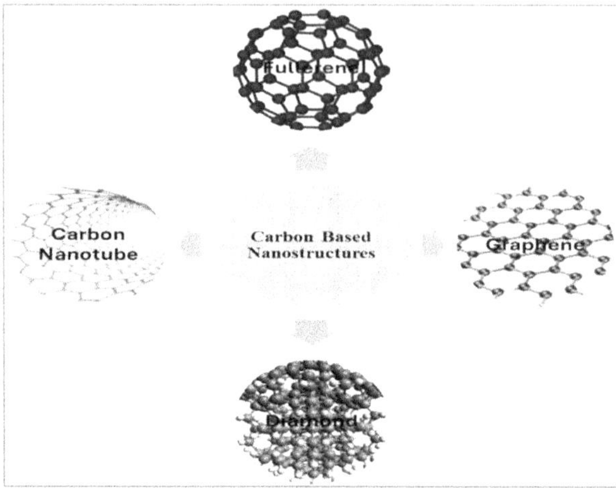

FIGURE 1.1 Illustration of the various cabon-based nanoparticles [13].

This chapter covers the major aspects of carbon-based sensors such as physical and chemical sensing performances, and their advantages, limitations, and applications in the determination of various analytes.

1.2 PROPERTIES OF SENSORS

Sensor is a device that is capable of detecting analytes in very low concentration. It must possess desirable properties for the effective recognition of the analytes, be environmentally friendly, have low recovery time and low cost, and must be easy to operate. The improvement in overall performances of sensors can be greatly achieved by using highly reactive carbonaceous materials. They operate in different mechanisms with respect to their nature of the functional groups. Sensors containing nitrogen vacancy defect (NVD) are promising candidates for the detection of biological species in low concentration. They contain delicate quantum structures and unaltered from external magnetic and electric fields. The NVD containing nanodiamond (ND) structures have different size and shape capable to detect various analytes, and they have different physical, chemical, and biological properties. Similar effects are seen in the case of fluorescent ND structures. They are studied for fluorescent applications. They have improved surface to volume ratio and high biocompatibility. Fullerene C60 can detect various biological species, such as DNA, glucose, adenosine triphosphate, etc. They are also being used as color-sensing materials. They have extensive π conjugation, ability to react with different molecules, high mechanical strength, etc.

On the other hand, CNTs possess unique features, with their surface area having electronic interactions of the functional groups, and are considered to be excellent materials for chemical and biological sensors. They have ultrathin surface, high charge mobility, and electrical properties, leading to their applications in semiconducting industries. Graphene also possesses high surface area, heterogeneous surface with high electron transfer rate, and mechanical and thermal stability, finding its applications in the field of quantifying gases aLso possess as high sensitity, selectivity, good recovery time etc. [14].

1.3 SENSOR PHYSICAL PROPERTIES

Physical properties of the sensors include humidity, strain, and pressure. Various sensors were developed for the establishment of sensors physical properties.

Humidity sensors are quite often used in semiconductor industries, pharmaceutical industries, food industries, etc. As reported by Leng *et al.* [15], graphene dispersed over graphene oxide (GO) on Nafion polymer acted as humidity sensors. The sensors showed 10% efficiency and long-term stability even after two months of their reuse. Li *et al.* [16] synthesized GO decorated with multi-walled carbon nano tubes (MWCNTs) for efficient humidity sensors. The electrodes were fabricated on Si wafer on the surface of the SiO_2. The sensitivity of the sensors was evaluated by using saturated salt solutions. Yao *et al.* [17] synthesized humidity sensors composed of quartz crystal microbalance decorated with MWCNTs and NDs. The sensor showed high efficiency and analysis was carried out by impedance analysis method. Recent studies carried out by Ding *et al.* [18] found that fullerene C60 dispersed with GO coated onto QSM acted as humidity sensors. The sensor showed reduced aggregation and formation of hydrophobic layers and was obtained through drop casting method. Zhao *et al.* [19] developed CNT decorated with cement sandwich composites for the effective determination of humidity.

Strain sensors commonly find their application in the measurement of piezoresistive and piezoelectric properties [20]. CNTs display significant strain-related properties. When subjected to external force, CNTs/composites display variations in electrical resistance or voltage. CNTs also show stronger electromechanical properties and are widely used for piezoresistive device applications [21].

1.4 SENSOR CHEMICAL PROPERTIES

One of the main applications of carbon-based sensors is to detect and quantify toxic gases such as NH_3 (ammonia), CH_4 (methane), carbon monoxide (CO), carbon dioxide (CO_2), hydrogen sulfide (H_2S), and nitrogen oxide (NO_x). Seekaew *et al.* [22] synthesized graphene decorated into Ni electrode system for the effective detection of NO_2 gas. Keshtkar *et al.* [23] synthesized SnO_2 decorated fullerene C60 quantum dots using hydrothermal method for the effective determination of H_2S gas at low temperatures. Kodu *et al.* [24] synthesized V_2O_5 decorated graphene by laser ablation method for the effective determination of NH_3 gas. Chen *et al.* [25] developed Li^+ decorated CNTs for the effective quantification of the CH_4 gas. Jaidev *et al.* [26] developed CNT-coated graphene structures decorated with Pt nanoparticles possessing high surface area for the effective detection of H_2 gas. Figure 1.2 presents a typical illustration of carbon-based sensors for the detection of various analytes.

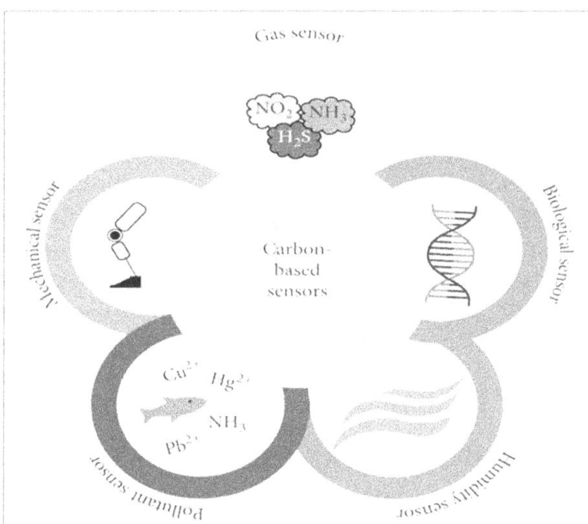

FIGURE 1.2 Illustration of the various sensor systems with different analytes. (Open aceess creative common attributions CC By 4.0 Ref [13].)

Various phenolic compounds such as monophenol and biphenol derivatives are quantified by using boron doped ND rods. The ND particles were dispersed over glassy carbon electrodes (GCE) for the development of sensors. Xuan and Park *et al.* [27] synthesized electrochemical (EC) miniatured sensors deposited with rGO on CNT composite for the detection of Au ions.

1.5 SENSING PERFORMANCES OF CARBON-BASED NANOMATERIALS

Carbon-based nanomaterials possess unique properties that make them ideal candidates for sensing applications. CNTs, for example, exhibit high electrical conductivity, mechanical strength, and large surface area, enabling sensitive detection of analytes through various sensing mechanisms. Graphene, a single layer of carbon atoms arranged in a two-dimensional honeycomb lattice, offers exceptional electrical and mechanical properties, making it suitable for ultrasensitive detection and biosensing. Carbon dots, on the other hand, are fluorescent carbon nanoparticles with tunable optical properties, enabling optical sensing and imaging applications. Sensitivity of some of the carbon-based nanoparticles these applications are reported in Refs. [28] and [29].

1.6 FUNDAMENTALS OF CARBON-BASED SENSORS

Fundamentals of carbon-based sensors lie in their utilization of carbon-based materials to interact with target analytes, inducing measurable changes in electrical, optical, or EC properties. CNTs, graphene, and carbon dots, among others, serve as sensing elements due to their exceptional properties, such as high surface area, electrical conductivity, and chemical reactivity. These materials enable sensitive detection across various analytes through diverse sensing mechanisms, including resistive, capacitive, optical, and EC mechanisms. The working principles are rooted in the adsorption or desorption processes occurring at the surface of carbon-based materials when exposed to the target analyte, leading to alterations in conductivity, capacitance, or optical properties, which are then quantified to determine the analyte concentration [30].

Through the exploitation of these fundamental principles, carbon-based sensors offer versatile solutions for a wide array of applications, including environmental monitoring, health care diagnostics, industrial process control, and security screening. Their sensitivity, selectivity, and rapid response make them indispensable tools in fields where precise detection and quantification of analytes are paramount. By continually refining the understanding of these fundamental principles and advancing fabrication techniques, carbon-based sensors are poised to play an increasingly pivotal role in addressing societal challenges and driving innovation in sensing technology.

Carbon-based sensors operate on the principle of interaction between the target analyte and the carbon-based sensing material, leading to measurable changes in electrical, optical, or EC properties. The choice of carbon material depends on factors such as sensitivity, selectivity, stability, and ease of fabrication. CNTs, for example, exhibit high surface area and sensitivity, making them suitable for gas-sensing applications, while graphene offers exceptional conductivity and mechanical strength, which is ideal for EC sensing [31]. This kind of work is reported in Bezzon *et al.*. [30]. Figure 1.3 presents the fundamentals of carbon nanosensors, which is opted for and modified from Rajagopalan *et al.* [32].

1.7 DESIGN CONSIDERATIONS FOR CARBON-BASED NANOSENSORS

The design of carbon-based nanosensors involves careful consideration of factors such as sensor geometry, surface functionalization, and sensing mechanism. The choice of carbon nanomaterial and sensor configuration depends on the specific application requirements, including sensitivity, selectivity, and detection limit. Surface functionalization techniques such as chemical modification

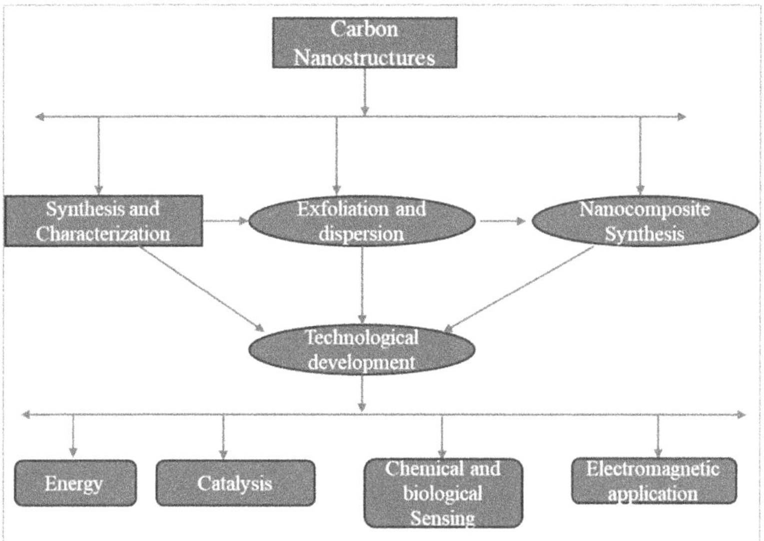

FIGURE 1.3 Pictorial illustration of the fundamentals of carbon nanosensors. (Open access creative common attributions CC by 4.0 Ref [32].)

and biomolecule immobilization can enhance sensor performance by increasing analyte binding affinity and reducing non-specific interactions. A simple and affordable method for determining the quantity of electroactive species in a solution, both quantitatively and qualitatively, is EC analysis. The advantages of electroanalytical techniques over other methods such as chromatography and spectroscopy include their low cost, convenience of use, sensitivity and dependability, short measurement time, and environmental friendliness. Scientists have created sensors and techniques for identifying electroactive analytes utilizing a range of materials, including nanomaterials, conductive polymers, and biomolecules, to enhance the sensitivity and selectivity [33]. These works are reported in Ref. [31]. Figure 1.4 illustrates the design of carbon-based sensors, which is adopted from Liang *et al.* [34] describing the carbon-based SERS biosensors.

1.8 FABRICATION TECHNIQUES FOR CARBON-BASED NANOSENSORS

Fabrication of carbon-based nanosensors involves various techniques, including chemical vapor deposition (CVD), solution-based methods, and nanolithography. CVD allows precise control over the growth of carbon nanomaterials, enabling the production of uniform and well-defined sensor structures. Solution-based methods such as inkjet printing and electrospinning offer scalability and cost-effectiveness for large-scale sensor manufacturing. Nanolithography techniques, including electron beam lithography and nanoimprint lithography, enable the fabrication of nanostructured sensor arrays with high spatial resolution [35–37], and these works are reported in Ref. [38]. Figure 1.5 shows the various steps involved in the fabrication process. Figure 1.6 displays the workflow of the fabrication of the carbon-based sensors.

During the fabrication process, working electrodes (WE) generally made with the circular design having a 12.6 mm^2 area and the counter electrode (CE) along with the connecting pads are made to fit 8 SATA connecting pins as shown in Figure 1.7 (a–c). However, depending on various samples under examination, modifications in the surfaces, their area, and electrolyte were carried out by Manjunatha *et al.* in their recent reports [39–43]. These sensors are commonly used for the EC determination of various analytes under the influence of temperature, interferents, and other reacting molecules [44].

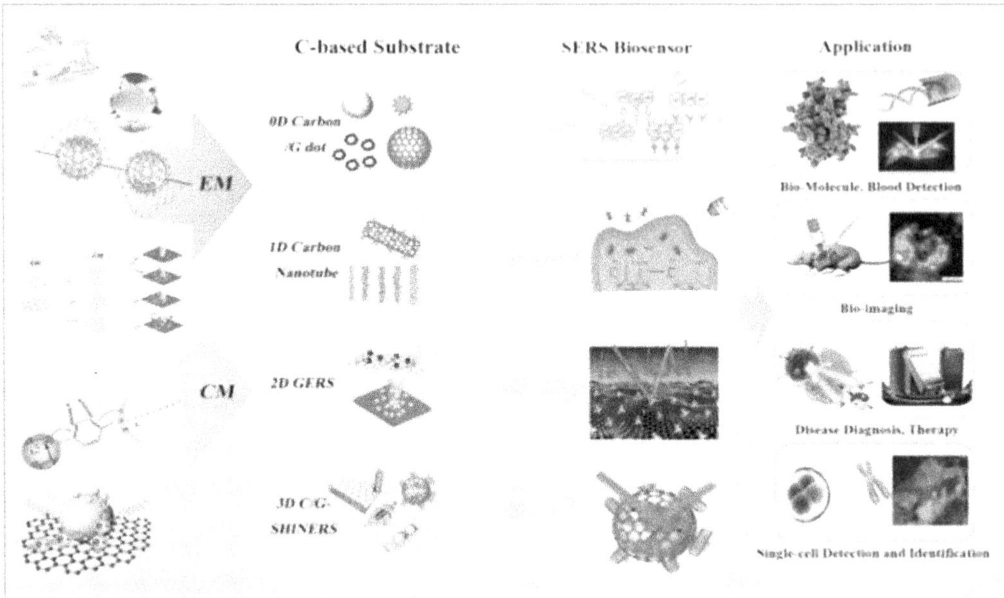

FIGURE 1.4 Illustration or designing of carbon-based sensors. (Open access creative common attributions CC by 4.0 Ref [34]).

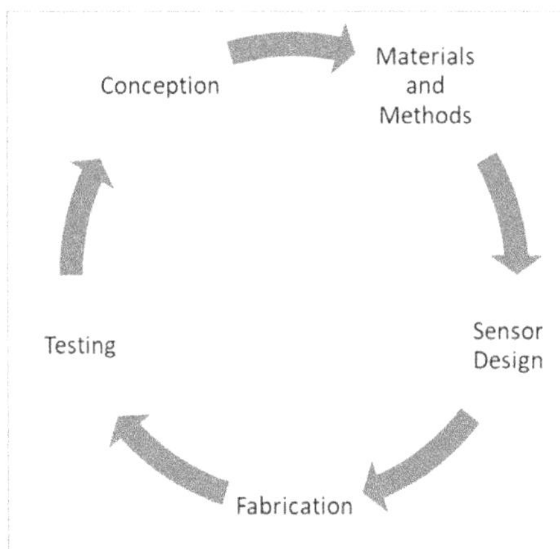

FIGURE 1.5 Illustration of the fabrication of carbon-based sensors (Open access creative common attributions CC by 4.0 Ref [38].)

After the thermal annealing process carried out on the electric pads, electric routes on the WE and CE, Ag paste was covered on the surface for effective establishment of the connections as well as interactions. The entire area was covered by carbon paste before depositing the Ag paste as shown in Figure 1.7 (c). Fabrication of the electrode is a laborious process and must be carried out with precision; otherwise, additional peaks are required to be obtained due to Ag--C interface. To optimize the process and enhance the sensor effectivity toward the analyte, only reference electrode (RE) is covered with the Ag paste and the rest (WE and CE) were covered with carbon paste (CH8). This method provides good choice for the measurement of electrochemical impedance spectroscopy (EIS) and cyclic voltamentery (CV) during EC interactions. A schematic illustration of the

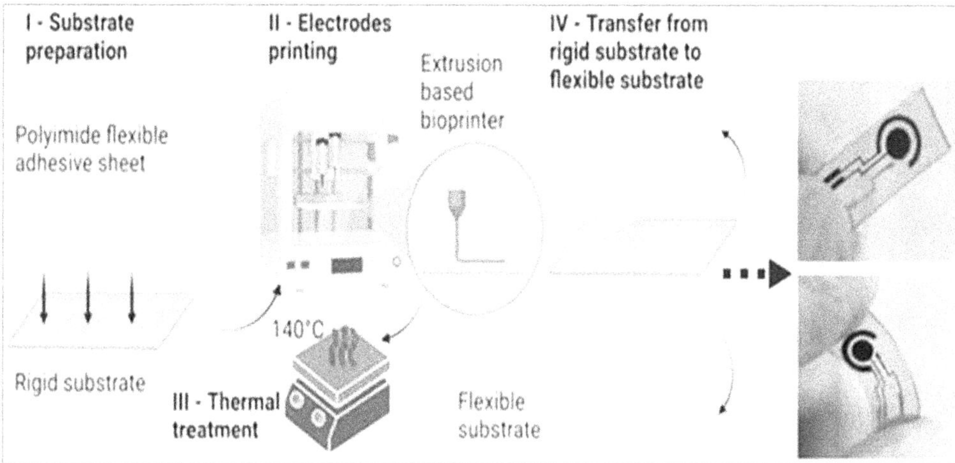

FIGURE 1.6 Illustration of workflow of fabrication of carbon-based sensors. (Open access creative common attributions CC by 4.0 Ref [38].)

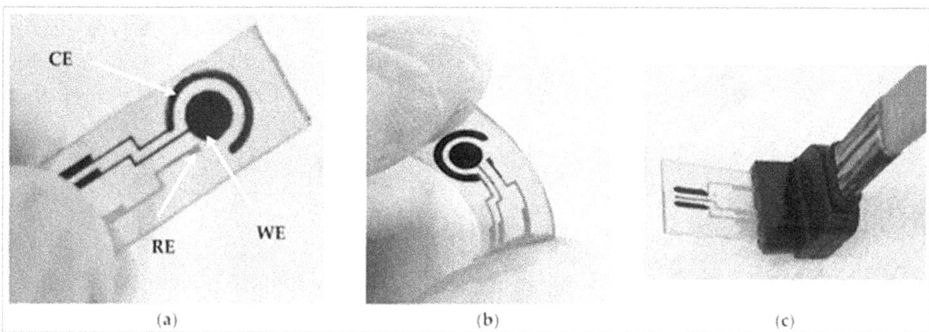

FIGURE 1.7 EC carbon-based sensors on flexible substrate. (a) Working and counter electrodes (b, c) preparation of Ag electric pads covered with WE/CE. (Open access creative common attributions CC by 4.0 Ref [38].)

carbon-based sensor fabrication process is presented in Figure 1.8 wherein the electrdoe surfaces are covered with low quantities of the materials under investigation and are less costly.

1.9 WORKING PRINCIPLES OF CARBON-BASED SENSORS

The working principle of carbon-based sensors revolves around the interaction between the target analyte and the carbon-based sensing material, leading to measurable changes in electrical, optical, or EC properties. In resistive carbon sensors, for example, the electrical resistance of the carbon material changes when it interacts with the target analyte, and this change in resistance can be attributed to various mechanisms, such as charge transfer, adsorption/desorption of analyte molecules, or changes in the electronic structure of the carbon material upon analyte binding. By measuring the magnitude of this resistance change, the concentration or presence of the analyte can be determined.

Similarly, capacitive carbon sensors rely on variations in capacitance resulting from the interaction between the analyte and the carbon-based sensing surface. When analyte molecules adsorb

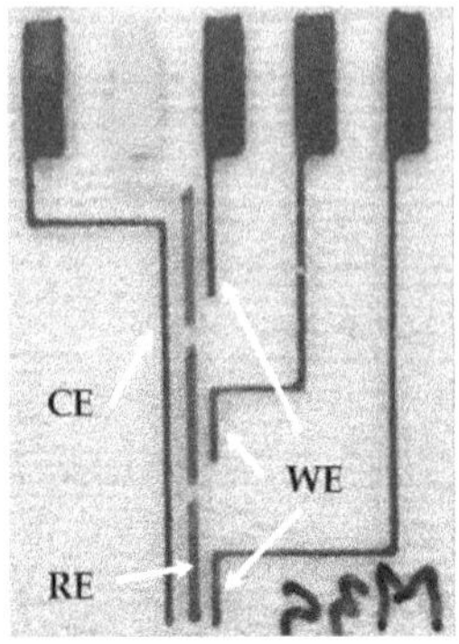

FIGURE 1.8 Electrochemical carbon-based array of electrodes, RE is made from Ag/AgCl. (Open access creative common attributions CC by 4.0 Ref [38].)

onto or desorb from the sensing surface, they alter the dielectric properties of the material, thereby changing the capacitance of the sensor. This change in capacitance is proportional to the concentration of the analyte and can be detected using appropriate readout circuitry. In EC carbon sensors, the interaction between the analyte and the carbon-based electrode leads to changes in the EC properties of the system. This may involve redox reactions occurring at the electrode surface, resulting in changes in current or potential that can be measured and correlated with the analyte concentration. Overall, the working principle of carbon-based sensors is based on the unique properties of carbon materials to facilitate sensitive and selective detection of target analytes through various physical and chemical interactions [45]. These works are reported in Ref. [35].

1.9.1 Significance of Carbon-Based Sensors

Carbon-based sensors hold immense significance due to their remarkable properties, including high surface area, electrical conductivity, mechanical strength, and chemical stability. These properties make carbon materials ideal candidates for sensing applications, enabling sensitive, selective, and reliable detection of various analytes. Moreover, carbon-based sensors offer advantages such as low cost, scalability, and compatibility with diverse substrates, making them attractive for widespread adoption in different industries and research fields. Carbon-based sensors are also used in the application of voltammetry field; in the detection of cancer cells, stress, and diseases; in the construction of voltammetry electrodes; and for food safety purpose [46]. The related works are reported in Ref. [45], describing the EC sensing applications of carbon based materials. Vinayak Adimule *et al.* illustrate the advanced sensor based on carbon nanomaterials and their significance [46]. Rajendrachari *et al.* significantly contributed the carbon nanomaterials-based EC sensors for the detection of food quality using voltammetric techniques [48–52]. Figure 1.9 illustrates the carbon materials used for the construction of EC voltammetry and it was reported by, Shashanka R. Vinayak A. *et al.* for the development of carbon-based structures and their EC sensing performances using voltammetric methods [53].

1.10 VARIOUS APPLICATIONS OF CARBON SENSORS

Carbon sensors have widespread applications in various sectors:

Environmental Monitoring: Carbon sensors play a crucial role in monitoring air and water quality, detecting pollutants, and assessing the impact of human activities on ecosystems.

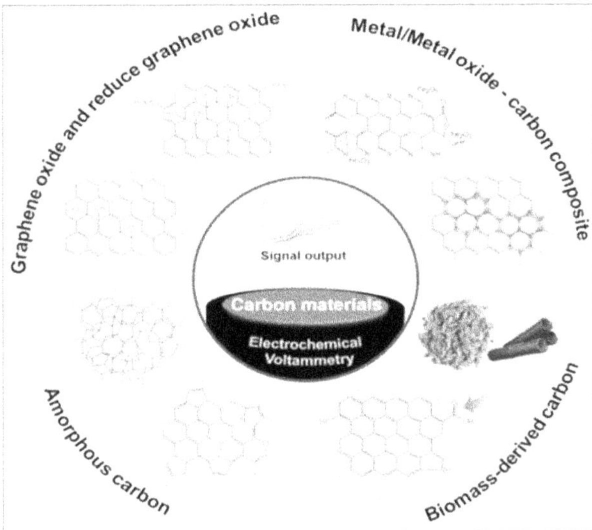

FIGURE 1.9 Carbon materials are commonly used to construct electrochemical voltammetry. (Open access creative common attributions CC by 4.0 Ref [53].)

Industrial Safety: In industrial settings, carbon sensors are used for detecting toxic gases, monitoring emissions, and ensuring workplace safety.

Health care: Carbon sensors are utilized in medical devices for monitoring respiratory gases, detecting biomarkers, and diagnosing diseases such as diabetes and metabolic disorders.

Automotive Industry: Carbon sensors are integrated into vehicles for monitoring exhaust gases, ensuring compliance with emission regulations, and optimizing engine performance.

Agriculture: In agriculture, carbon sensors aid in monitoring soil quality, assessing carbon sequestration in agricultural lands, and optimizing fertilizer usage [54]. Applications are reported in Refs. [55–56] involving graphene-based structures for the effective usage in e biosensor applications.

Water Treatment: Maintaining water quality is crucial nowadays because water contains many contaminants that are traced by analytical techniques using carbon nanosensors [57–58]. Figure 1.10 illustrates the applications of carbon nanosensors and it was opted and slightly modified from Holmannova *et al.* [59] describing the carbon NPs and their advanced biomedical applications.

1.10.1 Future Perspectives and Challenges

The field of carbon-based nanosensors is continuously evolving, driven by ongoing research and technological advancements. Future developments may involve the integration of carbon-based nanosensors with emerging technologies such as wearable devices, Internet of Things (IoT) platforms, and artificial intelligence (AI) algorithms for real-time monitoring and data analytics. However, challenges such as sensor stability, reproducibility, and scalability need to be addressed to realize the full potential of carbon-based nanosensors in practical applications and health care systems in the time of Covid-19, cooperation between and within the countries, and in the field of economics [60]. Figure 1.11 presents the work developed by Frantzidis *et al.* [61].

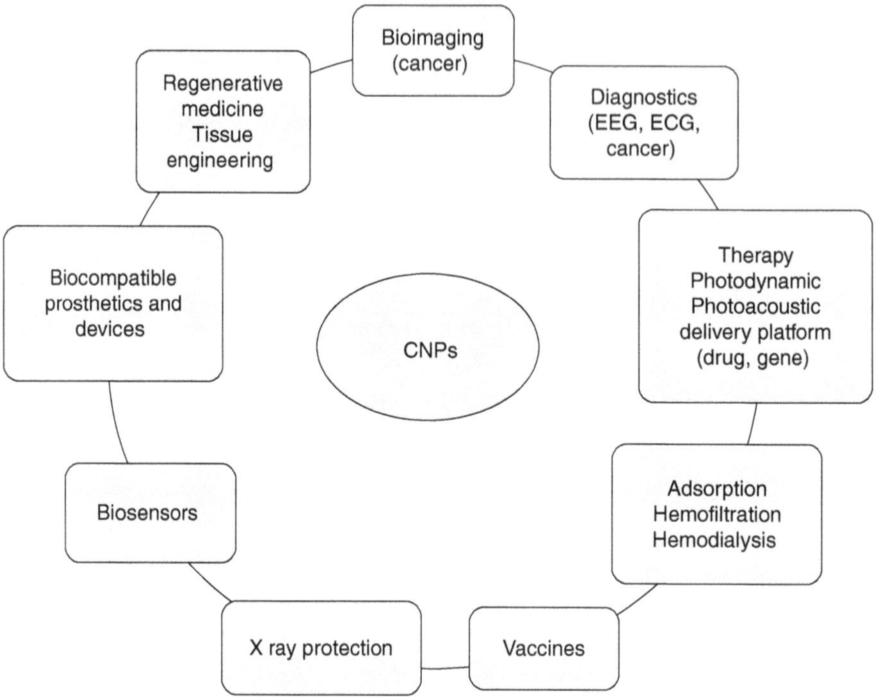

FIGURE 1.10 Illustration of applications of carbon nanosensors (open access creative common attributions CC by 4.0 Ref [59])

1.11 CONCLUSION

In this chapter, we have explored the fundamental principles and applications of carbon sensors. Carbon-based materials, such as CNTs and graphene, have shown remarkable potential in sensor technology due to their unique properties, including high surface area, excellent electrical conductivity, and chemical stability. We have also discussed various types of carbon sensors and examined their mechanisms, advantages, and limitations. The chapter also delved into the diverse applications of these sensors in fields such as environmental monitoring, health care, and industrial processes. The ability of carbon sensors to detect a wide range of analytes with high sensitivity and selectivity makes them invaluable tools in these areas. Furthermore, we explored the current challenges in the development and commercialization of carbon sensors, such as fabrication techniques, reproducibility, and integration into practical devices. Despite these challenges, ongoing research and advancements in nanotechnology continue to drive the evolution of carbon sensor technology. In conclusion, carbon sensors represent a promising and versatile class of sensing devices with significant potential to impact various industries. As research progresses, we can expect further improvements in their performance and broader adoption in real-world applications. The future of carbon sensors is bright, and continued innovation will likely lead to even more groundbreaking developments in this field.

AUTHORS' CONTRIBUTIONS

Dr Gangadhar Bagihalli and Bojappa Ginni wrote the draft of the chapter and were involved in taking permissions and making corrections. Dr Vinayak Adimule contributed to the grammatical and technical corrections of the chapter.

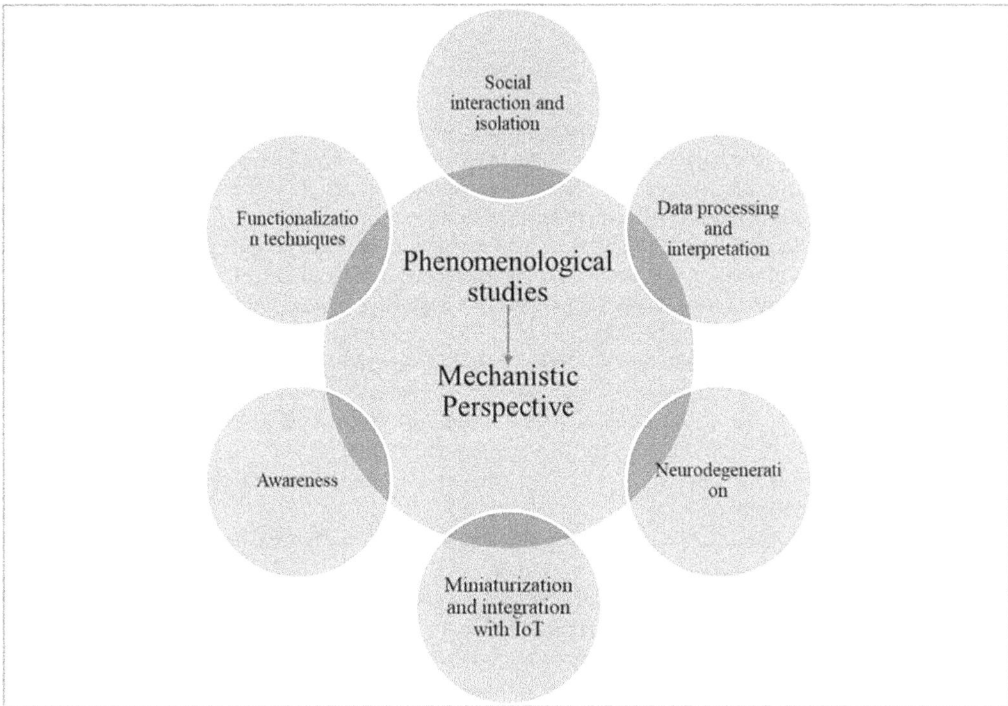

FIGURE 1.11 Visualization of the six primary challenges within the domain of cognitive neuroscience. (Open access creative common attributions CC by 4.0 Ref [61].)

ACKNOWLEDGMENTS

The authors are thankful to KLE Institute of Technology, Hubballi, Karnataka, India and Dr. MSSCET, KLE Tech University, Belagavi, Karnataka, India, for constant support and encouragement.

CONFLICT OF INTEREST

The authors declare that they do not have any conflict of interest.

FUNDING

The authors declare that they have not received any funding from any institutions or organizations.

REFERENCES

[1] Govardhan, K.R., Ramanathan, P. and Ganesapillai, M. 2020. Porous Carbon-Based Sensors and Their Applications. In Grace, A.N., Sonar, P., Bhardwaj, P., and Chakravorty, A. (Eds.), *Handbook of Porous Carbon Materials. Materials Horizons: From Nature to Nanomaterials*. Springer, Singapore. https://doi.org/10.1007/978-981-19-7188-4_14.

[2] Prinith, N.S. and Manjunatha, J.G. 2019. Surfactant modified electrochemical sensor for determination of Anthrone – A cyclic voltammetry. *Materials Science for Energy Technologies*, 2(3), pp. 408–416. https://doi.org/10.1016/j.mset.2019.05.004.

[3] Dara, A., Jafarbiglookarami, A., and Hafizi, A. (2022). Carbon-based materials for carbon capture and storage. In Gupta, R. (Ed.), *Handbook of Energy Materials*. Springer, Singapore. https://doi.org/10.1007/978-981-16-4480-1_46-1.

[4] Manjunatha, J.G. 2023. *Graphene-Based Sensors*. IOP Publishing. DOI: 10.1088/1674-1056/22/5/057701.

[5] Hernaez, M. 2020. Applications of graphene-based materials in sensors. *Sensors,* 20, p. 3196. DOI: 10.3390/s20113196.

[6] Vinayak, A., Sheetal, B. and Basappa, C.Y. 2024. Chapter 19 – Toxicity, safety, and disposal of surfactant-based electrochemical sensors and biosensor. In J.G. Manjunatha and C.M. Hussain (Eds.), *Surfactant Based Electrochemical Sensors and Biosensors*, pp. 473–488. ISBN 9780443154935. https://doi.org/10.1016/B978-0-443-15493-5.00010-5.

[7] Nandi, S.S., Suryavanshi, A., Vinayak A. and Maradur, S.R. 2020. Semiconductor current-voltage characteristics of some novel perovskite ionic nanocomposites of $Sr_{0.5}$, $Cu_{0.4}$, $Y_{0.1}$ and $Sr_{0.5}$, $Mn_{0.5}$ and their electronic sensor applications. In *AIP Conference Proceedings* (Vol. 2274, No. 1). AIP Publishing. DOI: 10.1063/5.0022453.

[8] Wang, X., Zhi, L. and Müllen, K. 2008. Transparent, conductive graphene electrodes for dye-sensitized solar cells. *Nano Letters*, 8(1), pp. 323–327. https://doi.org/10.1021/nl072838r.

[9] Zhang, H., Lv, X., Li, Y. and Wang, Y. 2019. Carbon dots: A small conundrum. *Trends in Chemistry*, 1(3), pp. 225–235. https://doi.org/10.1016/j.trechm.2019.02.003.

[10] Vinayak, A. and Rajendrachari, S. eds. 2022. *Optical and Sensor Characteristics of Nanocomposites*. Trans Tech Publications.

[11] Marculescu, C., et al. 2023. Customizable fabrication process for flexible carbon-based electrochemical biosensors. *Chemosensors*, 204. www.mdpi.com/2227-9040/11/4/204, ISSN 2227–9040.

[12] Manjunatha, J.G. and Uslu, B. 2022. Carbon-based composite materials for electrodes. *Materials*, 15, Art. 4908. https://doi.org/10.3390/ma15144908.

[13] Vinícius, D.N., et al. 2019. Carbon nanostructure-based sensors: A brief review on recent advances. *Advances in Materials Science and Engineering*, 2019, Art. 4293073. https://doi.org/10.1155/2019/4293073.

[14] Kaplan, A., Yuan, Z., Benck, J.D., Rajan, A.G., Chu, X.S., Wang, Q.H. and Strano, M.S. 2017. Current and future directions in electron transfer chemistry of graphene. *Chemical Society Reviews*, 46(15), pp. 4530–4571.

[15] Leng, X., Luo, D., Xu, Z. and Wang, F. 2018. Modified graphene oxide/Nafion composite humidity sensor and its linear response to the relative humidity. *Sensors and Actuators B: Chemical*, 257, pp. 372–381.

[16] Li, Y., Li, N., De Oliveira, N. and Wang, S. 2021. Implantable bioelectronics toward long-term stability and sustainability. *Matter*, 4(4), pp. 1125–1141.

[17] Yao, Y., Chen, Q., Li, Y.Q., Huang, X.H., Ling, W.W., Xie, Z.M., Wang, J.Q. and Chen, C.M. 2024. Nanodiamond/Ti_3C_2 MXene-coated quartz crystal microbalance humidity sensor with high sensitivity and high quality factor. *Rare Metals*, 43(6), pp. 2719–2729.

[18] Ding, X., Chen, X., Chen, X., Zhao, X. and Li, N. 2018. A QCM humidity sensor based on fullerene/graphene oxide nanocomposites with high quality factor. *Sensors and Actuators B: Chemical*, 266, pp. 534–542.

[19] Zhou, X., Shin, E., Wang, K.W. and Bakis, C.E. 2004. Interfacial damping characteristics of carbon nanotube-based composites. *Composites Science and Technology*, 64(15), pp. 2425–2437.

[20] Sanati, M., Sandwell, A., Mostaghimi, H. and Park, S.S. 2018. Development of nanocomposite-based strain sensor with piezoelectric and piezoresistive properties. *Sensors*, 18(11), p. 3789.

[21] Yan, Y., Sencadas, V., Zhang, J., Zu, G., Wei, D. and Jiang, Z. 2017. Processing, characterisation and electromechanical behaviour of elastomeric multiwall carbon nanotubes-poly (glycerol sebacate) nanocomposites for piezoresistive sensors applications. *Composites Science and Technology*, 142, pp. 163–170.

[22] Seekaew, Y. and Wongchoosuk, C. 2019. A novel graphene-based electroluminescent gas sensor for carbon dioxide detection. *Applied Surface Science*, 479, pp. 525–531.

[23] Keshtkar, S., Rashidi, A., Kooti, M., Askarieh, M., Pourhashem, S., Ghasemy, E. and Izadi, N. 2018. A novel highly sensitive and selective H_2S gas sensor at low temperatures based on SnO_2 quantum dots-C60 nanohybrid: Experimental and theory study. *Talanta*, 188, pp. 531–539.

[24] Kodu, M., Berholts, A., Kahro, T., Kook, M., Ritslaid, P., Seemen, H., Avarmaa, T., Alles, H. and Jaaniso, R. 2017. Graphene functionalised by laser-ablated V2O5 for a highly sensitive NH3 sensor. *Beilstein Journal of Nanotechnology*, 8(1), pp. 571–578.

[25] Chen, X., Huang, Z., Li, J., Wu, C., Wang, Z. and Cui, Y. 2018. Methane gas sensing behavior of lithium ion doped carbon nanotubes sensor. *Vacuum*, 154, pp. 120–128.
[26] Baro, M. and Ramaprabhu, S. 2018. Room temperature hydrogen gas sensing properties of mono dispersed platinum nanoparticles on graphene-like carbon-wrapped carbon nanotubes. *International Journal of Hydrogen Energy*, 43(33), pp. 16421–16429.
[27] Wang, X. and Zhang, X. 2013. Electrochemical co-reduction synthesis of graphene/nano-gold composites and its application to electrochemical glucose biosensor. *Electrochimica Acta*, 112, pp. 774–782.
[28] Huang, P., Lin, J., Wang, X., Wang, Z., Zhang, C., He, M., Wang, Q., Chen, F., Li, Z., Shen, G. and Cui, D. 2012. Light-triggered theranostics based on photosensitizer-conjugated carbon dots for simultaneous enhanced-fluorescence imaging and photodynamic therapy. *Advanced Materials (Deerfield Beach, Fla.)*, 24(37), p. 5104.
[29] Vinayak, A. 2024. Developments in inorganic and organic based nanostructured materials for electrochemical biosensing applications. In: J.G. Manjunatha (Ed.), *Novel Nanostructured Materials for Electrochemical Bio-Sensing Applications*, pp. 37–56. Elsevier. ISBN 9780443153341, https://doi.org/10.1016/B978-0-443-15334-1.00003-1.
[30] Swasti, S. and Ankit, K. S. 2020. Carbon nanotube-based sensors and their application. In S. Thomas, Y. Grohens, G. Vignaud, N. Kalarikkal and J. James (Eds.), *Nano-Optics*, pp. 265–291. Elsevier. https://doi.org/10.1016/B978-0-12-818392-2.00010-X.
[31] Leyla, K.S., et al. 2022. Recent trends on electrochemical carbon-based nanosensors for sensitive assay of pesticides. *Trends in Environmental Analytical Chemistry*, 34, Art. e00158, pp. 2214–1588.
[32] Rajagopalan, T., Poonam, S., Somnath, C. R. and Shantanu, B. From fundamentals to applications of carbon nanostructures: An overview. In R. Thiruvengadathan, S. C. Roy, P. Sundriyal and S. Bhattacharya (Eds.), *Carbon Nanostructures: Fundamentals to Applications*, pp. 1–14. AIP Publishing.
[33] Catalin, M., et al. 2022. A customizable fabrication process for flexible carbon-based electrochemical biosensors. *Trends in Environmental Analytical Chemistry*, 34, Art. e00158, pp. 2214–1588, https://doi.org/10.1016/j.teac.2022.e00158.
[34] Liang, X., Li, N., Zhang, R., et al. 2021. Carbon-based SERS biosensor: From substrate design to sensing and bioapplication. *NPG Asia Mater*, 13, p. 8. https://doi.org/10.1038/s41427-020-00278-5.
[35] Yagati, A.K. and Yoon, S.S. 2020. Graphene-based sensors: Fundamentals and applications. *Sensors*, 20(4), p. 1164. https://doi.org/10.3390/s20113196.
[36] Manjunatha, J.G. 2022. *Electrochemical Sensors Based on Carbon Composite Materials: Fabrication, Properties and Applications*. IOP Publishing.
[37] Manjunatha, J.G. 2020. Fabrication of efficient and selective modified graphene paste sensor for the determination of catechol and hydroquinone. *Surfaces*, 3, pp. 473–483. https://doi.org/10.3390/surfaces3030034.
[38] Marculescu, C., Preda, P., Burinaru, T., Chiriac, E., Tincu, B. and Matei, A. Brincoveanu, O. Pachiu, C. and Avram, M. 2023. Customizable fabrication process for flexible carbon-based electrochemical biosensors. *Chemosensors*, 11, p. 204. https://doi.org/10.3390/chemosensors11040204.
[39] Pushpanjali, P.A., Manjunatha, J.G., Amrutha, B.M. and Hareesha, N. 2021. Development of carbon nanotube-based polymer-modified electrochemical sensor for the voltammetric study of Curcumin. *Materials Research Innovations*, 25(7), pp. 412–420.
[40] Raril, C., Manjunatha, J.G., Ravishankar, D.K., Fattepur, S., Siddaraju, G. and Nanjundaswamy, L. 2020. Validated electrochemical method for simultaneous resolution of tyrosine, uric acid, and ascorbic acid at polymer modified nano-composite paste electrode. *Surface Engineering and Applied Electrochemistry*, 56, pp. 415–426.
[41] Tigari, G. and Manjunatha, J.G. 2020. Poly (glutamine) film-coated carbon nanotube paste electrode for the determination of curcumin with vanillin: An electroanalytical approach. *Monatshefte für Chemie-Chemical Monthly*, 151, pp. 1681–1688.
[42] Manjunatha, J.G., Raril, C., Hareesha, N., Charithra, M.M., Pushpanjali, P.A., Tigari, G., Ravishankar, D.K., Mallappaji, S.C. and Gowda, J. 2020. Electrochemical fabrication of poly (niacin) modified graphite paste electrode and its application for the detection of riboflavin. *Open Chemical Engineering Journal*, 14(1) pp. 90–98.

[43] Manjunatha Charithra, M. and Manjunatha, J.G. 2020. Electrochemical sensing of paracetamol using electropolymerised and sodium lauryl sulfate modified carbon nanotube paste electrode. *Chemistry Select*, 5(30), pp. 9323–9329.

[44] Hareesha, N., Manjunatha, J.G., Alothman, Z.A. and Sillanpää, M. 2022. Simple and affordable graphene nano-platelets and carbon nanocomposite surface decorated with cetrimonium bromide as a highly responsive electrochemical sensor for rutin detection. *Journal of Electroanalytical Chemistry*, 917, Art. 116388.

[45] Trong, D.N., My Thi Ngoc, N. and Lee, J.S. 2023. Carbon-based materials and their applications in sensing by electrochemical voltammetry journal. *Inorganics*, 11 (2), p. 81. https://doi.org/10.3390/inorganics11020081.

[46] Vinayak, A., Basappa, C.Y. and Adarsha, H.J.G. 2022. Advanced sensors based on carbon nanomaterials. In J.G. Manjunatha and C.M. Hussain (Eds.), *Carbon Nanomaterials-Based Sensors*, pp. 259–268. Elsevier. ISBN 9780323911740. https://doi.org/10.1016/B978-0-323-91174-0.00004.

[48] Rajendrachari, S., et al. 2022. Assessing the food quality using carbon nanomaterial based electrodes by voltammetric techniques. *Biosensors*, 12, p. 1173. https://doi.org/10.3390/bios12121173.

[49] Charithra, M.M. and Manjunatha, J.G. 2020. Enhanced voltammetric detection of paracetamol by using carbon nanotube modified electrode as an electrochemical sensor. *Journal of Electrochemical Science and Engineering*, 10(1), pp. 29–40. DOI: 10.5599/jese.717.

[50] Manjunatha, J.G. and Raril, C. 2018. Cyclic voltammetric investigation of caffeine at methyl orange modified carbon paste electrode. *Biomedical Journal of Scientific & Technical Research*, 9, pp. 1–6. https://doi.org/10.1007/s41664-019-00116-w.

[51] Manjunatha, J.G., et al. 2021. A highly responsive voltammetric methodology for the sensing of antihistamine drug cetirizine on the surface of cetrimonium bromide immobilized multi-walled carbon nanotube electrode. *Journal of Materials Science: Materials in Electronics*, 32. DOI: 10.1007/s10854-021-06751-3.

[52] Prinith, N.S. and Manjunatha, J.G. 2019. Surfactant modified electrochemical sensor for determination of Anthrone – A cyclic voltammetry. *Materials Science for Energy Technologies*, 2(3), pp. 408–416. https://doi.org/10.1016/j.mset.2019.05.004.

[53] Vinayak, A., et al. 2023. Electrochemical oxidation of methylene blue dye in wastewater using mechanically alloyed high entropy alloy modified carbon paste electrode using cyclic voltammetry. In *Materials Research Express*. IOP Publishing. DOI: 10.1088/2053-1591/acd3d7.

[54] Swasti, S. and Ankit, K. 2020. Carbon nanotube-based sensors and their application. In *Nano-Optics: Fundamentals, Experimental Methods, and Applications. Micro and Nano Technologies*, pp. 265–291. Elsevier. ISBN 9780128183922, https://doi.org/10.1016/B978-0-12-818392-2.00010-X.

[55] Maya, P., et al. 2022. Optical graphene for biosensor application: A review. *AMM*, 908, pp. 51–68. https://doi.org/10.4028/p-rs3qal.

[56] Vinayak, A., Nidhi, M. and Kalpana, S. 2023. Electro-analysis of hormones in a graphene-modified electrochemical sensor. In *Graphene-Based Sensors*, pp. 1–10. IOP Publishing.

[57] Manjunatha, J.G. and Hussain, C.M. 2022. *Carbon Nanomaterials-Based Sensors: Emerging Research Trends in Devices and Applications*. Elsevier. ISBN: 9780323911740.

[58] Vinayak, A. 2024. Developments in inorganic and organic based nanostructuredmaterials for electrochemical biosensing applications. In *Novel Nanostructured Materials for Electrochemical Bio-Sensing Applications*, pp. 37–56. Elsevier. ISBN 9780443153341, https://doi.org/10.1016/B978-0-443-15334-1.00003-1.

[59] Holmannova, D., et al. 2022. Carbon nanoparticles and their biomedical applications. *Applied Sciences*, 12, Art. 7865. https://doi.org/10.3390/app12157865.

[60] Amirhossein, T., Hajar, H. and Azam, R. 2021. Challenges, opportunities, and future perspectives. In Mohammad Hadi Dehghani, Rama Rao Karri, Sharmili Roy (eds.), *Environmental and Health Management of Novel Coronavirus Disease (COVID-19)*, pp. 443–477. Academic Press. https://doi.org/10.1016/B978-0-323-85780-2.00011-1.

[61] Frantzidis, C.A., et al. 2024. Editorial: New challenges and future perspectives in cognitive neuroscience. *Frontiers in Human Neuroscience*, 18, Art. 1390788. DOI: 10.3389/fnhum.2024.1390788.

2 Recent Advances in Development of Surfactant-Based Carbon Sensors

Kübra Turan, Nazlı Şimşek, Esra Ülker, and Gözde Aydoğdu Tığ

2.1 INTRODUCTION

As the name suggests, surfactants mean "substances that reduce surface tension" or "surface cleaning agents". The use of surfactants dates back to 2000 years.[1] Surfactants can be natural or synthetic. Surfactants are organic amphiphilic molecules with a hydrophobic head and a hydrophilic tail.[2] Surfactants are chemical substances that tend to accumulate on a surface or an interface. For a surfactant to be considered surface-active, it must possess the ability to adsorb at the interface between two distinct phases, such as electrode and solution, air and water, or oil and water.[3] A characteristic structure of the surfactant is formed by a polar head group, which may be zwitterionic, neutral, positive, or negative, and a hydrophobic tail consisting of one or more hydrocarbon chains with 6–22 carbon atoms.

Accordingly, surfactants can be classified as cationic, anionic, nonionic, and zwitterionic (Figure 2.1).[4] When anionic and cationic surfactants are used, negatively and positively charged surfaces are formed. Zwitterionic surfactants can create a negative or positive change depending on the pH value. Surfactants are used in oil refineries, soil and wastewater treatment, cosmetics, dyestuffs, pharmaceutical distribution, organic synthesis catalysts, pesticides, food processing, bioanalysis, electrochemical batteries, sensor design, etc. They play a role in many areas.

Surfactants reduce liquid surface tension between two immiscible phases by interacting with hydrophobic and hydrophilic solvents. With their distinct amphiphilic structure and self-assembly ability, surfactants are the key to extraction, adsorption, membrane filtration foaming, and precipitation. In a solution, surfactants reach a point where they form dense agglomerates, which are referred to as aggregate micelles. The point where clumping occurs is called the critical micelle concentration (CMC) point.[5] The essential properties of surfactants, namely adsorption at the interface and aggregation into micellar structures, are employed to advantage in electrochemistry. These substances can change the properties of the electrode surfaces in the electrode design. Surfactants are frequently used to increase reaction rates and control reaction pathways on the electrode surface.[6]

Carbon is an exemplary electrode material, exhibiting a combination of advantageous properties. These include cost-effectiveness, a wide potential window, workability, rich surface chemistry, relative inertness, and excellent electrochemical properties. The abundance of carbon relative to other popular electrode materials, such as platinum and gold, makes it arguably the most popular electrochemical substrate. The electrochemical characteristics of carbon are dependent on both microstructure and overall morphology. Carbon surfaces' robustness and relative inertness provide an ideal

DOI: 10.1201/9781003500322-2

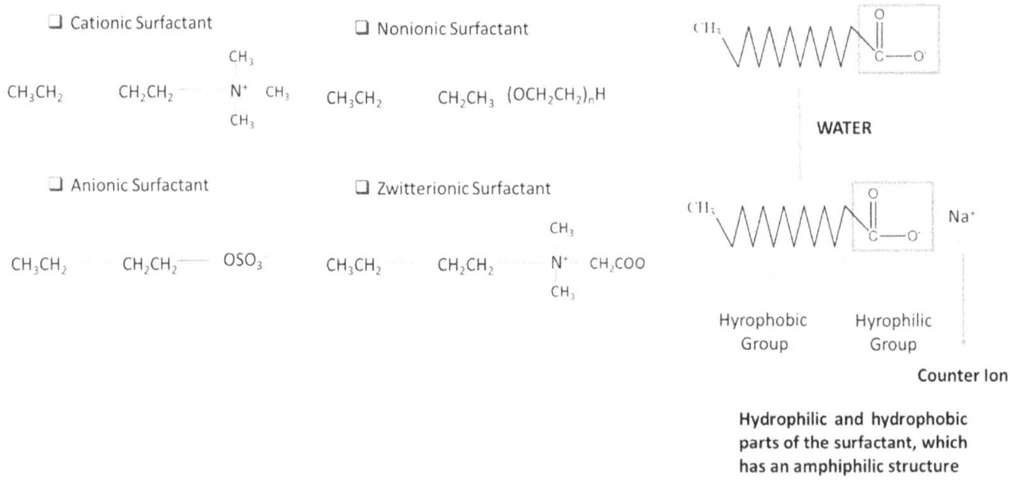

FIGURE 2.1 General classification and structure of surfactants.

platform for heterogeneous electron transfer.[7] Carbon-based electrodes are currently in widespread use in electroanalytical chemistry for various applications.[8] Combining carbon-based electrodes with surfactants is widely used in electrochemical studies to enhance surface properties and increase measurement sensitivity. As is well known, carbon-based electrodes typically exhibit a high surface area, and the surface properties of these electrodes can be further enhanced through surfactants.[9] Carbon sensors are devices or systems that measure or monitor various substances or compounds. They are based on carbon materials and are used in various applications. The term "carbon sensors" describes a variety of sensors, the specific characteristics of which depend on the particular application and the nature of the carbon-based material. The ongoing advancement of sensor technology is enabling the expansion of the capabilities and applications of carbon sensors across a range of industries and scientific disciplines. Surfactants can be used to prepare carbon sensors. They are added to the paste while preparing carbon paste electrodes (CPEs), dripped directly onto the carbon electrode surface, dried, and coated onto the carbon electrode surfaces by electrodeposition.

This chapter provides comprehensive information about surfactant-based carbon sensors, how they are prepared, and which analytes they are used to identify. In short, the reader will learn about the development of sensitive and selective electrochemical approaches using surfactant-based carbon sensors and the determination of some analytes. The carbon sensor applications presented here are classified into five categories, each discussed separately. Recent work on surfactant-based carbon sensors for detecting pharmaceuticals, food, pollutants, biomolecules, and other chemical and biochemical substances is also reviewed.

2.2 TYPES OF CARBON SENSORS/BIOSENSORS

The attractive electronic and structural properties of carbon, coupled with its superior electrochemical stability, render it particularly suitable for contemporary electrochemical applications.[10] Carbon electrodes have recently been employed and described for fabricating electrochemical sensors and biosensors.[11] The reliable, rapid, and in situ analysis of an ever-increasing range of biological species, environmental pollutants, heavy metals, and pharmaceutical compounds is now possible due to electrochemical sensing. The electrochemical approach has become the focal point among numerous transduction methods due to several factors, including its simplicity, low-cost instrumentation, miniaturization capability, and automation capability.[12] Currently, a significant proportion of analytical instruments are used in various settings, including environmental, food, pharmaceutical, and clinical laboratories, as well as many commercial point-of-care devices, and they operate using electrochemical sensors.[13] Carbon sensors are classified into the following categories: screen-printed electrode (SPE), glassy carbon electrode (GCE), CPE, and pencil graphite electrode (PGE).

2.2.1 Screen-Printed Electrode

In recent years, SPEs have been employed extensively to develop new electrochemical sensing platforms and enhance their performance due to their low cost, ease, and mass production speed using thick film technology.[14] SPEs display ductility when drawn in various shapes, are composed of multiple materials, and can be suitably modified utilizing a range of modifiers. Modifications can facilitate the utilization of SPEs based on metallic nanoparticles, which possess properties such as enhanced electrical conductivity and high surface areas. The modifications enhance the surface activity, which, as a result, improves reaction rates and detection sensitivity.[15] SPEs typically comprise three electrodes (working, counter, and reference electrodes) printed on diverse plastic or ceramic substrates. This configuration can be readily modified with many commercial or in-house-produced inks. The most commonly utilized pastes during the printing process of SPEs are silver ink and carbon ink. The surface of SPEs can be readily modified to suit many purposes about disparate analytes, thereby facilitating a range of enhancements. This versatility, coupled with its miniaturized dimensions and the potential for interfacing with portable devices, enables the determination of target analytes in situ with high specificity.[16]

2.2.2 Glassy Carbon Electrode

GCE is a hard and isotropic material. A considerable number of structural pores are responsible for its distinctive low density. Furthermore, GCE exhibits excellent impermeability to gases and liquids, high chemical stability, and good thermal and electrical conductivities.[17] GCEs have been extensively utilized as electrodes in electrochemical biosensors due to their cost-effectiveness, ease of microfabrication, simplicity of surface functionality, and superior mechanical and electrical properties. Moreover, the sensitivity and reproducibility of GCE can be enhanced through surface modification with a range of nanomaterials.

Moreover, the specificity of the electrodes can be enhanced through the functionalization of the working electrodes with a range of specific recognition molecules, including nucleic acids, antibodies, enzymes, and aptamers.[18] Another advantageous quality of GCE is that it can be manufactured from a range of renewable resources. As an illustration, furfuryl alcohol monomers employed in the synthesis of poly(furfuryl alcohol) resin can be derived from the processing of plant residues and lignocellulosic biomass, including rice husks, sugarcane bagasse, and corn cobs. Additionally, plant-derived precursors, such as sucrose, cellulose, and tannins, can also be utilized in GC production.[19]

2.2.3 Carbon Paste Electrode

CPEs have gained attention as ion-selective electrodes due to their advantageous characteristics compared to membrane electrodes. These include chemical inertness, robustness, renewability, a stable response, low ohmic resistance, the absence of an internal solution, and suitability for a range of sensing and detection applications.[20] Furthermore, CPEs are non-toxic and environmentally friendly electrodes. In recent years, scientists have been interested in developing new electrode materials that exhibit enhanced characteristics, including a wider potential window, a higher signal-to-noise ratio, improved mechanical stability, and resistance to passivation.[21] The characterization of materials is of great importance because electrode fouling is likely the most significant impediment to the more frequent application of electroanalytical methods in environmental analyses.[22].The issues associated with passivation are effectively circumvented through the expeditious restoration of the surface to its original state. CPEs are employed as electrodes or sensors for analysis in a range of environmental applications, including studies of DNA hybridization, in vivo measurements, and biosensors with immobilized enzymes in various medical applications. They are also utilized in pharmaceutical analysis, as recently outlined.[23]

2.2.4 Pencil Graphite Electrode

Pencil graphite tips and electrodes, when employed as working electrodes, are designated as PGEs. In addition to their affordability, PGEs offer a more straightforward and convenient alternative, eliminating the necessity for time-consuming electrode surface cleaning or polishing.[24] The application of disposable sensing materials, such as PGEs, represents a significant advancement in addressing the limitations of traditional gaseous carbon sensors.[25] The application of different voltammetric techniques for measuring various analytes from a diverse range of samples demonstrated the reproducibility of the signals produced by the PGEs and the precise definition of the voltammetric peaks. These electrodes offer good sensitivity and reproducibility, making them a viable, renewable, and economical tool.[26] Ag- and AgCl-doped pencil apparatus tips were utilized as reference electrodes in a paper-based electrode system in certain instances.[27] It has been demonstrated that PGEs provide an excellent working electrode in high-potential electrochemical sensing systems. PGEs can be readily modified directly with nanostructures, polymers, or a combination of both, analogous to that observed with precious metal electrodes such as gold or platinum.[28] PGEs have many applications in investigating environmental pollutants, including pesticides, pharmaceuticals, food additives, metal ions, and antioxidants.[29]

2.3 SURFACTANTS FOR SENSOR PREPARATION

2.3.1 Classification of Surfactants

The dissociated ions of surfactant molecules in an aqueous solution can be classified depending on the type of the surfactant. Surfactant molecules generally consist of a polar hydrophilic head and a hydrophobic tail with a long hydrocarbon chain. The hydrophilic polar end hydrogen bonds with water molecules and provides water solubility.[30] Some of the surfactants with different hydrophilic groups are shown in Table 2.1. The most valuable and general ranking of surfactants depends on the hydrophilic groups.[31]

TABLE 2.1
Commercially available surfactants with hydrophilic groups

Structure	General Class Name
R OSO_3^- M^+	Sulfate
R SO_3^- M^+	Sulfonate
R OPO_3^- M^+	Phosphate
R COO^- M^+	Carboxylate
R N^+RxHyX^-	Ammonium
R $OCH_2CH_2(OCH_2CH_2)_nOH$	Polyoxyethylene
R O $C_6H_7O(OH)_3$ O $C_6H_7(OH)_4$	Sucrose esters
R OCH_2 CH(OH) CH_2OH	Polyols
R–$(OCH_2CH[CH_2OH]CH_2)_n$-$OCH_2CH[CH_2OH]CH_2OH$	Polyglycidyl ester

2.3.1.1 Anionic Surfactants

In an aqueous environment, the hydrophilic end of an anionic surfactant carries a negative charge and interacts with positively charged ions such as Na^+, K^+, and Mg^{2+}. Examples include sulfates, sulfonates, phosphates, and carboxylates. They are widely used because they are nontoxic. They contain a saturated or unsaturated straight-chain aliphatic group ranging in length from C_{12} to C_{18}. The solubility of the surfactant in water is affected by the presence of double bonds.[32]

2.3.1.2 Cationic Surfactants

Surfactants that have a positively charged hydrophilic group during ion dissociation in aqueous media are called cationic surfactants. Positively charged cationic surfactants interact with negatively charged groups such as OH^-. Examples include amines and quaternary ammonium salts.[33] They are used in fabric softeners due to their softness and in household and bathroom cleaning due to their disinfecting properties.[34]

2.3.1.3 Nonionic Surfactants

The balance between hydrophilic and hydrophobic structures in the surfactant molecule provides the surface activity of nonionic surfactants. These substances do not dissociate into ions in aqueous solutions due to their hydrophilic groups, such as phenol, alcohol, ester, ether, or amide.[35] High molecules have high mass and are neutral molecules. Nonionic surfactants include phenol, alcohol, ester, ether, and amide. They resist hard water because they contain no negative or positive charge. They are used in laundry detergents, household cleaners, and hand dishwashing products because they can easily remove grease.

2.3.1.4 Zwitterionic Surfactants

Amphoteric surfactants contain both positively and negatively charged groups, and their properties are determined by the pH of the solution in which they are present. For this reason, the hydrophilic part of the molecule partially neutralizes the positive and negative charges. These surfactants are very mild and suitable for personal care products and household cleaners.[36] They are compatible with other classes of surfactants. These surfactants contain two distinct polar groups with opposing charges. In the majority of cases, the positive charge is ammonium, while the source of the negative charge may vary, including sulfonate, carboxylate, and sulfate. They possess excellent dermatological properties and are frequently utilized in dishwashing detergents and shampoos due to their high foaming capabilities.[37]

2.3.2 Sensor Preparation

Surface modifications are significant in electrochemical studies. There are three types of surface modifications: chemisorption, polymer film modifications, and covalent bonding.[38] A material used to fabricate the electrodes is essential for quantifying the electroactive molecule.[39] Various materials such as metal nanoparticles, multi-walled carbon nanotubes (MWCNT), carbon nanotubes (CNT), and carbon are used in surface modification studies of electrodes.[40] However, surfactants have recently been used in surface modifications of electrodes. Surfactants affect the electrochemical process and the solution/electrode interface by affecting the electrode surface used in carbon sensor design. The adsorption of the surfactant on the electrode surface can be both diffusion-controlled and kinetic.[41] One of the main reasons affecting the adsorption process is the structure of the molecule. The larger the length of the hydrophobic chain in the surfactant, the slower the adsorption.[42] These molecules reduce the detection limit of the sensor and increase its sensitivity.

Surfactants are used to prepare carbon sensors. They are added to the paste while preparing CPEs, dripped directly onto the carbon electrode surface, dried, and coated onto the carbon electrode surfaces by electrodeposition. On the surface of the carbon electrode, immobilized with surfactant, surfactants act as redox mediators and are used to determine many ions and molecules.

In a study by Manjunatha in 2020, a selective and sensitive electrochemical sensor for the determination of resorcinol and catechol based on a surfactant-modified GPE was designed. The modified electrode was characterized by differential pulse voltammetry (DPV), cyclic voltammetry (CV), and field emission scanning electron microscopy (FESEM). The anionic surfactant sodium dodecyl sulfate (SDS) was used and immobilized by dropping and drying on the GPE surface (Figure 2.2(a)). The results showed that SDS could be adsorbed on the surface of GPE and showed superior electrochemical oxidation with increasing peak current and decreasing peak potential.[43]

In another study by Patil et al. in 2022, a graphene (GR) and cationic surfactant cetyltrimethylammonium bromide (CTAB)-based electrochemical sensor (GR-CTAB/CPE) was designed for the electrochemical determination of the cholesterol-lowering drug atorvastatin (ATR). The morphological study of the electrode was performed using electronic scanning microscopy (SEM) technique and characterized by CV and DPV methods. The GR/CPE electrode was made of graphite powder, graphene, and paraffin oil homogeneously mixed and placed in a hollow polytetrafluoroethylene (PTFE) tube connected with a copper rod. Then 1.0 μL of CTAB cationic surfactant was dispersed on the CPE surface, and the electrode was immersed in pH 4.2 PBS for pretreatment, and CV scans were performed in the potential range of 0.7V–1.3V at a scan rate of 0.5 V/s (Figure 2.2(b)). The cationic surfactant CTAB was immobilized on the surface by electrodeposition. After the cationic surfactant CTAB ion was immobilized on the working surface, a working electrode interface parameter was developed, improving the electrochemical analysis of ATRV and increasing the sensitivity of the detection limit.[44]

In a study by Amrutha et al. in 2019, the determination of Alizarin red-S (ARS), an anthraquinone dye, using a carbon nanotube paste electrode (TX-100MCNTPE) modified with Triton X-100 (TX-100), a nonionic surfactant, was investigated using CV technique. For the preparation of TX-100MCNTPE, first 60% CNT powder and 40% silicone oil were pounded in a mortar and pestle and homogenized. Then the prepared paste was placed in a Teflon tube. A solution of 25×10^{-3} M TX-100 was prepared, and 10 μL was dropped onto the carbon nanotube electrode surface and immobilized for 5 min. Optimum conditions were determined with the prepared electrode, and according to the results, ARS shows a well-characterized irreversible oxidation peak at a scan rate of 0.1 V s^{-1} in a pH 6.5 solution. As a result of the scan rate studies, adsorption was found to be controlled. The prepared electrode showed an excellent linear operating range and a low limit of detection (LOD) for ARS. The fabricated sensor was found to be efficiently usable for the determination of ARS. The sensor showed superior electrocatalytic activity with TX-100MCNTPE obtained in contrast to carbon nanotube paste electrode (CNTPE) in the oxidation of ARS.[45]

Recently, Yang et al. designed a sensor using montmorillonite calcium and cationic surfactant CTAB. During the preparation of the electrode, 5.0 g of montmorillonite calcium was mixed with 3.0 g of CTAB in 50.0 mL of water and subjected to mechanical stirring at room temperature for 5 hours. The mixture was collected by filtration, washed several times with distilled water, and dried at 80 °C. To modify the electrode surface, 300.0 mg of CTAB-modified montmorillonite calcium (CTAB-MMT) was mixed with 700.0 mg of graphite powder and 250.0 μL of paraffin oil to obtain a homogeneous paste of CTAB-MMT and carbon and placed in a Teflon electrode. The resulting CTAB-MMT showed higher deposition efficiency toward 4-chlorophenol than unmodified MMT. At the CPE modified with CTAB-MMT, the oxidation peak current of 4-chlorophenol increases significantly. The prepared modified electrode was successfully applied to determine 4-chlorophenol in water samples.[46]

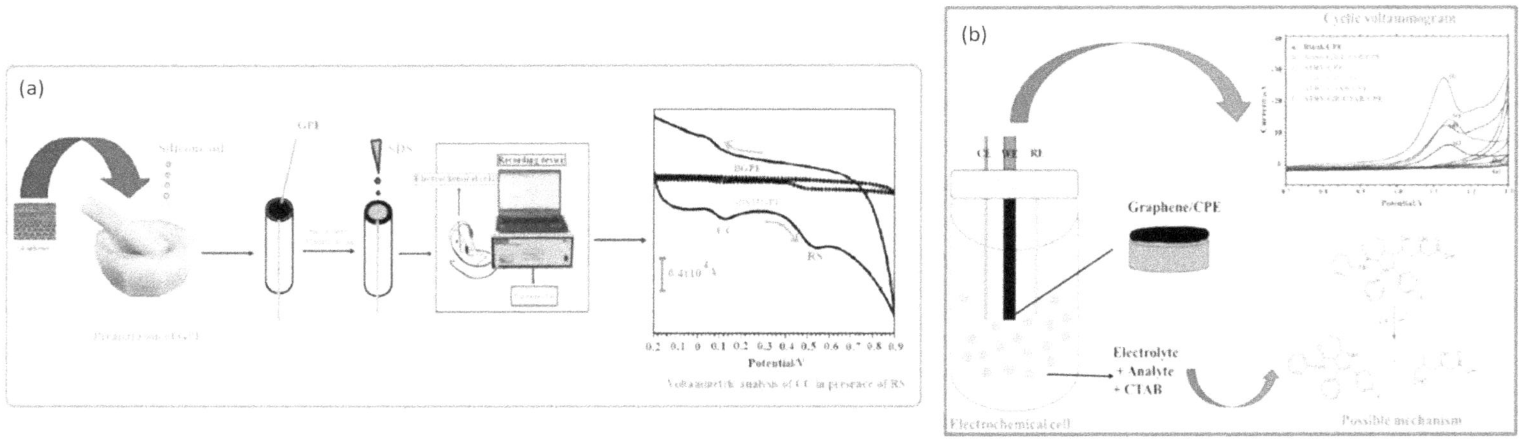

FIGURE 2.2 (a) Schematic representation of the preparation steps of the electrochemical sensor developed for catechol and resorcinol. (Reproduced with permission from Manjunatha, J. G. (2020). A surfactant enhanced graphene paste electrode as an effective electrochemical sensor for the sensitive and simultaneous determination of catechol and resorcinol. Chem. Data Coll. 25, 100331 © 2020 Elsevier Inc., with permission from Elsevier.) (b) Schematic representation of the preparation steps of a graphene sheet-based electrochemical sensor with cationic surfactant for the detection of atorvastatin. (Reprinted from: Patil, V. B., Malode, S. J., Tuwar, S. M., & Shetti, N. P. (2022). Graphene sheet-based electrochemical sensor with cationic surfactant for sensitive detection of atorvastatin. Sens. Int. 3, 100198. © 2022 Elsevier Inc., with permission from Elsevier.)

2.4 RECENT APPLICATIONS OF SURFACTANT-BASED CARBON SENSORS

The study of surfactants in electrochemistry has a history of over a century. Recently, there has been a notable increase in the use of surfactants in electroanalytical methods.[47] Because surfactants can influence the reaction mechanism of analytes on the electrode surface, they are capable to facilitate electron transfer due to their specific molecular structure, and have the ability to enhance electrode sensitivity.[48]

In electrochemical analysis using voltammetric methods, the introduction of surfactants onto the surface of the electrode results in the formation of an adsorptive film that facilitates the movement of electrons, thereby increasing the peak intensity and improving the redox potential.[49] The selection of modifiers is determined by their ability to minimize the likelihood of reduction and oxidation of electroactive species compared to unmodified electrodes. Surfactant modifiers such as CTAB, SDS, TX-100, poly(glycine), and nanoparticles exhibit high sensitivity to acidic and alkaline environments.[50] Due to their ability to reduce the size of nanoparticles, surfactants are used in nanoparticle synthesis.[51]

The fabrication of sensors entails the utilization of an array of materials, including those belonging to the categories of nano, micro, organic, inorganic, and polymer. This also includes surfactants. Surfactants (i.e., zwitterionic, anionic, or cationic) can be electrochemical sensors in amperometric or potentiometric configurations or combined with nanoparticles. Furthermore, surfactants can be integrated into the electrolyte solution in electrochemical sensing.[52]

Electrochemical sensing is a phenomenon that occurs at the electrode surface and analyte interface. The surface nature of the electrode has a significant impact on the sensing properties.[53] Combining carbon-based electrodes with surfactants is widely used in electrochemical studies to enhance surface properties and increase measurement sensitivity. As is well known, carbon-based electrodes typically exhibit a high surface area, and the surface properties of these electrodes can be further enhanced through surfactants. This allows reactants on the electrode surface to be adsorbed more effectively, thereby facilitating the monitoring of electrochemical reactions with greater precision. Surfactants possess the capacity to regulate molecular adsorption on the electrode surface. This can result in increased selectivity for specific analytes and the reduction of undesirable interactions on the electrode.[54]

The utilization of surfactants in electrochemical sensors and biosensors is on the rise due to their capacity to enhance the adsorption of the target analyte of interest and increase the rate of electron transfer. Numerous electrochemical studies have been conducted to analyze pharmaceuticals,[55] food, and food pathogens,[56] pesticides,[57] environmental pollutants,[58] and other chemical and biochemical substances by leveraging the distinctive characteristics of surfactants in aqueous media. A number of studies have been conducted in which surfactants and a variety of nanocomposites or nanomaterials have been employed with electrochemical sensors to enhance their sensitivity, reliability, and selectivity. The selection of surfactants for electrochemical sensors can be based on an evaluation of their properties, including their capacity to interact electrostatically with the sensor surface developed in the study and their compatibility with the structure of the material to be investigated. This evaluation can be carried out for anionic, cationic, or charge-free surfactants. The most commonly utilized surfactants in electrochemical studies are SDS, CTAB, and TX-100.[59]

Table 2.2 describes different types of sensors for the direct detection of various analytes using surfactant-based carbon sensors, as well as the electrode structure and fabrication, electrochemical technique, LODs, linear range, storage stability, sensitivity, real samples, and other important data related to these sensors. Table 2.2 also provides a comparison of these sensors.

2.4.1 Analysis of Pharmaceuticals

Surfactant-based carbon sensors are particularly advantageous for pharmaceutical analysis for several vital reasons. Surfactants improve the electrical conductivity of carbon sensors, resulting in

TABLE 2.2
Surfactant-based carbon sensors used for the detection of different analytes in recent applications

Analyte	Working electrode	Surfactant	Electrode modification	LOD	Lineer Range	Ref.
Rutin	CPE	CMB	CMB/GNPs-CBCPE	0.0027μM	0.1–7.0 μM	95
Tartrazin	CPE	SDS	SDSMCPE	52×10^{-7} M	2×10^{-5}–5×10^{-5} M 6×10^{-5}–1.1×10^{-4} M	96
L-tyrosine	CNTPE	SLS TX 100	STMCNTPE	15–160 μM	0.325 μM	97
Vanillin	GPE	CTAB	CTABMGPE	1.29 μM	4–15 μM 20–70 μM	98
Epinephrine	CPE	SDS	MCPE/pMWCNTs/SDS	4.57×10^{-8} M	1×10^{-7}–1×10^{-6} M 1×10^{-6}–1×10^{-4} M	99
Alloxan	CPE	CTAB	CTABMCPE	1.09 μM	5–80 μM	100
Bisphenol A	GCE	CTAB	AuNDs/CTAB/GCE	22 nM	0.025–10 μM	101
Bisphenol A	PGE	CTAB	Poly(CTAB)-MWCNTs	134 pM	2–808 nM	102
Evans blue	CNTPE	TX-100	TX100MCNTPE	5.06 μM	10–100 μM	103
Phenol	PGE	DTAB	DTAB-PGE	0.16 μg/mL	1–5 μg/mL	104
Hg(II) Cd(II)	GCE	CDT	CDTGCE	0.090 μg/L 0.084 μg/L	14–156 μg/L 12-90 μg/L	105
Estriol	CPE	SDS	SDS-PXAMCNTGCPE	1.9×10^{-7} M	10×10^{-6}–70×10^{-6} M	106
Paracetamol Adrenaline	CPE	TX-100	Cu-ZnO/TX-100/ MCPE Cu-ZnO/TX-100/ MCPE	4.6 μM 3.9 μM	10–100 μM 10–100 μM	107
Ciprofloxacin	CPE	SDS	SDSGr/CPE	0.029 μM	0.3–100 μM	108
Paracetamol, Diclofenac Tramadol	SPCE	SDS	aSPCE/SDS	1.49×10^{-8} 2.10×10^{-10} 1.71×10^{-9}	5.0×10^{-8}–2.0×10^{-5} 1.0×10^{-9}–2.0×10^{-7} 1.0×10^{-8}–2.0×10^{-7} 2.0×10^{-7}–2.0×10^{-10}	109
Morphine	GCE	CTAB	poly(CTAB)/GO	0.36 μM	50 nM–60 μM	110
Oxoplatin	GCE	DAN	DAN/AuNPs/GCE	14.5 pM	1–50 nM	111
Anthrone	CPE	OPE SLS	OPEMCPE SLSMCPE	18.8×10^{-7} M 10.6×10^{-7} M	4×10^{-6}–5×10^{-5} M 4×10^{-6}–5×10^{-5} M	112
Riboflavin	CPE	SLS	SLSMCNTPGCPE	11.6×10^{-9} M	0.2–0.8 μM 1–5 μM	113
8-hydroxy quinoline	CNTPE	SLS	SLSMCNTPE	1.1×10^{-7} M	2×10^{-5}–3.5×10^{-4} M	114
Phloroglucinol	CNTPE	OPE	OPEMCNTPE	0.71 μmol	10–90 μmol	115
Triclosan	CPE	SB16	AB/CNT/SB16	6.18 μM	0.01–2 μM	116
Aminotriazole	CPE	CTAB	TiO_2/CTAB-CPE	2.53 nM	0.01–0.65 μM	117

Note: PXAMCNTGCPE: sodium dodecyl sulfate and polymerized xanthacridinum modified carbon nanotube and graphite composite paste electrode; MCPE: modified carbon paste electrode; OPEMCNTPE: octyl phenol ethoxylate modified carbon nanotube paste electrode; OPE: octyl phenol ethoxylate; GNPs: graphene nanoplatelets; pMWCNTs: pristine multi-walled carbon nanotubes; AuNDs: Au nanodendrites; aSPCE: activated screen-printed carbon electrode; AB: acetylene black; SB16: cetyl sulfonyl betaine.

better sensitivity and selectivity. This is critical for detecting low concentrations of pharmaceutical compounds in complex matrices. They also stabilize the sensor surface, preventing fouling, and can improve the reproducibility of measurements.[60] Surfactant-based sensors are generally more cost-effective than other advanced sensor technologies, making them suitable for widespread use in research and industry.[61] Surfactants such as CTAB significantly improve drug detection performance when used to modify composite electrodes. They are known to increase the sensitivity of carbon sensors by increasing the electroactive surface area and improving the electron transfer rate. Therefore, surfactant-based carbon sensors are preferred for pharmaceutical analysis due to their higher sensitivity, selectivity, and cost-effectiveness.

MWCNTs represent the most common allotropic form of carbon, comprising cylindrical nanostructured material. These are widely utilized in the field of electrochemical analysis for the preparation of electrodes.[62] In a study by Bolat in 2020, a DNA biosensor based on a composite of poly(CTAB-MWCNTs) electrodeposited on graphite electrodes (PGEs) was created to detect DNA directly. This sensor was subsequently employed to determine the anticancer drug irinotecan (CPT-11). The biosensor has a porous structure comprising a surfactant. Consequently, the biosensor demonstrated enhanced selectivity and sensitivity for detecting CPT-11. Changes in the guanine oxidation signal depended on the interaction of irinotecan with DNA. Calibration graphs were obtained using the DPV method for surfactant-based electrochemical sensors and biosensors. An LOD of 1.03 μg mL^{-1}, corresponding to the linear concentration range of 2–500 μg mL^{-1}, was obtained for guanine. Furthermore, the concentration of CPT-11 in human serum samples was successfully quantified using this biosensor.[63]

A literature review reveals that surfactant-based sensors are typically prepared with CPE. CPEs are non-toxic and environmentally friendly electrodes. They have been widely considered as sensor materials for various reasons, including their suitability and stability in sensing applications and the fact that they do not require an internal solution.[64]

Manjunatha et al. in 2022 prepared zinc oxide nanoparticles (ZnO/NPs) and copper-doped ZnO nanoparticles (Cu-ZnO/NPs) using the co-precipitation method. The NPs and CTAB, TX-100 surfactants were successfully employed in CPE modification for sensor studies. The Cu-ZnO/CTAB and Cu-ZnO/TX-100 modified carbon paste electrodes (MCPEs) were prepared by immobilizing 10 μL of 0.1 mM CTAB and TX-100 surfactants on the surface of the Cu-ZnO/MCPE, followed by an incubation period of 5 min. The Cu-ZnO/TX-100/MCPE and Cu-ZnO/CTAB/MCPE sensors were employed to determine paracetamol (PA). The TX-100 modified electrode immobilized on the Cu-ZnO/MCPE surface exhibited enhanced redox properties, as evidenced by an increased peak current for PA compared to the Cu-ZnO/CTAB/MCPE. The Cu-ZnO/TX-100/MCPE demonstrated robust electrocatalytic activity toward the oxidation of PA, with an LOD of 4.6 μM and a limit of quantification (LOQ) of 16.2 μM, respectively. In conclusion, MCPE was utilized to detect analytes in pharmaceutical samples, yielding favorable recovery outcomes.[65]

Sawkar et al. in 2022 developed a novel SDS with a graphene-based electrochemical sensor (SDSGr/CPE) to detect and determine the pharmaceutical ciprofloxacin (CIP). The methodology involves the preparation of the paste within a Teflon tube, followed by adding SDS to the test solution and the subsequent fabrication of the electrodes. The surface morphology and topography of the sensor were evaluated using SEM and atomic force microscopy (AFM), and CV and EIS performed electrochemical characterization. The fabricated electrode demonstrated a detection limit of 2.9×10^{-8} M. The proposed sensor was shown to be effective for identifying CIP in both pharmaceutical and biological samples. The results revealed a high recovery rate for tablet analysis (93.8%) and urine analysis (93.3% to 98.7%). The stability of the electrode was evaluated through both daily and inter-day analyses. The results demonstrate the effectiveness of the method in determining CIP, even in the presence of other substances, and establish it as a rapid, reliable, and practical approach.[66]

Kozak et al. in 2021 proposed the first use of an electrochemically activated screen-printed carbon electrode modified with SDS (aSPCE/SDS) for the purpose of simultaneously determining PA, diclofenac (DF), and tramadol (TR). Following the activation process, the sensor was modified with SDS through immersion in a support electrolyte solution containing SDS. The differential pulse adsorptive stripping voltammetric (DPAdSV) procedure with SPCE/SDS enables the selective determination of low PA, DF, and TR concentrations. SEM, CV, and EIS examined the surface morphology and electrochemical behavior. The LODs obtained during the simultaneous determination of

PA, DF, and TR were 1.49×10^{-8} M, 2.10×10^{-10} M, and 1.71×10^{-9} M, respectively. The selectivity of aSPCE/SDS was evaluated by examining the impact of various organic and inorganic substances that are commonly present in biological and environmental samples on the responses of PA, DF, and TR. Ultimately, the DPAdSV procedure utilizing aSPCE/SDS was effectively employed to quantify PA, DF, and TR in pharmaceuticals, river water, and serum samples. The results demonstrate that the procedure offers a rapid, straightforward, cost-effective alternative to existing methods.[67]

The surfactant CTAB, which features a hydrophilic head and a hydrophobic tail, is critically important in enhancing the electrode's catalytic capacity. In another study, in 2020 Abraham et al. synthesized a composite of CTAB and graphene oxide (poly(CTAB)/GO) via in situ electropolymerization, and they employed it as a novel sensor for the detection of morphine (MO). An amount of 1 mg of GO and 0.176 g (0.045 M) of CTAB were dispersed in 10 mL of 0.1 M H_2SO_4 and sonicated for 30 min to form a black suspension. GCE was immersed in the suspension and subjected to 10 cycles at 100 mV s^{-1} over a potential range of –0.8 to 1.4 V (vs. Ag/AgCl). The resulting electrode was dried at room temperature and designated poly(CTAB)/GO/GCE. This study presents the initial observation of the sensitive electrooxidation of MO on a poly(CTAB)/GO sensor. The functionalized graphene sheet, modified with CTAB, exhibited a markedly enhanced electrochemical response to MO compared to the unmodified GCE. The wide linear range of 50–60 µM and the LOD of 0.36 µM demonstrate the proposed composite's characteristics in the MO analysis. The sensor exhibits excellent stability, anti-interference, and reproducibility properties. Furthermore, the sensor was tested on real samples. Furthermore, reducing the linear range to nanomolar levels enhances the effectiveness of the present work. The performance of the electrode indicates its potential for use as an effective sensing platform.[68]

In their study Zahid et al. in 2023 modified the GCE surface with a composite of surfactant 1-(2,4-initrophenyl)-dodecanoylthiourea (DAN) and gold nanoparticles (AuNPs) for the sensitive detection of oxoplatin (OXP), a widely used anticancer drug. The combination of DAN and NPs can act as an electrocatalyst for detecting analytes based on the unique interaction of the selected surfactant. The DAN/AuNPs/GCE platform, designed for the sensitive detection of oxoplatin, exhibits an LOD of 14.5 pM under optimal conditions. The sensor was characterized by electrochemical techniques, including spectroscopy and square wave voltammetry (SWV), CV, and EIS. OXP's response was also examined in an artificial serum sample. The modified electrode demonstrated an exceptional capacity for discriminating between the selected drug and various interfering agents at a concentration 100-fold higher than the drugs. Furthermore, it exhibited remarkable reproducibility, with an RSD of up to 3%. The efficacy of the developed sensors was assessed using serum samples, and the findings indicated promising recoveries, suggesting the potential practical applicability of the proposed sensors in the analysis of serum samples. The current research effectively addresses this field's previously identified research gap by performing voltammetric analysis to detect OXP at concentrations as low as picomolar levels.[69]

2.4.2 Analysis of Food

The use of surfactant-based carbon sensors offers several advantages in food analysis, particularly in enhancing the sensitivity, selectivity, and reliability of analytical measurements. Surfactants can modify the surface properties of carbon-based electrodes, such as graphene or CNTs, thereby enhancing their interaction with target analytes in food samples. By utilizing surfactant properties to modify carbon-based electrodes, researchers can develop sensors tailored toward specific food analytes, improve their analytical performance, and expand their applications in food analysis and safety assurance.

Sudan I is a food additive, as are chili powder and ketchup sauce, and is also used as a color additive. In their study, Mo et al. in 2010 created a unique composite-modified electrode containing a Gemini surfactant (cationic $C_{12}H_{25}N(CH_3)_2$-C_4H_8-$N(CH_3)_2C_{12}H_{25}Br_2$, C_{12}-C_4-C_{12}). The composite modified electrode was created by combining CNTs and hydrophilic ionic liquid to improve the detection of Sudan I. The electrochemical behavior of Sudan I was investigated using CV and linear scanning voltammetry (LSV). The resulting electrode demonstrated enhanced sensitivity for Sudan I compared to previous electrodes. In optimal conditions, the electrode exhibited a linear reaction for Sudan I between 0.05 and 2.0 μM, detecting levels as low as 0.03 μM. The method successfully detected Sudan I in powdered pepper samples and tomato sauce, with satisfactory recovery rates. The electrode demonstrated excellent stability and consistency, rendering it suitable for real-life applications, such as detecting Sudan I in food samples. Overall, the composite electrode exhibits the advantages of simple fabrication, stability, and high sensitivity, which could be advantageous for the analysis of food products.[70]

Rutin (RTN), or rutoside or vitamin P, is a highly potent bioflavonoid or plant pigment found in various plant sources, including fruits, vegetables, and plant extracts.[71] Hareesha et al. in 2022 developed a novel nanocomposite sensor for detecting RTN, a plant antioxidant. A CMB/GNPs-CBCPE sensor was prepared using graphene nanoplatelets (GNPs) and cetrimonium bromide (CMB)-modified CPE as a surfactant. The surfactants CMB, SDS, and octoxynol-9 (ONL-9) were added to the GNPs-CBCPE surface using the drop-casting methodology. A more sensitive surfactant and an optimal surfactant concentration were selected based on the electrochemical redox behavior of RTN on the GNPs-CBCPE surface, as determined by the CV method. The concentration of CMB required to modify the GNPs-CBCPE surface was determined to be 10.0 μL, which proved optimal in the context of the overall experimental investigation. This sensor, capable of measuring RTN in various samples, was successfully demonstrated for the first time using cost-effective preparation methods. The materials were subjected to extensive characterization using FESEM, EIS, and CV measurements, and the efficiency of the sensor was investigated using LSV, CV, and DPV electrochemical techniques. The sensor has been successfully applied to analyze RTN in various samples, including blood serum, urine, green tea, tomato, apple juice, and medical samples, with satisfactory results. This environmentally friendly sensor offers a cost-effective and efficient solution for rapid RTN detection in several fields, including food, pharmaceuticals, and the environment.[72]

Tartrazine (E102, TR) is one of the food industry's most frequently utilized azo dyes. It is a synthetic organic food dye commonly found in beverages, candies, dairy products, and bakery items. Given the azo groups and aromatic ring structures in the molecule, regulating its content is necessary to prevent potential harm to humans.[73] The use of SDS surfactant electrodes for the electrochemical detection of specific biomolecules has been extensively documented in the scientific literature. Raril and Manjunatha's research in 2021 aimed to examine the voltammetric behavior of tartrazine. An SDS-modified carbon paste electrode (SDSMCPE) was prepared by immobilizing 10 μL of SDS on the electrode surface for 5 min. The SDSMCPE sensor demonstrated excellent performance in TR oxidation, exhibiting high electrocatalytic activity and a low detection limit of 52×10^{-7} M. Compared to the bare CPE, the SDSMCPE exhibited enhanced sensitivity and selectivity. The modified electrode with surface immobilization of SDS demonstrated an increased ability to sense TR. The SDSMCPE sensor successfully detected TR in lemon yellow powder samples.[74]

Sweeteners, including vanillin (VN), are frequently incorporated into food products to enhance flavor. VN is a significant food flavoring substance utilized in many culinary preparations to augment the flavor of milk. Furthermore, it plays a pivotal role in the formulation of pharmaceuticals used to treat epilepsy.[75] Raril and Manjunatha present a novel approach for determining VN based on an ionic surfactant-modified GPE in their study. A CTAB-modified graphene paste electrode (CTABMGPE) was prepared by immobilizing the surfactant in BGPE. The optimal concentration of CTAB was determined to be 10 μL, and this concentration was subsequently employed in all subsequent experiments. The voltammetric behavior of VN on a CTABMGPEwas investigated using three analytical techniques: LSV, CV, and DPV. VN exhibits irreversible behavior, and the nature of the electrode response is observed to be adsorption controlled. Under optimized conditions, the detection limit for VN was found to be 1.29 μM. The developed sensor was also successfully used to determine VN in food samples, including biscuits and beverages, and it exhibited satisfactory recovery.[76]

2.4.3 Analysis of Pollutants

Surfactant-based carbon sensors offer many advantages in analyzing contaminants, thus making them valuable tools in environmental monitoring and pollution control efforts. The sensor studies demonstrated that surfactant coatings on CNTs enhanced the electrochemical response to heavy metals, resulting in improved detection limits and selectivity compared to unmodified electrodes. These studies emphasize the efficacy of surfactant-based carbon sensors in contaminant analysis and highlight their role in overcoming challenges related to sensitivity, selectivity, and interference in environmental samples.

Modifiers are selected based on their ability to reduce the likelihood of reduction and oxidation of electroactive species compared to a bare electrode. Various surfactant modifiers, including CTAB, SDS, TX-100, poly(glycine), and nanoparticles, are commonly employed due to their enhanced responsiveness to acidic and alkaline environments.[77] Amrutha et al. developed an electrochemical sensor for the specific and sensitive detection of Evans blue (EB) through surface modification of a carbon nanotube paste electrode (TX100MCNTPE) using a TX-100 surfactant. In the present study, immobilization was conducted by dispensing 10 μL of TX-100 onto the BCNTPE surface and allowing it to rest for 5 min. Subsequently, the excess surfactant was removed by washing with double-distilled water. The electrode morphology was characterized by FESEM. The anodic peak current of EB demonstrated a linear increase, observed with a 10-100 μM linear range, with an LOD of 5.06 μM and an LOQ of 16.87 μM. The operational sensor can simultaneously be employed to determine EB in the presence of TZ via CV. The modified sensor displays satisfactory linearity, reproducibility, and precision.[78]

In this study, Lien et al. present a novel approach for fabricating an electrochemical sensing platform, AuNDs/CTAB/GCE, to detect Bisphenol A (BPA). This platform employs the galvanostatic deposition of Au (nanodendrites, AuNDs) on a GCE coated with CTAB. In this context, the GCE-supported AuNDs and CTAB were demonstrated to function as electrochemical sensors with sensitive electrochemically active surface areas due to their hierarchical nanodendrite structure. Furthermore, the formation of nanodendrites on CTAB/GCE was completed in only 5 min using a simple one-step electrodeposition process. The addition of CTAB enhanced the sensitivity of the electrochemical sensors utilized to detect BPA. The electrochemical response of the sensor to BPA was investigated using CV and DPV. The sensor exhibited high reproducibility (RSD 4.9%, seven independent sensors) with a low detection limit of 22 nM and a linear range from 0.025 to 10 μM. The AuNDs/CTAB/GCE sensor exhibited high accuracy (96.60%–102.82%) in detecting BPA in real plastic water bottle samples.[79]

Phenol (PNL) and phenolic compounds are significant contaminants of water resources, as they are utilized to manufacture numerous industrial products. As a matter of environmental concern, Congur and Gül sought to develop a practical, robust, sensitive, selective, and cost-effective analytical instrument for monitoring PNL. A novel voltammetric sensor platform was fabricated for the selective, sensitive, and practical detection of PNL. To this end, a DTAB-PGE sensor modified with a dodecyl trimethylammonium bromide (DTAB) surfactant was fabricated for the first time in the PGEs literature. The DTAB-PGE-based electrochemical sensor platform has the potential to serve as a model for the development of miniaturized and microfluidic chip devices for the detection of a range of environmental and food contaminants.[80]

Zahid et al. developed a highly sensitive electrochemical GCE sensor for the detection of Cd(II) and Hg(II) using a novel surfactant, 1-cyclohexyl-3-dodecanoylthiourea (CDT), as a recognition layer. The use of CDT as a recognition layer represents the first step in developing an electrochemical sensor for the trace detection of Cd(II) and Hg(II) ions in water samples. Theoretical studies were conducted to calculate the interaction energy of CDT with Cd(II) and Hg(II), and the results were found to align with the experimental findings. The modified electrode exhibited a 10% enhancement in the current signals of the analytes relative to the BGCE. The RSD values obtained in the analysis of real water samples using the proposed method were less than 4%. The impact of diverse electrolytes on the sensor's functionality was also investigated. When conducted under suitable experimental conditions, linear calibration plots were obtained over wide concentration ranges, with detection limits of 0.084 μg/L and 0.090 μg/L for Cd(II) and Hg(II), respectively.[81]

2.4.4 Analysis of Biomolecules

Surfactant-based carbon sensors are highly effective for the analysis of biomolecules, exhibiting improved sensitivity, selectivity, and stability. Incorporating surfactants into carbon sensors enhances their electrochemical characteristics, namely, the surface area and electron transfer efficiency. This is important for detecting low concentrations of biomolecules, including proteins, enzymes, and nucleic acids. Moreover, surfactants facilitate the dissolution of hydrophobic biomolecules, enhancing their accessibility for detection and ensuring consistent measurements. These adaptable sensors can be applied to various fields, including clinical diagnostics, environmental monitoring, and food safety. Studies have demonstrated the potential of surfactant-modified electrodes for the sensitive detection of biomolecules and their applicability in developing biosensing technologies.[82]

Thomas et al. developed an amperometric sensor for the determination of epinephrine (EP) by modifying CPE with pristine multi-walled carbon nanotubes (pMWCNTs) using bulk modification, followed by the addition of SDS to the surface to optimize the sensor's potential application. This study demonstrates that MCPE/pMWCNTs/SDS exhibits sufficient sensitivity, selectivity, and stability, rendering it a viable sensor for EP determination. The addition of SDS significantly impacted the sensor's performance, acting as an effective supporting material. The presence of SDS on the surface was confirmed by energy-dispersive X-ray analysis. The presence of 100-fold excesses of serotonin, acetaminophen, folic acid, uric acid, tryptophan, tyrosine, and cysteine, as well as tenfold excesses of ascorbic acid and twofold excesses of dopamine, did not impede the quantification of EP at this electrode. The modified electrode is effective in analytical applications, as evidenced by its ability to quantify EP in blood serum and adrenaline tartrate injection.[83]

Pushpanjali and Manjunatha developed an electrochemical sensor designated sodium lauryl sulfate (SLS)-modified carbon nanotube paste electrode (SLSMCNTPE). The CNTPE was fabricated by mixing silicone oil with CNTs. Subsequently, the mixture was transferred into the fossa of the Teflon tube and kneaded until a smooth surface was achieved. An electrical connection was established using a copper wire, and the prepared electrode was designated as BCNTPE. The immobilization of the SLS solution onto the surface of the electrode resulted in the modification of BCNTPE, which enhanced the electrocatalytic response. The technique employed for the detection of 8-hydroxyquinoline (8HQ) was CV. The respective LOD and LOQ, as well as the linear range, were determined to be 1.1×10^{-7} M, 3.9×10^{-7} M, and 2×10^{-5} to 3.5×10^{-4} M[84].

Alloxan (AL) is a toxic glucose-like substance that induces diabetes mellitus by irreversibly damaging the β-cells of the pancreas, which are responsible for insulin production. In their study, Monnappa et al. employed a CPE immobilized with CTAB (CTABMCPE) to analyze AL in the presence of anthrone, as shown in Figure 2.3. The cationic surfactant CTAB, when combined with the substrate in convinced forms, was found to increase the redox current of AL. The electrode demonstrated excellent performance metrics and was employed to analyze AL in refined flour samples. The surface morphology analysis demonstrated an LOD value of 1.09 μM, thereby corroborating the effectiveness of the sensor. The recovery range for real sample analysis was 96.70% to 99.13%. The sensor demonstrated sufficient consistency, reliability, and durability, proving a valuable tool for determining AL in different samples. The modified electrode exhibited exceptional sensitivity and selectivity toward AL, thereby demonstrating the potential of surfactants to enhance

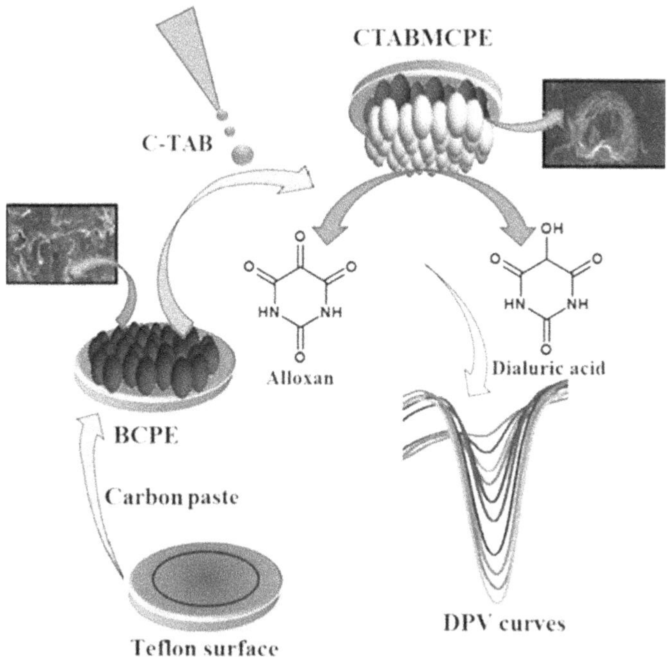

FIGURE 2.3 Schematic representation of the CTABMCPE sensor. (Reprinted from: Monnappa, A. B., Manjunatha, J. G., & Bhatt, A. S. (2020). Design of a sensitive and selective voltammetric sensor based on a cationic surfactant-modified carbon paste electrode for the determination of alloxan. ACS Omega, 5(36), 23481–23490. © 2020 ACS, with permission from ACS.)

AL performance. In conclusion, the electrode represents a rapid and cost-effective alternative to other techniques for detecting AL.[85]

Kumar and colleagues employed a mechanochemical approach to synthesize ZnO in a different study. ZnO/CTAB MCPE was prepared by immobilizing ZnO/CTAB on a bare CPE. A solution comprising 10 mL of ZnO and 25 mM CTAB was distributed on the CPE and incubated for 5 min. The same procedure was employed to prepare the ZnO/TX-100 MCPE, with TX-100 utilized in place of CTAB. The MCPEs containing ZnO/CTAB and ZnO/TX-100 were employed as dopamine (DA) electrochemical sensors. The impact of varying ZnO concentration, scan rate, and DA and ascorbic acid (AA) concentration was examined. Simultaneously, determining DA and AA at ZnO/TX-100 CPE was also conducted. The TX-100 modified CPE displays electrocatalytic characteristics, as evidenced by increased peak currents for DA. The determination of DA is less sensitive when CTAB-modified CPEs are employed. The high electrocatalytic behavior of the ZnO/TX-100 MCPE is anticipated to have significant applications in electroanalytical chemistry and biosensors.[86]

In another study, Prinith et al. in 2021 developed a novel sensor for the electrochemical oxidation of L-tyrosine (L-Ty) and DA using a carbon nanotube paste electrode (CNTPE) immobilized with composite surfactants sodium lauryl sulfate and TX-100. The surface of the prepared bare electrode was immobilized with 10 μL SLS. The electrode was then allowed to rest for 5 min and then rinsed with distilled water to remove the remaining SLS, labeled SLSMCNTPE. The same procedure was then performed with 10 μL of TX instead of SLS and named TXMCNTPE. The same procedures were performed in SLS and TX mixed surfactant (ST), and the modified electrode was named STMCNTPE. The STMCNTPE electrode showed an enhanced current signal with a detection limit of 3.75×10^{-7} M for L-Ty oxidation compared to other electrodes. The sensor was effectively utilized for L-Ty analysis in cow's milk and pharmaceuticals. It was found that an irreversible oxidation reaction with an adsorption-controlled mechanism characterizes the electrochemical behavior of L-Ty on STMCNTPE. In contrast, the oxidation of dopamine is reversible. The sensor showed high sensitivity and selectivity for L-Ty, even in the presence of DA, making it suitable for clinical diagnostic applications.[87]

Nanomaterials with zwitterionic surfactants have been demonstrated to enhance both biosensors' electrocatalytic activity and stability. In the subsequent study, Decarli and colleagues in 2022 developed a novel biosensor for the quantification of DA utilizing laccase (Lac) immobilized through physical adsorption on halide nanotubes, imidazolium zwitterionic surfactants, and 3-(1-alkyl-3-imidazolio) propane-sulfonate (HNT) in a CPE. The detection limit for DA was determined to be 0.252 μM for this biosensor. Considering the analytical and electrode performance, the produced biosensor exhibited a superior quality. Moreover, it was successfully employed for the detection of dopamine in pharmaceutical formulations. These findings demonstrate that the proposed modified sensor is an effective and innovative tool with the potential for practical clinical applications.[88]

2.4.5 Analysis of Other Chemical and Biochemical Substances

Surfactant-based carbon sensors are highly effective for the analysis of chemical and biochemical substances, exhibiting improved sensitivity, selectivity, and stability. Incorporating surfactants into carbon sensors enhances their electrochemical characteristics by increasing the surface area and facilitating superior electron transfer. This is of particular significance for detecting low concentrations of analytes, including toxins, enzymes, and small organic molecules.[89] The use of surfactant-modified electrodes for sensitively detecting toxic chemicals and biomolecules remains a highly active area of research.

In a study conducted by Prinith and Manjunatha, two promising electrochemical sensors were developed using an octyl phenol ethoxylate surfactant-modified carbon paste electrode (OPEMCPE) and a sodium lauryl sulfate-modified carbon paste electrode (SLSMCPE). The electrochemical response of Anthrone (ANT) was investigated by applying diverse voltammetric techniques. An amount of 10 mL of OPE was immobilized on the CPE surface via the drop-casting method for 5 min. The electrode was rinsed with distilled water to remove any residual surfactants that had not reacted. The resulting electrode was designated as OPEMCPE. The immobilization of 10 mL of SLS was conducted in accordance with the same procedure, resulting in the designation of the electrode as SLSMCPE. The modified sensors demonstrated excellent electrochemical performance, enhanced stability, rapid electron transfer rate, reproducibility, repeatability, and sensitivity exceeding the bare CPEs. The detection limit and quantification limit of anthrone used in CV were 18.8×10^{-7} M and 6.27×10^{-6} M on the OPEMCPE surface, while 10.6×10^{-7} M and 3.53×10^{-6} M on the SLSMCPE surface. Compared to the OPEMCPE electrode, the SLSMCPE electrode demonstrated a lower detection limit.[90]

In a recent study, Li et al. developed a novel electrochemical sensor for determining triclosan (TCS) using a different surfactant. A composite sensor was prepared using acetylene black (AB), CNT, and cetyl sulfonyl betaine (SB16). An amount of 10 μL of the SB16 solution (100 μM) was deposited onto the surface of the prepared electrode for 5 min. Subsequently, the electrode was washed with distilled water and dried, thus obtaining the AB/CNT/SB16-modified electrode. The effective combination of CNT, SB16, and AB enhanced the sensor's sensitivity and conductivity. The linear range of the prepared method for the detection of TCS was found to be from 0.01 μM to 2.0 μM, with a detection limit of 6.18 nM. The proposed sensor detected TCS in real samples, yielding an RSD of 0.56%–2.32%. These results align with those obtained through the high-performance liquid chromatography (HPLC) method. Consequently, the fabricated AB/CNT/SB16 sensor proved invaluable for detecting TCS in real samples, offering unparalleled selectivity, sensitivity, and accuracy.[91]

In a separate study, Tigari and colleagues in 2020 developed a novel sensor utilizing an anionic surfactant SLS-modified carbon nanotube and pencil graphite composite paste electrode (SLSMCNTPGCPE).[92] The surface modification of the electrode was conducted by immobilizing 10 mL of SLS surfactant on the carbon nanotube and pencil graphite composite paste electrode. The developed SLSMCNTPGCPE is a low-cost and highly active sensor successfully employed to determine riboflavin (RF) in the presence of DA. The fabricated sensor exhibits a linear current response to various RF concentrations of 0.2–0.8 mM and 1–5 mM with an LOD of 1.16×10^{-8} M by applying DPV. The sensor's stability, repeatability, reproducibility, and interference have been evaluated, and satisfactory results have been obtained. Furthermore, the sensor has been employed to investigate RF and DA simultaneously, yielding promising outcomes. The novel sensor has been used to estimate RF, demonstrating a notable enhancement in the analysis of Vitamin B-complex pills and natural food supplements. The presented sensing strategy has been successfully applied to pharmaceutical and food samples, demonstrating excellent recovery. Therefore, the presented sensor can be utilized for the routine analysis of RF in authentic samples. Furthermore, the electrode displays high stability, minimal interference effects, and excellent reproducibility while exhibiting environmentally friendly characteristics.[93]

Using pesticides for profit has given rise to substantial apprehensions about food safety and the environment and is regarded as one of numerous health concerns. Implementing sophisticated pesticide detection methodologies is essential in preventing pesticide-related hazards and safeguarding public health and the environment. Malode et al. in 2021 describe the electrochemical activity of aminotriazole (ATZ) by forming titanium dioxide nanoparticles (TiO_2) and a CTAB-based-CPE (TiO_2/CTAB-CPE). The square wave voltammetric technique investigated the herbicide ATZ in soil and water samples. The LOD and LOQ of ATZ were determined to be 2.53 nM and 8.44 nM, respectively. The proposed method was demonstrated to be a practical approach for sensitively determining ATZ herbicide in water and soil samples. The efficacy of the sensor in detecting ATZ in soil and water samples has been proved. Its applications include a low detection limit, high electrical conductivity, fast response, reproducibility, sensitivity, and electrocatalytic properties, as demonstrated by SWV and CV methods. The constructed TiO_2/CTAB-CPE sensor demonstrated an LOD of 2.53 nM for ATZ. From a rational standpoint, the sensor is straightforward to build, cost-effective, environmentally friendly, and aligns with the ecological standards of today's sensors. The method was proven effective and sensitive for the initial monitoring and determination of ATZ herbicide in real water and soil samples.[94]

2.5 CONCLUSION

The utilization of surfactants in electrochemical sensors and biosensors is on the rise due to their ability to facilitate the adsorption of the target analyte and accelerate electron transfer. Surfactants are organic compounds comprising hydrophilic and lipophilic groups, capable of markedly reducing the surface tension of a liquid. Surfactants have the capacity to modify and control the properties of electrode surfaces by forming a layer of molecules that can affect the catalytic activity of the electrode, adsorption of other species, mass transport, and electron transfer kinetics. Consequently, they are extensively utilized in the fabrication of electrochemical sensors. The use of surfactants in electroanalytical methods has increased in recent years. A variety of surfactants have been employed as electrode alternatives to enhance sensitivity. Including or excluding surfactants in an analytical method depends on the charge of the surfactant and nature of the analyte. The development of various platforms has demonstrated that the presence of trace amounts of surfactant markedly enhances the electrochemical responses of different analytes. Moreover, surfactants play a crucial role in maintaining the stability of the voltammetric response by preventing the contamination of the electrode surface. It has been demonstrated that surfactants can enhance the accumulation of specific electroactive molecules on the electrode surface, thereby improving the analytical signal and sensitivity of the proposed method. Combining carbon-based electrodes with surfactants is a widely utilized approach in electrochemical studies, as it enhances surface properties and elevates measurement sensitivity. The term "carbon sensors" is used to describe a variety of sensors, the specific designation of which depends on the particular application and the nature of the material being detected. Numerous research studies commonly utilize carbon sensors such as SPE, GCE, CPE, and PGE. It is widely acknowledged that carbon-based electrodes typically exhibit a high surface area. The use of surfactants can further enhance the surface properties of these electrodes. Subsequently, the CPEs are prepared and added directly to the paste, which is then dried and coated onto the carbon electrode surfaces through electrodeposition. Concurrently, using surfactant-based carbon sensors, many electrochemical studies have analyzed chemical and biochemical substances, including pharmaceuticals, food, food pathogens, pesticides, environmental pollutants, and biomolecules. Several surfactants have been employed in sensor studies, including CMB, SDS, SLS, CTAB, TX-100, CDT, DAN, OPE, SB16, and others. The most commonly used surfactants are SDS, CTAB, and TX-100. A literature review reveals that surfactant-based sensors are typically prepared with CPE.

ACKNOWLEDGMENTS

Kübra Turan was supported by the Scientific and Technological Research Council of Türkiye with a 2218-National Postdoctoral Research Fellowship Program scholarship in the 121C426 number project.

NOTES

1 Surbhi Mahender Saini, Titiksha Kumar Sagar, Yashi Dwivedi, and Sandeep Chandrashekharappa, "Brief Overview of Surfactants, Properties, Classification, Passivation, and Role in Chemistry." In *Surfactant Based Electrochemical Sensors and Biosensors*, edited by Jamballi G. Manjunatha and Chaudhery Mustansar Hussain, 3–20. Amsterdam: Elsevier, 2024. https://doi.org/10.1016/B978-0-443-15493-5.00018-X.

2 M. Aguirre-Ramírez, H. Silva-Jiménez, I. M. Banat, and M. A. Díaz De Rienzo, "Surfactants: Physicochemical Interactions with Biological Macromolecules." *Biotechnology Letters* 43, no. 3 (2021): 523–35. https://doi.org/10.1007/s10529-020-03054-1

3 R. Vittal, H. Gomathi, and Kang-Jin Kim, "Beneficial Role of Surfactants in Electrochemistry and in the Modification of Electrodes." *Advances in Colloid and Interface Science* 119, no. 1 (2006): 55–68. https://doi.org/10.1016/j.cis.2005.09.004.

4 Junmei Fan, Liping Ding, and Yu Fang, "Surfactant Aggregates Encapsulating and Modulating: An Effective Way to Generate Selective and Discriminative Fluorescent Sensors." *Langmuir* 35, no. 2 (January 15, 2019): 326–41. https://doi.org/10.1021/acs.langmuir.8b02111.

5 Imdad Ullah, Afzal Shah, Amin Badshah, Afzal Shah, Naseer Ali Shah, and Rico Tabor, "Surface, Aggregation Properties and Antimicrobial Activity of Four Novel Thiourea-Based Non-Ionic Surfactants." *Colloids and Surfaces A: Physicochemical and Engineering Aspects* 464 (January 2015): 104–9. https://doi.org/10.1016/j.colsurfa.2014.10.002.

6 Vittal, Gomathi, and Kim, "Beneficial Role of Surfactants."

7 Melinda Hersey, Shane N. Berger, Jordan Holmes, Alyssa West, and Parastoo Hashemi, "Recent Developments in Carbon Sensors for At-Source Electroanalysis." *Analytical Chemistry* 91, no. 1 (January 2, 2019): 27–43. https://doi.org/10.1021/acs.analchem.8b05151.

8 Pemmatte A. Pushpanjali, Jamballi G. Manjunatha, and Nagarajappa Hareesha, "An Overview of Recent Developments of Carbon-Based Sensors for the Analysis of Drug Molecules." *Journal of Electrochemical Science and Engineering* 11, no. 3 (2021): 161–77. https://doi.org/10.5599/JESE.999.

9 Hasret Subak, and Pınar Talay Pınar, "Electrochemical Oxidation of Ranitidine Using a Boron-Doped Diamond Electrode in the Presence of Anionic Surfactant: A Comprehensive Investigation." *Cumhuriyet Science Journal* 45, no. 2 (2024): 249–55.

10 Sunshine Holmberg, Maziar Ghazinejad, EunByul Cho, Derosh George, Brandon Pollack, Alexandra Perebikovsky, Regina Ragan, and Marc Madou, "Stress-Activated Pyrolytic Carbon Nanofibers for Electrochemical Platforms." *Electrochimica Acta* 290 (November 2018): 639–48. https://doi.org/10.1016/j.electacta.2018.09.013.

11 Raril Chenthattil, Jamballi G. Manjunatha, Doddarasinakere K. Ravishankar, Santosh Fattepur, Gurumallappa Siddaraju, and Lingappa Nanjundaswamy, "Validated Electrochemical Method for Simultaneous Resolution of Tyrosine, Uric

Acid, and Ascorbic Acid at Polymer Modified Nano-Composite Paste Electrode." *Surface Engineering and Applied Electrochemistry* 56, no. 4 (2020): 415–26. https://doi.org/10.3103/S1068375520040134
12 Hadi Beitollahi, Sayed Zia Mohammadi, Mohadeseh Safaei, and Somayeh Tajik, "Applications of Electrochemical Sensors and Biosensors Based on Modified Screen-Printed Electrodes: A Review." *Analytical Methods* 12, no. 12 (2020): 1547–60. https://doi.org/10.1039/c9ay02598g.
13 Zahra Taleat, Alireza Khoshroo, and Mohammad Mazloum-Ardakani, "Screen-Printed Electrodes for Biosensing: A Review (2008-2013)." *Microchimica Acta* 181, no. 9–10 (2014): 865–91. https://doi.org/10.1007/s00604-014-1181-1
14 Taleat, Khoshroo, and Mazloum-Ardakani, "Screen-Printed Electrodes for Biosensing."
15 Beitollahi et al., "Applications of Electrochemical Sensors."
16 Taleat, Khoshroo, and Mazloum-Ardakani, "Screen-Printed Electrodes for Biosensing."
17 Leonardo de Souza Vieira, A Review on the Use of Glassy Carbon in Advanced Technological Applications." *Carbon* 186 (022): 282–302. https://doi.org/10.1016/j.carbon.2021.10.022
18 P. Lakhera, V. Chaudhary, A. Jha, R. Singh, P. Kush, and P. Kumar, "Recent Developments and Fabrication of the Different Electrochemical Biosensors Based on Modified Screen Printed and Glassy Carbon Electrodes for the Early Diagnosis of Diverse Breast Cancer Biomarkers." *Materials Today Chemistry* 26 (2022): 101129. https://doi.org/10.1016/j.mtchem.2022.101129.
19 Vieira, "A Review on the Use of Glassy Carbon."
20 Hazem M. Abu Shawish, Mahmoud Elhabiby, Hadeel S. Abu Aziz, Salman M. Saadeh, and Ahmad Tbaza, "Determination of Trihexyphenidyl Hydrochloride Drug in Tablets and Urine Using a Potentiometric Carbon Paste Electrode." *Sensors and Actuators, B: Chemical* 235, no. 52 (2016): 18–26. https://doi.org/10.1016/j.snb.2016.05.058
21 Zorka Stanić and Stella Girousi, "Carbon Paste Electrodes in Potentiometry: The State of the Art and Applications in Modern Electroanalysis (a Review)." In *Sensing in Electroanalysis*, edited by Kalcher Kurt, Metelka Radovan, Švancara Ivan, and Vystřa Karel, Vol. 6. Pardubice: Univerzita Pardubice, 2011. http://dspace.upce.cz/handle/10195/42535
22 Abu Shawish et al., "Determination of Trihexyphenidyl Hydrochloride Drug in Tablets and Urine Using a Potentiometric Carbon Paste Electrode"; Stanić and Girousi, *Carbon Paste Electrodes in Potentiometry: The State of the Art and Applications in Modern Electroanalysis (a Review).*
23 Stanić and Girousi, "Carbon Paste Electrodes in Potentiometry."
24 Iulia Gabriela David, Dana-Elena Popa, and Mihaela Buleandra, "Pencil Graphite Electrodes: A Versatile Tool in Electroanalysis." *Journal of Analytical Methods in Chemistry* 2017, no. Cv (2017): 1905968. https://doi.org/10.1155/2017/1905968
25 Swati Sharma Annu, Rajeev Jain, and Antony Nitin Raja, "Review—Pencil Graphite Electrode: An Emerging Sensing Material." *Journal of The Electrochemical Society* 167, no. 3 (2020): 037501. https://doi.org/10.1149/2.0012003jes.
26 David, Popa, and Buleandra, "Pencil Graphite Electrodess."
27 Annu et al., "Review—Pencil Graphite Electrode."
28 Md Rajibul Akanda, Manzar Sohail, Md Abdul Aziz, and Abdel Nasser Kawde, "Recent Advances in Nanomaterial-Modified Pencil Graphite Electrodes for Electroanalysis." *Electroanalysis* 28, no. 3 (2016): 408–24. https://doi.org/10.1002/elan.201500374
29 Annu et al., "Review—Pencil Graphite Electrode."
30 Saini et al., "Brief Overview of Surfactants."
31 Drew Myers, *Surfactant Science and Technology*, edited by Drew Myers. Fourth. New Jersey, USA: John Wiley & Sons, 2020.
32 Ya R. Bazel, I. P. Antal, V. M. Lavra, and Zh A. Kormosh, "Methods for the Determination of Anionic Surfactants." *Journal of Analytical Chemistry* 69, no. 3 (2014): 211–36. https://doi.org/10.1134/S1061934814010043
33 Rui A. Gonçalves, Krister Holmberg, and Björn Lindman, "Cationic Surfactants: A Review." *Journal of Molecular Liquids* 375 (April 2023): 121335. https://doi.org/10.1016/j.molliq.2023.121335
34 Chengcheng Zhou, and Yilin Wang, "Structure–Activity Relationship of Cationic Surfactants as Antimicrobial Agents." *Current Opinion in Colloid & Interface Science* 45 (2020): 28–43. https://doi.org/10.1016/j.cocis.2019.11.009
35 Bengt Kronberg, Krister Holmberg, and Björn Lindman, "Types of Surfactants, Their Synthesis, and Applications." In *Surface Chemistry of Surfactants and Polymers*, 1–47. West Sussex, UK: Wiley, 2014. https://doi.org/10.1002/9781118695968.ch1
36 Guang-Guo Ying, "Fate, Behavior and Effects of Surfactants and Their Degradation Products in the Environment." *Environment International* 32, no. 3 (2006): 417–31. https://doi.org/10.1016/j.envint.2005.07.004
37 J Anandkumar, Manisha Agrawal, and Biju Prava Sahariah, "Surfactants and Bioremediation." In *Trends in Biological Processes in Industrial Wastewater Treatment*, 1–11. Philadelphia, USA: IOP Publishing, 2024. https://doi.org/10.1088/978-0-7503-5678-7ch1
38 Praveen Naik, Naveenchandra Pilicode, Kavya S. Keremane, Madhukara Acharya, and Airody Vasudeva Adhikari, "Synthesis, Optical, Electrochemical, and Computational Investigation of New Cyanopyridine-Centered Organic Dyads." *Optical Materials* 142 (August 2023): 114002. https://doi.org/10.1016/j.optmat.2023.114002

39 N. Hareesha, and J. G. Manjunatha, "Electro-Oxidation of Formoterol Fumarate on the Surface of Novel Poly(Thiazole Yellow-G) Layered Multi-Walled Carbon Nanotube Paste Electrode." *Scientific Reports* 11, no. 1 (2021): 1–14. https://doi.org/10.1038/s41598-021-92099-x
40 Maryam Mosleh, Sayed Mehdi Ghoreishi, Saeed Masoum, and Asma Khoobi, "Determination of Quercetin in the Presence of Tannic Acid in Soft Drinks Based on Carbon Nanotubes Modified Electrode Using Chemometric Approaches." *Sensors and Actuators B: Chemical* 272 (November 2018): 605–11. https://doi.org/10.1016/j.snb.2018.05.172
41 Isha Soni, and Gururaj Kudur Jayaprakash, "A Short Review on the Analysis of the Adsorptive Behavior of Surfactants at Carbon Paste Electrodes for Electrochemical Sensing." *Journal of Molecular Liquids* 388 (October 2023): 122737. https://doi.org/10.1016/j.molliq.2023.122737
42 N. Sandhyarani, "Surface Modification Methods for Electrochemical Biosensors." In *Electrochemical Biosensors*, edited by Ali A. Ensafi, 45–75. Amsterdam: Elsevier, 2019. https://doi.org/10.1016/B978-0-12-816491-4.00003-6
43 J. G. Manjunatha, "A Surfactant Enhanced Graphene Paste Electrode as an Effective Electrochemical Sensor for the Sensitive and Simultaneous Determination of Catechol and Resorcinol." *Chemical Data Collections* 25 (February 2020): 100331. https://doi.org/10.1016/j.cdc.2019.100331
44 Vinoda B. Patil, Shweta J. Malode, Suresh M. Tuwar, and Nagaraj P. Shetti, "Graphene Sheet-Based Electrochemical Sensor with Cationic Surfactant for Sensitive Detection of Atorvastatin." *Sensors International* 3 (2022): 100198. https://doi.org/10.1016/j.sintl.2022.100198
45 B. M. Amrutha, J. G. Manjunatha, A. S. Bhatt, C. Raril, and P. A. Pushpanjali, "Electrochemical Sensor for the Determination of Alizarin Red-S at Non-Ionic Surfactant Modified Carbon Nanotube Paste Electrode." *Physical Chemistry Research* 7, no. 3 (2019): 523–33. https://doi.org/10.22036/pcr.2019.185875.1636
46 Hongyu Yang, Xiaojiang Zheng, Wensheng Huang, and Kangbing Wu, "Modification of Montmorillonite with Cationic Surfactant and Application in Electrochemical Determination of 4-Chlorophenol." *Colloids and Surfaces B: Biointerfaces* 65, no. 2 (2008): 281–84. https://doi.org/10.1016/j.colsurfb.2008.04.013
47 Kübra Turan, Engin Er, Ahmet Uçar, and Gözde Aydoğdu Tığ, "Evaluations of Neurochemicals at Surfactant Modified Electrochemical Sensors and Biosensors." In *Surfactant Based Electrochemical Sensors and Biosensors*, 275–313. Elsevier, 2024. https://doi.org/10.1016/B978-0-443-15493-5.00016-6
48 Didem Nur Unal, Sercan Yıldırım, Sevinc Kurbanoglu, and Bengi Uslu, "Current Trends and Roles of Surfactants for Chromatographic and Electrochemical Sensing." *TrAC Trends in Analytical Chemistry* 144 (November 2021): 116418. https://doi.org/10.1016/j.trac.2021.116418
49 Rakesh R. Sawkar, Mahesh M. Shanbhag, Suresh M. Tuwar, Kunal Mondal, and Nagaraj P. Shetti, "Sodium Dodecyl Sulfate–Mediated Graphene Sensor for Electrochemical Detection of the Antibiotic Drug: Ciprofloxacin." *Materials* 15, no. 22 (2022): 7872. https://doi.org/10.3390/ma15227872
50 B. M. Amrutha, J. G. Manjunatha, S. Aarti Bhatt, and N. Hareesha, "Electrochemical Analysis of Evans Blue by Surfactant Modified Carbon Nanotube Paste Electrode." *Journal of Materials and Environmental Science* 10, no. 9 (2019): 882–90. http://www.jmaterenvironsci.com
51 K. G. Manjunatha, B. E. Kumara Swamy, G. K. Jayaprakash, S. C. Sharma, P. Lalitha, and K. A. Vishnumurthy, "Electrochemical Determination of Paracetamol at Cu Doped ZnO/Nanoparticle with TX-100-Surfactant MCPE: A Cyclic Voltammetric Technique." *Inorganic Chemistry Communications* 142 (August 2022): 109630. https://doi.org/10.1016/j.inoche.2022.109630
52 Ganeshlenin Kandasamy, and Dipak Maity, "Surfactant Sensors for Analysis of Toxic Chemicals." In *Advances in Surfactant Biosensor and Sensor Technologies*, edited by J. G. Manjunatha, 169–88. Cham: Springer Nature Switzerland, 2024. https://doi.org/10.1007/978-3-031-60832-2_9
53 Girish Tigari and J. G. Manjunatha, "Poly(Glutamine) Film-Coated Carbon Nanotube Paste Electrode for the Determination of Curcumin with Vanillin: An Electroanalytical Approach." *Monatshefte Fur Chemie* 151, no. 11 (2020b): 1681–88. https://doi.org/10.1007/s00706-020-02700-8
54 Subak and Pınar, "Electrochemical Oxidation of Ranitidine."
55 Anum Zahid, Afzal Shah, and Shamsa Munir, "Gold Nanoparticles and Surfactant-Based Electrochemical Sensing Platform for the Detection and Electroanalysis of the Anticancer Drug Oxoplatin." *ECS Advances* 2, no. 4 (2023): 040506. https://doi.org/10.1149/2754-2734/ad040d; Pinky Abraham, S. Renjini, T. E. Mary Nancy, and V. Anitha Kumary, "Electrochemical Synthesis of Thin-Layered Graphene Oxide-Poly(CTAB) Composite for Detection of Morphine." *Journal of Applied Electrochemistry* 50, no. 1 (2020): 41–50. https://doi.org/10.1007/s10800-019-01367-2; Jędrzej Kozak, Katarzyna Tyszczuk-Rotko, Magdalena Wójciak, and Ireneusz Sowa, "Electrochemically Activated Screen-Printed Carbon Sensor Modified with Anionic Surfactant (ASPCE/SDS) for Simultaneous Determination of Paracetamol, Diclofenac and Tramadol." *Materials* 14, no. 13 (June 26, 2021): 3581. https://doi.org/10.3390/ma14133581
56 Veera Erady, Ronald J. Mascarenhas, and Ashis K. Satpati, "Highly Efficient and Selective Quantification of Vanillin in Food, Beverages and Pharmaceuticals Using Surfactant Modified Carbon Paste Sensor." *Sensors International* 1 (2020): 100023. https://doi.org/10.1016/j.sintl.2020.100023

57 Shweta J. Malode, Keerthi Prabhu K., Nagaraj P. Shetti, and Kakarla Raghava Reddy, "Highly Sensitive Electrochemical Assay for Selective Detection of Aminotriazole Based on TiO2/Poly(CTAB) Modified Sensor." *Environmental Technology & Innovation* 21 (February 2021): 101222. https://doi.org/10.1016/j.eti.2020.101222
58 A. A. P. Khan, A. Khan, A. M. Asiri, M. M. Alam, M. M. Rahman, and M. Shaban, "Surfactant-Assisted Graphene Oxide/Methylaniline Nanocomposites for Lead Ionic Sensor Development for the Environmental Remediation in Real Sample Matrices." *International Journal of Environmental Science and Technology* 16, no. 12 (December 25, 2019): 8461–70. https://doi.org/10.1007/s13762-019-02447-8
59 Unal et al., "Current Trends and Roles of Surfactants."
60 S. Irem Kaya, Ahmet Cetinkaya, and Sibel A. Ozkan, "Surfactant Sensors for Pharmaceutical/Medical Applications." In *Surfactant-Based Sensors in Chemical and Biochemical Detection*, edited by J. G. Manjunatha, 23:163–92. Royal Society of Chemistry, 2023. https://doi.org/10.1039/BK9781837671182-00163
61 J. G. Manjunatha, *Advances in Surfactant Biosensor and Sensor Technologies*, edited by J. G. Manjunatha. Cham, Switzerland: Springer Nature Switzerland, 2024. https://doi.org/10.1007/978-3-031-60832-2.
62 P. A. Pushpanjali, J. G. Manjunatha, N. Hareesha, Edwin S. D. Souza, M. M. Charithra, and N. S. Prinith, "Voltammetric Analysis of Antihistamine Drug Cetirizine and Paracetamol at Poly(L-Leucine) Layered Carbon Nanotube Paste Electrode." *Surfaces and Interfaces* 24, no. April (2021): 101154. https://doi.org/10.1016/j.surfin.2021.101154.
63 Gulcin Bolat, "Investigation of Poly(CTAB-MWCNTs) Composite Based Electrochemical DNA Biosensor and Interaction Study with Anticancer Drug Irinotecan." *Microchemical Journal* 159 (December 2020): 105426. https://doi.org/10.1016/j.microc.2020.105426
64 Vinoda B. Patil, Davalasab Ilager, Suresh M. Tuwar, Kunal Mondal, and Nagaraj P. Shetti, "Nanostructured ZnO-Based Electrochemical Sensor with Anionic Surfactant for the Electroanalysis of Trimethoprim." *Bioengineering* 9, no. 10 (2022): 521. https://doi.org/10.3390/bioengineering9100521
65 Manjunatha et al., "Electrochemical Determination of Paracetamol."
66 Sawkar et al., "Sodium Dodecyl Sulfate–Mediated Graphene Sensor."
67 Kozak et al., "Electrochemically Activated Screen-Printed Carbon Sensor."
68 Abraham et al., "Electrochemical Synthesis of Thin-Layered Graphene Oxide-Poly(CTAB)."
69 Zahid, Shah, and Munir, "Gold Nanoparticles and Surfactant-Based Electrochemical Sensing Platform."
70 Zhirong Mo, Yafen Zhang, Faqiong Zhao, Fei Xiao, Gaiping Guo, and Baizhao Zeng. "Sensitive Voltammetric Determination of Sudan I in Food Samples by Using Gemini Surfactant–Ionic Liquid–Multiwalled Carbon Nanotube Composite Film Modified Glassy Carbon Electrodes." *Food Chemistry* 121, no. 1 (July 2010): 233–37. https://doi.org/10.1016/j.foodchem.2009.11.077
71 Fernanda Damin, Adriana Meinhart, Lucas Caldeirão, Milton Filho, Letícia da Silva, Livia Constant, José Teixeira-Filho, Roger Wagner, and Helena Godoy, "Chemoprevendetermination of Rutin in Fruits and Vegetables in Natura." *Journal of Food and Nutrition Research* 58, no. 4 (2019): 328–38
72 N. Hareesha, J. G. Manjunatha, Zeid A. Alothman, and Mika Sillanpää, "Simple and Affordable Graphene Nano-Platelets and Carbon Nanocomposite Surface Decorated with Cetrimonium Bromide as a Highly Responsive Electrochemical Sensor for Rutin Detection." *Journal of Electroanalytical Chemistry* 917 (July 2022): 116388. https://doi.org/10.1016/j.jelechem.2022.116388
73 P. Barciela, A. Perez-Vazquez, and M. A. Prieto, "Azo Dyes in the Food Industry: Features, Classification, Toxicity, Alternatives, and Regulation." *Food and Chemical Toxicology* 178, no. July (2023): 113935. https://doi.org/10.1016/j.fct.2023.113935
74 C. Raril, and J. G. Manjunatha, "Development of Sodium Dodecyl Sulfate Based Electrochemical Sensor for Tartrazine Determination." *Portugaliae Electrochimica Acta* 39, no. 1 (2021): 59–70. https://doi.org/10.4152/pea.202101059
75 Jianlian Liu, Chaojun Du, Shengyun Wang, and Zhaoyang Li, "Preparation and Electrochemical Sensing Properties of Vanillin Molecularly Imprinted Polymer." *International Journal of Electrochemical Science* 17, no. 7 (2022): 220715. https://doi.org/10.20964/2022.07.17
76 C. Raril and J. G. Manjunatha, "A Simple Approach for the Electrochemical Determination of Vanillin at Ionic Surfactant Modified Graphene Paste Electrode." *Microchemical Journal* 154, no. December 2019 (2020): 104575. https://doi.org/10.1016/j.microc.2019.104575
77 Girish Tigari, Jamballi G. Manjunatha, Chenthatill Raril, and Nagarajappa Hareesha, "Determination of Riboflavin at Carbon Nanotube Paste Electrodes Modified with an Anionic Surfactant." *ChemistrySelect* 4, no. 7 (2019): 2168–73. https://doi.org/10.1002/slct.201803191.
78 Amrutha et al., "Electrochemical Analysis of Evans Blue."
79 Nguyen Thi Lien, Le Quoc Hung, Nguyen Tien Hoang, Vu Thi Thu, Dau Thi Ngoc Nga, Pham Thi Hai Yen, Pham Hong Phong, and Vu Thi Thu Ha, "An Electrochemical Sensor Based on Gold Nanodendrite/Surfactant Modified Electrode for Bisphenol A Detection." *Journal of Analytical Methods in Chemistry* 2020 (December 29, 2020): 6693595. https://doi.org/10.1155/2020/6693595
80 Gulsah Congur and Ülküye Dudu Gül, "Phenol Monitoring in Water Samples Using an Inexpensive Electrochemical Sensor Based on Pencil Electrodes Modified with DTAB Surfactant." *Journal of Environmental Chemical Engineering* 9, no. 5 (2021): 105804. https://doi.org/10.1016/j.jece.2021.105804

81 Anum Zahid, Afzal Shah, Faiza Jan Iftikhar, Aamir Hassan Shah, and Rumana Qureshi, “Surfactant Modified Glassy Carbon Electrode as an Efficient Sensing Platform for the Detection of Cd (II) and Hg (II).” *Electrochimica Acta* 235 (2017): 72–78. https://doi.org/10.1016/j.electacta.2017.03.120
82 Manjunatha, *Adv. Surfactant Biosens. Sens. Technol.*; Kandasamy and Maity, “Surfactant Sensors for Analysis of Toxic Chemicals”; Kaya, Cetinkaya, and Ozkan, “Surfactant Sensors for Pharmaceutical/Medical Applications.”
83 Tony Thomas, Ronald J. Mascarenhas, Ozma J. D’Souza, Simon Detriche, Zineb Mekhalif, and Praveen Martis, “Pristine Multi-Walled Carbon Nanotubes/SDS Modified Carbon Paste Electrode as an Amperometric Sensor for Epinephrine.” *Talanta* 125 (2014): 352–60. https://doi.org/10.1016/j.talanta.2014.03.027
84 P. A. Pushpanjali and J. G. Manjunatha, “A Sensitive Novel Approach towards the Detection of 8-Hydroxyquinoline at Anionic Surfactant Modified Carbon Nanotube Based Biosensor: A Voltammetric Study.” *Physical Chemistry Research* 7, no. 4 (2019): 813–22. https://doi.org/10.22036/pcr.2019.198397.1663
85 Amrutha B. Monnappa, Jamballi G. Manjunatha, and Aarti S. Bhatt, “Design of a Sensitive and Selective Voltammetric Sensor Based on a Cationic Surfactant-Modified Carbon Paste Electrode for the Determination of Alloxan.” *ACS Omega* 5, no. 36 (September 15, 2020): 23481–90. https://doi.org/10.1021/acsomega.0c03517
86 Mohan Kumar, B. E. Kumara Swamy, Sathish Reddy, T. V. Sathisha, and J. Manjanna, “Synthesis of ZnO and Its Surfactant Based Electrode for the Simultaneous Detection of Dopamine and Ascorbic Acid.” *Analytical Methods* 5, no. 3 (2013): 735–40. https://doi.org/10.1039/C2AY26124C.
87 Nambudumada S. Prinith, Jamballi G. Manjunatha, and Nagarajappa Hareesha, “Electrochemical Validation of L-Tyrosine with Dopamine Using Composite Surfactant Modified Carbon Nanotube Electrode.” *Journal of the Iranian Chemical Society* 18, no. 12 (2021): 3493–3503. https://doi.org/10.1007/s13738-021-02283-z.
88 Nícolas Oliveira Decarli, Eduardo Zapp, Bruno Silveira de Souza, Edson Roberto Santana, João Paulo Winiarski, and Iolanda Cruz Vieira. “Biosensor Based on Laccase-Halloysite Nanotube and Imidazolium Zwitterionic Surfactant for Dopamine Determination.” *Biochemical Engineering Journal* 186 (August 1, 2022): 108565. https://doi.org/10.1016/j.bej.2022.108565
89 J. G. Manjunatha, *Surfactant-Based Sensors in Chemical and Biochemical Detection*, edited by J. G. Manjunatha. Vol. 23. Croydon, UK: Royal Society of Chemistry, 2023. https://doi.org/10.1039/9781837671182.
90 N. S. Prinith and J. G. Manjunatha, “Surfactant Modified Electrochemical Sensor for Determination of Anthrone A Cyclic Voltammetry.” *Materials Science for Energy Technologies* 2, no. 3 (December 2019): 408–16. https://doi.org/10.1016/j.mset.2019.05.004.
91 Hui Li, Ziwen Deng, Qingwu Tian, Limin Lun, Peng Zhao, Xiaomin Yang, Jingli Shen, Bing Jiang, Yusun Zhou, and Tingting Zhou, “Application of Carbon Nanotubes and Zwitterionic Surfactant-Modified Acetylene Black Electrode for the Determination of Triclosan in Household Commodities.” *International Journal of Environmental Analytical Chemistry* 102, no. 4 (March 16, 2022): 987–1000. https://doi.org/10.1080/03067319.2020.1730338
92 Girish Tigari and J. G. Manjunatha, “A Surfactant Enhanced Novel Pencil Graphite and Carbon Nanotube Composite Paste Material as an Effective Electrochemical Sensor for Determination of Riboflavin.” *Journal of Science: Advanced Materials and Devices* 5, no. 1 (2020c): 56–64. https://doi.org/10.1016/j.jsamd.2019.11.001
93 Tigari and Manjunatha, “A Surfactant Enhanced Novel Pencil.”
94 Malode et al., “Highly Sensitive Electrochemical Assay.”
95 Hareesha et al., “Simple and Affordable Graphene Nano-Platelets and Carbon Nanocomposite Surface.”
96 Raril and Manjunatha, “Development of Sodium Dodecyl Sulfate.”
97 Prinith, Manjunatha, and Hareesha, “Electrochemical Validation of L-Tyrosine.”
98 Raril and Manjunatha, “Simple Approach for the Electrochemical Determination of Vanillin.”
99 Thomas et al., “Pristine Multi-Walled Carbon Nanotubes/SDS Modified Carbon Paste Electrode.”
100 Monnappa, Manjunatha, and Bhatt, “Design of a Sensitive and Selective Voltammetric Sensor.”
101 Lien et al., “Electrochemical Sensor Based on Gold Nanodendrite/Surfactant.”
102 Gulcin Bolat, Yesim Tugce Yaman, and Serdar Abaci, “Highly Sensitive Electrochemical Assay for Bisphenol A Detection Based on Poly (CTAB)/MWCNTs Modified Pencil Graphite Electrodes.” *Sensors and Actuators B: Chemical* 255 (February 2018): 140–48. https://doi.org/10.1016/j.snb.2017.08.001
103 Amrutha et al., “Electrochemical Analysis of Evans Blue.”
104 Congur and Gül, “Phenol Monitoring in Water Samples.”
105 Zahid et al., “Surfactant Modified Glassy Carbon Electrode.”
106 Nagarajappa Hareesha and Jamballi G. Manjunatha, “Surfactant and Polymer Layered Carbon Composite Electrochemical Sensor for the Analysis of Estriol with Ciprofloxacin.” *Materials Research Innovations* 24, no. 6 (September 18, 2020): 349–62. https://doi.org/10.1080/14328917.2019.1684657
107 Manjunatha et al., “Electrochemical Determination of Paracetamol.”
108 Sawkar et al., “Sodium Dodecyl Sulfate–Mediated Graphene Sensor.”
109 Kozak et al., “Electrochemically Activated Screen-Printed Carbon Sensor.”
110 Abraham et al., “Electrochemical Synthesis of Thin-Layered Graphene Oxide-Poly(CTAB).”
111 Zahid, Shah, and Munir, “Gold Nanoparticles and Surfactant-Based Electrochemical Sensing Platform.”
112 Prinith and Manjunatha, “Surfactant Modified Electrochemical Sensor.”

113 Tigari and Manjunatha, "A Surfactant Enhanced Novel Pencil."
114 Pushpanjali and Manjunatha, "A Sensitive Novel Approach towards the Detection of 8-Hydroxyquinoline."
115 Girish Tigari and J. G. Manjunatha. "Optimized Voltammetric Experiment for the Determination of Phloroglucinol at Surfactant Modified Carbon Nanotube Paste Electrode." *Instruments and Experimental Techniques* 63, no. 5 (2020a): 750–57. https://doi.org/10.1134/S0020441220050139
116 Li et al., "Application of Carbon Nanotubes."
117 Malode et al., "Highly Sensitive Electrochemical Assay."

REFERENCES

Abraham, Pinky, S. Renjini, T. E. Mary Nancy, and V. Anitha Kumary. "Electrochemical Synthesis of Thin-Layered Graphene Oxide-Poly(CTAB) Composite for Detection of Morphine." *Journal of Applied Electrochemistry* 50, no. 1 (2020): 41–50. https://doi.org/10.1007/s10800-019-01367-2.

Abu Shawish, Hazem M., Mahmoud Elhabiby, Hadeel S. Abu Aziz, Salman M. Saadeh, and Ahmad Tbaza. "Determination of Trihexyphenidyl Hydrochloride Drug in Tablets and Urine Using a Potentiometric Carbon Paste Electrode." *Sensors and Actuators, B: Chemical* 235, no. 52 (2016): 18–26. https://doi.org/10.1016/j.snb.2016.05.058.

Aguirre-Ramírez, M., H. Silva-Jiménez, I. M. Banat, and M. A. Díaz De Rienzo. "Surfactants: Physicochemical Interactions with Biological Macromolecules." *Biotechnology Letters* 43, no. 3 (2021): 523–35. https://doi.org/10.1007/s10529-020-03054-1.

Akanda, Md Rajibul, Manzar Sohail, Md Abdul Aziz, and Abdel Nasser Kawde. "Recent Advances in Nanomaterial-Modified Pencil Graphite Electrodes for Electroanalysis." *Electroanalysis* 28, no. 3 (2016): 408–24. https://doi.org/10.1002/elan.201500374.

Amrutha, B. M., J. G. Manjunatha, A. S. Bhatt, C. Raril, and P. A. Pushpanjali. "Electrochemical Sensor for the Determination of Alizarin Red-S at Non-Ionic Surfactant Modified Carbon Nanotube Paste Electrode." *Physical Chemistry Research* 7, no. 3 (2019): 523–33. https://doi.org/10.22036/pcr.2019.185875.1636.

Amrutha, B. M., J. G. Manjunatha, S. Aarti Bhatt, and N. Hareesha. "Electrochemical Analysis of Evans Blue by Surfactant Modified Carbon Nanotube Paste Electrode." *Journal of Materials and Environmental Science* 10, no. 9 (2019): 882–90. http://www.jmaterenvironsci.com.

Ananda Murthy, H. C., Ararso Nagari Wagassa, C. R. Ravikumar, and H. P. Nagaswarupa. "Functionalized Metal and Metal Oxide Nanomaterial-Based Electrochemical Sensors." In *Functionalized Nanomaterial-Based Electrochemical Sensors*, edited by Chaudhery Mustansar Hussain and Jamballi G. Manjunatha, 369–92. Duxford: Elsevier, 2022. https://doi.org/10.1016/B978-0-12-823788-5.00001-6.

Anandkumar, J, Manisha Agrawal, and Biju Prava Sahariah. "Surfactants and Bioremediation." In *Trends in Biological Processes in Industrial Wastewater Treatment*, 1–11. Philadelphia, USA: IOP Publishing, 2024. https://doi.org/10.1088/978-0-7503-5678-7ch1.

Annu, Swati Sharma, Rajeev Jain, and Antony Nitin Raja. "Review—Pencil Graphite Electrode: An Emerging Sensing Material." *Journal of The Electrochemical Society* 167, no. 3 (2020): 037501. https://doi.org/10.1149/2.0012003jes.

Barciela, P., A. Perez-Vazquez, and M. A. Prieto. "Azo Dyes in the Food Industry: Features, Classification, Toxicity, Alternatives, and Regulation." *Food and Chemical Toxicology* 178, no. July (2023): 113935. https://doi.org/10.1016/j.fct.2023.113935.

Bazel, Ya R., I. P. Antal, V. M. Lavra, and Zh A. Kormosh. "Methods for the Determination of Anionic Surfactants." *Journal of Analytical Chemistry* 69, no. 3 (2014): 211–36. https://doi.org/10.1134/S10619 34814010043.

Beitollahi, Hadi, Sayed Zia Mohammadi, Mohadeseh Safaei, and Somayeh Tajik. "Applications of Electrochemical Sensors and Biosensors Based on Modified Screen-Printed Electrodes: A Review." *Analytical Methods* 12, no. 12 (2020): 1547–60. https://doi.org/10.1039/c9ay02598g.

Bolat, Gulcin. "Investigation of Poly(CTAB-MWCNTs) Composite Based Electrochemical DNA Biosensor and Interaction Study with Anticancer Drug Irinotecan." *Microchemical Journal* 159 (December 2020): 105426. https://doi.org/10.1016/j.microc.2020.105426.

Bolat, Gulcin, Yesim Tugce Yaman, and Serdar Abaci. "Highly Sensitive Electrochemical Assay for Bisphenol A Detection Based on Poly (CTAB)/MWCNTs Modified Pencil Graphite Electrodes." *Sensors and Actuators B: Chemical* 255 (February 2018): 140–48. https://doi.org/10.1016/j.snb.2017.08.001.

Congur, Gulsah, and Ülküye Dudu Gül. "Phenol Monitoring in Water Samples Using an Inexpensive Electrochemical Sensor Based on Pencil Electrodes Modified with DTAB Surfactant." *Journal of Environmental Chemical Engineering* 9, no. 5 (2021): 105804. https://doi.org/10.1016/j.jece.2021.105804.

Cserháti, Tibor, Esther Forgács, and Gyula Oros. "Biological Activity and Environmental Impact of Anionic Surfactants." *Environment International* 28, no. 5 (November 2002): 337–48. https://doi.org/10.1016/S0160-4120(02)00032-6.

Damin, Fernanda, Adriana Meinhart, Lucas Caldeirão, Milton Filho, Letícia da Silva, Livia Constant, José Teixeira-Filho, Roger Wagner, and Helena Godoy. "Chemoprevendetermination of Rutin in Fruits and Vegetables in Natura." *Journal of Food and Nutrition Research* 58, no. 4 (2019): 328–38.

David, Iulia Gabriela, Dana-Elena Popa, and Mihaela Buleandra. "Pencil Graphite Electrodes: A Versatile Tool in Electroanalysis." *Journal of Analytical Methods in Chemistry* 2017, no. Cv (2017): 1905968. https://doi.org/10.1155/2017/1905968.

Decarli, Nícolas Oliveira, Eduardo Zapp, Bruno Silveira de Souza, Edson Roberto Santana, João Paulo Winiarski, and Iolanda Cruz Vieira. "Biosensor Based on Laccase-Halloysite Nanotube and Imidazolium Zwitterionic Surfactant for Dopamine Determination." *Biochemical Engineering Journal* 186 (August 1, 2022): 108565. https://doi.org/10.1016/j.bej.2022.108565.

Erady, Veera, Ronald J. Mascarenhas, and Ashis K. Satpati. "Highly Efficient and Selective Quantification of Vanillin in Food, Beverages and Pharmaceuticals Using Surfactant Modified Carbon Paste Sensor." *Sensors International* 1 (2020): 100023. https://doi.org/10.1016/j.sintl.2020.100023.

Fan, Junmei, Liping Ding, and Yu Fang. "Surfactant Aggregates Encapsulating and Modulating: An Effective Way to Generate Selective and Discriminative Fluorescent Sensors." *Langmuir* 35, no. 2 (January 15, 2019): 326–41. https://doi.org/10.1021/acs.langmuir.8b02111.

Gonçalves, Rui A., Krister Holmberg, and Björn Lindman. "Cationic Surfactants: A Review." *Journal of Molecular Liquids* 375 (April 2023): 121335. https://doi.org/10.1016/j.molliq.2023.121335.

Hareesha, N., and J. G. Manjunatha. "Electro-Oxidation of Formoterol Fumarate on the Surface of Novel Poly(Thiazole Yellow-G) Layered Multi-Walled Carbon Nanotube Paste Electrode." *Scientific Reports* 11, no. 1 (2021): 1–14. https://doi.org/10.1038/s41598-021-92099-x.

Hareesha, N., J. G. Manjunatha, Zeid A. Alothman, and Mika Sillanpää. "Simple and Affordable Graphene Nano-Platelets and Carbon Nanocomposite Surface Decorated with Cetrimonium Bromide as a Highly Responsive Electrochemical Sensor for Rutin Detection." *Journal of Electroanalytical Chemistry* 917 (July 2022): 116388. https://doi.org/10.1016/j.jelechem.2022.116388.

Hareesha, Nagarajappa, and Jamballi G. Manjunatha. "Surfactant and Polymer Layered Carbon Composite Electrochemical Sensor for the Analysis of Estriol with Ciprofloxacin." *Materials Research Innovations* 24, no. 6 (September 18, 2020): 349–62. https://doi.org/10.1080/14328917.2019.1684657.

Hersey, Melinda, Shane N. Berger, Jordan Holmes, Alyssa West, and Parastoo Hashemi. "Recent Developments in Carbon Sensors for At-Source Electroanalysis." *Analytical Chemistry* 91, no. 1 (January 2, 2019): 27–43. https://doi.org/10.1021/acs.analchem.8b05151.

Holmberg, Sunshine, Maziar Ghazinejad, EunByul Cho, Derosh George, Brandon Pollack, Alexandra Perebikovsky, Regina Ragan, and Marc Madou. "Stress-Activated Pyrolytic Carbon Nanofibers for Electrochemical Platforms." *Electrochimica Acta* 290 (November 2018): 639–48. https://doi.org/10.1016/j.electacta.2018.09.013.

Ivanković, Tomislav, and Jasna Hrenović. "Surfactants in the Environment." *Archives of Industrial Hygiene and Toxicology* 61, no. 1 (March 1, 2010): 95–110. https://doi.org/10.2478/10004-1254-61-2010-1943.

Kandasamy, Ganeshlenin, and Dipak Maity. "Surfactant Sensors for Analysis of Toxic Chemicals." In *Advances in Surfactant Biosensor and Sensor Technologies*, edited by J. G. Manjunatha, 169–88. Cham: Springer Nature Switzerland, 2024. https://doi.org/10.1007/978-3-031-60832-2_9.

Kaya, S. Irem, Ahmet Cetinkaya, and Sibel A. Ozkan. "Surfactant Sensors for Pharmaceutical/Medical Applications." In *Surfactant-Based Sensors in Chemical and Biochemical Detection*, edited by J. G. Manjunatha, 23:163–92. Royal Society of Chemistry, 2023. https://doi.org/10.1039/BK9781837671182-00163.

Khan, A. A. P., A. Khan, A. M. Asiri, M. M. Alam, M. M. Rahman, and M. Shaban. "Surfactant-Assisted Graphene Oxide/Methylaniline Nanocomposites for Lead Ionic Sensor Development for the Environmental Remediation in Real Sample Matrices." *International Journal of Environmental Science and Technology* 16, no. 12 (December 25, 2019): 8461–70. https://doi.org/10.1007/s13762-019-02447-8.

Kozak, Jędrzej, Katarzyna Tyszczuk-Rotko, Magdalena Wójciak, and Ireneusz Sowa. “Electrochemically Activated Screen-Printed Carbon Sensor Modified with Anionic Surfactant (ASPCE/SDS) for Simultaneous Determination of Paracetamol, Diclofenac and Tramadol.” *Materials* 14, no. 13 (June 26, 2021): 3581. https://doi.org/10.3390/ma14133581.

Kronberg, Bengt, Krister Holmberg, and Björn Lindman. “Types of Surfactants, Their Synthesis, and Applications.” In *Surface Chemistry of Surfactants and Polymers*, 1–47. West Sussex, UK: Wiley, 2014. https://doi.org/10.1002/9781118695968.ch1.

Kumar, Mohan, B. E. Kumara Swamy, Sathish Reddy, T. V. Sathisha, and J. Manjanna. “Synthesis of ZnO and Its Surfactant Based Electrode for the Simultaneous Detection of Dopamine and Ascorbic Acid.” *Analytical Methods* 5, no. 3 (2013): 735–40. https://doi.org/10.1039/C2AY26124C.

Lakhera, P., V. Chaudhary, A. Jha, R. Singh, P. Kush, and P. Kumar. “Recent Developments and Fabrication of the Different Electrochemical Biosensors Based on Modified Screen Printed and Glassy Carbon Electrodes for the Early Diagnosis of Diverse Breast Cancer Biomarkers.” *Materials Today Chemistry* 26 (2022): 101129. https://doi.org/10.1016/j.mtchem.2022.101129.

Li, Hui, Ziwen Deng, Qingwu Tian, Limin Lun, Peng Zhao, Xiaomin Yang, Jingli Shen, Bing Jiang, Yusun Zhou, and Tingting Zhou. “Application of Carbon Nanotubes and Zwitterionic Surfactant-Modified Acetylene Black Electrode for the Determination of Triclosan in Household Commodities.” *International Journal of Environmental Analytical Chemistry* 102, no. 4 (March 16, 2022): 987–1000. https://doi.org/10.1080/03067319.2020.1730338.

Lien, Nguyen Thi, Le Quoc Hung, Nguyen Tien Hoang, Vu Thi Thu, Dau Thi Ngoc Nga, Pham Thi Hai Yen, Pham Hong Phong, and Vu Thi Thu Ha. “An Electrochemical Sensor Based on Gold Nanodendrite/Surfactant Modified Electrode for Bisphenol A Detection.” *Journal of Analytical Methods in Chemistry* 2020 (December 29, 2020): 6693595. https://doi.org/10.1155/2020/6693595.

Liu, Jianlian, Chaojun Du, Shengyun Wang, and Zhaoyang Li. “Preparation and Electrochemical Sensing Properties of Vanillin Molecularly Imprinted Polymer.” *International Journal of Electrochemical Science* 17, no. 7 (2022): 220715. https://doi.org/10.20964/2022.07.17.

Malode, Shweta J., Keerthi Prabhu K., Nagaraj P. Shetti, and Kakarla Raghava Reddy. “Highly Sensitive Electrochemical Assay for Selective Detection of Aminotriazole Based on TiO2/Poly(CTAB) Modified Sensor.” *Environmental Technology & Innovation* 21 (February 2021): 101222. https://doi.org/10.1016/j.eti.2020.101222.

Manjunatha, J. G. “A Surfactant Enhanced Graphene Paste Electrode as an Effective Electrochemical Sensor for the Sensitive and Simultaneous Determination of Catechol and Resorcinol.” *Chemical Data Collections* 25 (February 2020): 100331. https://doi.org/10.1016/j.cdc.2019.100331.

Manjunatha, J. G. *Surfactant-Based Sensors in Chemical and Biochemical Detection*, edited by J. G. Manjunatha. Vol. 23. Croydon, UK: Royal Society of Chemistry, 2023. https://doi.org/10.1039/9781837671182.

Manjunatha, J. G. *Advances in Surfactant Biosensor and Sensor Technologies*, edited by J. G. Manjunatha. Cham, Switzerland: Springer Nature Switzerland, 2024. https://doi.org/10.1007/978-3-031-60832-2.

Manjunatha, K. G., B. E. Kumara Swamy, G. K. Jayaprakash, S. C. Sharma, P. Lalitha, and K. A. Vishnumurthy. “Electrochemical Determination of Paracetamol at Cu Doped ZnO/Nanoparticle with TX-100-Surfactant MCPE: A Cyclic Voltammetric Technique.” *Inorganic Chemistry Communications* 142 (August 2022): 109630. https://doi.org/10.1016/j.inoche.2022.109630.

Melnyk, A., L. Wolska, and J. Namieśnik. “Coacervative Extraction as a Green Technique for Sample Preparation for the Analysis of Organic Compounds.” *Journal of Chromatography A* 1339 (April 2014): 1–12. https://doi.org/10.1016/j.chroma.2014.02.082.

Mo, Zhirong, Yafen Zhang, Faqiong Zhao, Fei Xiao, Gaiping Guo, and Baizhao Zeng. “Sensitive Voltammetric Determination of Sudan I in Food Samples by Using Gemini Surfactant–Ionic Liquid–Multiwalled Carbon Nanotube Composite Film Modified Glassy Carbon Electrodes.” *Food Chemistry* 121, no. 1 (July 2010): 233–37. https://doi.org/10.1016/j.foodchem.2009.11.077.

Monnappa, Amrutha B., Jamballi G. Manjunatha, and Aarti S. Bhatt. “Design of a Sensitive and Selective Voltammetric Sensor Based on a Cationic Surfactant-Modified Carbon Paste Electrode for the Determination of Alloxan.” *ACS Omega* 5, no. 36 (September 15, 2020): 23481–90. https://doi.org/10.1021/acsomega.0c03517.

Mosleh, Maryam, Sayed Mehdi Ghoreishi, Saeed Masoum, and Asma Khoobi. “Determination of Quercetin in the Presence of Tannic Acid in Soft Drinks Based on Carbon Nanotubes Modified Electrode Using

Chemometric Approaches." *Sensors and Actuators B: Chemical* 272 (November 2018): 605–11. https://doi.org/10.1016/j.snb.2018.05.172.

Myers, Drew. *Surfactant Science and Technology*, edited by Drew Myers. Fourth. New Jersey, USA: John Wiley & Sons, 2020.

Naik, Praveen, Naveenchandra Pilicode, Kavya S. Keremane, Madhukara Acharya, and Airody Vasudeva Adhikari. "Synthesis, Optical, Electrochemical, and Computational Investigation of New Cyanopyridine-Centered Organic Dyads." *Optical Materials* 142 (August 2023): 114002. https://doi.org/10.1016/j.optmat.2023.114002.

Patil, Vinoda B., Davalasab Ilager, Suresh M. Tuwar, Kunal Mondal, and Nagaraj P. Shetti. "Nanostructured ZnO-Based Electrochemical Sensor with Anionic Surfactant for the Electroanalysis of Trimethoprim." *Bioengineering* 9, no. 10 (2022): 521. https://doi.org/10.3390/bioengineering9100521.

Patil, Vinoda B., Shweta J. Malode, Suresh M. Tuwar, and Nagaraj P. Shetti. "Graphene Sheet-Based Electrochemical Sensor with Cationic Surfactant for Sensitive Detection of Atorvastatin." *Sensors International* 3 (2022): 100198. https://doi.org/10.1016/j.sintl.2022.100198.

Prinith, N. S., and J. G. Manjunatha. "Surfactant Modified Electrochemical Sensor for Determination of Anthrone A Cyclic Voltammetry." *Materials Science for Energy Technologies* 2, no. 3 (December 2019): 408–16. https://doi.org/10.1016/j.mset.2019.05.004.

Prinith, Nambudumada S., Jamballi G. Manjunatha, and Nagarajappa Hareesha. "Electrochemical Validation of L-Tyrosine with Dopamine Using Composite Surfactant Modified Carbon Nanotube Electrode." *Journal of the Iranian Chemical Society* 18, no. 12 (2021): 3493–3503. https://doi.org/10.1007/s13738-021-02283-z.

Pushpanjali, P. A., and J. G. Manjunatha. "A Sensitive Novel Approach towards the Detection of 8-Hydroxyquinoline at Anionic Surfactant Modified Carbon Nanotube Based Biosensor: A Voltammetric Study." *Physical Chemistry Research* 7, no. 4 (2019): 813–22. https://doi.org/10.22036/pcr.2019.198397.1663.

Pushpanjali, P. A., J. G. Manjunatha, N. Hareesha, Edwin S. D. Souza, M. M. Charithra, and N. S. Prinith. "Voltammetric Analysis of Antihistamine Drug Cetirizine and Paracetamol at Poly(L-Leucine) Layered Carbon Nanotube Paste Electrode." *Surfaces and Interfaces* 24, no. April (2021): 101154. https://doi.org/10.1016/j.surfin.2021.101154.

Pushpanjali, Pemmatte A., Jamballi G. Manjunatha, and Nagarajappa Hareesha. "An Overview of Recent Developments of Carbon-Based Sensors for the Analysis of Drug Molecules." *Journal of Electrochemical Science and Engineering* 11, no. 3 (2021): 161–77. https://doi.org/10.5599/JESE.999.

Raril, C., and J. G. Manjunatha. "A Simple Approach for the Electrochemical Determination of Vanillin at Ionic Surfactant Modified Graphene Paste Electrode." *Microchemical Journal* 154, no. December 2019 (2020): 104575. https://doi.org/10.1016/j.microc.2019.104575.

Raril, C., and J. G. Manjunatha. "Development of Sodium Dodecyl Sulfate Based Electrochemical Sensor for Tartrazine Determination." *Portugaliae Electrochimica Acta* 39, no. 1 (2021): 59–70. https://doi.org/10.4152/pea.202101059.

Raril, Chenthattil, Jamballi G. Manjunatha, Doddarasinakere K. Ravishankar, Santosh Fattepur, Gurumallappa Siddaraju, and Lingappa Nanjundaswamy. "Validated Electrochemical Method for Simultaneous Resolution of Tyrosine, Uric Acid, and Ascorbic Acid at Polymer Modified Nano-Composite Paste Electrode." *Surface Engineering and Applied Electrochemistry* 56, no. 4 (2020): 415–26. https://doi.org/10.3103/S1068375520040134.

Saini, Surbhi Mahender, Titiksha Kumar Sagar, Yashi Dwivedi, and Sandeep Chandrashekharappa. "Brief Overview of Surfactants, Properties, Classification, Passivation, and Role in Chemistry." In *Surfactant Based Electrochemical Sensors and Biosensors*, edited by Jamballi G. Manjunatha and Chaudhery Mustansar Hussain, 3–20. Amsterdam: Elsevier, 2024. https://doi.org/10.1016/B978-0-443-15493-5.00018-X.

Sandhyarani, N. "Surface Modification Methods for Electrochemical Biosensors." In *Electrochemical Biosensors*, edited by Ali A. Ensafi, 45–75. Amsterdam: Elsevier, 2019. https://doi.org/10.1016/B978-0-12-816491-4.00003-6.

Sawkar, Rakesh R., Mahesh M. Shanbhag, Suresh M. Tuwar, Kunal Mondal, and Nagaraj P. Shetti. "Sodium Dodecyl Sulfate–Mediated Graphene Sensor for Electrochemical Detection of the Antibiotic Drug: Ciprofloxacin." *Materials* 15, no. 22 (2022): 7872. https://doi.org/10.3390/ma15227872.

Soni, Isha, and Gururaj Kudur Jayaprakash. "A Short Review on the Analysis of the Adsorptive Behavior of Surfactants at Carbon Paste Electrodes for Electrochemical Sensing." *Journal of Molecular Liquids* 388 (October 2023): 122737. https://doi.org/10.1016/j.molliq.2023.122737.

Stanić, Zorka, and Stella Girousi. "Carbon Paste Electrodes in Potentiometry: The State of the Art and Applications in Modern Electroanalysis (a Review)." In *Sensing in Electroanalysis*, edited by Kalcher Kurt, Metelka Radovan, Švancara Ivan, and Vystřa Karel, Vol. 6. Pardubice: Univerzita Pardubice, 2011. http://dspace.upce.cz/handle/10195/42535.

Subak, Hasret, and Pınar Talay Pınar. "Electrochemical Oxidation of Ranitidine Using a Boron-Doped Diamond Electrode in the Presence of Anionic Surfactant: A Comprehensive Investigation." *Cumhuriyet Science Journal* 45, no. 2 (2024): 249–55.

Taleat, Zahra, Alireza Khoshroo, and Mohammad Mazloum-Ardakani. "Screen-Printed Electrodes for Biosensing: A Review (2008-2013)." *Microchimica Acta* 181, no. 9–10 (2014): 865–91. https://doi.org/10.1007/s00604-014-1181-1.

Thomas, Tony, Ronald J. Mascarenhas, Ozma J. D'Souza, Simon Detriche, Zineb Mekhalif, and Praveen Martis. "Pristine Multi-Walled Carbon Nanotubes/SDS Modified Carbon Paste Electrode as an Amperometric Sensor for Epinephrine." *Talanta* 125 (2014): 352–60. https://doi.org/10.1016/j.talanta.2014.03.027.

Tigari, Girish, and J. G. Manjunatha. "Optimized Voltammetric Experiment for the Determination of Phloroglucinol at Surfactant Modified Carbon Nanotube Paste Electrode." *Instruments and Experimental Techniques* 63, no. 5 (2020a): 750–57. https://doi.org/10.1134/S0020441220050139.

Tigari, Girish, and J. G. Manjunatha. "Poly(Glutamine) Film-Coated Carbon Nanotube Paste Electrode for the Determination of Curcumin with Vanillin: An Electroanalytical Approach." *Monatshefte Fur Chemie* 151, no. 11 (2020b): 1681–88. https://doi.org/10.1007/s00706-020-02700-8.

Tigari, Girish, and J. G. Manjunatha. "A Surfactant Enhanced Novel Pencil Graphite and Carbon Nanotube Composite Paste Material as an Effective Electrochemical Sensor for Determination of Riboflavin." *Journal of Science: Advanced Materials and Devices* 5, no. 1 (2020c): 56–64. https://doi.org/10.1016/j.jsamd.2019.11.001.

Tigari, Girish, Jamballi G. Manjunatha, Chenthatill Raril, and Nagarajappa Hareesha. "Determination of Riboflavin at Carbon Nanotube Paste Electrodes Modified with an Anionic Surfactant." *ChemistrySelect* 4, no. 7 (2019): 2168–73. https://doi.org/10.1002/slct.201803191.

Turan, Kübra, Engin Er, Ahmet Uçar, and Gözde Aydoğdu Tığ. "Evaluations of Neurochemicals at Surfactant Modified Electrochemical Sensors and Biosensors." In *Surfactant Based Electrochemical Sensors and Biosensors*, 275–313. Elsevier, 2024. https://doi.org/10.1016/B978-0-443-15493-5.00016-6.

Ullah, Imdad, Afzal Shah, Amin Badshah, Afzal Shah, Naseer Ali Shah, and Rico Tabor. "Surface, Aggregation Properties and Antimicrobial Activity of Four Novel Thiourea-Based Non-Ionic Surfactants." *Colloids and Surfaces A: Physicochemical and Engineering Aspects* 464 (January 2015): 104–9. https://doi.org/10.1016/j.colsurfa.2014.10.002.

Unal, Didem Nur, Sercan Yıldırım, Sevinc Kurbanoglu, and Bengi Uslu. "Current Trends and Roles of Surfactants for Chromatographic and Electrochemical Sensing." *TrAC Trends in Analytical Chemistry* 144 (November 2021): 116418. https://doi.org/10.1016/j.trac.2021.116418.

Vieira, Leonardo de Souza. "A Review on the Use of Glassy Carbon in Advanced Technological Applications." *Carbon* 186 (022): 282–302. https://doi.org/10.1016/j.carbon.2021.10.022.

Vittal, R., H. Gomathi, and Kang-Jin Kim. "Beneficial Role of Surfactants in Electrochemistry and in the Modification of Electrodes." *Advances in Colloid and Interface Science* 119, no. 1 (2006): 55–68. https://doi.org/10.1016/j.cis.2005.09.004.

Yang, Hongyu, Xiaojiang Zheng, Wensheng Huang, and Kangbing Wu. "Modification of Montmorillonite with Cationic Surfactant and Application in Electrochemical Determination of 4-Chlorophenol." *Colloids and Surfaces B: Biointerfaces* 65, no. 2 (2008): 281–84. https://doi.org/10.1016/j.colsurfb.2008.04.013.

Ying, Guang-Guo. "Fate, Behavior and Effects of Surfactants and Their Degradation Products in the Environment." *Environment International* 32, no. 3 (2006): 417–31. https://doi.org/10.1016/j.envint.2005.07.004.

Zahid, Anum, Afzal Shah, Faiza Jan Iftikhar, Aamir Hassan Shah, and Rumana Qureshi. "Surfactant Modified Glassy Carbon Electrode as an Efficient Sensing Platform for the Detection of Cd (II) and Hg (II)." *Electrochimica Acta* 235 (2017): 72–78. https://doi.org/10.1016/j.electacta.2017.03.120.

Zahid, Anum, Afzal Shah, and Shamsa Munir. "Gold Nanoparticles and Surfactant-Based Electrochemical Sensing Platform for the Detection and Electroanalysis of the Anticancer Drug Oxoplatin." *ECS Advances* 2, no. 4 (2023): 040506. https://doi.org/10.1149/2754-2734/ad040d.

Zhou, Chengcheng, and Yilin Wang. "Structure–Activity Relationship of Cationic Surfactants as Antimicrobial Agents." *Current Opinion in Colloid & Interface Science* 45 (2020): 28–43. https://doi.org/10.1016/j.cocis.2019.11.009.

3 Carbon Sensing Platforms for the Detection of Heavy Metals

D.M. Tejashwini, Ramachandra Naik, H.P. Nagaswarupa, and C.R. Ravikumar

3.1 SYNOPSIS OF ENVIRONMENTAL HEAVY METALS

Natural elements known as heavy metals have large atomic weights and densities that are at least five times higher than those of water. Lead (Pb), mercury (Hg), cadmium (Cd), arsenic (As), and chromium (Cr) are typical examples. Even at low concentrations, many heavy metals like copper (Cu) and zinc (Zn) are hazardous and provide serious health and environmental dangers. On the other hand, other heavy metals are necessary trace components for biological processes. There are two types of heavy metal sources in the environment: natural and man-made. Forest fires, rock weathering, and volcanic eruptions are examples of natural sources [1]. Human activities including mining, industrial processes (like smelting and battery production), agricultural practices (such as using pesticides and fertilizers), and inappropriate waste management are the main causes of heavy metal contamination. These metals linger in the environment, building up in living things, soils, and waterways, which may cause bioaccumulation and biomagnification in food chains. Because heavy metals are poisonous, non-biodegradable, and can lead to serious health problems like cancer, brain impairment, and organ failure, their presence in the environment is becoming a greater worry. These metals are persistent in the environment and accumulate in soils, aquatic environments, and living organisms. This can lead to bioaccumulation and biomagnification in food chains. The prevalence of heavy metals in the environment has become more concerning due to their poisonous nature, inability to biodegrade, and potential to cause major health issues such as cancer, brain damage, and organ failure [2].

Environmental detection of heavy metals is crucial because of its detrimental effects on ecosystems and human health. People may be exposed to heavy metals over extended periods of time if they are exposed to them in their food, water, or soil. These metals include arsenic, lead, cadmium, and mercury. This exposure is associated with serious health problems such as cancer, kidney damage, and neurological impairments. Heavy metal identification is essential for reducing risks and preventing pollution. Advanced detection techniques help ensure the safety of food and water supplies by enabling real-time monitoring. Effective cleanup strategies and legal actions are guided by the identification of pollution sources, which is made possible by early detection. The development of very sensitive, portable detection equipment is particularly critical for on-site monitoring, where areas at risk of contamination require immediate input [3].

DOI: 10.1201/9781003500322-3

3.2 CARBON-BASED MATERIALS AS PLATFORMS FOR SENSING

Activated carbon, graphene, and carbon nanotubes (CNTs) are common carbon compounds with unique properties that make them excellent sensing platforms. Their exceptional electrical conductivity, huge surface area, and chemical stability make it possible to detect a wide spectrum of analytes effectively. The large surface area improves target molecule adsorption, which raises sensitivity. The electrical conductivity of carbon compounds ensures effective electron transfer for electrochemical sensors. Additionally, because of their chemical stability, they can operate in a range of environments without deteriorating. Carbon materials can be functionalized with different chemical groups or nanoparticles to promote selectivity towards specific analytes, which would further improve their performance as sensing platforms. Because of their versatility and tunability, they can be utilized to detect gasses, biomolecules, and environmental pollutants, including heavy metals [4, 5].

There are various types of carbon materials, and each has its unique properties that make them valuable for use in sensing, electronics, and environmental monitoring applications. Graphene is a single layer of carbon atoms arranged in a hexagonal lattice with exceptional mechanical strength, electrical conductivity, and surface area. It is widely used in electrical devices and sensors. CNTs possess remarkable electrical and thermal conductivity, tensile strength, and chemical stability. CNTs are frequently used in nanocomposites, sensors, and energy storage systems. Carbon dots (CDs) are minuscule, fluorescent carbon nanoparticles possessing exceptional biocompatibility and tunable photoluminescence [6]. CDs are used in bioimaging, medication delivery, and environmental sensing. In filtration and adsorption applications, activated carbon is a highly porous carbon with a large surface area that is commonly used, particularly in gas adsorption and water purification. Fullerenes, which are spherical carbon molecules with unique electrical properties, are useful in photovoltaics, superconductors, and drug delivery [7].

Because of their special characteristics, carbon-based sensors offer significant advantages in the detection and sensing of heavy metal ions. Carbon compounds with a high surface area, such as activated carbon, graphene, and CNTs, enhance the adsorption of heavy metal ions and boost the sensitivity of detection. Their outstanding electrical conductivity, which promotes efficient electron transit, makes them ideal for electrochemical sensing applications. To improve carbon-based sensor's selectivity for specific heavy metals, it is easy to functionalize them with additional chemical groups or nanoparticles. This functionalization paves the way for the development of sensors that can detect even trace levels of contaminants in complex matrices [8]. Since carbon compounds do not deteriorate significantly under harsh environmental conditions, these sensors provide reliable endure monitoring choices. This is because carbon compounds are chemically stable. Carbon compounds are an inexpensive option for developing scalable and portable sensing devices since they are readily available and moderately priced. Since carbon-based sensors combine stability, affordability, sensitivity, and selectivity, they are particularly helpful for real-time heavy metal monitoring in food, water, and soil. In doing so, pollution is reduced and public health is maintained [9].

3.3 METHODS OF DETECTING HEAVY METALS USING CARBON SENSORS

3.3.1 Essential Mechanisms of Detection

Electrochemical sensing is a crucial detection technique for the recognition of heavy metals in carbon-based sensors. Examples of sensing materials whose electrical response is detected upon contact with target metal ions are graphene and CNTs. The working electrode of an electrochemical cell used in sensors is typically carbon. Heavy metal ions interact with the carbon electrode in the sample when they are present, altering electrical properties like voltage or current that are then measured [10]. The typical mechanism of electrochemical sensing involves metal ions contributing to or receiving electrons from redox reactions at the electrode surface. Because the resulting electrochemical signal is related to the concentration of heavy metal ions, quantification is achievable.

Carbon-based constituents are advantageous in this situation due to their large surface area, strong electrical conductivity, and ability to be functionalized to increase selectivity. These sensors may achieve exceptional sensitivity and selectivity for certain heavy metals, making them valuable for pollution detection and real-time environmental monitoring [11].

Optical sensing is a crucial technique for detecting heavy metals using carbon-based sensors, such as CDs and graphene oxide. This technique makes use of the interface between light and the carbon material to detect and analyse metal ions. Optical sensors commonly use absorbance, fluorescence, or Raman scattering to detect heavy metals. Because of their intense luminosity, carbon dots are utilized in fluorescence-based optical sensing. By detecting the change in fluorescence emission that heavy metal ions cause in the contact between the CDs and the metal ions, the amount of metal present in the mixture may be determined. Similarly, graphene oxide can be functionalized to interact with specific heavy metals, changing its optical properties such as absorbance or fluorescence [12].

The chemical interactions between carbon-based molecules and heavy metals are critical to the identification and removal of contamination effectively. Examples of carbon materials that exhibit distinct interactions with heavy metal ions are graphene, CNTs and CDs. The high surface area and many active sites of carbon materials promote heavy metal ion adsorption. The hydroxyl, carboxyl, and amino groups of these materials interact chemically and electrostatically with metal ions. It is feasible for graphene oxide to undergo redox interactions with heavy metals due to its carboxyl groups, which have the ability to bind to metal ions with great efficiency [13]. Stable complexes can be formed between functionalized carbon-containing materials and heavy metals. This interaction often requires the coordination of metal ions to the functional groups on the carbon surface in order to permit selective detection and separation. These chemical interactions support the effectiveness of carbon-based sensors in detecting and quantifying heavy metals in various environments [14].

3.3.2 Techniques for Signal Transduction

Conductometric sensing is a type of signal transduction employed by carbon-based sensors to identify heavy metals. This method measures alterations in the material's electrical conductivity by subjecting a sensor material to metal ions. The large surface area and strong electrical conductivity of carbon compounds such as graphene and CNTs make them particularly beneficial. In a conductometric sensor, the carbon material is integrated into a conducting electrode. The material's electrical conductivity is altered by the interaction of heavy metal ions with the sensor, which leads to the metals either contributing or absorbing electrons. This decrease in conductivity is precisely proportional to the concentration of metal ions in the sample [15]. Conductometric sensors may be made more sensitive by functionalizing carbon materials with certain chemical groups that increase their affinity for target metals. Because of this, detection is precise and sensitive even at low dosages. Conductometric sensors are valued for their ease of use, low cost, and ability to provide continuous, real-time monitoring of heavy metal pollution [16].

Carbon-based sensors that are used to detect heavy metals employ a signal transduction technique called voltammetric sensing. Using this method, one may measure a sensor's current response to applied voltage. Carbon-based materials, such as graphene, CNTs, and CDs, are utilized as electrodes due to their high surface area and superior electrical conductivity. A voltage that varies is delivered to the carbon electrode in a voltammetric sensor to cause redox reactions with heavy metal ions present in the sample. The current changes that arise from these reactions are measured, recorded, and examined to ascertain the metal ion concentration. By functionalizing the carbon compounds with certain ligands that boost their interaction with target metals, voltammetric sensors may be made more sensitive [17].

Carbon-derived sensors, fluorescence-based techniques are a widely utilized signal transduction approach for heavy metal detection. These methods are used to identify and measure metal ions by making use of the fluorescence characteristics of carbon materials, such as CDs or graphene oxide. Carbon materials are frequently modified with functional groups or nanoparticles that can precisely interact with target heavy metals in fluorescence-based detection. These functionalized carbon compounds bond to metal ions, which alter the fluorescence emission by either increasing or quenching the fluorescence. The heavy metal concentration in the material is then calculated by measuring these variations. Fluorescence-based techniques are very sensitive and selective, which makes them perfect for finding minute quantities of pollutants in intricate matrices. They also have the benefit of real-time, non-destructive examination, which is advantageous for biological applications and environmental monitoring [18].

3.4 TYPES OF CARBON-BASED SENSORS FOR HEAVY METALS

3.4.1 Graphene-Based Sensors

Due to graphene's remarkable mechanical strength, electrical conductivity, and large surface area, graphene-based sensors are very successful in detecting heavy metals. Graphene production and functionalization are essential for maximizing the performance of these sensors. Chemical vapour deposition (CVD) and chemical reduction of graphene oxide are simple procedures used to develop graphene. Large-area, high-quality graphene sheets are produced by CVD, but scalable manufacturing is better served by liquid-phase exfoliation and reduction techniques [19].

Various targeted heavy metal sensing is carried by using graphene-based sensors, as shown in Table 3.1. Using fluorescent CNTs to detect mercury ions through fluorescence quenching allowed for sensitive detection [20]. Using CNTs as electrodes in an electrochemical cell demonstrated high sensitivity for Pb detection in water [21]. Using Raman-active CNTs allowed for detecting cadmium with a high level of sensitivity and specificity in complex samples [22]. Using functionalized CNTs in the electrochemical device allowed for the uncovering of low levels of arsenic, making it suitable for monitoring safe drinking water [23]. Copper ions in food samples were detected by fluorescence changes [24].

3.4.2 Carbon Nanotube Sensors

The excellent electrical conductivity, vast surface area, and mechanical robustness of CNTs make them perfect for sensing applications. CNTs are available in two different forms: single-walled CNTs (SWCNTs) and multi-walled CNTs (MWCNTs), each with special benefits. MWCNTs offer improved mechanical strength and stability, but SWCNTs have a larger surface area and electrical conductivity. Their cylinder-shaped design facilitates effective charge transfer and material contact. CNTs can be functionalized with different chemical groups or nanoparticles to improve their sensing properties. Covalent bonding with organic molecules and non-covalent interactions with polymers and metal nanoparticles are two instances of functionalization techniques [25]. By strengthening the

TABLE 3.1
Heavy metal detection with the application

S No.	Targeted Heavy Metal	Application	Reference
1	Mercury	Environmental & biological monitoring	[20]
2	Lead	Water & soil monitoring	[21]
3	Cadmium	Industrial waste analysis	[22]
4	Arsenic	Drinking water	[23]
5	Copper	Food safety and environmental analysis.	[24]

TABLE 3.2
Heavy metal sensing with the detection limit

S No.	Heavy Metal	Detection Limit	Application	Reference
1	Arsenic	0.05 μg/L	Drinking water safety	[26]
2	Copper	1 μg/L	Food safety	[27]
3	Chromium	0.3 μg/L	Industrial wastewater	[28]
4	Nickel	0.2 μg/L	Environmental analysis	[29]
5	Zinc	1 μg/L	Soil and water testing	[30]

CNTs' affinity for the target compounds or changing their electrical characteristics, this customization enhances the CNTs' selectivity and sensitivity towards certain analytes.

Heavy metal sensing with the detection limit is presented in Table 3.2. Functionalized CNTs in an electrochemical sensor identified low amounts of arsenic, making them appropriate for monitoring safe drinking water [26]. Copper ions in food samples were identified by CNTs functionalized with fluorescent dyes through variations in fluorescence [27]. High-sensitivity chromium ion detection was achieved using CNT electrodes, which made them useful for monitoring industrial effluents [28]. Nickel ions were precisely recognized using CNT-based voltammetric sensors, which is helpful for environmental monitoring [29]. For the study of soil and water, conductometric sensors using CNTs were used to detect zinc ions through changes in conductivity [30].

3.4.3 Carbon Dot Sensors

Fluorescent nanostructures called CDs have special optical characteristics that make them useful for sensor applications. Processes like chemical oxidation, laser ablation, or hydrothermal carbonization are used to create CDs. To develop CDs, hydrothermal carbonization entails heating carbon-rich precursors in a sealed atmosphere. Whereas laser ablation creates CDs by irradiating a carbon target with a laser, chemical oxidation employs oxidizing chemicals to transform carbon sources into CDs [31]. Characterization like photoluminescence (PL) spectroscopy is used to measure fluorescence behaviour, ultraviolet-visible (UV-Vis) spectroscopy is utilized to measure optical potentials, and transmission electron microscopy (TEM) is used to study the morphology of CDs. CDs are modified with different chemical groups or nanoparticles to improve their sensing capabilities. Their selectivity is increased and their interaction with certain analytes is improved by surface modification. Covalent bonds with organic molecules and non-covalent interactions with polymers are examples of methods [32].

CD sensors have demonstrated significant potential for detecting heavy metals due to their high fluorescence, tunable emission properties, and biocompatibility. CDs functionalized with specific ligands have been used for Pb detection in water. The presence of Pb ions leads to fluorescence quenching, allowing sensitive and selective detection at low concentrations [33]. CDs functionalized with thiol groups have shown high sensitivity for Hg detection through fluorescence enhancement or quenching. The interface among Hg and thiol groups alters the fluorescence properties, facilitating detection. Fluorescent CDs with specific surface modifications have been used to detect Cd ions, where changes in fluorescence intensity correlate with Cd concentrations [34].

3.4.4 Composite Carbon Materials

Composite carbon materials, such as carbon–metal oxide hybrids, merge the assets of carbon resources with metal oxides to enhance sensing performance, particularly for heavy metal detection. Carbon–metal oxide composites integrate carbon materials (like graphene or CNTs) with metal oxides (such as zinc oxide, titanium dioxide, or iron oxide). This combination leverages the high conduction and huge exterior area of carbon materials with the catalytic and selective properties of metal oxides. Graphene-zinc oxide (ZnO) composites: These hybrids enhance sensitivity for detecting heavy metals through improved charge transfer and catalytic activity. ZnO nanoparticles on graphene have been used to detect Pb and Cd ions with high accuracy. CNT-TiO_2 hybrids offer enhanced photocatalytic activity and are used in sensors for detecting mercury and arsenic [34].

Synergistic effects in composite materials greatly improve sensing capacities by mixing carbon with metal oxides or other components. Improved perception skills such as enhanced sensitivity with higher surface areas and improved charge transport are provided by carbon-based composites, such as graphene-metal oxide hybrids, which enhance sensor sensitivity. In particular, increased electrochemical activity in graphene-ZnO composites enables the detection of trace heavy metals at lower limits. Specific chemical interactions can be introduced to improve selectivity by the integration of metal oxides with carbon materials. Since the photocatalytic qualities of TiO_2 are boosted by the conductive CNTs, CNT-TiO_2 composites exhibit selective detection of mercury and arsenic [35].

3.5 FABRICATION TECHNIQUES OF CARBON SENSING PLATFORMS

3.5.1 Synthesis of Carbon Nanomaterials

Different methods are used to synthesize carbon nanomaterials, and each has benefits of its own for creating materials with certain qualities, as displayed in Figure 3.1.

Techniques for manufacturing carbon nanomaterials are the following:

1. Chemical vapour deposition (CVD): CVD is the process of depositing carbon atoms onto a substrate by breaking down carbon-containing gases (such as methane) at high temperatures. This results in the development of nanomaterials like graphene and CNTs. As it can yield vast quantities of high-quality nanomaterials, this process is popular.
2. Hydrothermal Carbonization: To create carbon dots (CDs) and other nanomaterials, this process requires heating carbon-rich precursors in a sealed tank at high pressure and temperature. This method is preferred for being easy to use and due to its reasonable price.
3. Laser Ablation: To generate carbon nanomaterials such as graphene, a carbon target is exposed to a laser in an inert environment. Although this technique yields very high purity, small-scale synthesis is typically its application.
4. Chemical Oxidation: In this process, carbon sources are reduced to nanomaterials like graphene oxide, which may then be reduced even further to graphene, using oxidizing agents [36].

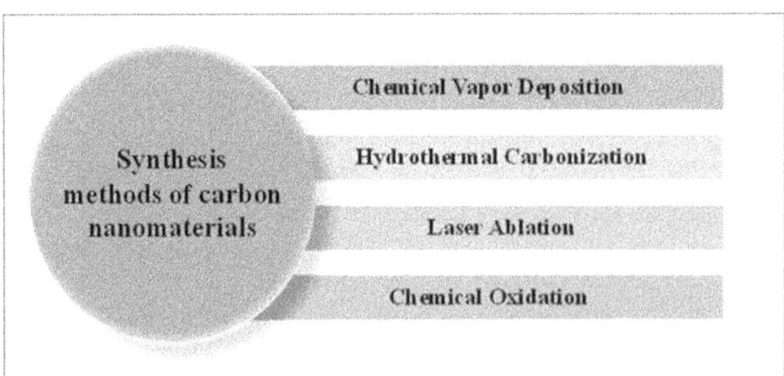

FIGURE 3.1 Synthesis methods of carbon nanomaterials.

5. Functionalization Techniques: Covalent functionalization uses covalent bonds. Functional groups are bonded to the carbon nanomaterial in this technique. Graphene oxide, for example, can be reduced and functionalized with hydroxyl, carboxyl, or amino groups to increase sensor sensitivity and improve its interaction with target analytes. Non-covalent functionalization is a non-covalent technique that relies on physical contacts including hydrogen bonds, van der Waals forces, and π-π stacking. For instance, to make CNTs more dispersible and perform better in sensing applications, they can be functionalized with polymers or surfactants. For nanoparticle decoration, which improves the catalytic and sensing capabilities of carbon materials, metal oxide or metal nanoparticles can be added. In particular, graphene with gold nanoparticle decoration has demonstrated increased sensitivity in finding the heavy metals [37].

3.5.2 Device Fabrication

Integrating carbon materials into sensing devices involves incorporating these materials into various sensor platforms to leverage their unique properties, such as high surface area, electrical conductivity, and optical characteristics. Electrochemical sensor utilizes carbon material's special qualities, such as their large surface area, electrical conductivity, and optical qualities, into different sensor platforms. Sensors that use electrochemistry are called electrochemical sensors that employ carbon-based materials as electrode materials, namely, graphene or CNTs. Their vast surface area and excellent conductivity improve sensitivity and charge transfer. Extremely sensitive electrochemical sensors for the determination of heavy metals like Pb and Hg have been developed using graphene-based electrodes. In optical sensors carbon nanoparticles are included to take use of their absorbance and fluorescence characteristics. Optical devices can be equipped with functionalized graphene or carbon dots to detect analytes through variations in absorbance or fluorescence. The conductometric sensor method employs carbon materials, namely, graphene and CNT, to monitor variations in electric conductivity caused by analyte contact. This technique has been used to find heavy metals and gases [38].

Achieving mobility and miniaturization is essential to creating useful and intuitive sensing devices. A variety of methods are used to create effective, small-sized sensors that are appropriate for field use. This method includes creating tiny sensor components on semiconductor substrates by photolithography, etching, and deposition. Wearable, portable sensors may be made by using flexible substrates like polyester (PET) films or polyimide. These flexible substrates are combined with carbon-based materials, such as graphene and CNTs, to create lightweight, conformable sensors. Lab-on-a-chip (LOC) devices are chips that combine sensors with microfluidic channels. These chips' microelectrodes and sensor components are made of carbon, which makes small, integrated sensing platforms for real-time analysis possible [39].

3.5.3 Surface Engineering for Selectivity and Sensitivity

Surface engineering of carbon-based sensors via functionalization with certain ligands or compounds improves the selectivity and sensitivity of the sensors for identifying target analytes, as illustrated in Figure 3.2. Covalent bonds bind certain ligands or molecules to the surface of carbon-based materials. For instance, adding thiol or amine groups to graphene can improve its selectivity for heavy metals like Pb or Hg by generating powerful chemical interactions that raise the sensitivity of detection. Carbon nanomaterial surfaces can be altered by non-covalent techniques like the adsorption of organic compounds or polymers. Target analytes can have specific binding sites introduced by using this method. For example, selective polymer-functionalized CNTs have been used to identify poisons in environmental samples. Sensing capabilities can be amended by uniting carbon materials with other nanomaterials, such as metal nanoparticles. The improved optical and catalytic

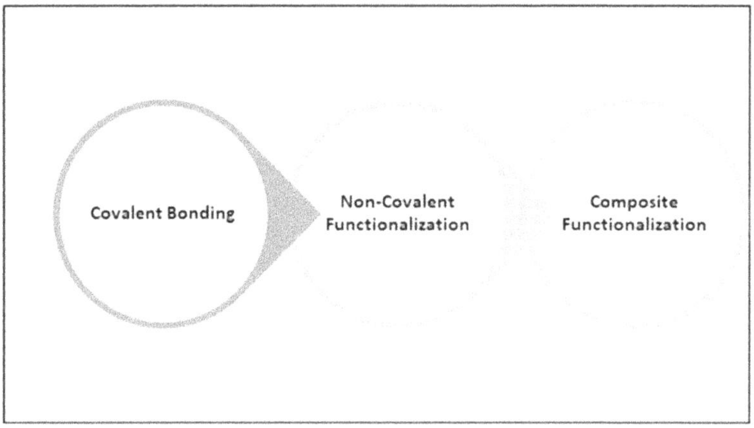

FIGURE 3.2 Various types of functionalization.

capabilities of CDs treated with gold nanoparticles result in higher sensitivity for detecting trace quantities of contaminants [40].

Improving the capacity of carbon-based sensors to detect target metals while reducing interference from other chemicals is known as "enhancing selectivity towards specific heavy metals". Strategies for Enhancement. Selective ligands that are highly affine to the target heavy metals are attached, significantly increasing the selectivity of the sensor. Using the molecular imprinting approach, carbon nanomaterials with pre-made cavities that fit the target metal ions are encased in a polymer matrix. By using this technique, the sensor's great selectivity for particular metals, such as Pb or Cd, is guaranteed. Selectivity can be further increased by mixing carbon materials with metal oxides or nanoparticles that particularly interact with the target metals [41].

3.6 ANALYTICAL PERFORMANCE OF CARBON-BASED SENSORS

3.6.1 Sensitivity and Detection Limits

The efficacy of carbon-based sensors in detecting tiny levels of target analytes can be greatly impacted by several important parameters that affect their sensitivity and detection limits, as displayed in Figure 3.3. Carbon nanomaterials, including graphene and CNT, have a high surface area that increases sensitivity by offering more active sites for interaction with the analytes of interest. By raising the affinity for target analytes, chemical modification of carbon materials with certain functional groups or molecules can improve sensitivity. Sensitivity is influenced by the electrical characteristics of carbon compounds. Higher electrical conductivity materials, like graphene, allow for greater charge transfer, which causes the sensor signal to alter more significantly when analyte binding occurs. The sensitivity of carbon nanomaterials is influenced by their size and structural characteristics, such as their porosity and dispersion [42].

Various techniques may be utilized to improve the sensitivity and performance of carbon-based sensors in order to detect minute quantities of target analytes, hence reducing the detection limits. Reducing detection limits can be achieved by enhancing the characteristics of carbon nanomaterials, like electrical conductivity and surface area. Selectivity and sensitivity can be increased by using sophisticated functionalization procedures to alter the carbon surface with highly selective ligands or receptors. Further lowering detection limits can be achieved by using signal amplification techniques, such as the use of electrochemical or optical enhancement techniques. Pre-concentration of analytes and more effective detection are made possible by the combination of carbon-based

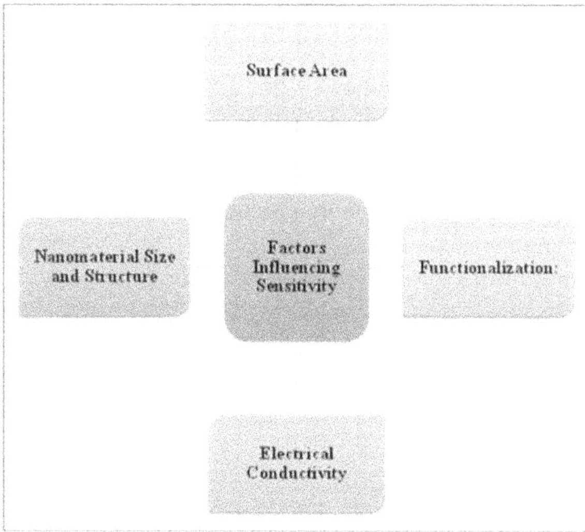

FIGURE 3.3 Different factors influencing the sensitivity.

sensors with microfluidic systems. By increasing the contact among the sensor and the analyte, this integration improves the sensor's capacity to detect lower concentrations [42].

3.6.2 Selectivity and Cross-Sensitivity

Developing selective carbon-based sensors presents a substantial problem due to cross-sensitivity, when a sensor reacts to ions other than the target analyte. This problem may result in lower detection accuracy and false positives for heavy metals. Cross-sensitivity to ions with comparable valence states or chemical characteristics may be seen in carbon-based sensors. Signal interference may arise due to the imprecise adsorption of different ions onto the sensor surface. High surface area carbon materials, such as graphene, can adsorb various ions, which can interfere with the sensor's capacity to discriminate between various analytes. Cross-sensitivity can result from changes in the temperature, ionic strength, and pH that affect how sensors respond. The reliability of detection may be impacted by this fluctuation, especially in complex environmental samples [43].

Enhancing selectivity in carbon-based sensors entails several crucial tactics to guarantee precise target analyte detection while reducing interference from other chemicals. Selectivity is increased by modifying the sensor surface to include particular functional groups or ligands that have a significant affinity for the target analytes. The process of molecular imprinting entails using polymer matrices that are specifically designed to fit the target analyte to create binding spots on the sensor surface. By offering precise recognition spots, molecularly imprinted polymers (MIPs) on carbon materials can greatly improve selectivity for certain heavy metals. To increase selectivity, composite materials can be made by mixing carbon with additional components that are chosen, such as metal oxides or nanoparticles. Enhancement of operational environment selectivity may be increased by modifying variables like pH, temperature, and ionic strength to optimize the sensor environment for the target analyte [44].

3.6.3 Stability and Reproducibility

Carbon-based sensor's long-term stability is critical for practical use since it ensures constant performance across time. Long-term sensor endurance is facilitated by the strong chemical and thermal stability of carbon resources like graphene and CNTs. Performance can be impacted by elements like oxidation or deterioration. For dependable long-term performance, surface functionalization must be stable. Over time, functional groups or ligands need to stay bonded and active. Sensor performance needs to be maintained despite variations in the surrounding environment, such as humidity and temperature. Long-term stability may be improved by using strong materials and creating sensors that can withstand environmental stresses [45]. The homogeneity of carbon nanomaterials utilized affects sensor repeatability. Sustainably produced materials such as graphene or CNT by high-quality synthesis guarantee that sensors function similarly. Reproducible sensor performance requires consistent manufacturing processes, such as accurate deposition and functionalization procedures. Differences in sensor properties and responses might result from variations in manufacturing procedures. To guarantee that sensors function consistently across a range of devices, the integration of carbon materials into diverse sensor platforms necessitates meticulous optimization. Reproducibility in sensor outputs can be achieved with the use of standardized integration processes [46].

3.7 REAL-WORLD APPLICATIONS OF CARBON-BASED HEAVY METAL SENSORS

3.7.1 Environmental Monitoring

Carbon-based sensors have proven essential in practical applications for heavy metal detection in environmental settings and water quality analysis. Portable systems for on-site water quality assessment are increasingly using graphene- and CNT-based sensors. They are appropriate for on-site detection of pollutants like Pb and Hg in water samples because of their high sensitivity and quick reaction times. Water bodies are monitored using carbon-based sensors in a diversity of outdoor circumstances. Their ease of use and quick results make them advantageous for regulating water quality in distant or resource-constrained places. Internet of Things (IoT) integration is frequently used to provide continuous data transfer and monitoring. Real-time analysis and remote control of water quality are made possible by this connection, which enhances response times to contamination events [47].

3.7.2 Industrial Applications

The sensitivity and versatility of carbon-based sensors make them a popular choice for identifying heavy metals in industrial effluents. These sensors, which include CNT, CD, and graphene oxide, provide practical ways to control and monitor metal contaminants in wastewater from industrial operations. Because CNTs have higher sensitivity for nickel and chromium, they can identify metal pollution in effluents early and allow for real-time monitoring. By enabling prompt remedial efforts, the integration of these sensors into industrial monitoring systems helps to minimize environmental effect while also ensuring compliance [48].

The distinct characteristics and versatility of carbon-based sensors have made them effective instruments for heavy metal detection. These sensors find traces of heavy metals in ambient samples by using different carbon allotropes such graphene, CNTs, and activated carbon. The large surface area of carbon-based sensors, which offers a wealth of active sites for the adsorption of heavy metal detection in metal ions, is one of its main benefits. As a result, metals at very low concentrations can be detected due to an increase in sensitivity. Because of their adaptability, carbon-based materials can have certain chemical groups functionalized on their surfaces to advance their selectivity for certain heavy metals. These sensors are appropriate for extensive environmental monitoring owing to their reputation for mechanical robustness, chemical stability, and affordability [49].

3.7.3 Biomedical and Food Safety Applications

The recognition of heavy metals in biological samples is being revolutionized by carbon-based sensors, improving applications in the fields of biomedicine and food safety. In blood, urine, and food items, graphene oxide and CNT sensors are especially useful for detecting traces of hazardous metals, including Cd, Pb, and Hg. For food safety and health problems to be avoided, precise finding of Pb in food samples is provided using CNTs. These sensors are valuable tools for food safety regulation compliance and health monitoring as they yield fast and accurate findings. This enhances environmental safety and public health outcomes [50]. The excellent sensitivity and quick detection of carbon-based sensors make them a popular choice for tracking heavy metal contamination in food items. In different food matrices, graphene oxide and CNT sensors are especially useful for finding traces of metals, including Pb, Cd, and Hg. CNTs shield the population from exposure to harmful metals; they provide accurate lead detection in a variety of food products. These sensors' capacity to deliver on-the-spot, real-time analysis improves food safety monitoring, lessens the need for conventional laboratory testing, and enables speedier reactions to contamination problems [51].

3.8 CHALLENGES AND FUTURE DIRECTIONS IN CARBON-BASED HEAVY METAL SENSING

3.8.1 Current Limitations

Despite their advances carbon-based heavy metal sensors still have several technical issues. A major drawback is the challenge of attaining high selectivity in the presence of intricate sample matrices, which might result in false positives and cross-sensitivity. A further concern is the stability and repeatability of these sensors in different environmental settings, which may have an impact on their accuracy and long-term performance. Issues with manufacturing consistency and scaling up make it difficult to integrate carbon-based sensors into affordable and portable devices. The development of novel materials and improved functionalization strategies are imperative in tackling these issues and augmenting the overall dependability of carbon-based heavy metal detection sensors [52].

Carbon-based sensors are promising, but they have material limits and manufacturing difficulties that reduce their usefulness. A major material problem is to achieve uniformity and quality consistently in carbon nanomaterials, like CNT and graphene, which can cause performance variations in sensors. One of the fabrication challenges is incorporating these materials into reliable and repeatable sensing devices. Another challenge is increasing the manufacturing of premium carbon-based sensors while preserving their price and functionality. In order to advance the actual deployment of carbon-based sensors in real-world situations, these materials and manufacturing problems must be addressed [53].

3.8.2 Innovations in Carbon Nanotechnology for Sensing

Novel synthesis techniques and newly developed materials have improved sensor performance appreciations to recent developments in carbon nanotechnology for sensing [54]. To increase selectivity and sensitivity for the heavy metal detection, new carbon nanomaterials are being created, such as doped graphene and hybrid CNTs. CVD and liquid-phase exfoliation are two examples of novel synthesis techniques that make it possible to produce high-quality carbon nanomaterials with specialized features for particular uses. By addressing issues with material homogeneity and functionalization, these developments produce more dependable and efficient sensors. For trace metal analysis, for example, the use of functionalized graphene quantum dots has demonstrated promise in obtaining high sensitivity and low detection limits. Carbon-based sensors are now much more feasible and widely used because of recent advancements in sensor integration and miniaturization. Microelectromechanical systems and nanofabrication techniques have made it possible to create high-performance compact sensors that are lighter and smaller, which makes them more appropriate for wearable and portable applications. These tiny sensors may be installed in a variety of contexts and provide real-time monitoring capabilities due to their integration with flexible electronics and wireless communication technologies [55].

3.8.3 Future Perspectives

Carbon-based sensing technologies have a bright future, with several possible advancements in the works. The main goal of innovations is probably to improve sensor performance by using sophisticated material engineering and integration methods. The creation of hybrid nanomaterials, which blend carbon with other elements to increase sensitivity and selectivity for a variety of analytes, is one emerging concept. More complex data analysis will be possible because of developments in artificial intelligence and machine learning, which will improve real-time detection and predictive maintenance. Furthermore, there is a rising interest in using IoT technology for wider applications by integrating these sensors into intelligent, networked systems for health and ecological monitoring. The aim of these developments is to improve the accuracy, adaptability, and applicability of sensors in a variety of domains, including industrial operations, biomedical diagnostics, and environmental monitoring [56].

Real-time monitoring systems are changing as a result of the IoT and carbon-based sensor integration. Real-time insights into environmental and health situations may be obtained through this interface, which enables continuous data gathering and analysis. Graphene- and CNT-based sensors are finding their ways into IoT frameworks for use in health diagnostics and air and water quality analyses among other uses. With the ability to wirelessly transfer data to centralized systems, these sensors allow for quick reaction to anomalies that are discovered [57].

3.9 CONCLUSION

The extraordinary sensitivity, adaptability, and real-time monitoring capabilities of carbon-based sensors have made them indispensable instruments for the detection and sensing of heavy metals. High surface areas and distinctive electrical characteristics of materials like graphene, CNTs, and carbon dots improve detection limits and selectivity for different heavy metals. In recent times, the ability of carbon-derived sensors has been greatly improved to detect heavy metals in an extensive range of applications. The discovery of materials with enhanced sensitivity and selectivity for detecting trace quantities of metals, such as graphene, CNT, and CD, is one of the major advancements. Recent developments have led to the integration of these sensors into a broad range of applications, including food safety and biomedical settings, industrial effluents, and environmental monitoring. Selectivity and stability were once constraints but advances in material synthesis and sensor miniaturization have increased their performance and usefulness. Their usefulness is further increased by the continued development of IoT integration, which offers real-time, actionable data that facilitates efficient environmental and health monitoring.

Improvements in functionalization and synthesis methodologies have made it possible to create sensors that are more precise and trustworthy. Real-time data collection and monitoring are made possible by the integration of these sensors with IoT technology, which enables prompt reactions to pollution. Applications of carbon-based sensors for the detection of heavy metal ion contamination include food safety, industrial effluents, industrial monitoring, and biological diagnostics. Carbon-based sensors will provide more dependable and affordable solutions to environmental problems as technology develops, enhancing sustainability and improving public health outcomes. The extraordinary sensitivity, adaptability, and real-time monitoring capabilities of carbon-based sensors have made them indispensable instruments for the detection of heavy metals. High surface areas and distinctive electrical characteristics of materials like graphene, CNTs, and CDs improve detection limits and selectivity for different heavy metals.

REFERENCES

1. Cimboláková, I., Uher, I., Laktičová, K. V., Vargová, M., Kimáková, T., & Papajová, I. (2019). Heavy metals and the environment. *Environmental Factors Affecting Human Health,* 1, 29.
2. Khalef, R. N., Hassan, A. I., & Saleh, H. M. (2022). Heavy metal's environmental impact. In Hosam Saleh & Amal I. Hassan (eds), *Environmental Impact and Remediation of Heavy Metals.* IntechOpen.
3. Zulkifli, S. N., Rahim, H. A., & Lau, W. J. (2018). Detection of contaminants in water supply: A review on state-of-the-art monitoring technologies and their applications. *Sensors and Actuators B: Chemical,* 255, 2657–2689.
4. Deshmukh, V. V., Tejashwini, D. M., Nagaswarupa, H. P., Naik, R., Al-Kahtani, A. A., & Kumar, Y. A. (2024). Sr and Fe substituted LaCoO3 nano perovskites: Electrochemical energy storage and sensing applications. *Journal of Energy Storage,* 89, 111724.
5. Tejashwini, D. M., Harini, H. V., Nagaswarupa, H. P., & Naik, R. (2024). Hybrid calcium magnesium ferrite: A multifaceted nanomaterial for energy, sensing, and environmental applications. *Results in Chemistry,* 7, 101456.
6. Pushpanjali, P. A., Manjunatha, J. G., Amrutha, B. M., & Hareesha, N. (2021). Development of carbon nanotube-based polymer-modified electrochemical sensor for the voltammetric study of Curcumin. *Materials Research Innovations,* 25(7), 412–420.
7. Raril, C., Manjunatha, J. G., Ravishankar, D. K., Fattepur, S., Siddaraju, G., & Nanjundaswamy, L. (2020). Validated electrochemical method for simultaneous resolution of tyrosine, uric acid, and ascorbic acid at polymer modified nano-composite paste electrode. *Surface Engineering and Applied Electrochemistry,* 56, 415–426.
8. Walekar, L., Dutta, T., Kumar, P., Ok, Y. S., Pawar, S., Deep, A., & Kim, K. H. (2017). Functionalized fluorescent nanomaterials for sensing pollutants in the environment: A critical review. *TrAC Trends in Analytical Chemistry,* 97, 458–467.
9. Deepa, H. M., Tejashwini, D. M., Harini, H. V., Nagaswarupa, H. P., Naik, R., & Basavaraju, N. (2024). Multifunctional applications of hybrid $CaMgTi_2O_4$ nanoparticles for electrochemical sensors and photocatalysis. *Sensing Technology*, 2(1), 2296890.
10. Huang, H., Chen, T., Liu, X., & Ma, H. (2014). Ultrasensitive and simultaneous detection of heavy metal ions based on three-dimensional graphene-carbon nanotubes hybrid electrode materials. *Analytica Chimica Acta,* 852, 45–54.
11. Liu, Y., Deng, Y., Dong, H., Liu, K., & He, N. (2017). Progress on sensors based on nanomaterials for rapid detection of heavy metal ions. *Science China Chemistry*, 60, 329–337.
12. Joshi, D. J., Koduru, J. R., Malek, N. I., Hussain, C. M., & Kailasa, S. K. (2021). Surface modifications and analytical applications of graphene oxide: A review. *TrAC Trends in Analytical Chemistry,* 144, 116448.
13. Manjunatha, J. G., Deraman, M., Basri, N. H., Nor, N. S. M., Talib, I. A., & Ataollahi, N. (2014). Sodium dodecyl sulfate modified carbon nanotubes paste electrode as a novel sensor for the simultaneous determination of dopamine, ascorbic acid, and uric acid. *Comptes Rendus. Chimie,* 17(5), 465–476.
14. Raju, C. V., Cho, C. H., Rani, G. M., Manju, V., Umapathi, R., Huh, Y. S., & Park, J. P. (2023). Emerging insights into the use of carbon-based nanomaterials for the electrochemical detection of heavy metal ions. *Coordination Chemistry Reviews,* 476, 214920.
15. Falina, S., Syamsul, M., Rhaffor, N. A., Sal Hamid, S., Mohamed Zain, K. A., Abd Manaf, A., & Kawarada, H. (2021). Ten years progress of electrical detection of heavy metal ions (HMIS) using various field-effect transistor (FET) nanosensors: A review. *Biosensors*, 11(12), 478.
16. Tejashwini, D. M., Harini, H. V., Nagaswarupa, H. P., Naik, R., Harlapur, S., & Basavaraju, N. (2024). Nanoferrites in photocatalytic wastewater treatment: Advancements, characterization, and environmental Implications. *Results in Chemistry*,7, 101247.
17. Charithra, M. M., & Manjunatha, J. G. (2020). Enhanced voltammetric detection of paracetamol by using carbon nanotube modified electrode as an electrochemical sensor. *Journal of Electrochemical Science and Engineering,* 10(1), 29–40.
18. Ralph, P. J., Smith, R. A., Macinnis-Ng, C. M. O., & Seery, C. R. (2007). Use of fluorescence-based ecotoxicological bioassays in monitoring toxicants and pollution in aquatic systems. *Toxicological and Environmental Chemistry*, 89(4), 589–607.

19. Bülbül, G., Hayat, A., & Andreescu, S. (2015). Portable nanoparticle-based sensors for food safety assessment. *Sensors*, 15(12), 30736–30758.
20. Selid, P. D., Xu, H., Collins, E. M., Face-Collins, M. S., & Zhao, J. X. (2009). Sensing mercury for biomedical and environmental monitoring. *Sensors*, 9(7), 5446–5459.
21. Manjunatha, J. G., Raril, C., Hareesha, N., Charithra, M. M., Pushpanjali, P. A., Tigari, G., ... & Gowda, J. (2020). Electrochemical fabrication of poly (niacin) modified graphite paste electrode and its application for the detection of riboflavin. *The Open Chemical Engineering Journal*, 14(1), 90–98.
22. Makwana, B. A., Vyas, D. J., Bhatt, K. D., Darji, S., & Jain, V. K. (2016). Novel fluorescent silver nanoparticles: Sensitive and selective turn off sensor for cadmium ions. *Applied Nanoscience,* 6, 555–566.
23. Kalluri, J. R., Arbneshi, T., Afrin Khan, S., Neely, A., Candice, P., Varisli, B., ... & Ray, P. C. (2009). Use of gold nanoparticles in a simple colorimetric and ultrasensitive dynamic light scattering assay: Selective detection of arsenic in groundwater. *Angewandte Chemie*, 121(51), 9848–9851.
24. Curulli, A. (2020). Nanomaterials in electrochemical sensing area: Applications and challenges in food analysis. *Molecules*, 25(23), 5759.
25. Hareesha, N., & Manjunatha, J. G. (2020). Fast and enhanced electrochemical sensing of dopamine at cost-effective poly (DL-phenylalanine) based graphite electrode. *Journal of Electroanalytical Chemistry,* 878, 114533.
26. Frisbie, S. H., & Mitchell, E. J. (2022). Arsenic in drinking water: An analysis of global drinking water regulations and recommendations for updates to protect public health. *PLoS One,* 17(4), e0263505.
27. Gravel, S., Roberge, B., Calosso, M., Gagné, S., Lavoie, J., & Labrèche, F. (2023). Occupational health and safety, metal exposures and multi-exposures health risk in Canadian electronic waste recycling facilities. *Waste Management,* 165, 140–149.
28. Shigeta, K., Fujita, A., Nakazato, T., & Tao, H. (2018). A robust method for the determination of Cr (VI) and Cr (III) in industrial wastewaters by liquid chromatography–inductively coupled plasma mass spectrometry combined with a chelating pretreatment with 2, 6-pyridinedicarboxylic acid. *Analytical Sciences*, 34(8), 925–932.
29. Baytak, S., & Türker, A. R. (2006). Determination of lead and nickel in environmental samples by flame atomic absorption spectrometry after column solid-phase extraction on Ambersorb-572 with EDTA. *Journal of Hazardous Materials,* 129(1–3), 130–136.
30. Beesley, L., Moreno-Jimenez, E., Clemente, R., Lepp, N., & Dickinson, N. (2010). Mobility of arsenic, cadmium and zinc in a multi-element contaminated soil profile assessed by in-situ soil pore water sampling, column leaching and sequential extraction. *Environmental Pollution*, 158(1), 155–160.
31. Hu, B., Wang, K., Wu, L., Yu, S. H., Antonietti, M., & Titirici, M. M. (2010). Engineering carbon materials from the hydrothermal carbonization process of biomass. *Advanced Materials*, 22(7), 813–828.
32. Mayes, A. G., & Whitcombe, M. J. (2005). Synthetic strategies for the generation of molecularly imprinted organic polymers. *Advanced Drug Delivery Reviews*, 57(12), 1742–1778.
33. Tejashwini, D. M., Nagaswarupa, H. P., Naik, R., Basavaraju, N., Alodhayb, A. N., Pandiaraj, S., ... & Kim, J. H. (2024). Synthesis, characterization, wastewater treatment & plant growth application of ZnFe2O4/Bi2O3 nanocomposite. *Desalination and Water Treatment*, 320, 100681.
34. Tejashwini, D. M., Harini, H. V., Nagaswarupa, H. P., Naik, R., Deshmukh, V. V., & Basavaraju, N. (2023). An in-depth exploration of eco-friendly synthesis methods for metal oxide nanoparticles and their role in photocatalysis for industrial dye degradation. *Chemical Physics Impact,* 7, 100355.
35. Deepa, H. M., Tejashwini, D. M., Nagaswarupa, H. P., Naik, R., & Basavaraju, N. (2024). Synthesis, characterisation, photocatalysis and Electrochemical studies of hybrid calcium magnesium titanate nanoparticles synthesised using co-precipitation method. *Sensing Technology,* 2(1), 2370407.
36. Tigari, G., Manjunatha, J. G., D'Souza, E. S., & Sreeharsha, N. (2021). Surfactant and polymer composite modified electrode for the sensitive determination of vanillin in food sample. *ChemistrySelect*, 6(11), 2700–2708.

37. Shabbir, M., Raza, Z. A., Shah, T. H., & Tariq, M. R. (2022). Recent progress in graphenes: Synthesis, covalent functionalization and environmental applications. *Journal of Nanostructure in Chemistry*, 12(6), 1033–1051.
38. Barhoum, A., Hamimed, S., Slimi, H., Othmani, A., Abdel-Haleem, F. M., & Bechelany, M. (2023). Modern designs of electrochemical sensor platforms for environmental analyses: Principles, nanofabrication opportunities, and challenges. *Trends in Environmental Analytical Chemistry,* 38, e00199.
39. Medina-Sánchez, M., Miserere, S., & Merkoçi, A. (2012). Nanomaterials and lab-on-a-chip technologies. *Lab on a Chip*, 12(11), 1932–1943.
40. Pushpanjali, P. A., & Manjunatha, J. G. (2020). Development of polymer modified electrochemical sensor for the determination of alizarin carmine in the presence of tartrazine. *Electroanalysis,* 32(11), 2474–2480.
41. Sobiech, M., Bujak, P., Luliński, P., & Pron, A. (2019). Semiconductor nanocrystal–polymer hybrid nanomaterials and their application in molecular imprinting. *Nanoscale*, 11(25), 12030–12074.
42. Aragay, G., Pons, J., & Merkoçi, A. (2011). Recent trends in macro-, micro-, and nanomaterial-based tools and strategies for heavy-metal detection. *Chemical Reviews,* 111(5), 3433–3458.
43. Barik, P., & Pradhan, M. (2022). Selectivity in trace gas sensing: Recent developments, challenges, and future perspectives. *Analyst*, 147(6), 1024–1054.
44. Zaidi, S. A. (2017). Molecular imprinting polymers and their composites: A promising material for diverse applications. *Biomaterials Science*, 5(3), 388–402.
45. Georgakilas, V., Tiwari, J. N., Kemp, K. C., Perman, J. A., Bourlinos, A. B., Kim, K. S., & Zboril, R. (2016). Noncovalent functionalization of graphene and graphene oxide for energy materials, biosensing, catalytic, and biomedical applications. *Chemical Reviews,* 116(9), 5464–5519.
46. Dkhar, D. S., Kumari, R., Malode, S. J., Shetti, N. P., & Chandra, P. (2023). Integrated lab-on-a-chip devices: Fabrication methodologies, transduction system for sensing purposes. *Journal of Pharmaceutical and Biomedical Analysis*, 223, 115120.
47. Yaroshenko, I., Kirsanov, D., Marjanovic, M., Lieberzeit, P. A., Korostynska, O., Mason, A., ... & Legin, A. (2020). Real-time water quality monitoring with chemical sensors. *Sensors*, 20(12), 3432.
48. Donga, C., Mishra, S. B., Abd-El-Aziz, A. S., & Mishra, A. K. (2021). Advances in graphene-based magnetic and graphene-based/TiO2 nanoparticles in the removal of heavy metals and organic pollutants from industrial wastewater. *Journal of Inorganic and Organometallic Polymers and Materials,* 31(2), 463–480.
49. Dahake, R. V., & Bansiwal, A. (2022). Disposable sensors for heavy metals detection: A review of carbon and non-noble metal-based receptors. *ChemistrySelect*, 7(42), e202202824.
50. Adley, C. C. (2014). Past, present and future of sensors in food production. *Foods*, 3(3), 491–510.
51. Vonnie, J. M., Rovina, K., Mariah, A. M. A., Erna, K. H., Felicia, W. X. L., & 'Aqilah, M. N. (2023). Trends in nanotechnology techniques for detecting heavy metals in food and contaminated water: A review. *International Journal of Environmental Science and Technology*, 20(7), 8041–8072.
52. Li, M., Shi, Q., Song, N., Xiao, Y., Wang, L., Chen, Z., & James, T. D. (2023). *Current trends in the detection and removal of heavy metal ions using functional materials*. Chemical Society Reviews, 52, 5827–5860.
53. Manjunatha, J. G., Swamy, B. K., Mamatha, G. P., Shankar, S. S., Gilbert, O., Chandrashekar, B. N., & Sherigara, B. S. (2009). Electrochemical response of dopamine at phthalic acid and triton X-100 modified carbon paste electrode: A cyclic voltammetry study. *International Journal of Electrochemical Science*, 4(10), 1469–1478.
54. Tejashwini, D. M., Harini, H. V., Nagaswarupa, H. P., Naik, R., & Chidananda, B. (2024). A comparative study of green and chemical approaches for photocatalytic activity of novel hybrid bismuth magnesium ferrites (BiMgFe2O4) nanoparticles for Acid Red-88 dye degradation. *Results in Chemistry,* 7, 101267.
55. Murthy, H. A., & Nagaswarupa, H. P. (2022). Novel trends for synthesis of carbon nanomaterial-based sensors. In Jamballi G. Manjunatha & Chaudhery Mustansar Hussain (eds), *Carbon Nanomaterials-Based Sensors* (pp. 29–42). Elsevier.

56. Harini, H. V., Tejashwini, D. M., Ishwarya, S., Deshmukh, V. V., Nagaswarupa, H. P., & Naik, R. (2024). Battery and fuel cell applications: Nanohybrid aluminates, ferrites, and vanadates. In H.P. Nagaswarupa, Mika E.T. Sillanpää, H.C. Ananda Murthy & Ramachandra Naik (eds), *Multifunctional Inorganic Nanomaterials for Energy Applications* (pp. 120–129). CRC Press.
57. Banerjee, A. N. (2022). Green syntheses of graphene and its applications in internet of things (IoT)—a status review. *Nanotechnology*, 33(32), 322003.

4 Carbon Sensor for Soil Analysis

Youssra El Hamdouni, Khalid Ait Ben Brahim, Mohamed Bendany, Najoua Labjar, Hamid Nasrellah, Hafssa Bensemlali, Halima Mortadi, and Souad El Hajjaji

4.1 INTRODUCTION

Soil analysis is a cornerstone of agriculture, natural resource management, and environmental protection [1,2]. As the primary substrate for plant growth, soils are crucial for food production and regulating biogeochemical cycles and water filtration [3]. A thorough understanding of the physical, chemical, and biological characteristics of soil is essential for optimizing agricultural practices, preventing land degradation, and preserving environmental quality [4]. With the challenges of climate change and the rising global population, soil analysis has become crucial for maintaining food security and ecosystem health [5]. A recent publication from the Food and Agriculture Organization highlights the pressing need for precise and continuous monitoring to avert further degradation of approximately 33% of the world's soils [6].

Traditional methods of soil analysis, while accurate, are characterized by complexity, high costs, and lengthy processing times, which limit their practicality for real-time and large-scale monitoring [7]. This situation has led to a growing demand for soil analysis technologies that are rapid, reliable, and adaptable for field use.

The advent of sensors has revolutionized various scientific domains, particularly precision agriculture and environmental management, by enabling real-time measurements. In soil analysis, these instruments enable real-time monitoring of essential parameters, including moisture levels, pH, nutrient concentrations, and the presence of contaminants [8]. This capability for real-time monitoring allows for informed decision-making, thereby enhancing agricultural efficiency and reducing environmental impacts [9]. However, traditional sensors often exhibit limitations, including bulkiness, high cost, and inadequate sensitivity to detect low concentrations of specific substances [10].

In response to these challenges, considerable focus has been placed on advancing new materials and technologies, with carbon-based materials demonstrating notable potential [11]. Carbon nanotubes (CNTs), graphene, and carbon nanofibers have distinctive physical and chemical characteristics that render them excellent materials for sensor manufacturing [12–14]. Their excellent electrical conductivity, extensive surface area, and ability to be functionalized enhance the development of sensors with high sensitivity and selectivity [15]. Moreover, the superior chemical and mechanical stability of these materials compared to conventional options enhances their durability and performance under challenging field conditions [16].

Sensors enhanced with carbon materials can detect various soil parameters, including nutrient levels, contaminants, and physical-chemical properties such as moisture and pH [17]. Their compact size and potentially lower cost make them practical real-time and large-scale soil monitoring tools

DOI: 10.1201/9781003500322-4

[18]. Furthermore, the integration of these sensors into sensor networks enables the collection and analysis of high-resolution data, optimizing soil management practices [19,20].

This chapter comprehensively examines recent advancements in carbon-modified sensors for soil analysis. Key areas of focus include the classification of sensors used in soil analysis, the various carbon-based materials employed, and the most effective functionalization methods to enhance sensor sensitivity and selectivity. Additionally, the chapter addresses sensor characterization techniques. It examines the possible uses of these sensors in diverse areas such as precision agriculture, environmental monitoring, and contaminant detection in industrial soils.

4.2 CLASSIFICATION OF CARBON SENSORS FOR SOIL ANALYSIS

Sensors are critical tools for measurement and detection in various scientific and industrial applications, including soil analysis [21,22]. They are generally classified into physical sensors and chemical sensors according to their functions and the types of signals they handle, as illustrated in Figure 4.1 [23,24].

Physical sensors are engineered to measure various physical parameters, including temperature and pressure, flow, humidity, and distance [25]. These sensors convert physical quantities into proportional electrical signals [26]. Common types include thermistors for temperature measurement, strain gauges for pressure sensing, and displacement sensors for distance measurements [27,28]. For instance, Gali et al. investigated the use of image sensors integrated with machine learning techniques to assess plant growth stages and detect diseases [29]. Although these sensors do not primarily use carbon materials, integrating carbon nanomaterials could potentially enhance their sensitivity and resolution [30].

Chemical sensors, on the other hand, are specialized in the detection and quantification of chemical substances [31]. These devices convert chemical data, such as the concentration of a particular substance, into an electrical or optical signal [32]. Electrochemical, optical, and gravimetric sensors represent major subcategories, each suited to specific applications in soil analysis [33]. For example, Neha et al. created an electrochemical sensor utilizing a carbon electrode enhanced with reduced graphene oxide functionalized with fluorenone to detect chlorpyrifos pesticide residues in soil [34]. Employing differential pulse voltammetry, the sensor was shown to be capable of detecting low concentrations of chlorpyrifos, highlighting the effectiveness of carbon materials in improving the sensitivity and specificity of sensors for soil contaminant monitoring.

Overall, integrating carbon materials into sensors used for soil analysis opens up new possibilities regarding precision, sensitivity, and durability, enabling large-scale real-time monitoring, which is essential for optimized soil management.

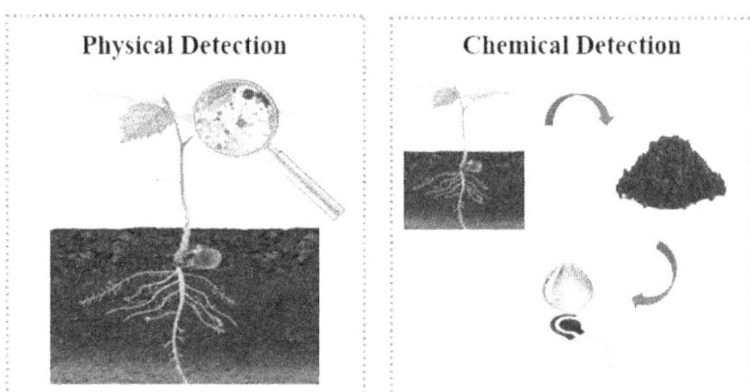

FIGURE 4.1 Classification of Physical and Chemical Sensors for Soil Analysis.

4.3 CARBON-BASED MATERIALS FOR SENSORS

The choice of sensor material is critical for achieving optimal measurement results. The appropriate surface modification of these sensors significantly influences their sensitivity and selectivity in detection [35]. Various materials or their combinations can be employed to modify electrodes, thereby increasing their effective surface area, lowering detection limits, and enhancing both responsiveness and specificity [36].

Given this scenario, various sensors employing carbon-based materials have been developed for soil analysis [37]. These materials encompass carbon nanoparticles, CNTs, carbon nanofibers, graphene and its derivatives, carbon quantum dots, and amorphous carbon [38]. Each material offers unique characteristics that make it suitable for detecting nutrients, contaminants, and other critical soil parameters [39]. Generally, carbon-based materials exhibit intrinsic advantages and distinctive properties such as ease of implementation, long-term stability, and high electrical conductivity [40,41]. A major advantage of these materials is their extensive electroactivity range, enabling the detection of a wide variety of species in both the cathodic and anodic regions [42]. Additionally, their unique morphology, large specific surface area, numerous adsorption sites, and rapid charge transfer kinetics enhance their effectiveness [43]. Their exceptional resistance to chemical corrosion and flexibility in carbon matrix deposition and operation further support large-scale production, making them highly promising for developing sensors aimed at environmental preservation and soil analysis [44]. However, they also present some limitations, including complex preparation processes and relatively high costs. The strengths and weaknesses of these materials are outlined in Table 4.1.

TABLE 4.1
Benefits and Limitations of Carbon-Based Materials for Soil Analysis

Type of Carbon	Advantages	Limitations	Ref.
Carbon Nanotubes	- Very high electrical conductivity, ideal for nanoelectronics and electrochemical sensors. - Large specific surface area and exceptional adsorption capacity. - Robust mechanical properties, especially for MWCNTs. - Nanometric structure allowing ultra-sensitive detection of target molecules.	- High production cost with challenges in large-scale synthesis. - Complexity of functionalization to improve selectivity. - Difficult dispersion in composite matrices.	[46]
Graphene	- Exceptional electrical conductivity, surpassing that of any other known material. - Large specific surface area, providing intense interaction with analytes. - Mechanical flexibility allows applications in portable devices. - Extremely high electron mobility, ideal for electronic devices.	- High production cost, especially for obtaining high-quality graphene. - Complex and technically demanding synthesis. - Frequent need for functionalization to adjust surface properties. - Agglomeration issues in composite matrices.	[47]
Graphite	- High thermal and chemical stability, resistant to extreme conditions. - Abundant and economical material. - Good electrical conductivity, useful for electrodes and other applications. - Natural lubrication due to its sheet-like structure. - Easy to process and handle for various industrial uses.	- Lower electrical conductivity compared to graphene. - Sheet structure limiting interactions with analytes. - Purity issues affecting performance in advanced applications. - Sensitivity to oxidation in certain environments.	[48]

(continued)

TABLE 4.1 (Continued)
Benefits and Limitations of Carbon-Based Materials for Soil Analysis

Type of Carbon	Advantages	Limitations	Ref.
Carbon Black	- Abundant and low-cost, with easy large-scale production. - Large specific surface area for interactions with analytes. - Improves the mechanical properties of composites when added to polymer matrices. - Good resistance to chemical corrosion.	- Moderate electrical conductivity compared to other carbon-based materials. - Difficult and non-uniform dispersion in matrices. - Amorphous structure limiting certain properties like conductivity and adsorption.	[49]

Moreover, the sensitivity and selectivity of these carbon-based materials can be enhanced through functionalization combined with other materials, creating composites with synergistically improved properties [45].

4.3.1 Carbon Nanotubes

CNTs are considered some of the most extraordinary and versatile nanomaterials, celebrated for their unique properties and diverse applications [49]. CNTs consist of tubular formations created from a hexagonal lattice of carbon atoms and are generally divided into two main types: single-walled carbon nanotubes (SWCNTs) and multi-walled carbon nanotubes (MWCNTs) [46]. SWCNTs are made from a single graphene sheet rolled into a cylindrical shape, whereas MWCNTs consist of several concentric graphene layers, each spaced about 0.34 nm apart [50]. Their tubular structure, combined with a large specific surface area and nanometric dimensions, provides exceptional adsorption capacity, which is crucial for detecting gas molecules, contaminants, or nutrients in soil [51]. Additionally, CNTs exhibit excellent electrical conductivity, which makes them highly suitable for applications in nanoelectronics and electrochemical sensors [52,53].

SWCNTs are utilized in ultra-sensitive detection devices due to their large specific surface area, which facilitates direct interaction with analytes and enhances sensor sensitivity and response speed [54]. MWCNTs, with their more complex structure, offer superior mechanical and chemical stability, making them suitable for demanding environments or applications requiring increased robustness [55,56].

However, challenges remain in the large-scale production of CNTs with precise control over their structure and properties. Ongoing research aims to develop more efficient synthesis methods that produce CNTs with homogeneous and reproducible properties. Additionally, functionalization, which involves attaching chemical groups to the CNTs' surface, is a common strategy to improve interactions with specific targets, thus increasing the sensors' selectivity and efficiency [50].

4.3.2 Graphene

Graphene, frequently hailed as a "remarkable material," consists of a single layer of carbon atoms arranged in a hexagonal pattern, serving as the foundation for various other carbon allotropes, including graphite, CNTs, and fullerenes. This two-dimensional (2D) substance is made up of carbon atoms linked by sp^2 bonds, imparting to graphene outstanding physical and chemical characteristics [57].

The most notable characteristic of graphene is its electrical conductivity, which exceeds that of any other known material [58]. This property, combined with extremely high charge carrier mobility, makes graphene highly suitable for electronic applications, including field-effect transistors, gas sensors, and energy storage devices. Furthermore, its exceptionally high specific surface area facilitates intense interaction with analytes, which is crucial for designing highly sensitive sensors [59].

Graphene is also renowned for its mechanical flexibility, allowing integration into flexible and transparent devices, thus enabling innovative applications in portable sensors and flexible electronics. Additionally, the 2D nature of graphene provides the maximum surface area for molecule adsorption, significantly enhancing the detection capabilities of graphene-based sensors [60].

Despite its promising properties, graphene presents challenges, particularly in achieving large-scale, low-cost production. Synthesis methods such as mechanical exfoliation, chemical vapor deposition (CVD), or chemical reduction of graphene oxide require precise control of reaction conditions to produce high-quality material with uniform properties [61]. Functionalization of graphene is also often necessary to improve compatibility with specific analytes or to modify its surface properties, thereby enhancing sensor efficiency and selectivity [57].

4.3.3 Graphite

Graphite is a three-dimensional (3D) crystalline form of carbon, consisting of multiple layers of graphene stacked together. The layers are interconnected through Van der Waals forces, which allow them to slide effortlessly against each other. This structure imparts graphite with natural lubricating properties and notable electrical conductivity, making it widely used in applications ranging from battery electrodes to pencils [62].

One of the most interesting properties of graphite is its thermal and chemical stability, making it highly resistant to extreme conditions, such as high temperatures and corrosive environments [63]. Furthermore, graphite is commonly employed in electrochemical sensors because of its capacity to catalyze certain reduction or oxidation reactions, which is essential for detecting various chemical species in soils. Graphite's physical and chemical properties—such as density, porosity, and conductivity—are markedly affected by interactions among the graphene layers [64].

However, like graphene, producing high-quality graphite with controlled properties remains a challenge. Research is ongoing to improve graphite purification and processing methods to maximize its performance in advanced applications. In addition, combining graphite with additional substances, including polymers or metals, is a commonly used strategy to enhance its mechanical properties or introduce new functionalities into graphite composites [62].

4.3.4 Carbon Black

Carbon black is a fine, non-crystalline carbon material created by the partial pyrolysis of carbon-rich sources such as hydrocarbons [65]. Its properties, such as specific surface area, pore size, and surface characteristics, are adjustable based on the conditions used during pyrolysis [66]. This capability to adjust both the physical and chemical traits of carbon black renders it exceptionally adaptable for a range of uses, including strengthening tires and improving composite materials [67].

In environmental sensors, carbon black is often used to improve the electrical performance of electrodes and to act as an active matrix for pollutant detection [68]. Its amorphous structure, consisting of nanometric particles, features a broad interface that improves engagement with target analytes [69]. However, carbon black has limitations, such as lower adsorption capacity compared to nanostructured materials like CNTs or graphene. To address these limitations, carbon black is often combined with additional materials, such as conductive polymers or metals, to improve its performance in detection devices [70].

For instance, incorporating carbon black into a polymer matrix can improve both the electrical conductivity and structural integrity of the composite, making it suitable for use in corrosive environments or under challenging conditions. Additionally, carbon black's ease of large-scale production makes it an economical and viable choice for developing environmental sensors [66].

4.4 SURFACE FUNCTIONALIZATION OF CARBON SENSORS

Surface modification of carbon sensors through recognition layers is essential for enhancing their performance and specificity in electrochemical applications. This modification is achieved through various chemical processes that create functional interfaces tailored to specific needs [71]. Techniques for immobilizing functional layers include methods such as embedding in polymer matrices or gels, creating cross-links through chemical reactions, forming self-assembled monolayers, establishing covalent connections, and applying electrodeposition processes [72]. The ability to customize the properties of carbon materials enhances their effectiveness in soil monitoring, with significant implications for precision agriculture and sustainable resource management [73,74].

Adsorption, whether chemical (chemisorption) or physical (physisorption), is a straightforward method for coating electrode surfaces. In this process, a solution with the modifying agent is spread over the electrode surface, and the solvent is then allowed to evaporate, leaving the modifying agent adhered to the surface [75]. This method is valued for its simplicity and rapidity. However, it can have limitations in terms of modifier stability and reproducibility, as the interactions between the modifier and the surface may not always be strong [76]. For example, glucose oxidase molecules are commonly adsorbed onto carbon electrodes for glucose detection. The effectiveness of this method depends heavily on preparation conditions and electrode properties [77].

Electrodeposition is another surface modification technique involving the deposition of functional materials onto the electrode substrate through electrolysis under an applied potential. This method facilitates the creation of hybrid films, with thickness and morphology controlled by parameters such as solvent, pH, temperature, and electrolyte composition [78,79]. For instance, the electrodeposition of gold nanoparticles (AuNPs) on CNT electrodes enhances conductivity and sensitivity for detecting toxic gases in soil, such as ammonia (NH_3) and sulfur dioxide (SO_2). These nanoparticles expand the effective surface of CNTs and promote electron movement, which is crucial for optimal electrochemical sensing [80].

Electropolymerization is an advanced technique that forms polymers on the electrode surface. This method is used to create porous polymer films with specific functions that enhance electron transfer and improve sensor sensitivity [81]. Films obtained through electropolymerization exhibit high selectivity due to their uniformity and strong adhesion to the electrode surface. For example, applying poly(ethylene glycol) (PEG) onto graphene has been utilized to target and detect organic contaminants selectively. PEG acts as a selective barrier that interacts specifically with target analytes while blocking interference from other soil compounds [82].

Covalent attachment is an additional method for surface modification, involving the introduction of reactive groups onto the electrode surface to facilitate the covalent bonding of modifying agents. This process generally involves oxidizing the surface to generate anchoring groups, such as carboxyl or amine groups, to which modifiers can bind [83]. This method offers high stability for the anchored modifiers and better measurement repeatability. For example, the functionalization of CNTs with carboxyl groups (-COOH) is a commonly used technique to enhance the detection of metal ions in soils. Studies have shown that CNTs functionalized with carboxyl groups exhibit increased affinity for lead ions (Pb^{2+}), allowing for more sensitive and selective detection compared to unfunctionalized CNTs [84]. Similarly, amine groups ($-NH_2$) can be used to detect organic compounds like pesticides, which is crucial for assessing soil quality in agriculture [85].

Different surface modification techniques each provide unique benefits related to ease of implementation, control over film structure, and stability of the functional layer. Selecting the appropriate method depends on the specific needs of the application and the intended characteristics of the sensor. Table 4.2 offers a detailed summary of these techniques, highlighting their benefits and drawbacks to assist in choosing the most appropriate method for specific electrochemical applications.

TABLE 4.2
Functionalization Techniques for Carbon-Based Sensors [86,87]

Method	Advantages	Limitations
Adsorption	- Simple and fast process. - Cost-effective. - Applicable to various types of carbon materials.	- Limited stability of the modifier. - Variability in repeatability. - Weak interactions with the surface.
Electrodeposition	- Controlled film thickness and morphology. - Enhances conductivity and sensitivity. - Versatile for different materials.	- Requires precise control of parameters. - Potential issues with uniformity. - Complex setup.
Electropolymerization	- High selectivity due to homogeneous films. - Strong adhesion to the electrode surface. - Enhanced sensitivity.	- Complex process. - Potentially high cost. - Limited to specific types of polymers.
Covalent Attachment	- High stability of modifiers. - Better repeatability. - Versatile functional group incorporation.	- Requires surface oxidation. - Potentially complex and time-consuming.

4.5 CHARACTERIZATION AND PERFORMANCE OF CARBON SENSORS

4.5.1 Characterization Methods for Sensors

Characterization methods are essential for the development and optimization of carbon sensors used in soil analysis. By employing a range of techniques such as electron microscopy, spectroscopy, and electrochemical methods, a detailed understanding of sensor properties and their performance under real-world conditions can be achieved [87,88]. Comprehensive characterization not only evaluates material and device quality but also informs necessary improvements to meet specific detection and analytical requirements.

Electron microscopy is essential for examining the structures of carbon sensors at microscopic and nanometric scales. Scanning electron microscopy (SEM) excels at imaging surface details of carbon electrodes and functional coatings with high resolution [89,90]. For instance, SEM has been used to observe the uniform distribution of biochar pores on modified electrodes, which enhances sensor sensitivity for detecting heavy metals in soils (Figure 4.2a) [91]. Transmission electron microscopy (TEM) provides a detailed perspective of internal structures, allowing for the observation of thin films and interfaces at an atomic scale. For example, TEM has been employed to analyze defects in carbon films deposited on electrodes, helping to optimize synthesis conditions for improved sensor performance [92].

Spectroscopic techniques provide critical insights into the chemical composition, electronic properties, and chemical interactions of sensors [93]. Fourier transform infrared spectroscopy (FTIR) is frequently employed to detect functional groups present on the surfaces of carbon electrodes [94]. For example, FTIR has confirmed the introduction of hydroxyl groups on carbon electrodes, which enhances their affinity for specific analytes (Figure 4.2b) [91]. Raman spectroscopy, which studies the vibrational modes of carbon atoms, provides information on the disorder and crystallinity of materials [95]. Variations in the D/G ratio in Raman spectra have been utilized to assess the effect of thermal treatments on carbon material structures, thereby improving sensor performance for detecting soil contaminants [96]. X-ray photoelectron spectroscopy (XPS) analyzes the valence states of elements on sensor surfaces, offering insights into the chemical elements present, their oxidation states, and their distribution. For example, XPS has been used to study changes in the oxidation states of elements after the functionalization of carbon electrodes with glycerol nanoparticles [97].

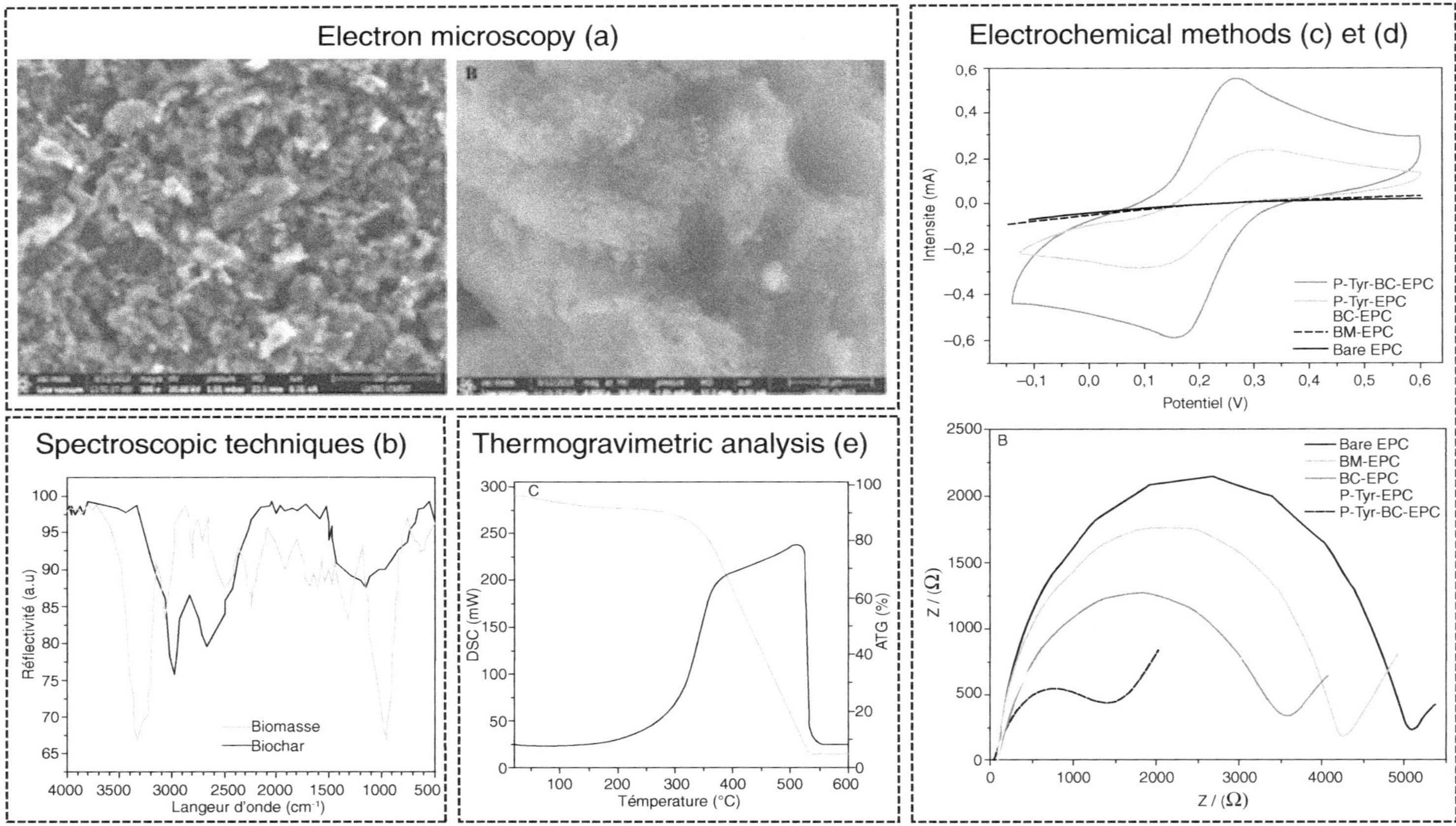

FIGURE 4.2 Characterization methods for Biochar-modified electrodes using SEM (a), FTIR (b), CV (c), EIS (d), and TGA (e).

Electrochemical methods are essential for evaluating conductivity properties, reactivity, and sensor performance [98]. Cyclic voltammetry (CV) generates current-potential curves, allowing for the analysis of redox reactions and charge storage capabilities [99]. For example, CV has demonstrated improved conductivity in electrodes functionalized with poly-tyrosine (Figure 4.2c [91]. Electrochemical impedance spectroscopy (EIS) is employed to investigate the resistance to charge transfer and the dynamics of reactions occurring at the electrode interface [100,101]. Nyquist plots obtained through EIS have facilitated the understanding of electrochemical processes and optimizing carbon sensor performance (Figure 4.2d) [91]. Chronoamperometry studies the real-time responses of electrodes to potential variations and has been used to assess the sensitivity of carbon electrodes to varying concentrations of soil contaminants [102].

Additional characterization techniques, such as atomic force microscopy (AFM) offer detailed imaging of surface topography, allowing for the measurement of roughness and nanoscale features. AFM has been used to assess the surface roughness of carbon electrodes following modifications, enhancing their performance [103]. Thermogravimetric analysis (TGA) evaluates the heat resistance of carbon materials. This method has been utilized to gauge the stability of functionalized electrodes under high temperatures, offering crucial insights into their application in extreme conditions (Figure 4.2e) [91].

Each characterization technique offers unique and complementary information, allowing for a comprehensive assessment of carbon sensors. Integrating these methods into the development process ensures optimal performance and precise adaptation to the specific needs of various applications.

4.5.2 Response to Environmental Conditions

The performance of carbon-based sensors, particularly in soil analysis, often undergoes significant changes due to environmental conditions such as temperature, humidity, pH, and other chemical substances. Understanding the impact of these variables is vital for maintaining the sensors' reliability and accuracy in various environmental settings [104].

Temperature is a key environmental factor that significantly impacts the performance of carbon sensors. Temperature variations can influence electrochemical reactions, electrical conductivity, and even the mechanical properties of materials [105]. For instance, while the conductivity of carbon materials generally rises with higher temperatures, this can also result in material degradation or trigger undesirable chemical reactions. Furthermore, temperature fluctuations can accelerate or decelerate redox reactions on carbon electrodes, causing variations in the measured signals [106]. To mitigate these effects, materials with high thermal stability are often employed, or thermal compensation mechanisms are integrated into measurement circuits. Regular calibrations based on temperature can also help correct data and improve accuracy under varying thermal conditions [107].

Humidity is another environmental factor influencing carbon sensors, potentially affecting both conductivity and surface properties [108]. High humidity can alter the conductivity of carbon electrodes by causing changes in the structure of functional layers or facilitating unwanted reactions. Excessive humidity can also lead to deposits or contamination on sensor surfaces, modifying electrode characteristics and disrupting measurements [109]. To address these issues, sensors can be protected with waterproof coatings or hydrophobic materials. Additionally, sensors equipped with dehydration or humidity compensation mechanisms can ensure stable performance despite changes in humidity levels [110].

pH variations and the presence of other chemical substances in the measurement environment can also influence sensor performance. Changes in pH can alter the electrochemical reactions occurring at the electrode surface, potentially causing inaccuracies in measurement [111]. Moreover, carbon electrodes may interact with ions in the solution, compromising the accuracy of the readings. Contaminating chemical substances in soil or samples can also affect the sensor's surface, leading to interference and degradation in performance [112]. To combat these issues, sensors can be made from materials resistant to pH fluctuations or designed with pH compensation mechanisms. Additionally, proper cleaning protocols and regular calibration can help maintain sensor reliability under varying chemical conditions [113].

4.5.3 Performance and Influencing Factors for Sensors

The validation of an electrochemical sensor involves establishing specific analytical requirements and confirming that the method used meets these criteria. The effectiveness of an electrode is closely connected to the characteristics of the materials used, the interaction between the sensor and analyte, its integration with the transducer, and the inherent properties of the electrochemical technique used [114]. To evaluate and validate the performance of sensors, several key criteria should be considered:

Selectivity, also known as specificity, denotes the sensor's capability to identify a particular analyte even when other interfering substances are present [115]. This attribute is crucial for analyzing complex mixtures, as it ensures the accurate detection of the target analyte amidst various other compounds [116].

Sensitivity is typically measured by the limit of detection, which represents the smallest concentration of an analyte that can be consistently identified, distinguishing it from non-detection [117]. The standard approach to calculating the limit of detection involves the 3σ criterion, setting the signal-to-noise ratio (S/N) at 3 [118]. Sensitivity is essential for detecting trace amounts of analytes, particularly in complex environmental samples such as soil.

Linearity, or calibration, evaluates how well a method generates a response that is directly proportional to the analyte concentration across a defined range [119]. It is crucial to choose a calibration range that is sufficiently broad to maintain result accuracy, as signal variation can increase with concentration. Generally, the calibration range should typically extend 1 to 1.5 logarithmic units beyond the detection limit, which corresponds to roughly 10 to 30 times this limit. It is beneficial to include a central value within this range for effective calibration curve development and analysis. For reliable results, it is recommended to use a minimum of seven distinct concentrations, including a blank, and to perform measurements in duplicate for thorough validation [116].

Precision refers to the method's capability to consistently reproduce measurements when conducted under identical conditions. This encompasses aspects such as instrument precision, repeatability, and reproducibility. While increasing the number of measurements can enhance precision, reproducibility often involves greater uncertainty compared to repeatability. Horwitz's research indicates that precision is also influenced by the analyte concentration range, with narrower ranges frequently showing higher relative standard deviations [120].

Accuracy measures the closeness of the average of an infinite number of results to a reference value. In practical terms, recovery rates between 95% and 105% are generally considered acceptable, indicating that the results closely align with the reference values [114].

Response time refers to the duration it takes for a sensor to achieve 90% of its final reading following exposure to the target analyte, especially when measurement conditions change rapidly. Response time is crucial for applications requiring quick and reliable detection of contaminants [121].

Sensor lifetime refers to the period during which a sensor maintains stable performance, ensuring long-term accuracy and reliability. Stability over time is vital for robust measurements, especially in field applications [116].

Evaluating these criteria is essential for validating the performance of carbon-based sensors. A thorough validation process involves repeating measurements, comparing results, and minimizing errors in calibration graphs. Incorporating appropriate statistical analyses ensures the scientific reliability of the method and confirms that the sensor meets the necessary performance standards.

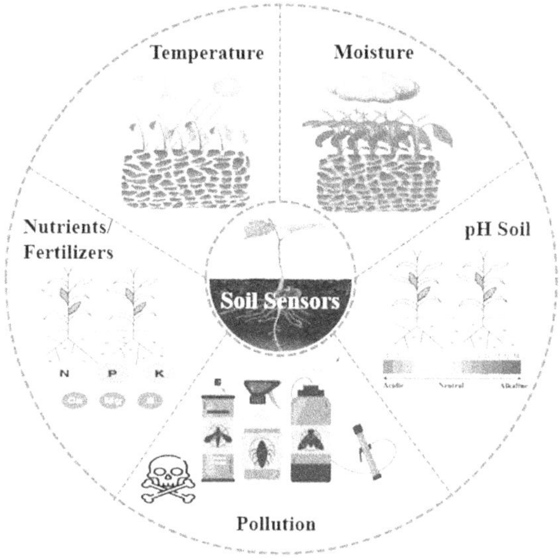

FIGURE 4.3 Applications of carbon sensors in soil analysis.

4.6 ADVANCED APPLICATIONS OF CARBON SENSORS IN SOIL ANALYSIS

Carbon-modified sensors represent a significant advancement in soil quality monitoring, offering highly accurate, real-time data crucial for assessing soil parameters such as pH, electrical conductivity, moisture content, and nutrient levels [21]. In modern agriculture, these sensors are essential tools because of their high sensitivity, rapid response times, robustness, and affordability [122,123]. They play a critical role in optimizing agricultural practices, particularly in areas like irrigation management, soil fertility assessment, and salinity monitoring [124]. By leveraging carbon-based sensors, farmers and agronomists can implement more sustainable and efficient practices, thereby enhancing crop productivity and soil health (Figure 4.3).

4.6.1 Monitoring Soil Quality

Soil pH is a fundamental indicator of nutrient availability and overall plant health. Advanced carbon-based sensors, such as those utilize oxidized graphene electrodes, offer unparalleled precision in pH measurement. These sensors exhibit enhanced sensitivity to pH variations, enabling real-time monitoring of soil acidity [125]. Recent studies highlight the efficacy of L-arginine-functionalized 3D macroporous graphene pH sensors in diverse agricultural environments, demonstrating their ability to monitor pH fluctuations in response to variations in fertilization and irrigation strategies [126].

Electrical conductivity, which reflects the concentration of dissolved salts and soil water retention capacity, can be accurately measured using the high conductivity properties of carbon materials [127]. For example, flexible graphene oxide-based antenna sensors have been engineered to effectively monitor salinity levels in soils, providing critical insights for the management of saline-affected lands. These sensors facilitate the timely application of remedial actions, contributing to improved soil health and crop productivity [128].

Soil moisture is another critical parameter for plant growth and water management. Carbon-modified sensors excel in this domain by utilizing absorbent materials or tracking changes in electrical conductivity to deliver precise and rapid moisture measurements [129,130]. Notably, modified carbon sensors have been demonstrated to provide real-time monitoring of soil moisture, offering essential data for optimizing irrigation systems and ensuring efficient water use [131].

In addition to basic soil parameters, carbon sensors are capable of detecting specific nutrients such as nitrates, phosphates, and potassium [132,133]. Functionalized carbon materials incorporating zinc oxide-multiwalled carbon nanotube-based sensors interact selectively with specific ions, enabling continuous monitoring of soil nutrient levels [134]. Electrochemical sensors utilizing functionalized Nafion-modified screen-printed carbon have been developed to measure potassium concentrations with high accuracy, enabling the fine-tuning of fertilization practices for enhanced crop yield and soil sustainability [135].

4.6.2 Detection of Environmental Contaminants

Detecting and quantifying environmental contaminants in soil is among the most important uses of sensors modified with carbon. These sensors, owing to their unique material properties, offer advanced solutions for the sensitive and specific monitoring of pollutants [136]. For example, hydrocarbons, often originating from gasoline spills or petroleum products, can be precisely measured using CNT-based electrochemical sensors [137]. Functionalized graphene oxide sensors have been successfully deployed to track the biodegradation of hydrocarbons, providing invaluable data on the environmental remediation process [138].

The persistence of pesticide and herbicide residues in soil poses significant environmental and health risks [139]. Carbon sensors are highly effective in detecting these residues at trace levels [140]. By measuring degradation products or monitoring changes in soil properties, carbon-based nanosensors can detect even the smallest concentrations of these chemicals, facilitating monitoring of their degradation over time [141].

Heavy metals, though non-biodegradable, can undergo biological transformations that alter their toxicity [142]. Electrochemical sensors can monitor these transformations by measuring changes in concentration and chemical state [143]. Carbon fiber electrode have been specifically designed to track the presence and transformation of heavy metals in soils, providing critical data for assessing soil contamination and the effectiveness of remediation strategies [144].

4.6.3 Monitoring of Soil Degradation Processes

Monitoring soil degradation processes, including erosion, organic matter loss, salinization, and chemical contamination, is essential for maintaining soil health and ensuring sustainable land use [145]. Carbon-modified sensors offer a robust means of tracking these degradation indicators. For instance, soil organic matter, a key indicator of soil fertility, can be monitored by measuring total organic carbon content or soil respiration rates [146]. Functionalized graphene by polydopamine electrochemical sensors has been developed to measure total organic carbon with high precision, yielding valuable insights into the state of soil organic matter degradation [147].

Soil salinization, which is a major issue in arid areas, can be effectively monitored with conductivity sensors that utilize a composite electrode made from a β-NiOx hybrid film and MWCNTs. These sensors provide precise measurements of dissolved salt concentrations, allowing for the accurate tracking of salinization processes and timely intervention [148].

Although direct detection of soil erosion is more challenging, carbon-modified sensors can effectively monitor changes in soil properties indicative of erosion, such as soil density and structure [149,150]. These sensors integrate advanced measurement techniques to detect physical and chemical changes in the soil, offering a comprehensive approach to erosion monitoring [151].

4.7 CONCLUSION AND FUTURE PERSPECTIVES

The development and application of modified carbon sensors mark a transformative advance in soil analysis, offering a compelling alternative to traditional methods that are often hindered by complexity, high costs, and time-consuming processes. These advanced sensors, which utilize materials such as graphene, carbon nanofibers, and CNTs, are known for their outstanding sensitivity, selectivity, and stability. Their ability to continuously monitor critical soil parameters—including moisture, pH, nutrient levels, and contaminants—positions them as invaluable tools for optimizing agricultural practices, enhancing environmental protection, and improving resource management.

The remarkable properties of carbon-based materials, such as excellent electrical conductivity, expansive specific surface area, and superior chemical and mechanical stability, greatly enhance the performance of these sensors under various challenging field conditions. Moreover, the potential for incorporating these sensors into sensor networks facilitates high-resolution data acquisition, which promotes data-driven decision-making to improve agricultural productivity and support environmental sustainability.

Looking forward, several promising avenues for research and development emerge. Advancing sensor materials by exploring innovative carbon composites and hybrid materials could further enhance performance metrics such as sensitivity and selectivity. Moreover, integrating these sensors with artificial intelligence (AI) can revolutionize data analysis, offering sophisticated tools for precision agriculture and comprehensive environmental monitoring [152].

Field testing and scalability are critical next steps in validating sensor performance under various environmental conditions, ensuring these technologies become more accessible and cost-effective [153]. Furthermore, developing regulatory standards and protocols is vital to ensure data reliability and consistency across various applications, which will support the broader adoption and integration of carbon sensors in soil analysis [154].

REFERENCES

[1] A. Rakshit, S. Ghosh, S. Chakraborty, V. Philip, and A. Datta, "Soil Analysis for the Advancement of Soil-Based Production Systems," *Rhizosphere*, vol. 15, p. 100216, 2020, doi: 10.1016/j.rhisph.2020.100216.

[2] M. Raghavendra, M. P. Sharma, A. Ramesh, A. Richa, S. D. Billore, and R. K. Verma, "Soil Health Indicators: Methods and Applications," *Soil Anal. Recent Trends Appl.*, pp. 221–253, 2020, doi: 10.1007/978-981-15-2039-6_13.

[3] J. Telo da Gama, "The Role of Soils in Sustainability, Climate Change, and Ecosystem Services: Challenges and Opportunities," *Ecologies*, vol. 4, no. 3, pp. 552–567, 2023, doi: 10.3390/ecologies4030036.

[4] N. K. Fageria, "Soil Quality vs. Environmentally-Based Agricultural Management Practices," *Commun. Soil Sci. Plant Anal.*, vol. 33, no. 13–14, pp. 2301–2329, 2002, doi: 10.1081/CSS-120005764.

[5] R. Lal, "Climate Strategic Soil Management," *Challenges*, vol. 5, no. 1, pp. 43–74, 2014, doi: 10.3390/challe5010043.

[6] L. Montanarella *et al.*, "World's Soils Are under Threat," *Soil*, vol. 2, no. 1, pp. 79–82, 2016, doi: 10.5194/soil-2-79-2016.

[7] J. Wang, X. Zhang, L. Xiao, and T. Li, "Survey for Soil Sensing with IOT and Traditional Systems," *Network*, vol. 3, no. 4, pp. 482–501, 2023, doi: 10.3390/network3040021.

[8] H. Yin, Y. Cao, B. Marelli, X. Zeng, A. J. Mason, and C. Cao, "Soil Sensors and Plant Wearables for Smart and Precision Agriculture," *Adv. Mater.*, vol. 33, no. 20, pp. 1–24, 2021, doi: 10.1002/adma.202007764.

[9] A. Garciá-Miranda Ferrari, P. Carrington, S. J. Rowley-Neale, and C. E. Banks, "Recent Advances in Portable Heavy Metal Electrochemical Sensing Platforms," *Environ. Sc.: Water Res. Technol.*, vol. 6, no. 10. 2020. doi: 10.1039/d0ew00407c.

[10] Y. Wu, R. D. Tilley, and J. J. Gooding, "Challenges and Solutions in Developing Ultrasensitive Biosensors," *J. Am. Chem. Soc.*, vol. 141, no. 3, pp. 1162–1170, 2019, doi: 10.1021/jacs.8b09397.

[11] S. Nasir, M. Z. Hussein, Z. Zainal, and N. A. Yusof, "Carbon-Based Nanomaterials/Allotropes: A Glimpse of Their Synthesis, Properties and Some Applications," *Materials (Basel).*, vol. 11, no. 2, pp. 1–24, 2018, doi: 10.3390/ma11020295.

[12] Z. Wang, S. Wu, J. Wang, A. Yu, and G. Wei, "Carbon Nanofiber-Based Functional Nanomaterials for Sensor Applications," *Nanomaterials*, vol. 9, no. 7, 2019, doi: 10.3390/nano9071045.

[13] T. Liyanage, A. Z. Qamar, and G. Slaughter, "Application of Nanomaterials for Chemical and Biological Sensors: A Review," *IEEE Sens. J.*, vol. 21, no. 11, pp. 12407–12425, 2021, doi: 10.1109/JSEN.2020.3032952.

[14] S. Sharma, S. Shekhar, S. Gautam, B. Sharma, A. Kumar, and P. Jain, "Carbon-Based Nanomaterials as Novel Nanosensors," *Nanofabrication Smart Nanosensor Appl.*, pp. 323–347, 2020, doi: 10.1016/B978-0-12-820702-4.00014-3.

[15] S. Morais, "Special Issue: Carbon-Based Nanomaterials for (bio)sensors Development," *Nanomaterials*, vol. 11, no. 9, pp. 10–11, 2021, doi: 10.3390/nano11092430.

[16] J. A. Puértolas and S. M. Kurtz, "Evaluation of Carbon Nanotubes and Graphene as Reinforcements for UHMWPE-Based Composites in Arthroplastic Applications: A Review," *J. Mech. Behav. Biomed. Mater.*, vol. 39, pp. 129–145, 2014, doi: 10.1016/j.jmbbm.2014.06.013.

[17] M. Singh, R. S. Patkar, M. Vinchurkar, and M. S. Baghini, "Cost Effective Soil pH Sensor Using Carbon-Based Screen-Printed Electrodes," *IEEE Sens. J.*, vol. 20, no. 1, pp. 47–54, 2020, doi: 10.1109/JSEN.2019.2941695.

[18] O. Kanoun, R. Ramalingame, A. Al-Hamry, A. Bouhamed, J. R. Bautista-Quijano, and D. Rejandran, "Potential of Carbonaceous Materials for the Realization of High Performance Sensors," *Proc. 17th Int. Multi-Conference Syst. Signals Devices, SSD 2020*, 2020, doi: 10.1109/SSD49366.2020.9364173.

[19] A. Bah, S. K. Balasundram, and M. H. A. Husni, "Sensor Technologies for Precision Soil Nutrient Management and Monitoring," *Am. J. Agric. Biol. Sci.*, vol. 7, no. 1, pp. 43–49, 2012, doi: 10.3844/ajabssp.2012.43.49.

[20] Y. Mekonnen, S. Namuduri, L. Burton, A. Sarwat, and S. Bhansali, "Review—Machine Learning Techniques in Wireless Sensor Network Based Precision Agriculture," *J. Electrochem. Soc.*, vol. 167, no. 3, p. 037522, 2020, doi: 10.1149/2.0222003jes.

[21] M. Nadporozhskaya, N. Kovsh, R. Paolesse, and L. Lvova, "Recent Advances in Chemical Sensors for Soil Analysis: A Review," *Chemosensors*, vol. 10, no. 1, 2022, doi: 10.3390/chemosensors10010035.

[22] M. Bajaj, P. Swami, A. Singla, V. Mittal, J. Chawla, and S. Jabin, "Analysis of Contamination of Soil by Sensor Monitoring Systems: A Green Technology," *Ecol. Environ. Conserv.*, vol. 29, no. 02, pp. 787–792, 2023, doi: 10.53550/eec.2023.v29i02.041.

[23] A. De Marcellis and G. Ferri, "Physical and Chemical Sensors," *Analog Circuits Signal Process.*, pp. 1–35, 2011, doi: 10.1007/978-90-481-9828-3_1.

[24] R. S. K. Muthumanickam, "Electrochemical Treatment of Dye Wastewater Using Nickel Foam Electrode," *J. Electrochem. Sci. Eng.*, vol. 11, no. 3, pp. 209–215, 2021, doi: http://dx.doi.org/10.5599/jese.1011.

[25] H. Cao *et al.*, "Recent Advances in Physical Sensors Based on Electrospinning Technology," *ACS Mater. Lett.*, vol. 5, no. 6, pp. 1627–1648, Jun. 2023, doi: 10.1021/acsmaterialslett.3c00144.

[26] C. Frenette *et al.*, "A Comprehensive Review on Advancements in Sensors for Air Pollution Applications," *Food Webs*, vol. 951, p. e00276, 2024, doi: 10.1016/j.scitotenv.2024.175696.

[27] W. S. Jones, "Electromechanical Transducers in Hostile Environments," *Proc. Inst. Electr. Eng.*, vol. 126, no. 11, pp. 1029–1052, 1979, doi: DOI: 10.1049/piee.1979.0189.

[28] H. H. Busta, J. E. Pogemiller, and B. J. Zimmerman, "The Field Emitter Triode as a Displacement/Pressure Sensor," *J. Micromechanics Microengineering*, vol. 3, no. 2, pp. 49–56, 1993, doi: 10.1088/0960-1317/3/2/003.

[29] B. Gali, T. Jaina, G. Katpally, G. Kankipati, and C. Rajyalakshmi, "Image Based Plant Disease Detection Using Computer Vision," *2024 4th Int. Conf. Adv. Electr. Comput. Commun. Sustain. Technol. ICAECT 2024*, 2024, doi: 10.1109/ICAECT60202.2024.10468857.

[30] T. Zhou *et al.*, "A Rapid Detection Method for Soil Organic Matter Using a Carbon Dioxide Sensor In Situ," *Meas. J. Int. Meas. Confed.*, vol. 208, p. 112471, 2023, doi: 10.1016/j.measurement.2023.112471.

[31] P. Wang and Q. Liu, "Chemical Sensors and Measurement," *Adv. Top. Sci. Technol. China*, pp. 117–197, 2011, doi: 10.1007/978-3-642-19525-9_4.

[32] C. A. Buckner *et al.*, "Introductory Chapter: What Is Chemical Sensor?," *Intech*, vol. 11, p. 13, 2016, doi: http://dx.doi.org/10.5772/64626.

[33] H. J. Kim, K. A. Sudduth, and J. W. Hummel, "Soil Macronutrient Sensing for Precision Agriculture," *J. Environ. Monit.*, vol. 11, no. 10, pp. 1810–1824, 2009, doi: 10.1039/b906634a.

[34] N. Neha, K. Ranjeet, Jyoti, S. Rana, and K. Navneet, "Fluorenone Decorated RGO Based Electrochemical Sensor for Rapid Detection of Chlorpyrifos in Vegetable and Soil Samples," *J. Electrochem. Soc.*, vol. 170, no. 5, p. id.057508, 10 pp., 2023, doi: 10.1149/1945-7111/acd3b7.
[35] I. G. Munteanu and C. Apetrei, "A Review on Electrochemical Sensors and Biosensors Used in Assessing Antioxidant Activity," *Antioxidants*, vol. 11, no. 3, 2022, doi: 10.3390/antiox11030584.
[36] Q. WU, H. M. BI, and X. J. HAN, "Research Progress of Electrochemical Detection of Heavy Metal Ions," *Chinese J. Anal. Chem.*, vol. 49, no. 3, pp. 330–340, 2021, doi: 10.1016/S1872-2040(21)60083-X.
[37] P. S. Adarakatti and S. K. Kempahanumakkagari, "Modified Electrodes for Sensing," *Electrochemistry*, vol. 15, pp. 58–95, 2019, doi: 10.1039/9781788013895-00058.
[38] J. Wei *et al.*, "Highly Sensitive and Selective Electrochemical Detection of Pb(II) in Serum via an α-Fe_2O_3/NiO Heterostructure: Evidence from Theoretical Calculations and Adsorption Investigation," *Sensors Actuators, B Chem.*, vol. 344, p. 130295, 2021, doi: 10.1016/j.snb.2021.130295.
[39] M. Mehmandoust, G. Li, and N. Erk, "Biomass-Derived Carbon Materials as an Emerging Platform for Advanced Electrochemical Sensors: Recent Advances and Future Perspectives," *Industrial & Engineering Chemistry Research*, vol. 62, pp. 4628–4635, 2023, doi: 10.1021/acs.iecr.2c03058.
[40] H. Li, H. an Xi, S. Zhu, Z. Wen, and R. Wang, "Preparation, Structural Characterization, and Electrochemical Properties of Chemically Modified Mesoporous Carbon," *Microporous Mesoporous Mater.*, vol. 96, no. 1–3, pp. 357–362, 2006, doi: 10.1016/j.micromeso.2006.07.021.
[41] B. B. Arnold and G. W. Murphy, "Studies on the Electrochemistry of Carbon and Chemically-Modified Carbon Surfaces," *J. Phys. Chem.*, vol. 65, no. 1, pp. 135–138, 1961, doi: 10.1021/j100819a038.
[42] Y. Zhang, Y. Liu, Y. Bai, Y. Liu, and E. Xie, "Boosting the Electrochemical Properties of Carbon Materials as Bipolar Electrodes by Introducing Oxygen Functional Groups," *RSC Adv.*, vol. 10, no. 58, pp. 35295–35301, 2020, doi: 10.1039/d0ra06888h.
[43] L. Cui, J. Wu, and H. Ju, "Biosensors and Bioelectronics Electrochemical Sensing of Heavy Metal Ions with Inorganic, Organic and Bio-materials," *Biosens. Bioelectron.*, vol. 63, pp. 276–286, 2015, doi: 10.1016/j.bios.2014.07.052.
[44] P. Ramnani, N. M. Saucedo, and A. Mulchandani, "Carbon Nanomaterial-Based Electrochemical Biosensors for Label-Free Sensing of Environmental Pollutants," *Chemosphere*, vol. 143, pp. 85–98, 2016, doi: 10.1016/j.chemosphere.2015.04.063.
[45] Q. Wang *et al.*, "Recent Advances in Electrochemical Sensors for Antibiotics and Their Applications," *Chinese Chem. Lett.*, vol. 32, no. 2, pp. 19–29, 2021, doi: 10.1016/j.cclet.2020.10.025.
[46] S. Rathinavel, K. Priyadharshini, and D. Panda, "A Review on Carbon Nanotube: An Overview of Synthesis, Properties, Functionalization, Characterization, and the Application," *Mater. Sci. Eng. B*, vol. 268, p. 115095, 2021, doi: 10.1016/j.mseb.2021.115095.
[47] L. Mohapatra, D. Cheon, and S. H. Yoo, "Carbon-Based Nanomaterials for Catalytic Wastewater Treatment: A Review," *Molecules*, vol. 28, no. 4, 2023, doi: 10.3390/molecules28041805.
[48] P. Prasertying *et al.*, "Modified Pencil Graphite Electrode as a Low-Cost Glucose Sensor," *J. Sci. Adv. Mater. Devices*, vol. 5, no. 3, pp. 330–336, 2020, doi: 10.1016/j.jsamd.2020.07.004.
[49] S. Islam, R. Malha, J. Mandli, and A. Ourari, "Carbon Black-Modified Electrodes as Sensitive Tools for the Electrochemical Detection of Nitrite and Nitrate," *Electroanalysis*, vol. 25, pp. 1–9, 2013, doi: 10.1002/elan.201300257.
[50] M. Sajid, M. K. Nazal, M. Mansha, A. Alsharaa, S. M. S. Jillani, and C. Basheer, "Chemically Modified Electrodes for Electrochemical Detection of Dopamine in the Presence of Uric Acid and Ascorbic Acid: A Review," *TrAC – Trends Anal. Chem.*, vol. 76, pp. 15–29, 2016, doi: 10.1016/j.trac.2015.09.006.
[51] S. S. Ogunsola, M. E. Oladipo, P. O. Oladoye, and M. Kadhom, "Carbon Nanotubes for Sustainable Environmental Remediation: A Critical and Comprehensive Review," *Nano-Structures and Nano-Objects*, vol. 37, p. 101099, 2024, doi: 10.1016/j.nanoso.2024.101099.
[52] P. A. Pushpanjali, J. G. Manjunatha, N. Hareesha, E. S. D. Souza, M. M. Charithra, and N. S. Prinith, "Voltammetric Analysis of Antihistamine Drug Cetirizine and Paracetamol at Poly(L-Leucine) Layered Carbon Nanotube Paste Electrode," *Surfaces Interfaces*, vol. 24, p. 101154, 2021, doi: 10.1016/j.surfin.2021.101154.
[53] J.-S. Ye and F.-S. Sheu, "Functionalization of CNTs: New Routes Towards the Development of Novel Electrochemical Sensors," *Curr. Nanosci.*, vol. 2, no. 4, pp. 319–327, 2006, doi: 10.2174/157341306778699329.

[54] N. Hareesha and J. G. Manjunatha, "Electro-Oxidation of Formoterol Fumarate on the Surface of Novel Poly(Thiazole Yellow-G) Layered Multi-walled Carbon Nanotube Paste Electrode," *Sci. Rep.*, vol. 11, no. 1, pp. 1–14, 2021, doi: 10.1038/s41598-021-92099-x.
[55] A. G. Crevillén, M. Ávila, M. Pumera, M. C. González, and A. Escarpa, "Food Analysis on Microfluidic Devices Using Ultrasensitive Carbon Nanotubes Detectors," *Anal. Chem.*, vol. 79, no. 19, pp. 7408–7415, 2007, doi: 10.1021/ac071247i.
[56] P. Vichchulada, L. D. Lipscomb, Q. Zhang, and M. D. Lay, "Incorporation of Single-Walled Carbon Nanotubes into Functional Sensor Applications," *J. Nanosci. Nanotechnol.*, vol. 9, no. 4, pp. 2189–2200, 2009, doi: 10.1166/jnn.2009.SE11.
[57] W. Boumya, N. Taoufik, M. Achak, and N. Barka, "Chemically Modified Carbon-Based Electrodes for the Determination of Paracetamol in Drugs and Biological Samples," *J. Pharm. Anal.*, vol. 11, no. 2, pp. 138–154, 2021, doi: 10.1016/j.jpha.2020.11.003.
[58] C. Leow, P. B. Kreider, S. Sommacal, P. Kluth, and P. Compston, "Electrical and Thermal Conductivity in Graphene-Enhanced Carbon-Fibre/PEEK: The Effect of Interlayer Loading," *Carbon N. Y.*, vol. 215, p. 118463, 2023, doi: 10.1016/j.carbon.2023.118463.
[59] J. Paek *et al.*, "Stretchable Electronic Devices Using Graphene and Its Hybrid Nanostructures," *FlatChem*, vol. 3, pp. 71–91, 2017, doi: 10.1016/j.flatc.2017.06.002.
[60] P. Joshi, R. Mishra, and R. J. Narayan, "Biosensing Applications of Carbon-Based Materials," *Curr. Opin. Biomed. Eng.*, vol. 18, p. 100274, 2021, doi: 10.1016/j.cobme.2021.100274.
[61] X. J. Lee *et al.*, "Review on Graphene and Its Derivatives: Synthesis Methods and Potential Industrial Implementation," *J. Taiwan Inst. Chem. Eng.*, vol. 98, pp. 163–180, 2019, doi: 10.1016/j.jtice.2018.10.028.
[62] S. Michalkiewicz, A. Skorupa, and M. Jakubczyk, "Carbon Materials in Electroanalysis of Preservatives: A Review," *Materials (Basel).*, vol. 14, no. 24, 2021, doi: 10.3390/ma14247630.
[63] M. Ahmed, H. Ahmed, P. Adebayo, and R. Carter, "Mechanical Properties and Application of Graphite and Graphite-based Nanocomposite: A Review," *Chem. Mater. Res.*, vol.15, no. 1, 2023, doi: 10.7176/cmr/15-1-02.
[64] X. Zhu *et al.*, "Oxygen-Defects Functionalized Graphite Nanoplatelets as Electrode Materials for Electrochemical Sensing," *J. Electrochem. Soc.*, vol. 166, no. 15, pp. B1400–B1407, 2019, doi: 10.1149/2.0271915jes.
[65] L. Wang *et al.*, "Preparation of Carbon Black from Rice Husk by Hydrolysis, Carbonization and Pyrolysis," *Bioresour. Technol.*, vol. 102, no. 17, pp. 8220–8224, 2011, doi: 10.1016/j.biortech.2011.05.079.
[66] M. R. Pacheco, S. C. Barbosa, R. F. N. Quadrado, A. R. Fajardo, and D. Dias, "Glassy Carbon Electrode Modified with Carbon Black and Cross-Linked Alginate Film: A New Voltammetric Electrode for Paraquat Determination," *Anal. Bioanal. Chem.*, vol. 411, no. 15, pp. 3269–3280, 2019, doi: 10.1007/s00216-019-01769-3.
[67] S. Greenough, M. J. Dumont, and S. Prasher, "The Physicochemical Properties of Biochar and Its Applicability as a Filler in Rubber Composites: A Review," *Mater. Today Commun.*, vol. 29, 2021, doi: 10.1016/j.mtcomm.2021.102912.
[68] C. Gibi, C. H. Liu, S. C. Barton, S. Anandan, and J. J. Wu, "Carbon Materials for Electrochemical Sensing Application – A Mini Review," *J. Taiwan Inst. Chem. Eng.*, vol. 154, p. 105071, 2024, doi: 10.1016/j.jtice.2023.105071.
[69] F. Arduini, S. Cinti, V. Mazzaracchio, V. Scognamiglio, A. Amine, and D. Moscone, "Carbon Black as an Outstanding and Affordable Nanomaterial for Electrochemical (Bio)sensor Design," *Biosens. Bioelectron.*, vol. 156, p. 112033, 2020, doi: 10.1016/j.bios.2020.112033.
[70] T. A. Silva, F. C. Moraes, B. C. Janegitz, O. Fatibello-Filho, and D. Ganta, "Electrochemical Biosensors Based on Nanostructured Carbon Black: A Review," *J. Nanomater.*, vol. 2017, 2017, doi: 10.1155/2017/4571614.
[71] D. Bizzotto, I. J. Burgess, T. Doneux, T. Sagara, and H. Z. Yu, "Beyond Simple Cartoons: Challenges in Characterizing Electrochemical Biosensor Interfaces," *ACS Sensors*, vol. 3, no. 1, pp. 5–12, 2018, doi: 10.1021/acssensors.7b00840.
[72] A. Sassolas, L. J. Blum, and B. D. Leca-Bouvier, "Immobilization Strategies to Develop Enzymatic Biosensors," *Biotechnol. Adv.*, vol. 30, no. 3, pp. 489–511, 2012, doi: 10.1016/j.biotechadv.2011.09.003.

[73] A. A. Bhojiya, "The potential Contribution of Nanocarbon to Fostering Sustainable Agriculture for Future Generations," *Nano Carbons*, vol. 1, no. 1, p. 228, 2023, doi: 10.59400/n-c.v1i1.228.

[74] M. J. Riding, F. L. Martin, K. C. Jones, and K. T. Semple, "Carbon Nanomaterials in Clean and Contaminated Soils: Environmental Implications and Applications," *Soil*, vol. 1, no. 1, pp. 1–21, 2015, doi: 10.5194/soil-1-1-2015.

[75] N. D. J. Yates, M. A. Fascione, and A. Parkin, "Methodologies for 'Wiring' Redox Proteins/Enzymes to Electrode Surfaces," *Chem. – A Eur. J.*, vol. 24, no. 47, pp. 12164–12182, 2018, doi: 10.1002/chem.201800750.

[76] D. Mandler and S. Kraus-Ophir, "Self-Assembled Monolayers (SAMs) for Electrochemical Sensing," *J. Solid State Electrochem.*, vol. 15, no. 7–8, pp. 1535–1558, 2011, doi: 10.1007/s10008-011-1493-6.

[77] M. Florescu and C. M. Christopher, "Development and Evaluation of Electrochemical Glucose Enzyme Biosensors Based on Carbon Film Electrodes," *Talanta*, vol. 65, no. 2 SPEC. ISS., pp. 306–312, 2005, doi: 10.1016/j.talanta.2004.07.003.

[78] M. Willander, P. Klason, L. L. Yang, S. M. Al-Hilli, Q. X. Zhao, and O. Nur, "ZnO Nanowires: Chemical Growth, Electrodeposition, and Application to Intracellular Nano-sensors," *Phys. Status Solidi Curr. Top. Solid State Phys.*, vol. 5, no. 9, pp. 3076–3083, 2008, doi: 10.1002/pssc.200779232.

[79] G. Tigari and J. G. Manjunatha, "Poly(glutamine) Film-Coated Carbon Nanotube Paste Electrode for the Determination of Curcumin with Vanillin: An Electroanalytical Approach," *Monatshefte Chem.*, vol. 151, no. 11, pp. 1681–1688, 2020, doi: 10.1007/s00706-020-02700-8.

[80] A. Machín, M. Cotto, J. Duconge, C. Morant, F. I. Petrescu, and F. Márquez, "Sensitive and Reversible Ammonia Gas Sensor Based on Single-Walled Carbon Nanotubes," *Chemosensors*, vol. 11, no. 4, pp. 1–16, 2023, doi: 10.3390/chemosensors11040247.

[81] C. Vedhi, G. Selvanathan, P. Arumugam, and P. Manisankar, "Electrochemical Sensors of Heavy Metals Using Novel Polymer-Modified Glassy Carbon Electrodes," *Ionics (Kiel).*, vol. 15, no. 3, pp. 377–383, 2009, doi: 10.1007/s11581-008-0277-1.

[82] D. R. Rout and H. M. Jena, "Polyethylene Glycol Functionalized Reduced Graphene Oxide Coupled with Zinc Oxide Composite Adsorbent for Removal of Phenolic Wastewater," *Environ. Res.*, vol. 214, no. P3, p. 114044, 2022, doi: 10.1016/j.envres.2022.114044.

[83] J. Hong, Y. Wang, L. Zhu, and L. Jiang, "An Electrochemical Sensor Based on Gold-Nanocluster-Modified Graphene Screen-Printed Electrodes for the Detection of β-Lactoglobulin in Milk," *Sensors (Switzerland)*, vol. 20, no. 14, pp. 1–12, 2020, doi: 10.3390/s20143956.

[84] H. M. Marwani, J. Ahmed, and M. M. Rahman, "Development of a Toxic Lead Ionic Sensor Using Carboxyl-Functionalized MWCNTs in Real Water Sample Analyses," *Sensors*, vol. 22, no. 22, 2022, doi: 10.3390/s22228976.

[85] S. Boher, R. Ullah, M. Tuzen, and T. A. Saleh, "Metal Doped Nanocomposites for Detection of Pesticides and Phenolic Compounds by Colorimetry: Trends and Challenges," *OpenNano*, vol. 13, p. 100168, 2023, doi: 10.1016/j.onano.2023.100168.

[86] Y. Hui, Z. Huang, M. E. E. Alahi, A. Nag, S. Feng, and S. C. Mukhopadhyay, "Recent Advancements in Electrochemical Biosensors for Monitoring the Water Quality," *Biosensors*, vol. 12, no. 7, pp. 1–31, 2022, doi: 10.3390/bios12070551.

[87] R. Rejithamol and S. Beena, "Carbon Paste Electrochemical Sensors for the Detection of Neurotransmitters," *Front. Sensors*, vol. 3, pp. 1–9, 2022, doi: 10.3389/fsens.2022.901628.

[88] S. Bilge, L. Karadurmus, A. Sınağ, and S. A. Ozkan, "Green Synthesis and Characterization of Carbon-Based Materials for Sensitive Detection of Heavy Metal Ions," *TrAC – Trends Anal. Chem.*, vol. 145, 2021, doi: 10.1016/j.trac.2021.116473.

[89] A. Knapek, F. Mika, J. Prasek, and P. Majzlikova, "SEM Characterization of Carbon Nanotubes Based Active Layers of Chemical Sensors," *Proc. 2014 37th Int. Spring Semin. Electron. Technol. ISSE 2014*, pp. 361–364, 2014, doi: 10.1109/ISSE.2014.6887625.

[90] C. Raril, J. G. Manjunatha, D. K. Ravishankar, S. Fattepur, G. Siddaraju, and L. Nanjundaswamy, "Validated Electrochemical Method for Simultaneous Resolution of Tyrosine, Uric Acid, and Ascorbic Acid at Polymer Modified Nano-Composite Paste Electrode," *Surf. Eng. Appl. Electrochem.*, vol. 56, no. 4, pp. 415–426, 2020, doi: 10.3103/S1068375520040134.

[91] A. S. Y. El Hamdouni, S. El Hajjaji, T. s Szabo, L. Trif, I. Felhosi, K. Abbi, N. Labjar, L. Harmouche, "Biomass Valorization of Walnut Shell into Biochar as a Resource for Electrochemical Simultaneous

Detection of Heavy Metal Ions in Water and Soil Samples: Preparation, Characterization, and Applications," *Arab. J. Chem.*, vol. 15, 2022, doi: 10.1016/j.arabjc.2022.104252.

[92] X. Ke, C. Bittencourt, and G. van Tendeloo, "Possibilities and Limitations of Advanced Transmission Electron Microscopy for Carbon-Based Nanomaterials," *Beilstein J. Nanotechnol.*, vol. 6, no. 1, pp. 1541–1557, 2015, doi: 10.3762/bjnano.6.158.

[93] E. Mazzotta, S. Rella, A. Turco, and C. Malitesta, "XPS in Development of Chemical Sensors," *RSC Adv.*, vol. 5, no. 101, pp. 83164–83186, 2015, doi: 10.1039/c5ra14139g.

[94] V. Ţucureanu, A. Matei, and A. M. Avram, "FTIR Spectroscopy for Carbon Family Study," *Crit. Rev. Anal. Chem.*, vol. 46, no. 6, pp. 502–520, 2016, doi: 10.1080/10408347.2016.1157013.

[95] M. S. Dresselhaus, M. A. Pimenta, P. C. Eklund, and G. Dresselhaus, "Raman Scattering in Fullerenes and Related Carbon-Based Materials," *Verlag Berlin Heidelb.*, pp. 314–364, 2000, doi: 10.1007/978-3-662-04221-2_10.

[96] Q. Chen *et al.*, "Raman Spectroscopic Quantification of Graphene Oxide in Soil: Transport, Surficial Enrichment and Environmental Effects," *Environ. Pollut.*, vol. 357, p. 124412, 2024, doi: 10.1016/j.envpol.2024.124412.

[97] A. Simpson, R. R. Pandey, C. C. Chusuei, K. Ghosh, R. Patel, and A. K. Wanekaya, "Fabrication Characterization and Potential Applications of Carbon Nanoparticles in the Detection of Heavy Metal Ions in Aqueous Media," *Carbon N. Y.*, vol. 127, pp. 122–130, 2018, doi: 10.1016/j.carbon.2017.10.086.

[98] J. Baranwal, B. Barse, G. Gatto, G. Broncova, and A. Kumar, "Electrochemical Sensors and Their Applications: A Review," *Chemosensors*, vol. 10, no. 9, 2022, doi: 10.3390/chemosensors10090363.

[99] V. Climent and J. M. Feliu, "Cyclic Voltammetry," *Ref. Modul. Chem. Mol. Sci. Chem. Eng.*, pp. 48–74, 2015, doi: 10.1016/B978-0-12-409547-2.10764-4.

[100] A. B. Monnappa, J. G. Manjunatha, and A. S. Bhatt, "Design of a Sensitive and Selective Voltammetric Sensor Based on a Cationic Surfactant-Modified Carbon Paste Electrode for the Determination of Alloxan," *ACS Omega*, vol. 5, no. 36, pp. 23481–23490, 2020, doi: 10.1021/acsomega.0c03517.

[101] H. L. Haijiang Wang and X.-Z. Yuan, "Electrochemical Impedance Spectroscopy," *PEM Fuel Cell Diagnostic Tools*, pp. 37–70, 2011, doi: 10.1201/b11100-5.

[102] N. G. Yasri, A. J. Halabi, G. Istamboulie, and T. Noguer, "Chronoamperometric Determination of Lead Ions Using PEDOT:PSS Modified Carbon Electrodes," *Talanta*, vol. 85, no. 5, pp. 2528–2533, 2011, doi: 10.1016/j.talanta.2011.08.013.

[103] S. Bellucci, G. Gaggiotti, M. Marchetti, F. Micciulla, R. Mucciato, and M. Regi, "Atomic Force Microscopy Characterization of Carbon Nanotubes," *J. Phys. Conf. Ser.*, vol. 61, no. 1, pp. 99–104, 2007, doi: 10.1088/1742-6596/61/1/021.

[104] M. Nankali *et al.*, "Highly Stretchable and Sensitive Strain Sensors Based on Carbon Nanotube-Elastomer Nanocomposites: The Effect of Environmental Factors on Strain Sensing Performance," *J. Mater. Chem. C*, vol. 8, no. 18, pp. 6185–6195, 2020, doi: 10.1039/d0tc00373e.

[105] Y. Zou, J. D. Clark, and A. A. May, "A Systematic Investigation on the Effects of Temperature and Relative Humidity on the Performance of Eight Low-Cost Particle Sensors and Devices," *J. Aerosol Sci.*, vol. 152, p. 105715, 2021, doi: 10.1016/j.jaerosci.2020.105715.

[106] L. Xiaowei, R. Jean-Charles, and Y. Suyuan, "Effect of Temperature on Graphite Oxidation Behavior," *Nucl. Eng. Des.*, vol. 227, no. 3, pp. 273–280, 2004, doi: 10.1016/j.nucengdes.2003.11.004.

[107] B. Yuan, Z. Tang, P. Zhang, and F. Lv, "Thermal Calibration of Triaxial Accelerometer for Tilt Measurement," *Sensors*, vol. 23, no. 4, pp. 1–16, 2023, doi: 10.3390/s23042105.

[108] J. M. Tulliani, B. Inserra, and D. Ziegler, "Carbon-Based Materials for Humidity Sensing: A Short Review," *Micromachines*, vol. 10, no. 4, pp. 1–30, 2019, doi: 10.3390/mi10040232.

[109] B. del Moral *et al.*, "Temperature and Humidity Influence on the Strain Sensing Performance of Hybrid Carbon Nanotubes and Graphite Cement Composites," *Constr. Build. Mater.*, vol. 284, p. 122786, 2021, doi: 10.1016/j.conbuildmat.2021.122786.

[110] G. Korotcenkov, N. P. Simonenko, E. P. Simonenko, V. V. Sysoev, and V. Brinzari, "Paper-Based Humidity Sensors as Promising Flexible Devices, State of the Art, Part 2: Humidity-Sensor Performances," *Nanomaterials*, vol. 13, no. 1381, pp. 1–67, 2023, https://doi.org/10.3390/nano13081381.

[111] A. Slesinski, S. Sroka, K. Fic, E. Frackowiak, and J. Menzel, "Operando Monitoring of Local pH Value Changes at the Carbon Electrode Surface in Neutral Sulfate-Based Aqueous Electrochemical

Capacitors," *ACS Appl. Mater. Interfaces*, vol. 14, pp. 37782–37792, 2022, https://doi.org/10.1021/acsami.2c09920.
[112] P. Chen, J. Wang, Y. Xue, and C. Wang, "Comprehensive Review of Advanced Sensors for Monitoring Emerging Contaminants in Water," *Water Res.*, vol. 24, p. 122297, 2024, doi: 10.1016/j.watres.2024.122297.
[113] I. Abimbola, A. I. Alina, and Q. Ibrahim, "Challenges and Opportunities in Calibrating Low-Cost Environmental Sensors," *Sensors*, vol. 24, no. 3650, 2024, doi: 10.3390/s24113650.
[114] E. Bodoki, R. Săndulescu, and L. Roman, "Method Validation in Quantitative Electrochemical Analysis of Colchicine Using Glassy Carbon Electrode," *Cent. Eur. J. Chem.*, vol. 5, no. 3, pp. 766–778, 2007, doi: 10.2478/s11532-007-0034-8.
[115] T. Verbić, Z. Dorkó, and G. Horvai, "Selectivity in analytical chemistry," *Rev. Roum. Chim.*, vol. 58, no. 7–8, pp. 569–575, 2012, http://web.icf.ro/rrch/.
[116] S. A. Ozkan, J.-M. Kauffmann, and P. Zuman, *Electroanalysis in Biomedical and Pharmaceutical Sciences*. 2015. Electroanalysis in Biomedical and Pharmaceutical Sciences.
[117] D. A. Armbruster and T. Pry, "Limit of Blank, Limit of Detection and Limit of Quantitation," *Clin. Biochem. Rev.*, vol. 29, pp. S49–52, 2008, [Online]. Available: pmcid: PMC2556583
[118] K. S. Galhardo, T. R. L. Dadamos, R. J. N. Bettencourt da Silva, and S. A. S. Machado, "Development and Validation of an Advanced Electrochemical Sensor for the Fast and Cheap Determination of Hydrochlorothiazide in Urine Samples Using the Monte-Carlo Method for Uncertainty Evaluation," *Talanta*, vol. 215, p. 120883, 2020, doi: 10.1016/j.talanta.2020.120883.
[119] E. Desimoni and B. Brunetti, "Presenting Analytical Performances of Electrochemical Sensors. Some Suggestions," *Electroanalysis*, vol. 25, no. 7, pp. 1645–1651, 2013, doi: 10.1002/elan.201300150.
[120] A. C. Power and A. Morrin, "Electroanalytical Sensor Technology," *Electrochemistry*, p. 210, 2013, http://dx.doi.org/10.5772/51480.
[121] C. Maccà, "Response Time of Ion-Selective Electrodes. Current Usage Versus IUPAC Recommendations," *Anal. Chim. Acta*, vol. 512, pp. 183–190, 2004, doi: 10.1016/j.aca.2004.03.010.
[122] P. Parameswari *et al.*, "Nanotechnology-Based Sensors for Real-time Monitoring and Assessment of Soil Health and Quality: A Review," *Asian J. Soil Sci. Plant Nutr.*, vol. 10, no. 2, pp. 157–173, 2024, doi: 10.9734/ajsspn/2024/v10i2272.
[123] K. Y. Chen, J. Kachhadiya, S. Muhtasim, S. Cai, J. Huang, and J. Andrews, "Underground Ink: Printed Electronics Enabling Electrochemical Sensing in Soil," *Micromachines*, vol. 15, no. 5, 2024, doi: 10.3390/mi15050625.
[124] A. Mandal *et al.*, "Impact of Agricultural Management Practices on Soil Carbon Sequestration and Its Monitoring Through Simulation Models and Remote Sensing Techniques: A Review," *Crit. Rev. Environ. Sci. Technol.*, vol. 52, no. 1, pp. 1–49, 2022, doi: 10.1080/10643389.2020.1811590.
[125] M. Siddiqui, Mohd Salman search by orcid; Aslam, "Highly Stable and Reusable 3D Graphene-Quinizarin Voltammetric pH Sensor," *J. Electrochem. Soc.*, vol. 170, no. 4, p. id.047511, 14 pp, 2023, doi: 10.1149/1945-7111/accba7.
[126] M. S. Siddiqui and M. Aslam, "Three-Dimensional Conductometric Network Based on Reduced Graphene Oxide for Soil pH Sensors," *ACS Appl. Nano Mater.*, vol. 6, no. 19, pp. 17376–17386, Oct. 2023, doi: 10.1021/acsanm.3c02033.
[127] X. Deng *et al.*, "A Method of Electrical Conductivity Compensation in a Low-Cost Soil Moisture Sensing Measurement Based on Capacitance," *Meas. J. Int. Meas. Confed.*, vol. 150, p. 107052, 2020, doi: 10.1016/j.measurement.2019.107052.
[128] V. P. Singh, K. Kandasamy, and M. R. Rahman, "A Flexible and Biodegradable Graphene Oxide Antenna Sensor for Monitoring Subsoil Health," *ACS Appl. Nano Mater.*, vol. 7, no. 13, pp. 15223–15231, Jul. 2024, doi: 10.1021/acsanm.4c02013.
[129] B. Kashyap and R. Kumar, "Sensing Methodologies in Agriculture for Monitoring Biotic Stress in Plants due to Pathogens and Pests," *Inventions*, vol. 6, no. 2, pp. 14095–14121, 2021, doi: 10.3390/INVENTIONS6020029.
[130] D. F. Cao, B. Shi, G. Q. Wei, S. E. Chen, and H. H. Zhu, "An Improved Distributed Sensing Method for Monitoring Soil Moisture Profile Using Heated Carbon Fibers," *Meas. J. Int. Meas. Confed.*, vol. 123, pp. 175–184, 2018, doi: 10.1016/j.measurement.2018.03.052.

[131] M. Wang, X. Li, L. Chen, S. Hou, G. Wu, and Z. Deng, "A Modified Soil Water Content Measurement Technique Using Actively Heated Fiber Optic Sensor," *J. Rock Mech. Geotech. Eng.*, vol. 12, no. 3, pp. 608–619, 2020, doi: 10.1016/j.jrmge.2019.11.003.
[132] M. A. Ali, L. Dong, J. Dhau, A. Khosla, and A. Kaushik, "Perspective—Electrochemical Sensors for Soil Quality Assessment," *J. Electrochem. Soc.*, vol. 167, no. 3, p. 037550, 2020, doi: 10.1149/1945-7111/ab69fe.
[133] J. V. Sinfield, D. Fagerman, and O. Colic, "Evaluation of Sensing Technologies for on-the-go Detection of Macro-nutrients in Cultivated Soils," *Comput. Electron. Agric.*, vol. 70, no. 1, pp. 1–18, 2010, doi: 10.1016/j.compag.2009.09.017.
[134] A. A. Kumar, S. K. N. Kumar, and R. E. Fernandez, "Real Time Sensing of Soil Potassium Levels Using Zinc Oxide-Multiwall Carbon Nanotube-Based Sensors," *IEEE Trans. Nanobioscience*, vol. 20, no. 1, pp. 50–56, 2021, doi: 10.1109/TNB.2020.3027863.
[135] S. Bhandari, U. Singh, and S. Kumbhat, "Nafion-modified Carbon Based Sensor for Soil Potassium Detection," *Electroanalysis*, vol. 31, no. 5, pp. 813–819, 2019, doi: 10.1002/elan.201800583.
[136] S. R. Benjamin and E. J. Miranda Ribeiro Júnior, "Graphene-Based Electrochemical Sensors for Detection of Environmental Pollutants," *Curr. Opin. Environ. Sci. Heal.*, vol. 29, p. 100381, 2022, doi: 10.1016/j.coesh.2022.100381.
[137] I. R. Comnea-Stancu, J. (Koos) F. van Staden, and R.-I. Stefan-van Staden, "Review—Trends in Recent Developments in Electrochemical Sensors for the Determination of Polycyclic Aromatic Hydrocarbons from Water Resources and Catchment Areas," *J. Electrochem. Soc.*, vol. 168, no. 4, p. 047504, 2021, doi: 10.1149/1945-7111/abf260.
[138] H. Cui *et al.*, "Electrochemical Sensor for the Detection of 1-Hydroxypyrene Based on Composites of PAMAM-Regulated Chromium-Centered Metal-Organic Framework Nanoparticles and Graphene Oxide," *ACS Omega*, vol. 6, no. 46, pp. 31184–31195, 2021, doi: 10.1021/acsomega.1c04765.
[139] Y. Li *et al.*, "Recent Advances of Biochar-Based Electrochemical Sensors and Biosensors," *Biosensors*, vol. 12, no. 6, 2022, doi: 10.3390/bios12060377.
[140] C. J. Venegas, S. Bollo, and P. Sierra-Rosales, "Carbon-Based Electrochemical (Bio)sensors for the Detection of Carbendazim: A Review," *Micromachines*, vol. 14, no. 9, 2023, doi: 10.3390/mi14091752.
[141] L. Karadurmus, A. Cetinkaya, S. I. Kaya, and S. A. Ozkan, "Recent Trends on Electrochemical Carbon-Based Nanosensors for Sensitive Assay of Pesticides," *Trends Environ. Anal. Chem.*, vol. 34, p. e00158, 2022, doi: 10.1016/j.teac.2022.e00158.
[142] M. K. Abd Elnabi *et al.*, "Toxicity of Heavy Metals and Recent Advances in Their Removal: A Review," *Toxics*, vol. 11, no. 7, 2023, doi: 10.3390/toxics11070580.
[143] B. Li *et al.*, "Recent Advance of Nanomaterials Modified Electrochemical Sensors in the Detection of Heavy Metal Ions in Food and Water," *Food Chem.*, vol. 440, p. 138213, 2024, doi: 10.1016/j.foodchem.2023.138213.
[144] A. R. Dhanapal, M. Thiruvengadam, J. Vairavanathan, B. Venkidasamy, M. Easwaran, and M. Ghorbanpour, "Nanotechnology Approaches for the Remediation of Agricultural Polluted Soils," *ACS Omega*, vol. 9, no. 12, pp. 13522–13533, 2024, doi: 10.1021/acsomega.3c09776.
[145] R. Lal, "Restoring Soil Quality to Mitigate Soil Degradation," *Sustain.*, vol. 7, no. 5, pp. 5875–5895, 2015, doi: 10.3390/su7055875.
[146] K. Rani, "Estimation of Soil Organic Carbon by Remote Sensing," *Curr. Res. Soil Sci.*, no. July, pp. 77–92, 2022, doi: 10.22271/ed.book.1736.
[147] J. Liu, T. Yang, J. Xu, and Y. Sun, "A New Electrochemical Detection Technique for Organic Matter Content in Ecological Soils," *Front. Chem.*, vol. 9, pp. 1–7, 2021, doi: 10.3389/fchem.2021.699368.
[148] M. L. M. Firmino *et al.*, "Sensor Based on B-NiOx Hybrid Film/Multi-walled Carbon Nanotubes Composite Electrode for Groundwater Salinization Inspection," *Chem. Eng. J.*, vol. 323, pp. 47–55, 2017, doi: 10.1016/j.cej.2017.04.077.
[149] D. I. Anguiano *et al.*, "Electrochemical Detection of Iron in a Lixiviant Solution of Polluted Soil Using a Modified Glassy Carbon Electrode," *Int. J. Electrochem.*, vol. 2012, pp. 1–6, 2012, doi: 10.1155/2012/739408.
[150] D. Ilager, H. Seo, S. S. Kalanur, N. P. Shetti, and T. M. Aminabhavi, "A Novel Sensor Based on WO3·0.33H2O Nanorods Modified Electrode for the Detection and Degradation of Herbicide, Carbendazim," *J. Environ. Manage.*, vol. 279, p. 111611, 2021, doi: 10.1016/j.jenvman.2020.111611.

[151] S. Senanayake, B. Pradhan, A. Huete, and J. Brennan, “A Review on Assessing and Mapping Soil Erosion Hazard Using Geo-informatics Technology for Farming System Management,” *Remote Sens.*, vol. 12, no. 24, pp. 1–25, 2020, doi: 10.3390/rs12244063.
[152] Z. Zhang, X. Liu, H. Zhou, S. Xu, and C. Lee, “Advances in Machine-Learning Enhanced Nanosensors: From Cloud Artificial Intelligence Toward Future Edge Computing at Chip Level,” *Small Struct.*, vol. 5, no. 4, 2024, doi: 10.1002/sstr.202300325.
[153] H. Lu, E. Campbell, D. E. Campbell, C. Wang, and H. Ren, “Field Test of Several Low-Cost Particulate Matter Sensors in High and Low Concentration Urban Environments,” *Atmos. Environ.*, vol. 23, no. 1994, pp. 248–258, 2017, doi: 10.1016/j.hal.2017.06.001.Submit.
[154] D. Arrouays *et al.*, “On Soil Carbon Monitoring Networks,” *Soil Carbon*, pp. 59–68, 2014, doi: 10.1007/978-3-319-04084-4.

5 Analysis of Textile Dyes Pollutants Using Carbon Sensors

Hülya Silah and Bengi Uslu

5.1 INTRODUCTION

Worldwide, worsening ecological contamination is of main public worry, venturing the healthiness of the public and future generations (S. Khan et al., 2019). With the progress of industrialization, there is an enhancement in the quantity of organic pollutants directly emptied into environmental mediums by various industries (such as textile, pharmaceutical, food, agricultural, leather, plastic, cosmetic, and domestic industries) and most of them are hazardous to living things (Kang et al., 2019). The wastes of these industrial types are harmful to the entire environment, and the most important type of pollution they cause can be described as water pollution.

Water makes up most of the world, although only 0.3% is utilizable for humans and animals. The enormous oceans, icecaps, and other non-drinkable water sources make up the remaining 99.7% (Periyasamy, 2024).

Textile dyes are one of the most important pollutants in wastewater. More than 3,600 textile dyes are being manufactured by industries today. In order to make easier the dyeing procedures properly, the textile dye added with many auxiliary chemical compounds must be dissolved within a sufficient quantity of aqueous solution. In addition, the number of chemicals used by this sector throughout all production activities, including all dyeing and printing processes, reaches 8,000 (Kant, 2012; Jahan et al., 2023). Textile dyes and accompanying chemicals, in addition to inadmissible views and hazardous effects of them after their degradation, may pollute the nearby soils, sediments, and surface waters, causing a crucial environmental contamination problem worldwide (Figure 5.1) (Yaseen & Scholz, 2019). The vogue and textile industries are generally found faulty with harmful environmental effects (e.g., generation of various wastes, consumption of sources, especially carbon footprint) during its procurement chain operations (Akter et al., 2022; Niinimäki et al., 2020).

Also, recently, the progression of technology and globalization, coupled with the diversification fashion options of humans, has remarkably subscribed to the augmentation of textile industry and garment manufacturing (Ganesan et al., 2024; Akter et al., 2022; Akhtar et al., 2022). This situation causes an increase in the demand for the fashion industry and therefore an increase in the number of dyes and auxiliary chemicals used.

Textile industry exhausts huge amounts of water every day and is liable for discharging noteworthy quantities of textile dyes and other polluting materials into water sources (Barreto et al., 2010). As a result, it can be said that large quantities of wastewater are created from the textile, paper, cosmetics, leather, rubber, printing, and other industries everyday (Wang et al., 2022).

Water contamination is ascribed to the degradation of water quality owing to human beings and their anthropogenic activities (Shetti et al., 2019). Textile dyes and residues are quite stable in the aquatic system. Because of their color, they inhibit light penetration into the water bodies. This situation causes some problems such as gas solubility and phytoplankton's photosynthesis. As a

DOI: 10.1201/9781003500322-5

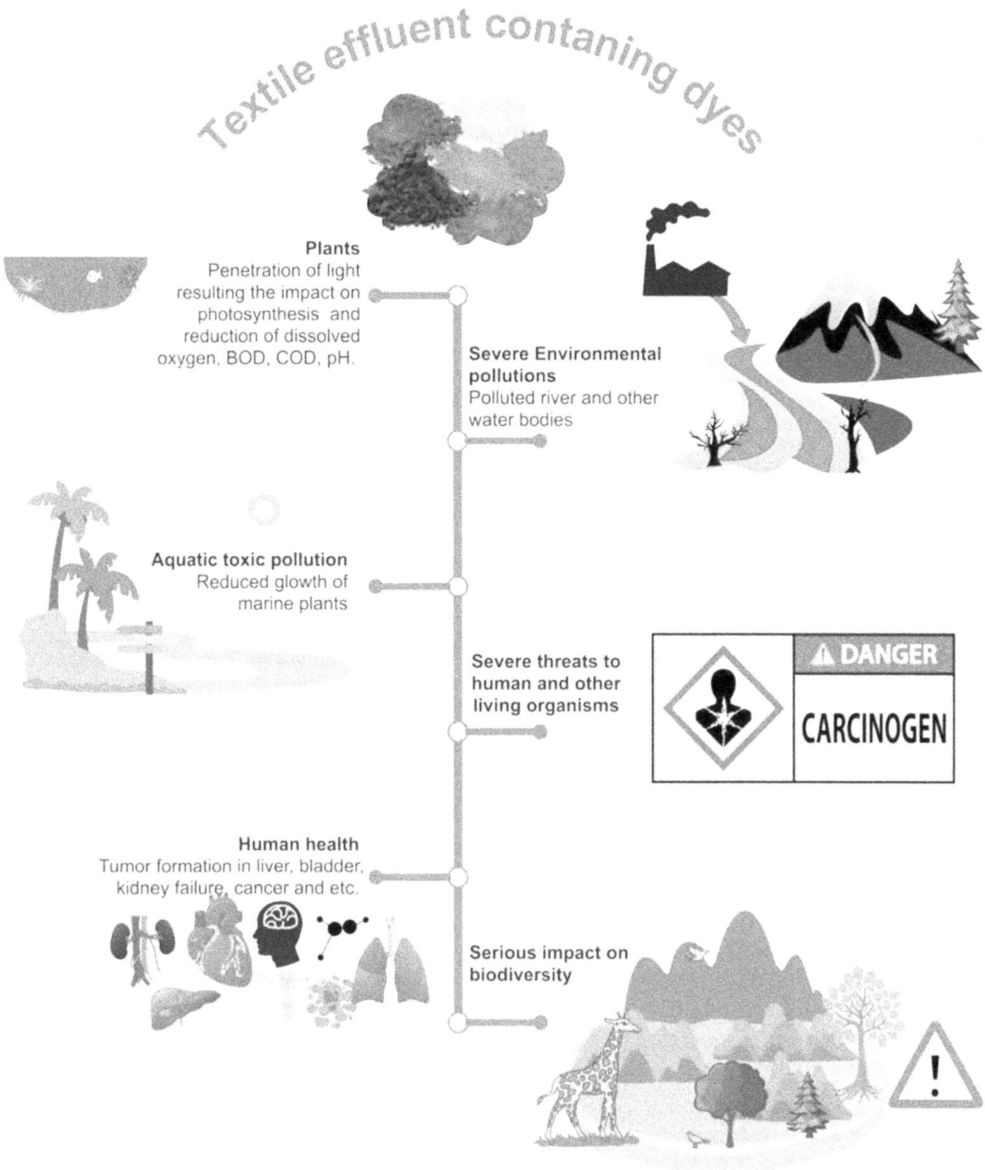

FIGURE 5.1 Impacts of dye-containing wastewater from textiles on the environment and health hazards (Periyasamy, 2024).

consequence, dissolved oxygen is decreased for aquatic living beings (Hayat, Shah, Hakeem et al., 2022; Arya et al., 2020; Ozturk & Silah, 2020).

Dye molecules, pigments and their degradation products in environmental mediums, and as a consequence, their mixing into the food chain, give rise to some toxic effects such as mutagenicity, carcinogenicity, and teratogenicity problems, affecting kidney, liver, brain, central nervous, and reproductive systems. Also, analysis and breakdown of these toxic dye molecules utilizing sensitive analytical methods for lower quantity assist many industries for manufacturing, variation, and determination of product structures. Consequently, the need for the development of new analytical methodologies for rapid, sensitive, and selective determination of toxic dyes and their degradation products has emerged (Shetti et al., 2019; Zhou et al., 2019; Manjunatha, 2019; Erkmen et al., 2024). Several analytical methods have been utilized to analyze and measure the residual amounts of textile dyes. When the literature studies of previous years are examined, it is seen that the UV-Vis spectroscopy technique stands out among the methods used for the determination of dyes in various environmental environments (Şahin et al., 2007; Barreto et al., 2010). However, in addition to this, in the literature other techniques can be seen. These are as follows: high-performance liquid chromatography (HPLC) (Novotná et al., 1999); liquid chromatography (LC) (Surowiec et al., 2006); gas chromatography (Abo-Ayad et al., 2024); surface-enhanced Raman spectroscopy (SERS) (M. Yao et al., 2023); also immunoassay methods such as ELISA (Ju et al., 2008); some techniques based on real-time time-of-flight mass spectrometry (Djelal et al., 2017); mass spectrometry such as laser desorption/ionization (LDI)-MS (Ribechini et al., 2013); matrix-assisted laser desorption/ionization (Dunn & Allison, 2007); time-of-flight secondary ion mass spectrometry (Zhou et al., 2012); and liquid microjunction surface sampling probe mass spectrometry (ElNaggar & Van Berkel, 2011).

Among these methods, electrochemical methods are chosen because of their portability, quick responsiveness, uncomplicated handling, very high sensitivity and selectivity (Hayat, Shah, Hakeem et al., 2022). Also, electrochemical methods are appropriate instruments for researching procedures and investigating kinetics (Dhrithi & Swamy, 2024).

Carbon-based materials have gained momentum in recent years due to their significant characteristics such as chemical stability, high optical absorptivity, biocompatibility, and low toxicity (Prasad et al., 2019). This chapter summarizes recent progress and developments in electrochemical textile dye sensing, dependent on carbon-based materials and their typical enforcements.

5.2 TEXTILE DYES WIDELY USED IN THE INDUSTRY

Textile dyes are primarily constituted of organic compounds such as cyclic aromatic structures and inorganic matters and are described as having high molecular mass (Khosropour et al., 2019).

The chemical structure of a dye is made up of a diversity of chemical groups, auxochromes, and chromophore groups as shown in Figure 5.2 (Jahan et al., 2023). The complicated and stable chemical structure of the textile dyes causes the degradation of these dyes (with existence in wastewater and other complicated substrates) arduous (Wang et al., 2022). Chromophore structures are made up of diverse functional groups (e.g., -NO_2, -N=N-, = O, O=(C_6H_4), and -C=O) and are liable to ensuring color features of the dyes. Auxochrome groups affect the solubility of dyes via sharing or donation of electrons, and also increase fiber affinity. These groups are generally formed as -COOH, -SO_3H, -OH, and-NH_3 (Sharma et al., 2021; Alharbi et al., 2022).

Depending on the chromophore groups, textile dyes can be categorized into different groups, such as azo, xanthene, oxazine, diphenylmethane, triphenylmethane, cyanine, nitro, indigoid, phthalein, anthraquinone, quinone-imine, nitro, nitroso, and diarylmethane (Benkhaya, Rabet et al., 2020; Rauf et al., 2011). Table 5.1 provides a general classification of dyes according to their chemical structures and application areas.

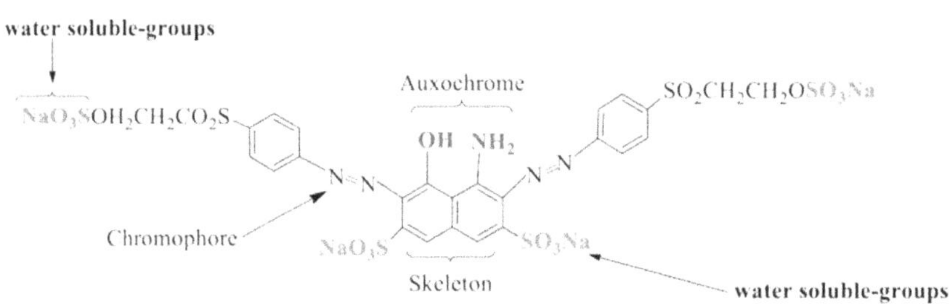

FIGURE 5.2 A general example of dye structures based on azo dyes (Benkhaya, M'rabet, et al., 2020).

TABLE 5.1
A general classification of dyes according to their chemical structures and their application areas

Group	Sample	Applications
Acidic dyes	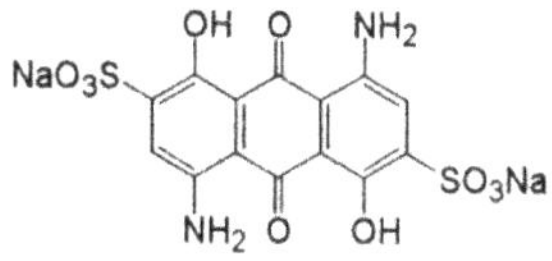 Acid Blue 45	Polyamide fibers, silk, acrylics, wool, leather, cosmetics, food, nylon, also paper printing ink
	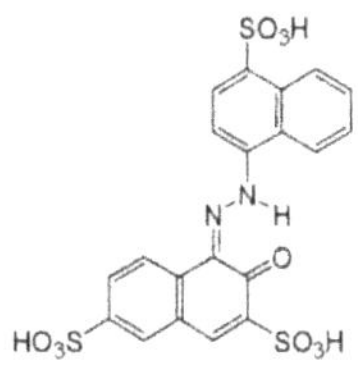 Acid Red 27	
Basic dyes	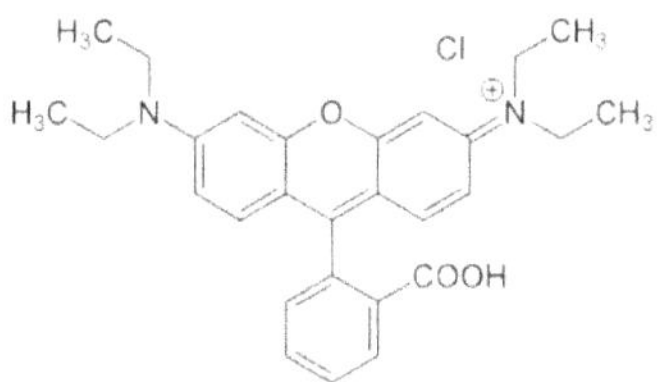 Rhodamine	Silk, polyester, wool, modified nylon, leather, straw tannin mordant cotton, inks, pharmaceuticals, paper, wood
	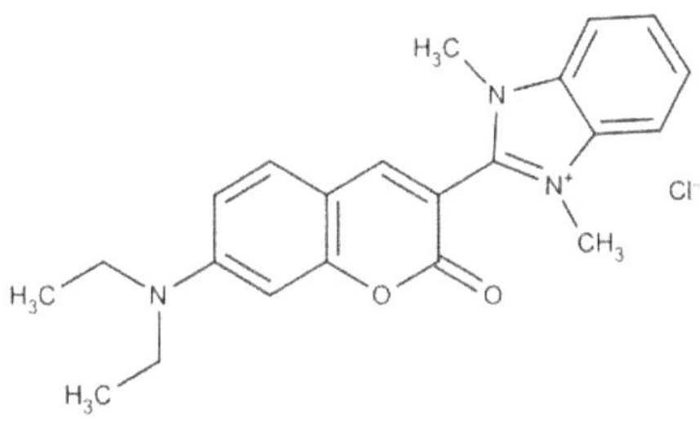 Basic Yellow 40	
Azo dyes	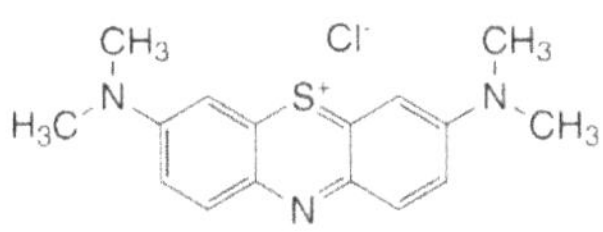 Methylene Blue	Rayon, cotton, acetate, cellulosic materials, polyester, plastic, detergents
	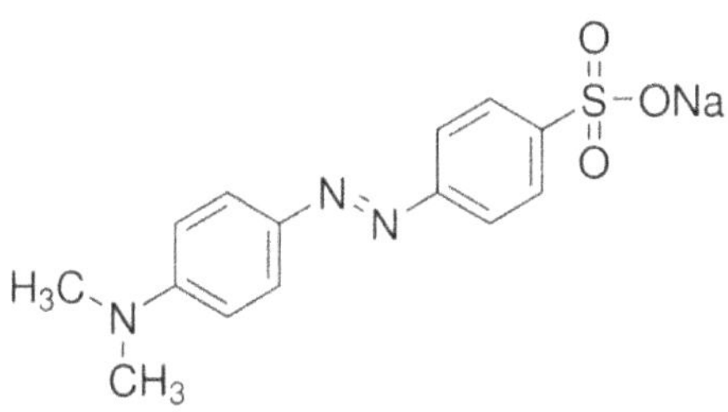 Methylene orange	

(continued)

TABLE 5.1 (Continued)
A general classification of dyes according to their chemical structures and their application areas

Group	Sample	Applications
Dispers dyes	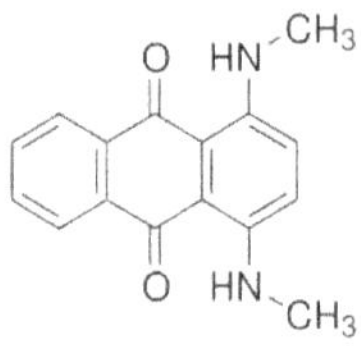 Disperse Blue	Acrylic fiber, acetate, cellulose, polyamide, nylon, polyester, polyester-cotton, cotton, and plastic
	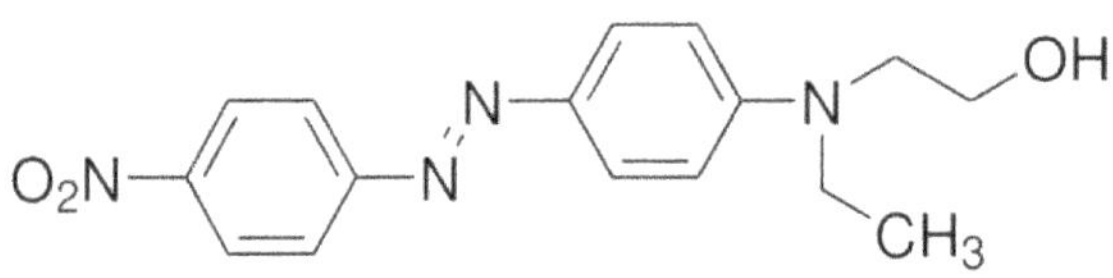 Disperse Red	
Reactive dyes	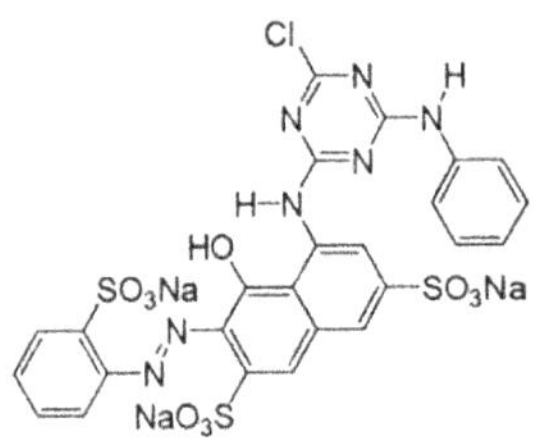 Reactive Red 3	Cellulosic, cotton, nylon, silk, wool, yarn, and multi-color printing process
	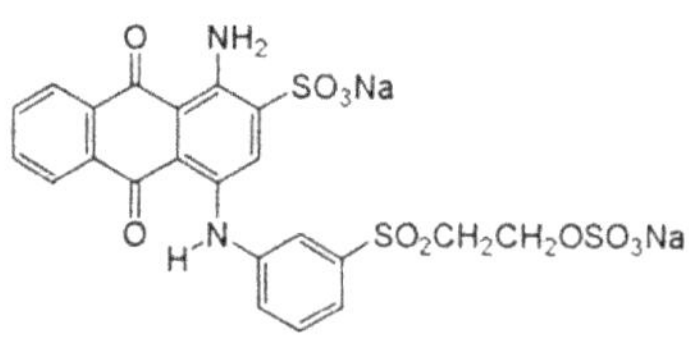 Reactive Blue 19	
Mordant dyes	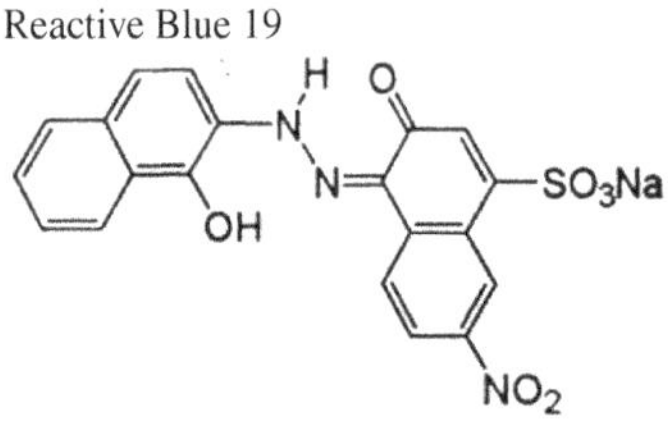 Mordant Black 11	Fiber, cotton, anodized aluminum, wool, leather, fur and hair
	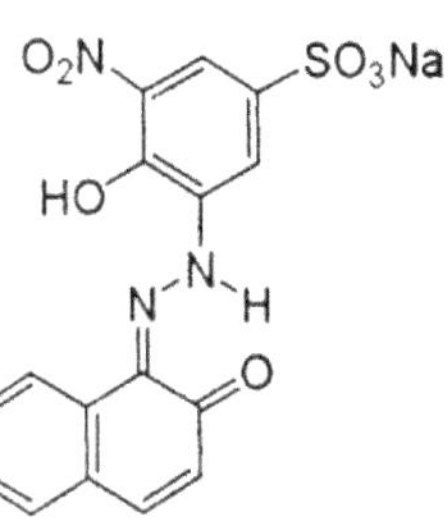 Mordant Black 17	

Source: Adapted from IARC (n.d., p. 66).

Azo dyes, which constitute an important and large group of textile dyes, are defined by double bonds between nitrogen and nitrogen (–N=N–). These are generally attached to two moieties of which at least one (but generally both) is (are) aromatic groups (especially naphthalene or benzene rings). Synthetic azo dye structures are very persistent and carcinogenic in nature due to the existence of –N=N– bond. In terms of application, the most commonly used azo dyes include Metanil Yellow, Direct Blue 15, Acid Orange 7, Orange II, Disperse Orange 1, Acid Orange 20, Methyl Orange, Reactive Orange 16, Methyl Red, Reactive Red 2, Reactive Black 5, Reactive Brilliant Red, Remazol Brilliant Blue R, Remazol Brilliant Orange 3R, Congo Red, Basic Blue 41, Direct Blue 160, Basic Yellow 15, Disperse Red 1, Acid Violet 7, etc. (Benkhaya, M'rabet et al., 2020; Sarkar et al., 2017; Rauf et al., 2011).

5.3 CARBON-BASED SENSORS IN ELECTROCHEMISTRY

Sensor systems are electronic apparatus with the capability to determine small differences in input energy and convert it into the same or another kind of energy as output in the form of signals (Kangmennaa et al., 2024; Chaithra et al., 2020). There are diversified sensors present like piezoelectric sensors, Fiber Bragg Grating sensors, metallic/semiconductor strain gauges, field effect transistor sensors, Raman shift-based sensors, piezoresistive sensors, capacitive sensors/apparatus, linear variable differential transducers, ultrasonic sensors, acoustic sensors, and accelerometers that are being employed to accomplish many aims in daily life (Rao et al., 2024).

Sensor systems as instrumental devices in especially analytical chemistry are utilized for the translation of chemical data into an electronic signal, employing two major elements, such as a transducer and a receptor connected in series (Ganesan & Paramathevar, 2020; Silah et al., 2022). On the surface of recognition layer, the interplay with the analyte species implies, and the transduction part turns into some chemical alterations coming into the open from this interplay into electrical signals (Vrabelj & Finšgar, 2022).

Due to their advantages of having very low limit of detection (LOD) values, rapidity, and cheap sensing devices, electrochemical sensor systems—the market leaders—are maybe the most generally utilized sort of sensors. Potentiometric, impedimetric, amperometric, electrogenerated and also photoelectrochemical chemiluminescence are some of the various kinds of electrochemical sensors (Thakur & Kumar, 2023).

Carbon-based materials have become one of the most researched and used materials in the field of nanotechnology in recent years due to their superior properties and advantages (Bezzon et al., 2019). Carbon-based electrodes are mostly exploited in electrochemical analysis because of their specific properties, which comprise tunable porosity, very large surface area, high electroconductivity, temperature resistance, chemical stability, nanostructuring probabilities, and convenience of producing at low cost (Pisoschi et al., 2021). Production procedures can be straightforward, yielding a desirable quantity of carbonaceous material with low densification faults. Additionally, these materials can be thought about as an option to recently pricey electronic components, supplying perfect performance and being considered an eco-friendly material. When considered from the perspective of sensor systems and electrochemical studies, it can be seen that carbon materials are quite suitable materials for these studies because they increase the electroactive surface area, are chemically inert, facilitate and accelerate electron transfer between analytes and converters, reduce overpotential, have relatively wide potential ranges, and support the adsorption of analyte molecules on the electrode surface (Bezzon et al., 2019; Power et al., 2018; Silah et al., 2022).

A trustworthy, rapid, and low-cost technique utilizes electrochemical sensor systems, mainly those modified with carbon-based materials with very good analytical efficiency (Venegas et al., 2023).

Carbonaceous materials generally take the form of tubes, sheets, hollow spheres, nanoparticles, and ellipsoids (Saleem & Zaidi, 2020; Silah, Nimal, et al., 2024). The carbon-based materials most frequently implemented in these studies included graphene, single- and multi-walled carbon nanotubes (MWCNT), and also their derivatives, reduced graphene oxide, graphene oxide, and carbon material mixes. The electrochemical investigation of textile dyes extensively makes use of pencil graphite electrodes (PGE), carbon paste electrodes (CPE), graphite electrodes, screen-printed carbon electrodes (SPCE), and glassy carbon electrodes (GCE) modified with various nanomaterials (Celik et al., 2024; Reddy Prasad et al., 2019).

5.4 APPLICATIONS OF CARBON SENSORS IN TEXTILE DYE DETERMINATION

Electrochemical methods are emerging as an alternative for accurate, rapid, selective, sensitive, and economical measurement. They address several disadvantages associated with conventional methods, particularly spectroscopy for textile dye determination. These drawbacks include prolonged analysis and sample preparation times, the use of non-eco-friendly and costly chemical reagents, uncertain reaction times, and the requirement for expensive, space-consuming equipment (Nimal, Selcuk, et al., 2022; Pisoschi et al., 2021; Chiorcea-Paquim et al., 2020).

Electrochemical methods which have sensor systems using carbon-based modified electrodes are generally applied in the determination of various analytes. When literature studies on the electrochemical determination of textile dyes were examined, it was found that electrochemical methods are mostly actualized using carbon-based sensors (Table 5.2). Depending on the mode of potential alteration, voltammetric methods implicate diversified kinds such as cyclic voltammetry (CV), linear sweep voltammetry (LSV), differential pulse voltammetry (DPV), square wave voltammetry (SWV), differential pulse adsorptive stripping voltammetry, and square wave adsorptive stripping voltammetry (Manjunatha et al., 2021).

In a study conducted by Ganash et al., the surface of the CPE was coated with gold nanoparticles using the electrodeposition technique and this modified electrode was successfully used for the sensitive and selective determination of Congo red dye (Figure 5.3). In the study, the surface characterization of the prepared electrode was investigated using scanning electron microscopy, transmission electron microscopy, Energy dispersive X-ray analysis, and Fourier transform infrared spectroscopy techniques. In the electrochemical analysis of Congo red dye, SWV technique was preferred and analyses were carried out in acetate buffer solution at pH 6.5. With the prepared electrode, the limit of detection value for Congo red was calculated as 0.7 μM, and the linear range was 1–30 μM and 50–200 μM. Congo red determination was also carried out in real samples and the recovery value as a result of the analysis in the wastewater sample was calculated as 103.29% (Ganash et al., 2023).

In a study conducted in 2022, Hayat and his colleagues designed a new electrode for the sensitive detection of the amount of methylene blue, an important textile dye in the textile industry, in aqueous environments and modified it using MWCNTs functionalized with amine on the GCE surface. The electrochemical parameters and properties of the prepared sensor were investigated by applying CV, EIS, and SWV. As a result of the modification, it was shown that the use of amine groups provided a negative charge to the MWCNT structure and enriched the electrode surface in terms of electrons. As a result of the electrochemical studies carried out with SWV, a linear working range of 0.01–0.5 μM and an LOD value of 0.21 nM were calculated for methylene blue. In addition, a mechanism was proposed for the oxidation of methylene blue dye on the NH_2-fMWCNTs/GCE surface (Figure 5.4) in their study (Hayat, Shah, Nisar, et al., 2022).

Acid Green 25, which is one of the pollutants in the textile sector and has an organic sulfonic acid structure, finds applications in the leather and cosmetics industry as well as in the textile industry, such as being used as a temporary hair dye. However, Acid Green 25 dye and its residues containing effluent can cause skin and eye irritation. In addition, this textile dye is toxic to many aquatic species (Jain & Gogate, 2018). In 2015, De Oliveira and his working group aimed to develop a voltammetric based carbon sensor for the selective and sensitive determination of this dye. The developed method includes the use of GCE electrode modified with activated MWCNT in H_2SO_4. The developed voltammetric method is based on the reduction of the anthraquinone group,

TABLE 5.2
Some of the selected studies based on electrochemical methods using carbon sensors for the determination of textile dyes

Dye	Method	Sensor Material	Linearity Range	LOD	Applications	Reference
Acid Blue 40	SWV	RuO_2-rGO/CMWCNTs-GCE	0.05–50.0 μM	0.01 μM	Wastewater from cloth dyeing	Khosropour et al., 2019
Acid Blue 29	SWV	RuO_2-rGO/CMWCNTs-GCE	0.5–75.0 μM	0.1 μM	Wastewater from cloth dyeing	Khosropour et al., 2019
Acid Blue 29	DPV	IL/CMWCNT-GCE	0.10–25 μM	0.06 μM	Wastewater samples	Rezaei et al., 2017
Acid Blue 113	CV	BentMCPE	1.00×10^{-9}–10.00×10^{-9} M	7.93×10^{-10} M	N.S.	Acar et al., 2015
Acid Green 25	SWV	CNTME	1.0×10^{-7}–7.0×10^{-6} M	2.7×10^{-9} M	Water treatment station sample	De Oliveira et al., 2015
Acid Green 25	DPV	IL/$CoFe_{12}O_{19}$/GO/CPE	0.08 – 900 μM	0.067 μM	Textile industry wastewater samples	Darabi & Shabani-Nooshabadi, 2020
Acid Yellow 17	CV	BentMCPE	1.00×10^{-9} –10.00×10^{-9} M	5.73×10^{-10} M	N.S.	Acar et al., 2015
Alizarin Red-S	SWV	RuNPs@CoNPs/Gr-Chi/PGE	2.5 μM – 2550 μM	0.277 μM	Tap water, Buckingham canal water, and textile industry wastewater samples	Kumari et al., 2024
Basic Brown G	SWV	CPE	0.004–0.2 μM 0.2–2.0 μM	1.2 nM	Wastewater and river water samples	Zheng et al., 2019
Basic Red 13	DPV	AgNP/GO/GCE	0.5 – 100.0 μM	2.0 μM	Wastewater samples	Jafari et al., 2017
Calmagite	CV	ZnO/MCPE	0.1 to 0.6×10^{-4} M	0.3834×10^{-8} M	Industrial wastewater samples	Dhrithi & Swamy, 2024
Chrysoidine	DPV	DNA/MWCNTs-PDDA/PGE	0.05–15.00 μg/mL	0.03 μg/mL	Industrial wastewater samples	Ensafi et al., 2014
Congo Red	SWV	AuNPs/CPE	1–30 and 50–200 μM	0.07 μM	Wastewater samples	Ganash et al., 2023
Congo Red	DPV	GO/GCE	0.01 μM–0.2 μM	2.4×10^{-7} M	Water and soil samples	Shetti et al., 2019
Disperse Red 1	EIS	Ethanolamine/Ab anti-DR1/GCE	8.40–100 nM	2.52 nM	Water samples	Rocha et al., 2016
Lanaset Red 2B	CV	BentMCPE	1.00×10^{-9}–10.00×10^{-9} M	7.08×10^{-10} M	N.S.	Acar et al., 2015
Lanaset Red G	CSVW	BGPE	5×10^{-9}–3×10^{-8} M	0.65 nM	N. S.	Ortaboy, 2017
Methylene Blue	DPV	MIP-CPE	1–50 μg/mL	36.4×10^{-6} M	N.S.	Soysal et al., 2013
Methylene Blue	SWV	NH_2-fMWCNTs/GCE	0.01–0.5 μM	0.21 nM	N. S.	Hayat, Shah, Nisar, et al., 2022
Methyl Green	SWAdASV	mag-MIP/CPE	9.9×10^{-8}-1.8×10^{-6} M	1.0×10^{-8} M	River water and industrial wastewater	S. Khan et al., 2019
Methyl Orange	CV	MG-MCPE	1–4 mM	0.080 μM	N.S.	Rajendrachari et al., 2023
Methyl Orange	DPV	Fe_3O_4/SWNTs/GCE	0.025–4000 μM	4 nM	Lake water	Kang et al., 2019

(continued)

TABLE 5.2 (Continued)
Some of the selected studies based on electrochemical methods using carbon sensors for the determination of textile dyes

Dye	Method	Sensor Material	Linearity Range	LOD	Applications	Reference
Mordant Yellow 10	DPV	IL/CMWCNT-GCE	0.6–30 µM	0.5 µM	Wastewater samples	Rezaei et al., 2017
Orange II	SWV	NH_2-*f*MWCNTs/ZnO/GCE	0.03–0.09 µM	0.57 nM	Industrial wastewater sample	Irfan et al., 2022
p-nitro phenol	DPV	GO/PVI/PdNPs-CPE	10–300 µM	2.20 µM	Synthetic textile wastewater samples	D'Almeida & Buoro, 2021
Reactive Red 120	DPV	MWCNT/GCE	N.S.	0.3 µg/mL	Dyeing process samples	Valarmathi et al., 2013
Reactive Yellow 84	DPV	MWCNT/GCE	N.S.	0.6 µg/mL	Dyeing process samples	Valarmathi et al., 2013
Rhodamine B	DPV	Fe_3O_4/SWNTs/GCE	0.1–250 µM	25 nM	Lake water	Kang et al., 2019
Rhodamine B	DPV	f-MWCNT/GCE	0.05–100 µM	2.6 nM	Sauce samples	K. Li et al., 2022
Rhodamine B	SWV	COOH-MWCNTs/NH_2-fMWCNTs	0.1 nM–0.05 µM	57.4 pM	Tap water and industrial wastewater samples	Hayat,Shah, Hakeem, et al., 2022
Safranin T	SWV	COOH-fMWCNTs/Fe-ZnO/GCE	0.08–80 µM	2.3 nM	N. S.	Sadiq et al., 2023
Turquoise blue 15	CSV	GCE	5.0×10^{-8} M–1.0×10^{-6} M	8.34×10^{-9} M	Tap water and in textile industry effluent samples	Osugi et al., 2003

Note: **Ab:** antibodies, **BentMCPE:** bentonite modified carbon paste electrode, **BGPE:** graphite paste electrode modified with natural Na-bentonite, **Chi:** Chitosan, **CMWCNTs:** carboxylated multiwalled carbon nanotubes, **CNTME:** carbon nanotube-modified electrode, **CPE:** carbon paste electrode, **CSV:** cathodic stripping voltammetry, **CSWV:** cathodic square wave voltammogram, **CV:** cyclic voltammetry, **DPV:** differential pulse voltammetry, **DR1:** disperse red 1, **EIS:** electrochemical impedance spectroscopy, **f-MWCNT/GCE**: carboxyl-functionalized multi-walled carbon nanotube, **GCE:** glassy carbon electrode, **GO:** graphene oxide, **Gr:** graphene, **IL:** ionic liquid, **LSV:** linear sweep voltammetry, **mag-MIP:** magnetic molecularly imprinted, polymer, **MG-MCPE:** metallic glass-modified carbon paste electrode, **MIP:** molecular imprinted polymer, **NP:** Nanoparticle, **N.S.:** Not statement, **PDDA:** Poly(diallyldimethylammonium chloride), **PGE:** pencil graphite electrode, **PVI:** poly-vinylimidazole, **rGO:** reduced graphene oxide, **SWAdASV:** square-wave adsorptive anodic stripping voltammetry, **SWNTs:** single-walled carbon nanotubes, **SWV:** square wave voltammetry.

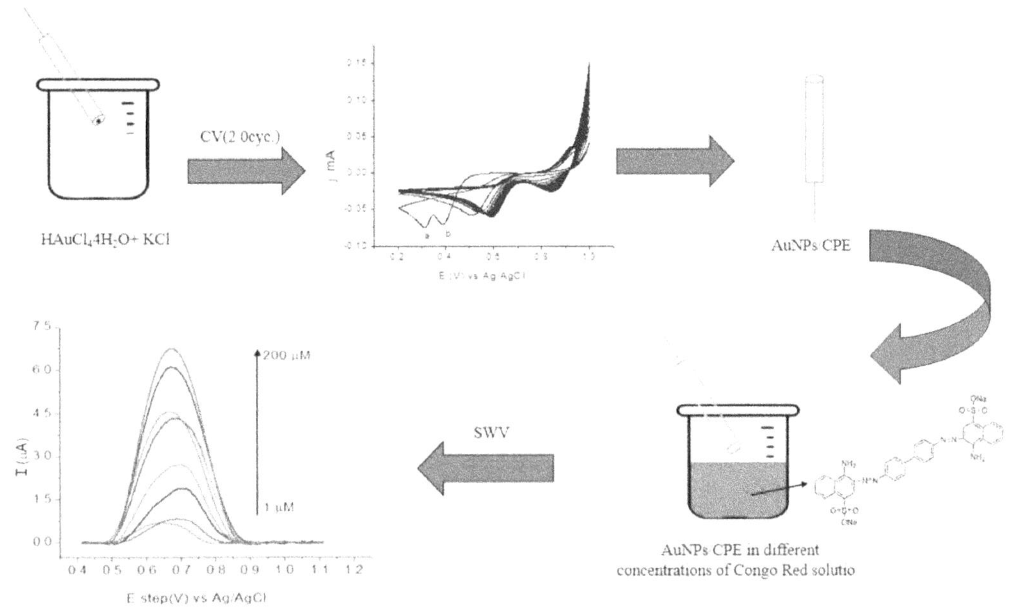

FIGURE 5.3 Procedure steps of the electrochemical determination of Congo red (Ganash et al., 2023).

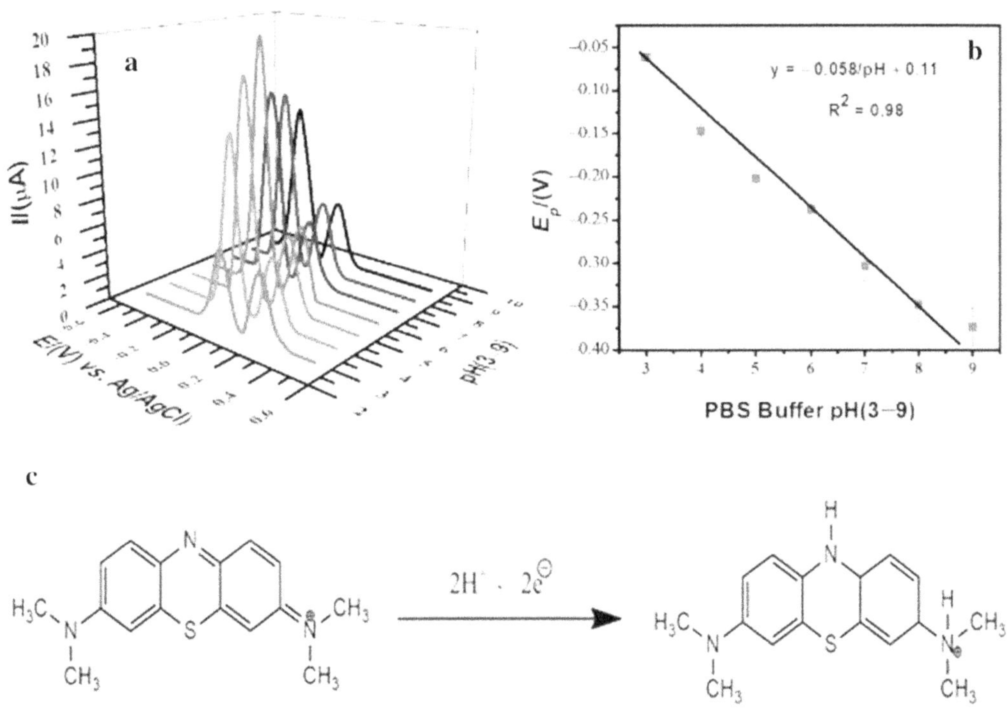

FIGURE 5.4 (a) SWV of methylene blue obtained by using NH_2-fMWCNTs/GCE as an indicator electrode. (b) Ep–pH plot of methylene blue. (c) Proposed oxidation mechanism of methylene blue on the NH_2-fMWCNTs/GCE (Hayat, Shah, Nisar, et al., 2022).

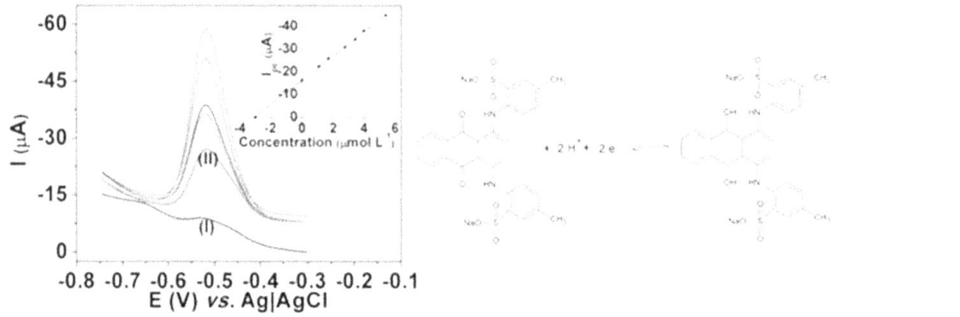

FIGURE 5.5 SWV obtained under optimized conditions with increasing concentrations of Acid Green 25 and Mechanism of Acid Green 25 reduction at the carbon nanotube-modified electrode (De Oliveira et al., 2015).

which is present in the structure of the Acid Green 25 dye molecule also acts as a chromophore, in the Britton-Robinson (B-R) buffer solution (pH 4.0) containing two electrons. The reduction process is defined as reverse and the relevant mechanism is given in Figure 5.5. SWV was used as the electrochemical method in the study and the linear working range for Acid Green 25 dye was found to be from 1.0×10^{-7} to 7.0×10^{-6} M and the observability value was calculated as 2.7×10^{-9} M (De Oliveira et al., 2015).

In recent years, it has been observed that nanomaterials and quantum dots are widely used in electrode design, especially in electrochemical studies. These nanomaterials can be used in different types, such as synthetic polymer nanomaterials, polymeric nanoparticles, natural nanomaterials, inorganic nanomaterials, and lipid nanomaterials. It has been stated in many studies that these nanomaterials facilitate and increase electron transfer (Ji et al., 2023; Jin et al., 2020; Uslu, 2021). In a study conducted by Irfan et al. in 2022, they modified the GCE surface using an MWCNT structure functionalized with amine groups and ZnO nanoparticles as nanoparticles in order to determine Orange II, a widely used textile dye with high toxic values, in different sample environments. In the study, it was stated that ZnO nanoparticles facilitated electron transfer between the analyte and the electrochemical transducer. Again, it was shown that the positive charging of the surface of the MWCNT structure modified with amine groups increased the adsorption of the negatively charged Orange II molecule in acidic solution on the electrode surface. The study showed that the modification process shifted the oxidation potential to negative values and caused a significant increase in the peak current. In the electrochemical Orange II analysis performed using SWV, the linear working range was found to be 0.03–0.09 μM, while the LOD value was calculated as 0.57 nM. In addition, a mechanism for the oxidation of Orange II on the prepared modified electrode surface was proposed in this study (Irfan et al., 2022).

5.5 CONCLUSION

Carbon sensors have attracted the notice of the world of science in consequence of their unique physical and chemical properties, such as ideal electrical conductivity, perfect heat conductivity, high mechanical strength, and also huge chemical stability. Carbon sensor systems are a rapid analysis apparatus for qualitative/quantitative determination of textile dyes. However, there are still some topics that need to be overcome regarding carbon sensor system fssor electrochemical applications in the determination of textile dyes.

- Although remarkable research has been channeled and reported to carbon sensor systems, a systematic comprehension of the development of nano-engineered smart carbon composite/

sensors for on-site determination is incomplete. There is a gap between laboratory studies and real area implementation in structural components (Rao et al., 2024).

- Commercial and industrial production of carbon sensors is not yet at the desired level and improvements are needed at this stage (Raya et al., 2022).
- Although carbon sensor-based electrochemical methods are extraordinarily sensitive, selective, and low cost, there is even an exigency to advance their monitoring performance. It is feasible that research will keep going into the surface modification of electrode with many nanomaterials, quantum dots, nanoparticles, conducting polymers, etc., with regard to the improvement of selectivity/sensitivity, and the response of electrochemical biosensors (Silah, Nimal, et al., 2024).
- Nowadays, the use of self-powered sensor systems and especially disposable electrodes is gaining importance in determination processes of different analytes and these sensor systems provide extraordinary advantages in the low-cost, simple, and on-site determination of these analytes in non-established laboratories. Therefore, it is necessary to focus on research on the use of such sensor systems in the analysis of textile dyes in different environmental environments (Silah, Erkmen, et al., 2024; Sivaranjanee et al., 2022).
- It is important that simultaneous qualitative and quantitative determination of different types of pollutants in environmental samples is effectively performed in situ by carbon-based electrochemical sensors modified with nanomaterials. Since the samples taken from environmental matrices, especially industrial wastewaters, as a result of today's industrial activities, are quite complex due to the large number of pollutant types of different origins and contain a large number of analytes.
- The ordinarily exploited analytical methods for the monitoring of textile dyes and their degradation products are spectroscopic methods. Consequently, the combined electrochemical based sensor systems and the spectroscopic methods-based devices can be generated for textile dye determination in soil, water, food products, etc.
- One of the most important advantages that direct researchers from many fields to work on biosensors is the versatility of this field of study. Biosensor-based systems are the product of multidisciplinary work and include collaborative initiatives from engineering, chemistry, physics, biology, microbiology, electricity, biotechnology, electronics, etc. As a result, it can be said that future discoveries on biosensors will be possible with the cooperation of the mentioned disciplines.

REFERENCES

Abo-Ayad, Z. A., Hussein, M. F., Zayed, M. A., & Abdelraheem, O. H. (2024). GC–MS identification of DR31 textile dye degradation products during its efficient electrochemical removal from wastewater media. *Journal of Molecular Liquids*, *400*, 124408. https://doi.org/10.1016/j.molliq.2024.124408

Acar, E. T., Ortaboy, S., Hisarlı, G., & Atun, G. (2015). Sensitive determination and electro-oxidative polymerization of azodyes on a carbon paste electrode modified with bentonite. *Applied Clay Science*, *105–106*, 131–141. https://doi.org/10.1016/j.clay.2014.12.035

Akter, M. M. K., Haq, U. N., Islam, M. M., & Uddin, M. A. (2022). Textile-apparel manufacturing and material waste management in the circular economy: A conceptual model to achieve sustainable development goal (SDG) 12 for Bangladesh. *Cleaner Environmental Systems*, *4*, 100070. https://doi.org/10.1016/j.cesys.2022.100070

Akhtar, W. H., Watanabe, C., Tou, Y., & Neittaanmäki, P. (2022). A new perspective on the textile and apparel industry in the digital transformation era. *Textiles*, *2*(4), 633–656. https://doi.org/10.3390/textiles 2040037

Alharbi, H. A., Hameed, B. H., Alotaibi, K. D., Al-Oud, S. S., & Al-Modaihsh, A. S. (2022). Recent methods in the production of activated carbon from date palm residues for the adsorption of textile dyes: A review. *Frontiers in Environmental Science*, *10*. https://doi.org/10.3389/fenvs.2022.996953

Arya, M. C., Bafila, P. S., Mishra, D., Negi, K., Kumar, R., & Bughani, A. (2020). Adsorptive removal of Remazol Brilliant Blue R dye from its aqueous solution by activated charcoal of Thuja orientalis leaves: an eco-friendly approach. *SN Applied Sciences*, *2*(2). https://doi.org/10.1007/s42452-020-2063-2

Barreto, W. J., Barreto, S. R. G., Scarminio, I. S., & Inoue, F. (2010). Quantification of textile dyes in industrial effluent using UV-Vis spectrophotometry combined with principal components regression. *Analytical Letters*, *43*(5), 814–822. https://doi.org/10.1080/00032710903486278

Benkhaya, S., Rabet, S. M., & Harfi, A. E. (2020). A review on classifications, recent synthesis and applications of textile dyes. *Inorganic Chemistry Communications*, *115*, 107891. https://doi.org/10.1016/j.inoche.2020.107891

Benkhaya, S., M'rabet, S., & Harfi, A. E. (2020). Classifications, properties, recent synthesis and applications of azo dyes. *Heliyon*, *6*(1), e03271. https://doi.org/10.1016/j.heliyon.2020.e03271

Bezzon, V. D. N., Montanheiro, T. L. A., De Menezes, B. R. C., Ribas, R. G., Righetti, V. a. N., Rodrigues, K. F., & Thim, G. P. (2019). Carbon nanostructure-based sensors: A brief review on recent advances. *Advances in Materials Science and Engineering*, *2019*, 1–21. https://doi.org/10.1155/2019/4293073

Çelik, M., Bozal Palabiyik, B., Kanbes Dindar, C., & Uslu, B. (2024). Voltammetric determination of metrafenone using an iridium oxide nanoparticle – Ionic liquid nanocomposite modified glassy carbon electrode. *Analytical Letters*. https://doi.org/10.1080/00032719.2024.2318653

Chaithra, K. P., Bhat, V. S., Akshaya, K. B., Maiyalagan, T., Hegde, G., Varghese, A., & George, L. (2020). Unique host matrix to disperse PD nanoparticles for electrochemical sensing of Morin: Sustainable engineering approach. *ACS Biomaterials Science & Engineering*, *6*(9), 5264–5273. https://doi.org/10.1021/acsbiomaterials.0c00758

Chiorcea-Paquim, A., Enache, T. A., De Souza Gil, E., & Oliveira-Brett, A. M. (2020). Natural phenolic antioxidants electrochemistry: Towards a new food science methodology. *Comprehensive Reviews in Food Science and Food Safety*, *19*(4), 1680–1726. https://doi.org/10.1111/1541-4337.12566

D'Almeida, S. R., & Buoro, R. M. (2021). Determination of P-Nitrophenol in synthetic textile wastewater samples using a graphene oxide/palladium nanoparticles modified carbon paste electrode. *Electroanalysis*, *33*(6), 1623–1632. https://doi.org/10.1002/elan.202060539

Darabi, R., & Shabani-Nooshabadi, M. (2020). Determination of Acid Green 25 in the presence of Brilliant Blue in hair dye and wastewater using voltammetric sensor amplified with reduced graphene oxide/CoFe12O19. *Environmental Technology & Innovation*, *19*, 101020. https://doi.org/10.1016/j.eti.2020.101020

De Oliveira, R., Hudari, F., Franco, J., & Zanoni, M. (2015). Carbon Nanotube-Based electrochemical sensor for the determination of anthraquinone hair dyes in wastewaters. *Chemosensors*, *3*(1), 22–35. https://doi.org/10.3390/chemosensors3010022

Dhrithi, H. R., & Swamy, B. E. K. (2024). Electrochemical investigation of environmental pollutant, azo dye (calmagite) using modified zinc oxide carbon paste electrode. *Sensing Technology*, *2*(1). https://doi.org/10.1080/28361466.2024.2369531

Djelal, H., Cornée, C., Tartivel, R., Lavastre, O., & Abdeltif, A. (2017). The use of HPTLC and Direct Analysis in Real Time-Of-Flight Mass Spectrometry (DART-TOF-MS) for rapid analysis of degradation by oxidation and sonication of an azo dye. *Arabian Journal of Chemistry*, *10*, S1619–S1628. https://doi.org/10.1016/j.arabjc.2013.06.003

Dunn, J. D., & Allison, J. (2007). The detection of multiply charged dyes using matrix-assisted laser desorption/ionization mass spectrometry for the forensic examination of pen ink dyes directly from paper. *Journal of Forensic Sciences*, *52*(5), 1205–1211. https://doi.org/10.1111/j.1556-4029.2007.00535.x

ElNaggar, M. S., & Van Berkel, G. J. (2011). Liquid microjunction surface sampling probe fluid dynamics: Characterization and application of an analyte plug formation operational mode. *Journal of the American Society for Mass Spectrometry*, *22*(10). https://doi.org/10.1007/s13361-011-0209-6

Ensafi, A. A., Jamei, H. R., Heydari-Bafrooei, E., & Rezaei, B. (2014). Development of a voltammetric procedure based on DNA interaction for sensitive monitoring of chrysoidine, a banned dye, in foods and textile effluents. *Sensors and Actuators B Chemical*, *202*, 224–231. https://doi.org/10.1016/j.snb.2014.05.001

Erkmen, C., Silah, H., & Uslu, B. (2024). Electrochemical biosensing for determination of toxic dyes. In *Elsevier eBooks* (pp. 215–241). https://doi.org/10.1016/b978-0-443-15334-1.00009-2

Ganash, A., Alshammari, S., & Ganash, E. (2023). Development of a novel electrochemical sensor based on gold nanoparticle-modified carbon-paste electrode for the detection of Congo red dye. *Molecules*, *28*(1), 19. https://doi.org/10.3390/molecules28010019

Ganesan, M., & Paramathevar, N. (2020). Quantum dots as nanosensors for detection of toxics: A literature review. *Analytical Methods*, *12*(35), 4254–4275. https://doi.org/10.1039/D0AY01293A

Ganesan, S., Kokulnathan, T., Sumathi, S., & Palaniappan, A. (2024). Efficient photocatalytic degradation of textile dye pollutants using thermally exfoliated graphitic carbon nitride (TE–g–C3N4). *Scientific Reports*, *14*(1). https://doi.org/10.1038/s41598-024-52688-y

Hayat, M., Shah, A., Hakeem, M. K., Irfan, M., Haleem, A., Khan, S. B., & Shah, I. (2022). A designed miniature sensor for the trace level detection and degradation studies of the toxic dye Rhodamine B. *RSC Advances*, *12*(25), 15658–15669. https://doi.org/10.1039/d2ra01722a

Hayat, M., Shah, A., Nisar, J., Shah, I., Haleem, A., & Ashiq, M. N. (2022). A novel electrochemical sensing platform for the sensitive detection and degradation monitoring of methylene blue. *Catalysts*, *12*(3), 306. https://doi.org/10.3390/catal12030306

IARC. (n.d.). Some aromatic amines, organic dyes and related exposures. In *Monographs on the evaluation of carcinogenic risks to humans* (Vol. 99).

Irfan, M., Shah, A., Iftikhar, F. J., Hayat, M., Ashiq, M. N., & Shah, I. (2022). Electrochemical sensing platform based on functionalized Multi-Walled carbon nanotubes and metal oxide for the detection and degradation studies of Orange II dye. *ACS Omega*, *7*(36), 32302–32312. https://doi.org/10.1021/acsomega.2c03641

Jafari, S., Dehghani, M., Nasirizadeh, N., & Akrami, H. R. (2017). Voltammetric determination of Basic Red 13 during its sonoelectrocatalytic degradation. *Microchimica Acta*, *184*(11), 4459–4468. https://doi.org/10.1007/s00604-017-2482-y

Jahan, R. A., Hassan, M. M., Rana, A. A., & Karim, M. M. (2023). Adsorption of anionic and cationic dyes from textile effluents by activated carbon prepared from sawdust and fish scale. *Advances in Chemical Engineering and Science*, *13*(03), 189–202. https://doi.org/10.4236/aces.2023.133014

Jain, S. N., & Gogate, P. R. (2018). Efficient removal of Acid Green 25 dye from wastewater using activated Prunus Dulcis as biosorbent: Batch and column studies. *Journal of Environmental Management*, *210*, 226–238. https://doi.org/10.1016/j.jenvman.2018.01.008

Ji, C., Li, J., Mei, J., Su, W., Dai, H., Li, F., & Liu, P. (2023). Advanced nanomaterials for the diagnosis and treatment of renal cell carcinoma. *Advanced NanoBiomed Research*, 3(2). https://doi.org/10.1002/anbr.202200079

Jin, C., Wang, K., Oppong-Gyebi, A., & Hu, J. (2020). Application of nanotechnology in cancer diagnosis and therapy – A mini-review. *International Journal of Medical Sciences*, *17*(18), 2964–2973. https://doi.org/10.7150/ijms.49801

Ju, C., Tang, Y., Fan, H., & Chen, J. (2008). Enzyme-linked immunosorbent assay (ELISA) using a specific monoclonal antibody as a new tool to detect Sudan dyes and Para red. *Analytica Chimica Acta*, *621*(2), 200–206. https://doi.org/10.1016/j.aca.2008.05.055

Kang, N., Ji, L., Zhao, J., Zhou, X., Weng, X., Li, H., Zhang, X., & Yang, F. (2019). Uniform growth of Fe3O4 nanocubes on the single-walled carbon nanotubes as an electrosensor of organic dyes and the study on its catalytic mechanism. *Journal of Electroanalytical Chemistry*, *833*, 70–78. https://doi.org/10.1016/j.jelechem.2018.11.012

Kangmennaa, A., Forkuo, R. B., & Agorku, E. S. (2024). Carbon-based electrode materials for sensor application: A review. *Sensing Technology*, *2*(1). https://doi.org/10.1080/28361466.2024.2350174

Kant, R. (2012). Textile dyeing industry an environmental hazard. *Natural Science*, *04*(01), 22–26. https://doi.org/10.4236/ns.2012.41004

Khan, S., Wong, A., Zanoni, M. V. B., & Del Pilar Taboada Sotomayor, M. (2019). Electrochemical sensors based on biomimetic magnetic molecularly imprinted polymer for selective quantification of methyl green in environmental samples. *Materials Science and Engineering C*, *103*, 109825. https://doi.org/10.1016/j.msec.2019.109825

Khosropour, H., Rezaei, B., & Ensafi, A. A. (2019). A selective and sensitive detection of residual hazardous textile dyes in wastewaters using voltammetric sensor. *Microchemical Journal*, *146*, 548–556. https://doi.org/10.1016/j.microc.2019.01.055

Kumari, K. G. B., Brahman, P. K., Lavanya, A. L., Satyadev, T. N. V. S. S., & Reddy, T. V. (2024). Ultra-thrifty/cost-effective approach to determine Alizarin Red S (ARS) in natural water supplies through ruthenium-cobalt modified pencil graphite as minuscule electrode. *International Journal of Environmental & Analytical Chemistry*, 1–19. https://doi.org/10.1080/03067319.2024.2330015

Li, K., Xiao, Y., Zhao, Y., Xia, Y., Ding, J., He, Q., Ling, J., & Li, G. (2022). A metal-free voltammetric sensor for sensitive determination of Rhodamine B using carboxyl-functionalized carbon nanomaterials. *Inorganic Chemistry Communications*, *145*, 110025. https://doi.org/10.1016/j.inoche.2022.110025

Manjunatha, J. G. (2019). Electrochemical polymerised graphene paste electrode and application to catechol sensing. *The Open Chemical Engineering Journal*, *13*(1), 81–87. https://doi.org/10.2174/1874123101913010081.

Manjunatha, J. G., Pushpanjali, P. A., & Hareesha, N. (2021). An overview of recent developments of carbon-based sensors for the analysis of drug molecules. *Journal of Electrochemical Science and Engineering*. https://doi.org/10.5599/jese.999

Niinimäki, K., Peters, G., Dahlbo, H., Perry, P., Rissanen, T., & Gwilt, A. (2020). The environmental price of fast fashion. *Nature Reviews Earth & Environment*, *1*(4), 189–200. https://doi.org/10.1038/s43017-020-0039-9

Nimal, R., Selcuk, O., Kurbanoglu, S., Shah, A., Siddiq, M., & Uslu, B. (2022). Trends in electrochemical nanosensors for the analysis of antioxidants. *TrAC Trends in Analytical Chemistry*, *153*, 116626. https://doi.org/10.1016/j.trac.2022.116626

Novotná, P., Pacáková, V., Bosáková, Z., & Stulík, K. (1999). High-performance liquid chromatographic determination of some anthraquinone and naphthoquinone dyes occurring in historical textiles. *Journal of Chromatography A*, *863*(2), 235–241. https://doi.org/10.1016/s0021-9673(99)00980-2

Ortaboy, S. (2017). Electrochemistry and sensitive determination of a metal complex azo dye using graphite paste electrode modified with Na-bentonite. *JOTCSA*,*4*(3), 931–952.

Osugi, M. E., Carneiro, P. A., & Zanoni, M. V. B. (2003). Determination of the phthalocyanine textile dye, reactive turquoise blue, by electrochemical techniques. *Journal of the Brazilian Chemical Society*, *14*(4), 660–665. https://doi.org/10.1590/s0103-50532003000400025

Ozturk, G., & Silah, H. (2020). Adsorptive removal of remazol brilliant blue R from water by using a macroporous polystyrene resin: Isotherm and kinetic studies. *Environmental Processes*, *7*(2), 479–492. https://doi.org/10.1007/s40710-020-00429-4

Periyasamy, A. P. (2024). Recent advances in the remediation of Textile-Dye-Containing wastewater: prioritizing human health and sustainable wastewater treatment. *Sustainability*, *16*(2), 495. https://doi.org/10.3390/su16020495

Pisoschi, A. M., Pop, A., Iordache, F., Stanca, L., Bilteanu, L., & Serban, A. I. (2021). Antioxidant Determination with the Use of Carbon-Based Electrodes. *Chemosensors*, *9*(4), 72. https://doi.org/10.3390/chemosensors9040072

Power, A. C., Gorey, B., Chandra, S., & Chapman, J. (2018). Carbon nanomaterials and their application to electrochemical sensors: A review. *Nanotechnology Reviews*, *7*(1), 19–41.

Rajendrachari, S., Basavegowda, N., Vinaykumar, R., Narsimhachary, D., Somu, P., & Lee, M. (2023). Electrocatalytic determination of methyl orange dye using mechanically alloyed novel metallic glass modified carbon paste electrode by cyclic voltammetry. *Inorganic Chemistry Communications*, *155*, 111010. https://doi.org/10.1016/j.inoche.2023.111010

Rao, R. K., Gautham, S., & Sasmal, S. (2024). A comprehensive review on carbon nanotubes based smart nanocomposites sensors for various novel sensing applications. *Polymer Reviews*, 1–64. https://doi.org/10.1080/15583724.2024.2308889

Rauf, M., Meetani, M., & Hisaindee, S. (2011). An overview on the photocatalytic degradation of azo dyes in the presence of TiO2 doped with selective transition metals. *Desalination*, *276*(1–3), 13–27. https://doi.org/10.1016/j.desal.2011.03.071

Raya, I., Kzar, H. H., Mahmoud, Z. H., Ahmed, A. a. A., Ibatova, A. Z., & Kianfar, E. (2022). A review of gas sensors based on carbon nanomaterial. *Carbon Letters*, *32*(2), 339–364. https://doi.org/10.1007/s42823-021-00276-9

Reddy Prasad, P., E. B. Naidoo, and N. Y. Sreedhar. 2019. Electrochemical preparation of a novel type of C-dots/ZrO2 nanocomposite onto glassy carbon electrode for detection of organophosphorus pesticide. *Arabian Journal of Chemistry*, *12*(8), 2300–2309. http://doi.org/10.1016/j.arabjc.2015.02.012

Rezaei, B., Khosropour, H., & Ensafi, A. A. (2017). A modified electrode using carboxylated multiwalled carbon nanotubes and 1-butyl-2,3-dimethylimidazolium hexafluorophosphate ionic liquid for a simultaneous hazardous textile dye sensor. *Analytical Methods*, *9*(2), 267–275. https://doi.org/10.1039/c6ay01884j

Ribechini, E., Pérez-Arantegui, J., & Colombini, M. P. (2013). Positive and negative-mode laser desorption/ionization-mass spectrometry (LDI-MS) for the detection of indigoids in archaeological purple. *Journal of Mass Spectrometry*, *48*(3), 384–391. https://doi.org/10.1002/jms.3169

Rocha, C. G., Ferreira, A. A., & Yamanaka, H. (2016). Label-free impedimetric immunosensor for detection of the textile azo dye Disperse Red 1 in treated water. *Sensors and Actuators B Chemical*, *236*, 52–59. https://doi.org/10.1016/j.snb.2016.05.040

Sadiq, M. U., Shah, A., Nisar, J., & Shah, I. (2023). Photoelectrocatalytic Detection and Degradation studies of a hazardous textile dye Safranin t. *Nanomaterials*, *13*(15), 2218. https://doi.org/10.3390/nano13152218

Saleem, H., & Zaidi, S. J. (2020). Developments in the application of nanomaterials for water treatment and their impact on the environment. *Nanomaterials*, *10*(9), 1764. https://doi.org/10.3390/nano10091764

Sarkar, S., Banerjee, A., Halder, U., Biswas, R., & Bandopadhyay, R. (2017). Degradation of synthetic azo dyes of textile industry: A sustainable approach using microbial enzymes. *Water Conservation Science and Engineering*, *2*(4), 121–131. https://doi.org/10.1007/s41101-017-0031-5

Sharma, J., Sharma, S., & Soni, V. (2021). Classification and impact of synthetic textile dyes on Aquatic Flora: A review. *Regional Studies in Marine Science*, *45*, 101802. https://doi.org/10.1016/j.rsma.2021.101802

Shetti, N. P., Malode, S. J., Malladi, R. S., Nargund, S. L., Shukla, S. S., & Aminabhavi, T. M. (2019). Electrochemical detection and degradation of textile dye Congo red at graphene oxide modified electrode. *Microchemical Journal*, *146*, 387–392. https://doi.org/10.1016/j.microc.2019.01.033

Silah, H., Demir, E., Yıldırım, S., & Uslu, B. (2022). Carbon nanomaterial-based sensors for the development of sensitive sensor platform. In *Elsevier eBooks* (pp. 191–246). https://doi.org/10.1016/b978-0-323-91174-0.00009-3

Silah, H., Erkmen, C., Yıldırım, S., & Uslu, B. (2024). Future trends of nanomaterials in environmental analysis. In *Elsevier eBooks* (pp. 455–470). https://doi.org/10.1016/b978-0-12-820643-0.00022-5

Silah, H., Nimal, R., Shah, A., & Uslu, B. (2024). Nanostructure materials for wastewater treatment. In *Elsevier eBooks* (pp. 331–374). https://doi.org/10.1016/b978-0-12-820643-0.00017-1

Sivaranjanee, R., Kumar, P. S., Saravanan, R., & Govarthanan, M. (2022). Electrochemical sensing system for the analysis of emerging contaminants in aquatic environment: A review. *Chemosphere*, *294*, 133779. https://doi.org/10.1016/j.chemosphere.2022.133779

Soysal, M., Muti, M., Esen, C., Gençdağ, K., Aslan, A., Erdem, A., & Karagözler, A. E. (2013). A novel and selective methylene blue imprinted polymer modified carbon paste electrode. *Electroanalysis*, *25*(5), 1278–1285. https://doi.org/10.1002/elan.201200569

Surowiec, I., Quye, A., & Trojanowicz, M. (2006). Liquid chromatography determination of natural dyes in extracts from historical Scottish textiles excavated from peat bogs. *Journal of Chromatography A*, *1112*(1–2), 209–217. https://doi.org/10.1016/j.chroma.2005.11.019

Şahin, S., Demir, C., & Güçer, Ş. (2007). Simultaneous UV–vis spectrophotometric determination of disperse dyes in textile wastewater by partial least squares and principal component regression. *Dyes and Pigments*, *73*(3), 368–376. https://doi.org/10.1016/j.dyepig.2006.01.045

Thakur, A., & Kumar, A. (2023). Recent trends in nanostructured carbon-based electrochemical sensors for the detection and remediation of persistent toxic substances in real-time analysis. *Materials Research Express*, *10*(3), 034001. https://doi.org/10.1088/2053-1591/acbd1a

Uslu, B. (2021). Electroanalytical applications of Quantum Dot-Based biosensors. In *Elsevier eBooks*. https://doi.org/10.1016/c2019-0-04090-3

Valarmathi, M., Gomathi, A., & Manisankar, P. (2013). Development of multiwalled carbon nanotube based electrochemical sensor for reactive AZO dyes. *Advanced Materials Research*, *678*, 321–325. https://doi.org/10.4028/www.scientific.net/amr.678.321

Venegas, C. J., Bollo, S., & Sierra-Rosales, P. (2023). Carbon-based electrochemical (bio)sensors for the detection of carbendazim: A review. *Micromachines*, *14*(9), 1752. https://doi.org/10.3390/mi14091752

Vrabelj, T., & Finšgar, M. (2022). Recent progress in non-enzymatic electroanalytical detection of pesticides based on the use of functional nanomaterials as electrode modifiers. *Biosensors*, *12*(5), 263. https://doi.org/10.3390/bios12050263

Wang, X., Jiang, J., & Gao, W. (2022). Reviewing textile wastewater produced by industries: Characteristics, environmental impacts, and treatment strategies. *Water Science & Technology*, *85*(7), 2076–2096. https://doi.org/10.2166/wst.2022.088

Yao, M., Liu, J., Narengaowa, N., Li, Y., & Zhao, F. (2023). In situ detection of natural dyes in archaeological textiles by SERS substrates immobilized on the fiber. *Microchemical Journal*, *192*, 108939. https://doi.org/10.1016/j.microc.2023.108939

Yaseen, D. A., & Scholz, M. (2019). Textile dye wastewater characteristics and constituents of synthetic effluents: A critical review. *International Journal of Environmental Science and Technology*, *16*(2), 1193–1226. https://doi.org/10.1007/s13762-018-2130-z

Zheng, L., Feng, X., & Zhang, C. (2019). Electrochemical oxidation and voltammetric determination of basic brown G at a carbon paste electrode. *International Journal of Electrochemical Science*, *14*(11), 10565–10575. https://doi.org/10.20964/2019.11.38

Zhou, Y., Lu, J., Zhou, Y., & Liu, Y. (2019). Recent advances for dyes removal using novel adsorbents: A review. *Environmental Pollution*, *252*, 352–365. https://doi.org/10.1016/j.envpol.2019.05.072

Zhou, C., Li, M., Garcia, R., Crawford, A., Beck, K., Hinks, D., & Griffis, D. P. (2012). Time-of-flight-secondary ion mass spectrometry method development for high-sensitivity analysis of acid dyes in nylon fibers. *Analytical Chemistry*, *84*(22), 10085–10090. https://doi.org/10.1021/ac3025569

6 Enhanced Carbon Sensors for Determination of Hormones

Nida Aydogdu Ozdogan, Seyda Nur Samanci, and Ersin Demir

6.1 INTRODUCTION

The discovery of hormones began with the groundbreaking work of Arnold Berthod and Ernest Starling, who laid the foundation for understanding testosterone and secretin, respectively. This was followed by the discovery of numerous hormones, including secretin, gastrin, cholecystokinin, ghrelin, and obestatin, in the gastrointestinal tract [1]. The term "hormone" was first introduced by Starling in 1905 [2], and the concept of chemical regulation in physiology took a major step forward with the isolation and synthesis of secretin in the 1960s [3]. These discoveries have significantly advanced our understanding of the role of hormones in maintaining bodily homeostasis. In Latin, *hormaein* means stimulation (activation) [4]. Hormones are organic compounds synthesized in ductless endocrine glands, classically known as endocrine organs, such as the pituitary gland, adrenal gland, thyroid gland, parathyroid gland, and gonads. Hormones play a crucial role in various physiological processes, including immune system regulation, energy metabolism, and the development and homeostasis of organs [5,6]. The endocrine system affects the target cells by secreting chemicals into the circulatory system. Endocrinology studies the endocrine glands and hormones secreted from these glands. Although hormones are similar to enzymes, they do not initiate chemical reactions like enzymes, but they affect the rate of the reaction. The human body produces a wide variety of hormones, each with its own unique function. Some of the most well-known hormones include testosterone, estrogen, insulin, thyroid hormones, and cortisol. Hormones generally regulate metabolism, the immune system (stimulation and inhibition), growth (stimulation and inhibition), restructuring, and cell life. The tissues on which hormones act are called target tissue. Contrary to the general rule, some hormones also act locally. For example, acetylcholine is a type of hormone that acts locally and is released from nerve endings. Secretin and cholecystokinin released from the duodenal wall are such hormones. Other hormones are usually released into the blood from a special gland and then transported via the blood to the target tissue where they will have an effect [4].

There are various classification criteria for hormones owing to their place of formation, mechanism of action, the chemical structure, or exerted final effect. Classification of hormones according to their solubility is divided into hydrophilic and lipophilic hormones. Water-soluble hormones, amino acid derivatives, peptide, and protein hormones fall into this group. Water-soluble hormones reach target tissues more easily through the carrier system. Lipophilic hormones are lipid-soluble hormones. Steroid hormones and thyroid hormones fall into this group. Because of their chemical makeup, hormones can be broadly categorized into four primary groups, including peptide and protein hormones, amino acid derivative hormones, steroid hormones, and fatty acid analogs. The classification of hormones according to their chemical structures is shown in Figure 6.1.

DOI: 10.1201/9781003500322-6

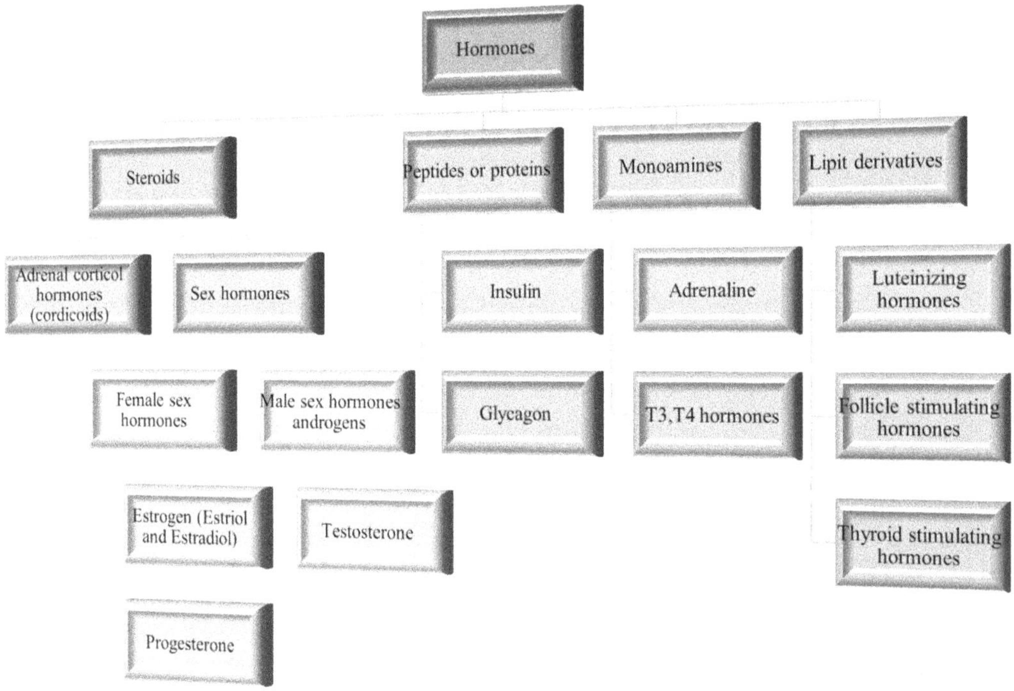

FIGURE 6.1 Classification of animal hormones.

Recently, nanotechnology has become popular in the field of sensors. Various studies have proven that the use of such technologies and nanosized materials can have positive effects on the performance of sensors. The fact that nanosized materials have a number of new and interesting physical and chemical properties has defined a new research field in condensed-matter physics in recent years. Carbon-based nanomaterials have been developed as the most discussed, researched, and applied synthetic nanomaterials owing to their impressive properties. Carbon materials offer many advantages over traditional materials, such as large surface area, versatility, and functionality. Carbon-based nanomaterials, including carbon nanotubes (CNTs), graphene, graphene oxide (GO), fullerenes, and nanodiamonds, are particularly suitable for environmental monitoring, food safety control, and healthcare because of their remarkable properties, such as high mechanical stability and high durability [7]. Carbon-based nanomaterials are increasingly used in sensor design owing to their advantageous intrinsic properties, such as high electrical and thermal conductivity, chemical stability, optical properties, large specific surface area, biocompatibility, and easy functionalization [8]. Figure 6.2 shows different carbon materials and the production methods of these materials.

A range of methods have been developed for the determination of hormones in various samples. Kang and friends utilized homogeneous ionic liquid microextraction and high-performance liquid chromatography to extract and determine hormones in cosmetics, achieving high precision and sensitivity [10]. Zhou and Chai employed ultra-performance liquid chromatography-electrospray ionization tandem mass spectrometry for the simultaneous characterization of 14 hormones in urine, demonstrating high extraction efficiency and precision [11]. López-García and friends developed a method for the determination of steroid hormones and their metabolites in meat samples using ultra-high performance liquid chromatography coupled with Orbitrap high-resolution mass spectrometry, achieving suitable recovery, repeatability, and reproducibility [12]. The disadvantages of commonly used methods are long analysis time, pretreatment, need for expert analyzers, use of a lot of organic solvents, and expensive devices. Because of these disadvantages, scientists are

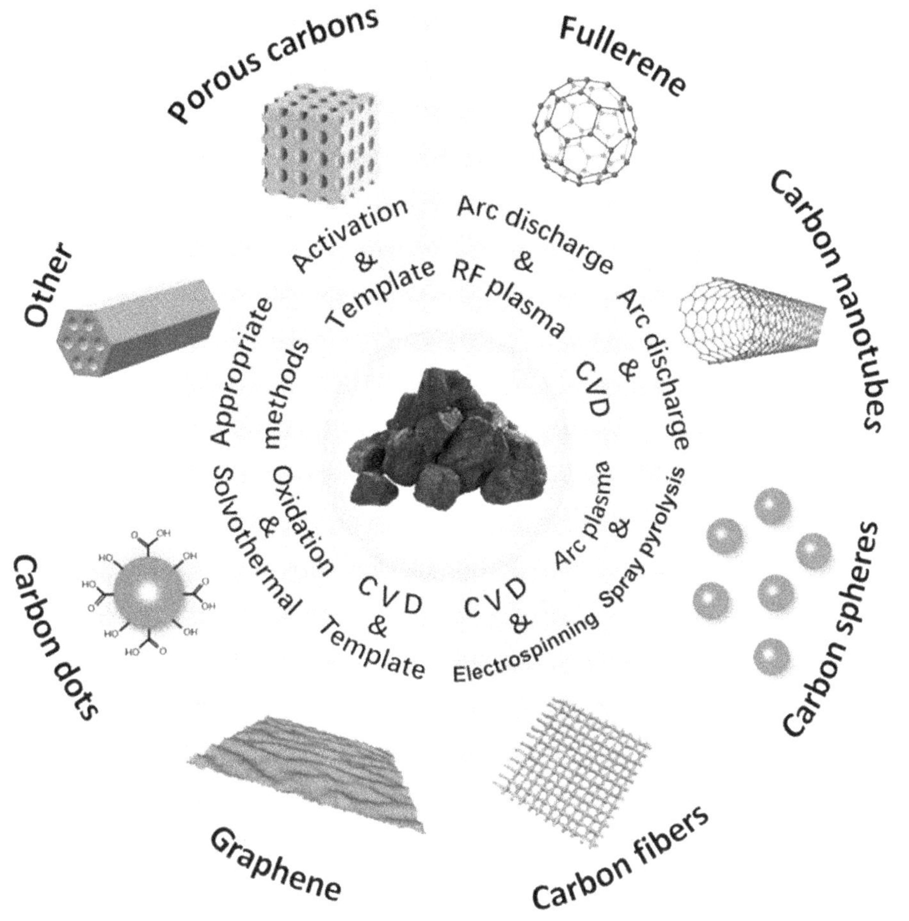

FIGURE 6.2 Different carbon materials and production techniques [9].

turning to nature-friendly, applicable, low-cost, and sensitive methods. Electrochemical methods are well-established analytical techniques that are widely used for the determination of pharmaceutical compounds. These methods, which are considered particularly fast, accurate, and selective with high sensitivity, have attracted considerable attention in recent years for the detection of electroactive analytes. Sample preparation for electrochemical measurements is generally quick and does not require expensive procedures. Furthermore, the mechanisms of analyte reactions on the surface of the working electrode and their redox properties, which can be investigated using cyclic voltammetry, can provide additional information about the interaction of the compound with cells and tissues. This chapter presents advances in the electrochemical determination of hormones with carbon sensors over the last five years. It is planned that this research will shed light on future studies by reporting on recent studies.

6.2 CARBON MATERIALS IN SENSOR TECHNOLOGY: PRINCIPLES AND TYPES

Over the past decade, nanomaterials have emerged as viable alternatives for molecular sensors and biosensors. Carbon nanomaterials (CNMs) are among the most discussed, researched, and applied synthetic nanomaterials. This is because of their diverse intrinsic electronic, magnetic, and optical

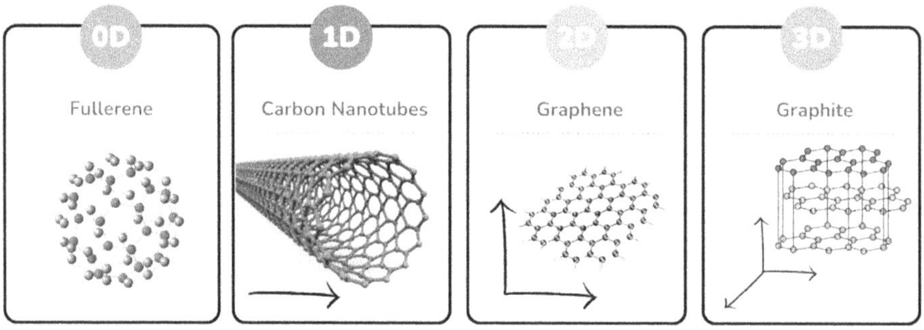

FIGURE 6.3 Carbon materials according to the number of dimensions.

properties, chemical versatility, ease of manipulation, biocompatibility, and performance as a chemically robust platform. This last feature distinguishes them from many other nanomaterials, whose stability can be compromised by factors such as temperature, high ionic strength, and solvent stability [13].

Carbon shows great structural diversity, existing in allotropic forms such as diamond, graphene, amorphous carbon, C60, and single-walled carbon nanotubes (SWNTs). These materials can be conveniently divided into zero-dimensional (0D) nanoparticles, one-dimensional (1D) nanotubes, and two-dimensional (2D) layered materials according to the number of dimensions larger than 100 nm (Figure 6.3).

Carbon materials, including graphite, CNTs, graphene, and their derivatives, are widely used as transducer elements in sensor technologies. Carbon-based sensors have gained significant attention in recent years owing to their superior analytical performance and unique properties [14]. These sensors, including CNTs, graphene, graphene oxide, fullerenes, and nanodiamonds, offer high sensitivity, selectivity, and biocompatibility, making them suitable for a wide range of applications such as environmental monitoring, food safety control, and healthcare [7]. The advantageous properties of carbon-based nanomaterials include high electrical and thermal conductivity, chemical stability, and large specific surface area, further enhancing their potential for use in sensor and biosensor development [8].

6.3 APPLICATIONS OF CARBON SENSORS IN HORMONE DETECTION

The variety and level of hormones in the body, which constitute an important component for health individuals, is of great importance. The fact that hormones are involved in various physiological processes, especially immune system regulation, energy metabolism and organ development, has caused research to focus on hormones. The human body synthesizes various hormones, each with its own specific function, and therefore, the determination of the amount of these hormones is also important. It is necessary to determine the levels at which these hormones should be in order to fulfill their functional duties.

Numerous analytical methods have been applied for the qualitative and quantitative determination of hormones. Among these methods, especially chromatography, spectrophotometry, and electrophoresis have some deficiencies, so scientists are looking for alternative analytical methods. The first alternative method that comes to mind is electroanalytical. Electrochemistry is fast, highly selective, has excellent sensitivity, does not require pre-treatment, is portable, miniaturized, requires minimal organic solvent, and is an environmentally friendly analysis technique. Therefore, electrochemical determinations of numerous analytes have been carried out in the last two decades. In the literature a wide range of electrochemical sensors have been developed for the detection of

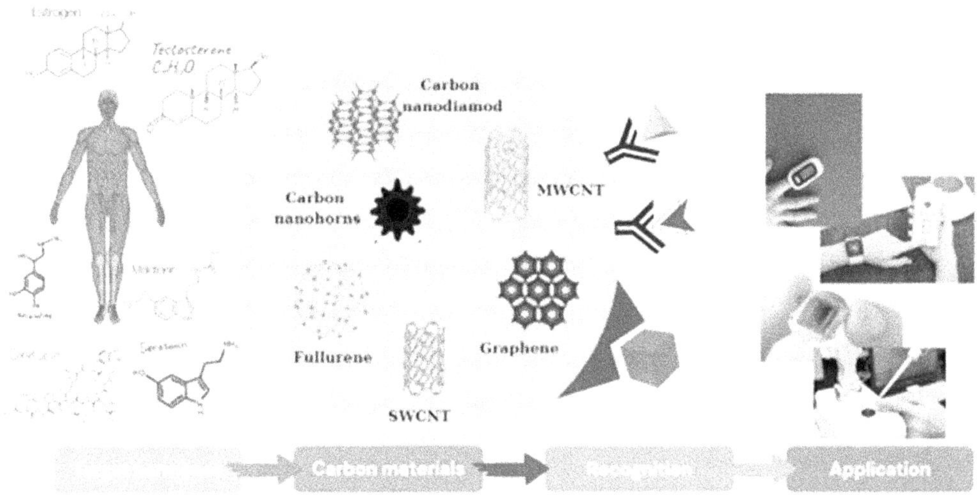

FIGURE 6.4 Schematic illustration of the determination of hormones with the aid of carbon-based materials.

hormones. The most preferred nanomaterial among these sensors is carbon and its derivatives such as diamond, graphene, amorphous carbon, C60, and carbon nanotubes (SWNTs and MWCNTs), glassy carbon (GCE), and graghite (Figure 6.4). The main reason for this is the superior versatility of carbonaceous materials, such as high conductivity, stable structures, and unique catalytic properties. At the same time, more sensitive and selective hybrid sensitizers have been developed by combining different carbon nanomaterials. The main aim here is to improve the sensitivity and catalytic properties of the sensors, thus allowing them to analyze hormones with high selectivity and accuracy even in complex matrix environments. It also leads to the discovery of new generation carbon sensors that can easily compete with classical analytical methods and even have superior performance. In recent years, the perspective of the sensor world has changed with the increasing importance of on-site analysis. This is also valid for hormones. Therefore, nanosensors provide great contributions to the in vitro and in vivo qualitative and quantitative analysis of hormones. In addition, the fact that both chemical wastes are very low and that very few organic solvents are used increases the importance of carbon sensors even more. In terms of green chemistry, this type of nanodevelopment is seen as an extremely vital development. In addition, the development of integrated biosensors and instant data acquisition allows hormones to be closely monitored from biological samples.

In this section, the advantages of voltammetric, amperometric and electrochemical impedance spectroscopic methods performed using carbon sensors for hormone detection are emphasized. Additionally, unique methodologies and innovations in new-generation composite carbon electrochemical sensors modified with different nanomaterials for the detection of hormones were discussed in detail. Moreover, the advantages of electroanalytical methods based on different hormone types analyzed by electrochemical techniques, indicator electrode types based on carbonaceous material used in electrochemical techniques, working ranges of electrochemical methods, limits of determination (LOD and LOQ), sample classification used in analytical applications, and recovery studies of carbon sensors are indicated in Table 6.1. Considering all these advantages, it would be a more appropriate step to focus on the status of nanosensors, their future place and perspective.

In their study published in 2022, Antolin Jesila Jesu Amalraj and her colleagues constructed a hybrid sensor with a glassy carbon electrode (GCE) surface doped with platinum (Pt) nanoparticles and selenium (Se) rods [15]. They found that this composite sensor exhibited a unique conductive and catalytic performance with the synergistic effect between Pt and Se nanomaterials. Then, the structural and surface morphology of the developed carbon composite nanosensor were investigated.

Then, the analytical performance of the constructed Pt/Se/GCE sensor for the determination of 17β-estradiol (E2) and methanol was investigated. The linear scanning voltammetry (LSV) and amperometry electrochemical techniques found that the Pt/Se/GCE sensor exhibited effective performance in the anodic reaction of estradiol. The recommended carbon-based hybrid sensors achieved a wide working range from 0.05 to 85.5 μM, and a very low limit of detection (LOD) of 11.12 nM with outstanding sensitivity. The optimum obtained supporting electrolyte pH 7.0 river and tap water samples were used for determination of analytes with high recoveries at low detection limit levels. Also, the oxidation reaction (MOR) of methanol on the Pt/Se/GCE sensor was investigated by cyclic voltammetry (CV) and chronoamperometry. As a result, the electrochemical determinations of 17β-estradiol (E2) and methanol on the developed carbon nanosensor (Pt/Se/GCE) were performed with high selectivity, recovery, accuracy and in a short time [15].

Rapid, sensitive and accurate determination of insulin is extremely important and necessary for the diagnosis of diabetes, a chronic disease. In the study conducted by Bhardwaj et al. (2023), a glassy carbon-based hybrid sensor was developed for the determination of insulin, which plays a regulatory role in the blood glucose level in the body [16]. For this purpose, they constructed the AgNP@ONP/GCE composite sensor and elucidated the electrode structure and surface morphology by electrochemical impedance spectroscopy (EIS), (CV, high-resolution transmission electron microscopy (HRTEM), and scanning electron microscopy (SEM). Then, they carried out the determination of insulin in aqueous samples using CV, linear scanning voltammetry (LSV), chronoamperometry, UV-visible and fluorescence spectroscopy methods on the proposed AgNP@ONP/GCE. They developed a method with a wide working range of 0–225 nM on carbon nanomaterials and a very trace detection limit of 2 nM. They also proved that the AgNP@ONP/GCE sensor exhibits high sensitivity and excellent reproducibility for insulin determination. They detected an important hormone such as insulin with high selectivity and accuracy even in complex matrix environments. The insulin detection of this sensor is schematically shown in Figure 6.5. As a result, insulin amount determination, which is an important data in terms of diabetes diagnosis, was carried out with a less costly, pre-processing-free, fast and environmentally friendly alternative method compared to traditional analytical methods by containing carbon nanomaterials [16].

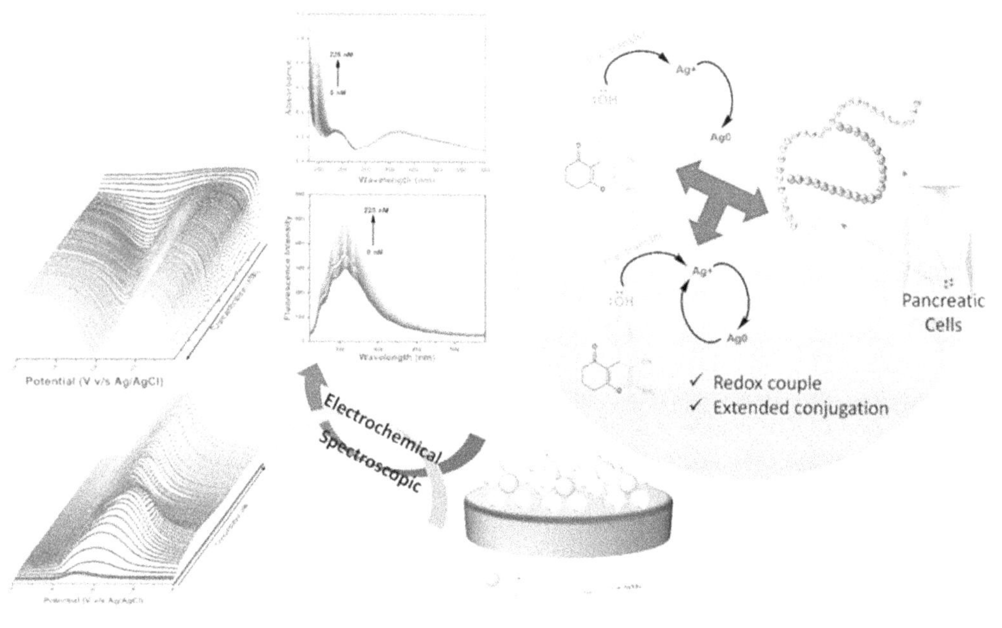

FIGURE 6.5 Schematic representation of the carbon nanosensor developed for insulin determination [16].

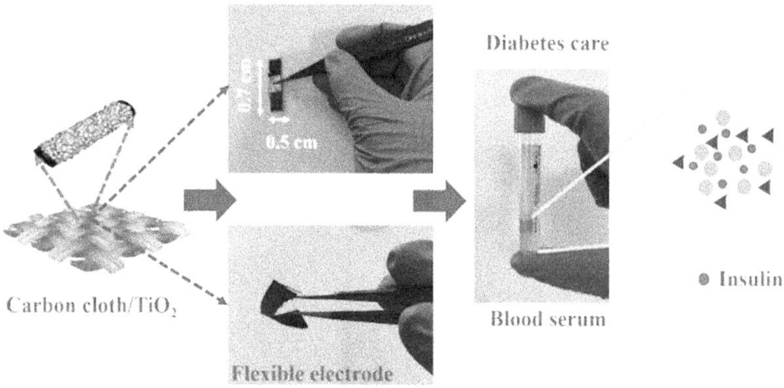

FIGURE 6.6 Schematic representation of flexible, portable, and miniaturized carbon-based nanosensor for inuslin determination [17].

Diabetes is among the most important chronic diseases in the world. The online monitoring of insulin hormone, which has an important role in the diagnosis and treatment of this disease, is very important in the treatment of the disease. For this, the production, commercialization and even miniaturization of insulin-specific biosensors are required. In particular, online (on-site) analysis systems are also needed. In their study conducted in 2024, Ramya and colleagues developed a nanofiber-supported carbon fabric (CC)-supported titanium dioxide (TiO_2) flexible nanofiber sensor (flexible TiO_2/CC) for insulin detection [17]. Despite the flexibility and mechanical stability of carbon fabric (CC), wearable and implantable properties, this material needs to be developed or modified with other nanomaterials due to its limiting properties. For this purpose, TiO_2 nanofibers increased both the surface area of the carbon material and improved its electron transfer ability (Figure 6.6). Thus, they improved the sensitivity and catalytic properties of the flexible nanofiber sensor. Then, the surface morphology and structural properties of the proposed flexible nanosensor were investigated by CV, field-emission scanning electron microscopy (FE-SEM), energy-dispersive X-ray spectroscopy (EDAX) methods. Chronoamperometry (CA) and CV techniques stabilized the analytical performance of the proposed flexible TiO_2/CC nanosensor by optimizing the parameters such as pH, electrode resistance and scanning speed. Then, a very low and wide linear detection scale from 7 pM to 1000 pM was determined on the flexible TiO_2/CC nanosensor. In addition, LOD and LOQ values were calculated as 2.7 pM and 8.7 pM, respectively. These trace level detection limits show that they can easily compete with traditional analysis methods. Insulin was analyzed in human serum samples with high recovery and low relative error by the proposed flexible and portable carbon nanosensor. In conclusion, they successfully determined insulin selectively, with high accuracy and precision using a miniaturized, flexible and portable carbon-based nanosensor [17]. This study is seen as a promising study, especially for the development of commercialization of nanosensors.

In particular, due to the expertise required, time-consuming analysis times, and expensive equipment required for the determination of testosterone, scientists have turned to alternative analytical methods. In their study conducted in 2023, Sanchez-Almirola and colleagues developed an electrochemical detection sensor (Figure 6.7) based on molecularly imprinted polymers (MIPs) for the determination of testosterone (a hormone that regulates body functions) [18]. For this purpose, they produced a MIP-based Poly(oPD)/SPE miniature carbon nanosensor by coating o-phenylenediamine (oPD) on a screen-printed electrode (SPE) with cyclic voltammetry. They determined testosterone in a wide working range of 1 ng/dL (34.7 pM)–25 ng/dL (866.8 pM) on the Poly(oPD)/SPE miniature carbon sensor and obtained the limit of detection (LOD) value as 1 ng/dL (34.7 pM). They performed analytical applications on complex matrix samples such as saliva taken from a healthy individual and successfully analyzed testosterone with high recovery and low relative error. As a result, they

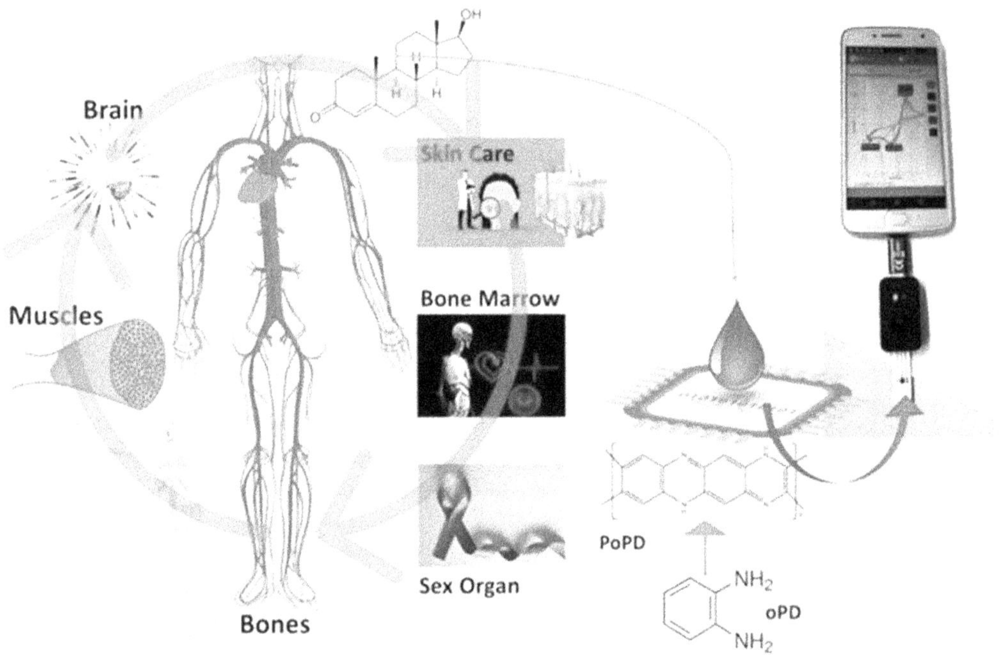

FIGURE 6.7 General schematic representation of the miniature analysis kit for testosterone testing [18].

constructed a miniaturized, selective and sensitive carbon-based molecularly imprinted polymer nanosensor for the determination of testosterone, a hormone of utmost importance for human health [18]. This work is also understood to be promising for flexible, wearable sensor designs. Moreover, it paves the way for the development of biomarkers integrated into the human body and may even contribute to the development of a smartphone online analysis system using a PalmSens M-P with Bluetooth capabilities and remote control.

Melatonin (MT) is a pineal gland hormone class and is an important biomarker in biological functions such as sleep regulation, depression, and prostate cancer. In the scientific article published by Jayaraman and his colleagues in 2023, they modified the disposable screen printed carbon electrode (SPCE) surface with carbon spheres (Ru-CS) rich in C and O and containing a high level of ruthenium (Ru) with a catalytic effect for the determination of Melatonin (MT), which is extremely important for human health [19]. This modification is schematically shown in Figure 6.8. The non-covalent interaction between the nanomaterial and the ion-molecule affinity accelerated the MT electrode reaction and ensured that it was oxidized with less positive potential. The surface morphology and microstructural features of the developed miniature carbonaceous Ru-CS/SPCE were elucidated by FE-SEM, transmission electron microscopy (TEM), high-angle annular dark-field scanning transmission electron microscopy (HAADF-STEM), X-ray photoelectron spectroscopy (XPS), and energy-dispersive X-ray (EDX) analysis. Then, the calibration graph of different concentrations of Melatonin was obtained on Ru-CS/SPCE and a wide working range of 1.25–20.0 μM was obtained. The LOD value of 0.27 μM was determined with the carbonaceous nanosensor having a large surface and catalytic effect. Furthermore, the interference effect of organic compounds such as norepinephrine (NE), dopamine, ascorbic acid, tyrosine and epinephrine on MT determination was investigated amperometrically on CS/SPCE and it was found that none of the interference types had a significant effect on MT determination. Finally, they successfully implemented the analytical application for MT detection on CS/SPCE in pharmaceutical samples. As a result, they detected melatonin hormone rapidly, simply, selectively and sensitively using miniature carbon nanosensor [19].

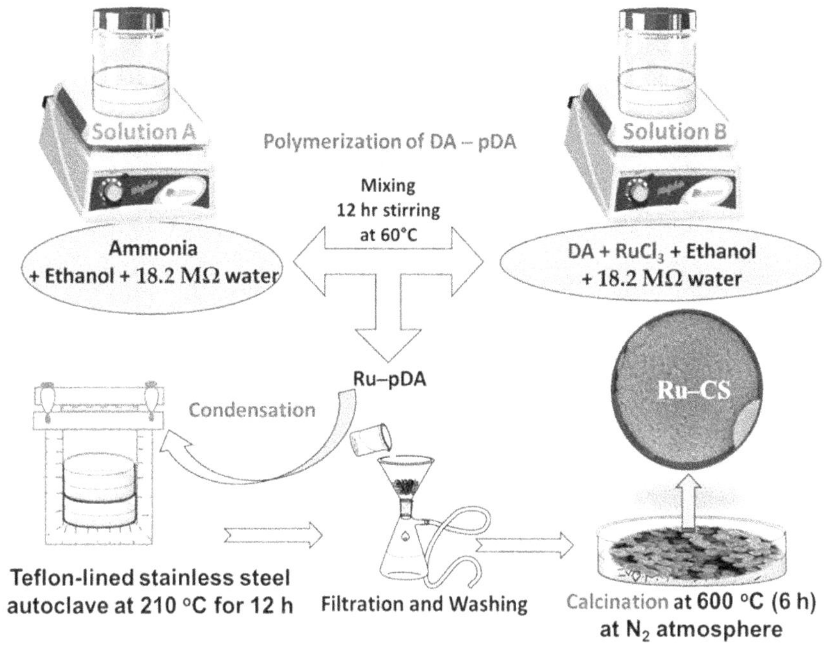

FIGURE 6.8 Schematic representation of the preparation of Ru-CS/SPCE nanosensor [19].

Due to some deficiencies in the sensors developed for the 17-estradiol hormone, a nanomarker that received full marks has not yet been developed. Therefore, there is always a need for new, cheap, portable and commercializable nanosensors. In their scientific study published in 2021, Supchocksoonthorn and colleagues developed a simple, cheap, non-disposable and portable nanosensor containing polyaniline (PANI) and carbon dots (CDs) synthesized from iota-carrageenan for the analysis of 17-estradiol in natural samples [20]. For this purpose, poly(vinylidene fluoride) was used as a binder on a GCE and a CDs-PANI/GCE composite nanosensor was produced. Later, the surface topography of the developed carbon nanosensor was illuminated with field-emission scanning electron microscope (FE-SEM) equipped with energy dispersive X-ray spectroscopy (EDS). It was observed that CDs-PANI/GCE composite nanosensor significantly increased the 17-estradiol anodic signal compared to the bare electrode (Figure 6.9). Under optimum conditions, a wide working range of 1–100 nM was obtained for 17-estradiol hormone in phosphate buffer solution (PBS) at pH 1.0 with LSV and the LOD values calculated as 43 nM. It was also stated that the nanosensor could be used hundreds of times at a reproducible level in this working range. This shows that the stability of CDs-PANI/GCE composite nanosensor is quite good and its shelf life is satisfactory. In addition, it was shown that it is highly selective against 17-estradiol in the presence of some organic species with the recommended CDs-PANI/GCE. Moreover, the accuracy and precision test of the LSV method and the developed carbon nanosensor was carried out on human serum sample and different types of waters such as drinking water, tap water, and river water samples. Finally, they produced a sensitive, cheap, simple, and selective new generation carbon nanosensor for the determination of 17-estradiol, an extremely important hormone [20].

Yang and his team synthesized nanoporous carbon/nickel oxide (OMC-NiO) material using the hard templating method in their 2021 study [21] (Figure 6.10). Thus, they produced a more regular mesostructure and nanostructure with a high surface area. Then, they coated this nanomaterial on the GCE surface using the drop-dry technique with the help of an infrared lamp. They illuminated the

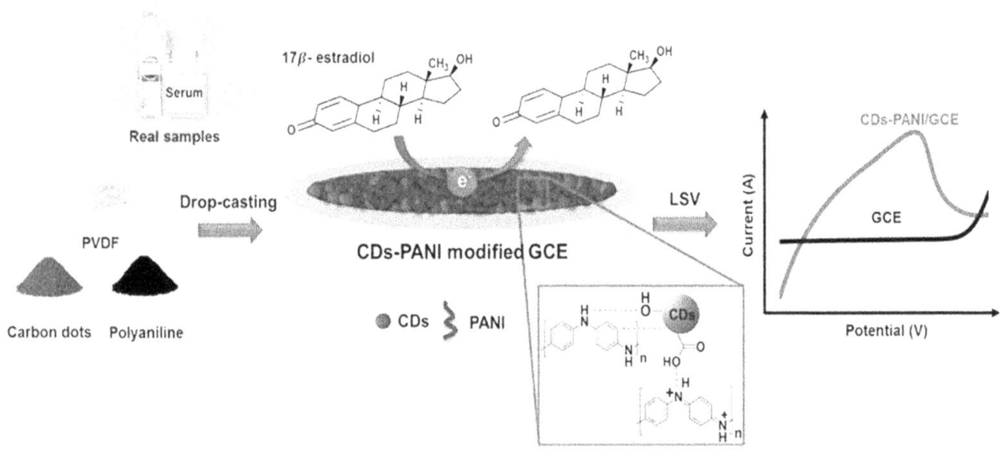

FIGURE 6.9 Schematic representation of the constructed new generation carbon nanosensor for the determination of 17-estradiol [20].

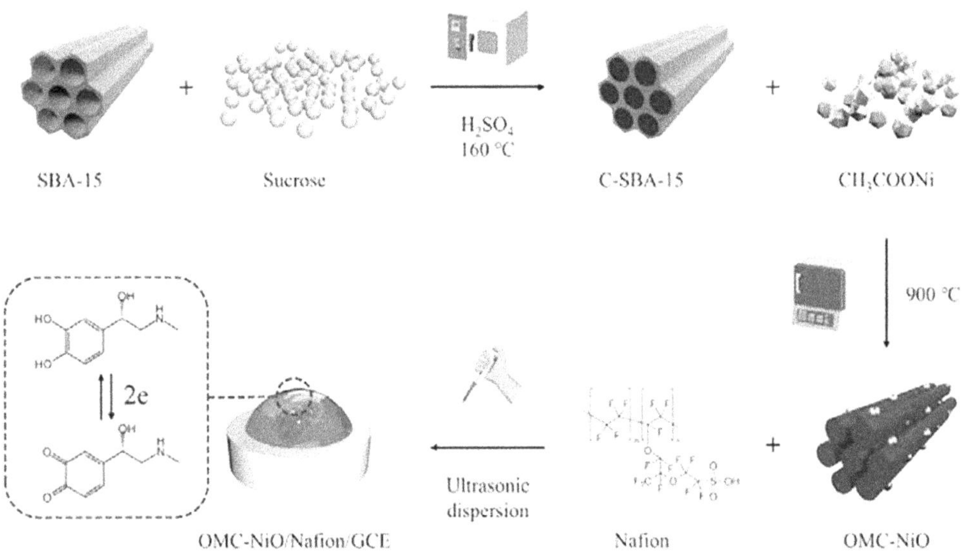

FIGURE 6.10 Schematic representation of the constructed carbon nanosensor as OMC-NiO/Nafion/GCE for epinephrine determination [21].

OMC-NiO/GCE surface topology and structural features with small-angle X-ray diffraction (XRD), pore size distribution, TEM, and cyclic voltammetry (CV). They also investigated the electrochemical behavior of epinephrine qualitatively and quantitatively using the developed sensitive carbon indicator electrode as OMC-NiO/GCE. They obtained a dynamic working range between 0.8 μM and 50 μM on OMC-NiO/GCE with optimized differential pulse voltammetry (DPV) and calculated the LOD value (S/N ×3) as 85 nM. They also elucidated the oxidation reaction mechanism of epinephrine with CV technique. Moreover, they investigated the interference effect of some organic substances on the surface of OMC-NiO/GCE on the determination of epinephrine and found that foreign substances did not cause a significant tension. This shows that the carbon hybrid electrode recommended for the determination of epinephrine is highly selective. In the last stage of the study, they successfully performed the analysis of epinephrine with high recovery and low relative error in spiked blood samples. As a result, the analysis of epinephrine, an important substance for healthy individuals, was carried out simply, cheaply, sensitively, and selectively with high accuracy and precision [21].

Priyanka and Latha, in their scientific study conducted in 2021, developed a carbon material-based sensor based on $MnCr_2O_4$ nanoparticle modified carbon paste electrode ($MnCr_2O_4$/MCPE) and used it as a sensor in the electrochemical quantitative analysis of NE [22] (Figure 6.11). They synthesized the $MnCr_2O_4$ nanocomposite material used in the production of modified carbon electrode by hydrothermal method. They elucidated the surface topology of the synthesized nanocomposite with scanning electron microscope (SEM), infrared (IR), and XRD spectroscopy techniques. In addition, they carried out CV studies to determine the redox properties of the $MnCr_2O_4$/MCPE nanosensor and showed that the modified carbon electrode has a more active surface area than the bare electrode. The most ideal support electrolyte for qualitative and quantitative analysis of NE was obtained in 0.2 M phosphate buffer with pH 7.4. Under the optimum electrochemical method, a linear working range was obtained for the determination of NE between 0.3 μM and 4.5 μM using the $MnCr_2O_4$/MCPE carbon sensor and the LOD value was calculated as 34 nM. In addition, in order to demonstrate the accuracy and precision of the method, they performed the analytical application of NE in pharmaceutical samples and successfully analyzed NE with very high recovery of 98.5 to 99.5. Consequently, a simple, rapid, sensitive, and selective $MnCr_2O_4$ nanoparticle modified carbon material sensor ($MnCr_2O_4$/MCPE) was developed for the determination of NE [22].

In the study conducted by Jayaraman et al. in 2023, a new generation electrochemical sensor based on Cu nanomaterial loaded carbon nanospheres (Cu-CNSs) was designed [23]. First, CNSs were synthesized using dark brown sugar, a natural or cheap carbon source. Then, to improve the catalytic property of this material, they made a composite with Cu nanomaterial. Finally, the constructed Cu-CNSs were modified on screen-printed carbon electrode (SPCE) and used as markers for the simultaneous determination of dopamine (DA) and melatonin (MT) and are schematized in Figure 6.12. In addition, the surface morphology and structural properties of the developed carbon-based nanosensor were elucidated by FE-SEM coupled with EDS, TEM, CV and electrochemical impedance spectroscopy (EIS) techniques. They investigated the electrochemical properties of both

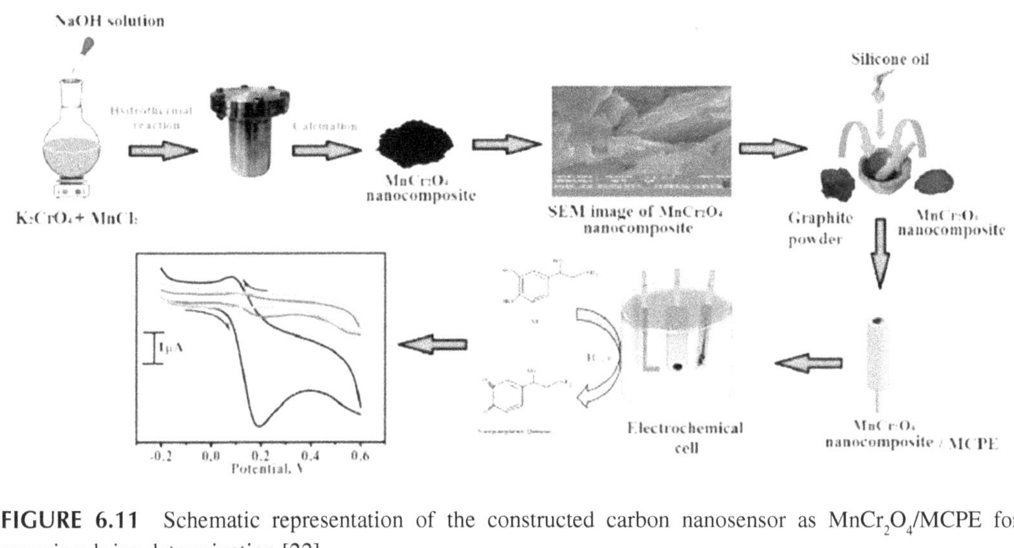

FIGURE 6.11 Schematic representation of the constructed carbon nanosensor as $MnCr_2O_4$/MCPE for norepinephrine determination [22].

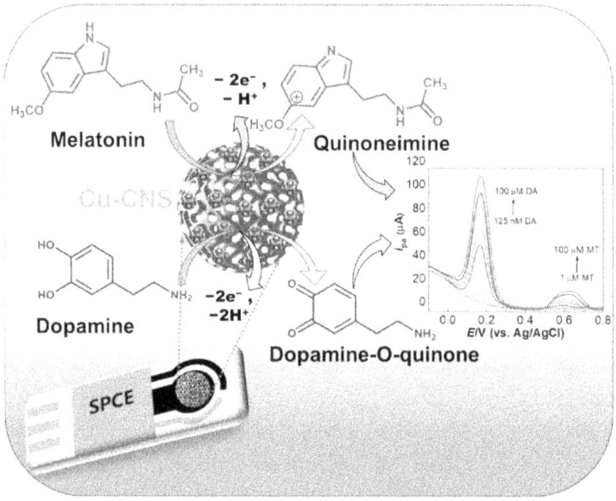

FIGURE 6.12 Schematic representation of the developed carbon-based Cu-CNS/SPCE sensor for the simultaneous determination of dopamine (DA) and melatonin (MT) [23].

hormones on Cu-CNS/SPCE by differential pulse voltammetry (DPV) and determined them in natural samples. They obtained a very wide working range of 0.125–100 μM for DA and 1.0–100 μM for MT on the proposed carbonaceous Cu-CNS/SPCE. Moreover, they calculated the LOD values for both substances as 0.34 μM and 0.33 μM (S/N = 3) on Cu-CNS/SPCE by DPV, respectively. They also investigated the interference effects on the simultaneous determination of DA and MT substances of organic substances, such as ascorbic acid (AA), glucose, uric acid (UA), urea, and creatinine with a new generation carbon nanosensor, and showed that the interference species did not affect the quantitative determination of both substances. In the last step of the study, simultaneous determination of DA and MT in human serum samples on Cu–CNS/SPCE indicator electrode was carried out with high recovery, extraordinary selectivity, and excellent accuracy. Consequently, simultaneous determination of DA and melatonin MT was successfully achieved in a fast, simple, selective, and sensitive manner with carbon-based nanosensor containing carbon nanospheres and metal nanoparticles.

6.4 DISCUSSION AND CONCLUSION

Public health has been the most important issue in the world for years. The state allocates a serious budget every year for the training and raising of healthy individuals. Raising healthy individuals directly affects not only the economy but also the sociocultural impact of the country. The fact that hormones play a role in physiological processes, especially in the regulation of the immune system, energy metabolism, and organ development, has caused research in recent years to focus on the determination of hormones. In addition, since hormones play an active role in the diagnosis and treatment of diseases, the qualitative and quantitative analyses of these substances are of great importance. In addition, it is necessary to determine the levels at which hormones should be in order to fulfill their functional duties. Therefore, the analyses of hormones in biological samples are gaining importance and becoming necessary day by day.

Although many analytical methods have been applied for the qualitative and quantitative determination of hormones in real samples, we can say that there is no simple, fast, and cheap fully validated technique yet. Scientists are always looking for alternatives due to the shortcomings of traditional analytical methods such as chromatography, spectrophotometry, and electrophoresis,

TABLE 6.1
Variety of carbonaceous materials and analytical performances of electrochemical methods developed for hormone determination

Analyte	Electrode	Method	Linear Range (μM)	LOD (μM)	LOQ (μM)	Sample	%Recovery	Ref
Cortisol	BSA/C-AB/PSE/GO/GCE	CV	2.5×10^{-5}–5.0	8.8×10^{-7}	-	Saliva, urine, sweat samples	100.24–101.54	[24]
Cortisol	PANI-GO/GCE	CV	100–1000	30	-	Human saliva samples	-	[25]
Cortisol	RGO/ AbC/SPE	Amperometry	0.00028–0.55	0.00028	-	-	-	[26]
Dexamethasone	P(L-Cys)/GCE	CV	10 – 100	0.0055	0.016	Human Serum samples	96.49–98.85	[27]
Dopamine	e-CS/MWCNTs/GCE	DPV	20–1000	12	-	Human Saliva samples	119.78	[28]
Dopamine	Cu–CNS/SPCE	DPV	0.125–20	0.34	-	Human serum samples	95.2–102.9	[23]
Dopamine	GO-BAMB-$Co(OH)_2$/GCE	DPV	3–20 and 25–100	0.4	1.5	Urine samples	95.37–97.75	[29]
Dopamine	Poly(BTCA)/GCE	DPV	0.1–10	0.028	0.094	Urine and blood samples	98.0–102.3	[30]
Dopamine	Porous carbon sensor	DPV	0.8–400	0.1	-	-	-	[31]
Dopamine	GO/GCE	DPV	1.0–15	0.27	-	Urine samples	98.6–102.4	[32]
Dopamine	SWCNT/PET	DPV	1.5–30	0.51	-	-	-	[33]
Dopamine	Au/RGO/GCE	DPV	6.8–41	1.4	-	Serum Sample	97.5–102.0	[34]
Dopamine	RGO/PAMAM/MWCNT/AuNP/GCE	DPV	10–320	3.3	-	-	-	[35]
Dopamine	3D N, P-doped carbon	DPV	2–200	0.6	-	-	-	[36]
Dopamine	N-rGO/GCE	DPV	0.5–150	0.41	-	-	-	[37]
Dopamine	PTh/GPE	LSV	10–180	1	-	Urine and blood samples	95.6–102.9	[38]
Dopamine	FeS_2/CoS_2NHs/GCE	LSV	1–24000	1.1	-	Human serum samples	98.6–104.5	[39]
Dopamine	BiOCl-Fe_2O_3/GCE	LSV	100–800 μL	-	-	Biological samples	-	[40]
Dopamine	SPCE/PVA/ZnO	CV	100–600	120.8	-	Human Serum samples	92.5–99.8	[41]
Dopamine	$Ag_3Fe(CN)_6$-graphene/GCE	Chronoamperometry	0.4–28	0.03	-	Human urine samples	96–107	[42]
Dopamine	Graphene nanobelts/GCE	Amperometry	2 –202	0.58	1.95	-	-	[43]
Epinephrine	OMC-NiO/GCE	DPV	0.8–50.0	0.085	0.37	Blood serum sample	94.5–108.0	[21]
Epinephrine	MWCNT-PANI-TiO_2/Au	DPV	0.077 – 4.9	0.16	-	Pharmaceutical samples	99.5–101.3	[44]
Epinephrine	MWCNT/Fe_3O_4/2,3-Nc/GCE	DPV	20–48	12.3	-	-	-	[45]
Epinephrine	AuNP/cysteic acid/GCE	DPV	0.4–11.7	0.008	-	-	-	[46]
Epinephrine	S-MCF/GCE	DPV	0.1–12	0.04	-	Pharmaceutical samples	98–108	[47]
Epinephrine	Gr/SnO_2/Au com/GCE	DPV	0.5–10	0.05	-	Human urine samples	95.4–104.5	[48]

(continued)

TABLE 6.1 (Continued)
Variety of carbonaceous materials and analytical performances of electrochemical methods developed for hormone determination

Analyte	Electrode	Method	Linear Range (μM)	LOD (μM)	LOQ (μM)	Sample	%Recovery	Ref
Epinephrine	AuNPs/poly(trisamine)/GCE	DPV	3.9–61.8	0.9	-	Human blood samples	98.4–101.3	[49]
Epinephrine	GCB/GCE	SWV	0.1–0.5	0.0036	0.0118	Blood serum samples	97.05–102.4	[50]
Epinephrine	CNTs/PGE	CSWV	2.5–250	0.873	2.65	Pharmaceutical samples	98.31–98.78	[51]
Epinephrine	PSAF/CPE	LSV	10–80	0.61	1.58	Pharmaceuticals samples	97.53–100.2	[52]
Estriol	N-MWCNT/GONRs/GE	Amperometry	0.1–20 μg·mL^{-1}	2 ng·mL^{-1}	-	Clinical and water samples	93.7–105.4 and 94.2–106.2	[53]
Estriol	Lac/rGO/Sb_2O_5/GCE	Chronoamperometry	0.025–1.03	0.011	-	Human urine	100.1 ± 1	[54]
Ethinylestradiol	SPCE-Ni	SWV	0.23–30	0.052	0.171	Pharmaceutical samples	95–106.6	[55]
Follicle-Stimulating Hormone	rGO/MWCNTs/Thi/AuNP/SPE	DPV	1 mIU/mL to 250 mIU/mL	0.05 mIU/mL	-	Human Serum	94.0–109.8	[56]
Follicle-Stimulating Hormone	r-GO/Thi/AuNPs/SPE	DPV	1–100 mIU/mL	1 mIU/mL	-	Blood Sample	-	[57]
Follicle-Stimulating Hormone	$NiCo_2O_4$/rGO and MIP	EIS	1 × 10^{-7}–1 μM	1 × 10^{-7}	-	Human serum	-	[58]
Human chorionic gonadotrophin	Anti-hCG/AuNPs/Cys/AuNps/GE	DPV	0.001–0.2, 0.2–60.7	0.3 ng mL^{-1}	-	Human serun	97–105	[59]
Human chorionic gonadotrophin	Graphene immunosensor	Amperometry	0.62–5.62 ng/mL	0.62 ng/mL		-	-	[60]
Hydrocortisone	P(L-Cys)/GCE	CV	10–100	0.011	0.033	Human Serum samples	99.05–103.88	[27]
Insulin	Aptamer/cDNA-Gated amine functionalized MSTF/GCE	DPV	0.01–0.35	0.003	-	Blood Sample	96.28–108.05	[61]
Insulin	NiO/Co_3O_4 CuAl LDH@MWCNT/CPE	DPV	2.0×10^{-5}–0.07	1.73×10^{-5}	-	Human Serum	98.0–106.0	[62]
Insulin	SiO_2NPs-Nafion/GCE	DPV	0.01–0.05	0.0031	-	Bovine insulin	97.2–103.5	[63]
Insulin	AuNPs/PANI/GE	DPV	0.001 ng ml^{-1} to 1000 ng ml^{-1}	0.017 pg ml^{-1}	-	Human serum	99–104	[64]
Insulin	MIP cryogel/f- MWCNTs/AuE	SWV	5 × 10^{-8} –1.4 × 10^{-6}	33 × 10^{-8}	-	Blood serum samples	-	[65]
Insulin	NiNPS/methylebenzoate/PGE	SWV	0.01–0.6	0.0083	-	Human Serum samples	98.4–103.2	[66]

Insulin	Nafion-SiO_2NPs/ GCE	CV	0.1–0.8	0.03	-	Bovine insulin	97.2–103.5	[63]
Insulin	AgNP@ONP/GCE	CV	0–0.225	0.002	-	Human serum samples	98.3–100.02	[16]
Insulin	CC/TiO_2/FE	CV	7×10^{-6} –0.001	2.7×10^{-6}	8.7×10^{-6}	Human serum samples	-	[17]
Insulin	CC/TiO_2/FE	Chronoamperometry	7×10^{-6}–10^{-4}	4.5×10^{-6}	1.49×10^{-5}	Human serum	96.32–98.31	[17]
Insulin	$Co(OH)_2$ NPs/CCE	FIA	0.0005–0.0015	0.00011	-	Human samples	95.2–104%	[67]
Insulin	GCE/CQDs	Amperometry	0.04–02	0.0022	-	Blood samples	93.2–104.1	[68]
Insulin	GCE/CQD-Chit/GLA/APT/BSA	EIS	5×10^{-7}–1×10^{-5}	0.0001	-	Human blood serum	96.8–105.6	[69]
Insulin	Aptamer/Au-polymer film modified PGE	EIS	0.001–1	0.0002	-	Plasma sample	88.9–103	[70]
Melatonin	e-CS/MWCNTs/GCE	DPV	20–1000	22	-	Human Saliva samples	80.13	[28]
Melatonin	Cu–CNS/SPCE	DPV	1.0–100	0.33	-	Human serum	96.1–99.4	[23]
Melatonin	Fe_2O_3/C–Fe-gC_3N_4/SPCE	DPV	0.1–233.1	0.004	-	Urine Samples	96.86–98.69	[71]
Melatonin	MIP/GCE	DPV	0.097–200	0.029	-	Urine and saliva samples	102–103	[72]
Melatonin	FeCo@CNFs/GCE	DPV	0.05–250	0.0027	-	Pharmaceutical samples	98.6–101.4	[73]
Melatonin	$ZnFe_2O_4$/CPE	DPV	6.5–145	3	-	Blood serum samples	-	[74]
Melatonin	Gr–Fe_3O_4/CPE	DPV	0.02–5.8	0.0084	-	Pharmaceutical samples	-	[75]
Melatonin	SnO_2–Co_3O_4@ rGO/IL/CPE	DPV	0.02–6.0	0.04	-	Urine samples	96.7–104.7	[76]
Melatonin	Gr-AV	SWV	10–100	0.49	-	Serum, Urine, and pharmaceutical Samples	89.8–110	[77]
Melatonin	Ru–CS/SPCE	Chronoamperometry	1.25–20	0.27	-	Pharmaceuticals	95.2–102.2	[19]
Melatonin	AuNPs/PDA/GCE	EIS	1×10^{-6}–18×10^{-5}	3.2×10^{-7}	-	Urea samples	99.0-107.0	[78]
Norepinephrine	$MnCr_2O_4$/MCPE	CV	0.3–4.5	0.034	0.113	Pharmaceuticals samples	98.5–99.5	[22]
Parathyroid	MB-APBA-HRP-Ab-PE-PTH/SPCE	SWV	10–300 pg ml^{-1}	11.56 pg ml^{-1}	-	Serum Sample	85.64–105.73	[79]
Progesterone	Sn/GCE	DPV	5.0–80.0	0.12	1.47	Pharmaceutical samples	97.89–103.5	[80]
Progesterone	Fe_3O_4@GQD/f-MWCNTs/GCE	DPAdSV	0.01–3.0	2.18	-	Pharmaceutical Samples	101.7–102.2	[81]
Progesterone	GQDs-PSSA/GO/GCE	DPV	0.001–6.0	0.00031	0.00087	Pharmaceutical Samples	99–101.6	[82]
Progesterone	GO-IMZ/GCE	SWV	0.22–14.0	0.064	0.21	Pharmaceutical samples	-	[83]
Progesterone	AuNPs/5-Amino-2-mercaptobenzimidazole /rGO/SPCE	SWV	0.0009–27	0.28	0.93	Serum samples	98.76–100.78	[84]
Progesterone	P(L-Cys)/GCE	CV	10 – 100	0.0008	0.0023	Human Serum samples	-	[27]

(continued)

TABLE 6.1 (Continued)
Variety of carbonaceous materials and analytical performances of electrochemical methods developed for hormone determination

Analyte	Electrode	Method	Linear Range (μM)	LOD (μM)	LOQ (μM)	Sample	%Recovery	Ref
Progesterone	AuNP/MnSB/GCE	CV	35–800	0.0114	-	Milk samples	92.3–104.2	[85]
Progesterone	GCE/AuNP/MnSB	CV	35–800	0.0114	-	Milk samples	92.3–104.2	[85]
Progesterone	WO3 NBs@GR/GCE	Amperometry	0.025– 1792.5	0.0042	-	Human serum	97.0–98.36	[86]
Serotonin	e-CS/MWCNTs/GCE	DPV	9–1000	10	-	Human Saliva samples	79.19	[28]
Testosterone	rGO-MOFs/GCE	DPV	0.03–4.125	0.0001	-	Human serum samples	97.80–99.05	[87]
Testosterone	rGO/GCE	AdSV	0.002–0.21	0.0001	-	Urine and plasma samples	98.1–104.2%	[88]
Testosterone	CuONP/PGE	SW-AdSV	0.005–0.2	0.0046	0.0153	Pharmaceutical samples	98.3%–103.0%	[89]
Testosterone	GCE	SW-AdSV	0.01–0.07	0.0018	0.0039	Urine samples	100.03	[90]
Testosterone	PoPD-MIPs/SPCE	CV	1–25 ng/dL	1 ng/dL	-	Human Serum samples	-	[18]
Testosterone	NCD/silicon wafer/MIP/carbon layer	EIS	0.005–0.02	0.005	-	Human urine and saliva	-	[91]
Testosterone	GCE/SA/BSA/BiNb16	EIS	0.05 – 5 ng mL^{-1}	0.045 ng mL^{-1}	-	Human serum	90%–112%	[92]
Testosterone	MIP/GO	EIS	1.0×10^{-9}–1.0	1×10^{-9}	-	Human serum sample	98.6–104.2	[93]
β-Estradiol	Ionic Liquid/GCE	DPV	0.1–10	0.05	-	Human sesum	103	[94]
β-Estradiol	MC/ SiO_2NPGCE/	DPV	0.5–10	0.37	-	Water	-	[95]
β-Estradiol	RGO/CuTthP/GCE	DPV	0.1–1.0	0.0053	-	River water	96.6–98.3	[96]
β-Estradiol	Biochar nanoparticles /GCE	DPV	0.05–20	0.0113	-	Tap water	103.4–107.2	[97]
β-Estradiol	wMC/GCE	DPV	0.05–10 10–80	0.0083 nM	.-	Milk and river water	92–108.4	[98]
β-Estradiol	NH_2-SWCNT/NMB/AuNP	DPV	5pG	10 pg–500 ng	-	Human serum	-	[99]
β-Estradiol	Fe_3O_4-MIP/SPCE	DPV	0.05–10.0	0.02	0.09	River water	95.6–97	[100]
β-Estradiol	$NTiO_2$TP/GCE	LSV	9.9–14	0.17	-	Tap water and urine samples	91–114 87–99	[101]
β-Estradiol	CDs/PANI/ GCE	LSV	0.001–100	0.043	-	Water and human serum	94.4–103.7	[20]
β-Estradiol	Pt/Se/GCE	Amperometry	0.05–85.5	0.0111	0.036	River and tap water	-	[15]

which are frequently used in hormone determinations, such as requiring long pre-treatment, using too many organic solvents, requiring expertise, requiring expensive equipment, and long analysis times. In this case, the first alternative method that comes to mind is electroanalytical techniques. The reason is that electrochemical techniques are quite fast, selective, have excellent sensitivity, do not require pre-treatment, are portable, miniaturized, require minimum organic solvent, and are environmentally friendly. For this reason, electrochemical determinations of a large number of hormones have been carried out in the last 20 years. Among the electrochemical methods, carbon electrodes are the most preferred material. We can count this nanomaterial as carbon and its derivatives such as diamond, graphene, amorphous carbon, C60 and carbon nanotubes (SWNTs and MWCNTs), GCE, and graphite. This is mainly because carbonaceous materials can exhibit superior versatility, such as high conductivity, stable structures, and unique catalytic properties. At the same time, they can be recombined with carbon nanomaterials and become more sensitive and selective hybrid sensitizers. When various nanomaterials such as nanoparticles, polymers, and organic substances, are added to this, excellent conductivity or extraordinary catalytic properties are added. Thus, the sensitivity and catalytic properties of the sensors are improved, allowing them to analyze hormones with high selectivity and accuracy, even in complex matrix environments. This has also led to the discovery of a generation of carbon sensors that can easily compete with classical analytical methods for hormone determination and even exhibit superior performance. In terms of green chemistry, the development of carbon-based nanomaterials is seen as an extremely vital development. In addition, the development of these integrated biosensors allows for on-site and online data acquisition, and close monitoring of hormones from biological samples. It is understood that voltammetric, amperometric, and electrochemical impedance spectroscopic methods performed using carbon sensors recommended for hormone detection are highly selective, sensitive, and inexpensive analytical methods. Hormones analyzed by electrochemical techniques using carbon nanosensors were analyzed with wide working ranges, low detection limits (LOD and LOQ), high recovery and high selectivity in analytical applications. As a future perspective of enhanced carbon sensors for determination of hormones, although there is a lot of variety, carbon sensors need to be developed for the determination of harmonics in in vivo and in vitro samples. This will be possible with the development of different nanomaterials and the advancement of technology. Thus, the development of more sensitive, more selective, on-site, and portable analysis kits will contribute. It will especially lead to the production of integrated analysis systems in the body. In this sense, there will always be a need for the development of more simple, cheap, sensitive, and selective sensors for the development of commercial component analysis kits.

ACKNOWLEDGEMENTS

Seyda Nur Samanci thanks the financial support from the Scientific and Technological Research Council of Türkiye (TUBITAK) under the BIDEB/2211-A Ph.D. and ARDEB/1004 Ph.D. Scholarship Programs.

ABBREVIATIONS

APBA	Aminophenyl boronic acid
BCNPs	Nano sized biochar particles
CC	Nanofiber-enhanced carbon cloth
CCE	Carbon ceramic electrode
CNFs	Carbon nanofibers
CNT	Carbon nanotube
CSWV	Cyclic square wave voltammetry
CuTthP	Cu(II)-meso-tetra(thien-2-yl)porphyrin

CV	Cyclic voltammetry
Da	Dalton
EIS	Electrochemical impedance spectroscopy
FE	Flexible electrode
FIA	Flow injection amperometry
GCE	Glassy carbon electrode
GE	Gold electrode
Gr-AV	Graphite/automotive varnish ink
GSs	Graphene sheets
Hcg	Human serum chorionic gonadotropin
HRP	Horseradish peroxidase
IL	Ionic liquid
LIG	Laser-induced graphitization
MBs	Magnetic beads
MC	Mesoporous carbon
MCF	Mesoporous carbon foam
MCPE	Modified carbon paste electrode
MIP	Molecularly imprinted polymer
MSTF	Mesoporous silica thin-film
MWCNT	Multi-walled carbon nanotube
NMB	New methylene blue
NPEDMA	N-phenylethylene diamine methacrylamide
NPG	Nanoporous gold
N-TiO_2	Nitrogen-doped titanium dioxide nanoparticles
OMC-NiO	Ordered mesoporous carbon/nickel oxide
ONPs	Organic nanoparticles
P(L-Cys)	Poly(L-cysteine)
PAMAM	Poly(amido-amine)
PANI	Polyaniline
PET	Poly(ethylene terephthalate)
PoPD	Poly(o-phenylenediamine)
PSAF	Poly (Safranin)
PTH	Parathyroid hormone
PVA	Polyvinyl alcohol
RDGCE	Rotating-disc Glassy Carbon Electrode
rGO	Reduced graphene oxide
Ru–CS	Ruthenium-incorporated carbon spheres
SiC	Silicon carbide nanoparticles
SPCE	Screen-printed carbon electrode
SWCNTS	Single-walled carbon nanotubes
TES	Three-electrode system
wMC	Wrinkled mesoporous carbon

REFERENCES

[1] P. Ceranowicz, Z. Warzecha, A. Dembinski, Peptidyl hormones of endocrine cells origin in the gut--their discovery and physiological relevance, *J. Polish Physiol. Soc.* 66 (2015) 11–27. https://doi.org/https://pubmed.ncbi.nlm.nih.gov/25716961/.

[2] L. Eaton, College looks back to discovery of hormones, *BMJ* 330 (2005) 1466.3. https://doi.org/10.1136/bmj.330.7506.1466-b.

[3] O.B.S.J.H. Henriksen, Secretin, its discovery, and the introduction of the hormone concept, *Scand. J. Clin. Lab. Invest.* 60 (2000) 463–472. https://doi.org/10.1080/003655100448446.
[4] T. Ası, Hormonlar, in: *Tablolarla Biyokim.* Cilt II, Ankara, 1999: pp. 71–111.
[5] D.S. Reich, C.F. Lucchinetti, P.A. Calabresi, Multiple sclerosis, *N. Engl. J. Med.* 378 (2018) 169–180. https://doi.org/10.1056/NEJMra1401483.
[6] A. Westphal, R. Mrowka, New insights into the astonishing diversity of hormone functions, *Acta Physiol.* 224 (2018). https://doi.org/10.1111/apha.13188.
[7] S. Sharma, S. Shekhar, S. Gautam, B. Sharma, A. Kumar, P. Jain, Carbon-based nanomaterials as novel nanosensors, *Nanofabrication Smart Nanosensor Appl.* (2020) 323–347. https://doi.org/10.1016/B978-0-12-820702-4.00014-3.
[8] S. Morais, Special issue: Carbon-based nanomaterials for (bio)sensors development, *Nanomaterials* 11 (2021) 2430. https://doi.org/10.3390/nano11092430.
[9] H. Li, X. He, T. Wu, B. Jin, L. Yang, J. Qiu, Synthesis, modification strategies and applications of coal-based carbon materials, *Fuel Process. Technol.* 230 (2022) 107203. https://doi.org/10.1016/j.fuproc.2022.107203.
[10] M. Kang, S. Sun, N. Li, D. Zhang, M. Chen, H. Zhang, Extraction and determination of hormones in cosmetics by homogeneous ionic liquid microextraction high-performance liquid chromatography, *J. Sep. Sci.* 35 (2012) 2032–2039. https://doi.org/10.1002/jssc.201200206.
[11] Y. Zhou, Z. Cai, Determination of hormones in human urine by ultra-high-performance liquid chromatography/triple-quadrupole mass spectrometry, *Rapid Commun. Mass Spectrom.* 34 (2020). https://doi.org/10.1002/rcm.8583.
[12] M. López-García, R. Romero-González, A. Garrido Frenich, Determination of steroid hormones and their metabolite in several types of meat samples by ultra high performance liquid chromatography—Orbitrap high resolution mass spectrometry, *J. Chromatogr. A* 1540 (2018) 21–30. https://doi.org/10.1016/j.chroma.2018.01.052.
[13] F.R. Baptista, S.A. Belhout, S. Giordani, S.J. Quinn, Recent developments in carbon nanomaterial sensors, *Chem. Soc. Rev.* 44 (2015) 4433–4453. https://doi.org/10.1039/C4CS00379A.
[14] G. Speranza, Carbon nanomaterials: Synthesis, functionalization and sensing applications, *Nanomaterials* 11 (2021) 967. https://doi.org/10.3390/nano11040967.
[15] A. Jesila, J. Amalraj, N. Murthy, S.-F. Wang, Pt nanoparticle-decorated Se rods for electrochemical detection of 17β-estradiol and methanol oxidation, *ACS Appl. Nano Mater* 5 (2022) 1944–1957. https://doi.org/10.1021/acsanm.1c03590.
[16] G. Bhardwaj, R. Kaur, S. Saini, N. Kaur, N. Singh, Electrochemical and spectroscopic detection of insulin using AgNPs-decorated organic derivative, *Colloids Surfaces A Physicochem. Eng. Asp.* 679 (2023) 132511. https://doi.org/10.1016/J.COLSURFA.2023.132511.
[17] K. Ramya, K. Amreen, S. Goel, Nanofiber enhanced carbon cloth based flexible Miniaturized electrodes for real-time and specific insulin detection in blood serum, *Microchem. J.* 204 (2024) 111029. https://doi.org/10.1016/J.MICROC.2024.111029.
[18] J. Sanchez-Almirola, A. Gage, R. Lopez, D. Yapell, M. Mujawar, V. Kamat, A. Kaushik, Label and bio-active free electrochemical detection of testosterone hormone using MIP-based sensing platform, *Mater. Sci. Eng. B* 296 (2023) 116670. https://doi.org/10.1016/J.MSEB.2023.116670.
[19] S. Jayaraman, T. Rajarathinam, H.-G. Jang, D. Thirumalai, J. Lee, H.-J. Paik, S.-C. Chang, Ruthenium-anchored carbon sphere-customized sensor for the selective amperometric detection of melatonin, *Biosensors* 13 (2023) 936. https://doi.org/10.3390/bios13100936.
[20] P. Supchocksoonthorn, M.C. Alvior Sinoy, M.D.G. de Luna, P. Paoprasert, Facile fabrication of 17β-estradiol electrochemical sensor using polyaniline/carbon dot-coated glassy carbon electrode with synergistically enhanced electrochemical stability, *Talanta* 235 (2021) 122782. https://doi.org/10.1016/J.TALANTA.2021.122782.
[21] X. Yang, P. Zhao, Z. Xie, M. Ni, C. Wang, P. Yang, Y. Xie, J. Fei, Selective determination of epinephrine using electrochemical sensor based on ordered mesoporous carbon / nickel oxide nanocomposite, *Talanta* 233 (2021) 122545. https://doi.org/10.1016/j.talanta.2021.122545.
[22] S.R. Priyanka, K.P. Latha, MnCr2O4 nanocomposite modified carbon paste electrode based electrochemical sensor for determination of Norepinephrine: A cyclic voltammetry study, *Chem. Data Collect.* 35 (2021) 100769. https://doi.org/10.1016/j.cdc.2021.100769.

[23] S. Jayaraman, T. Rajarathinam, S.-C. Chang, Disposable sensor with copper-loaded carbon nanospheres for the simultaneous determination of dopamine and melatonin, *Chemosensors* 11 (2023) 254. https://doi.org/10.3390/chemosensors11040254.
[24] N. Olgaç, E. Karakuş, Y. Şahin, L. Liv, Electrochemical biosensing of cortisol in a hormone tablet and artificial bodily fluids, *Diam. Relat. Mater.* 132 (2023) 109622. https://doi.org/10.1016/J.DIAMOND.2022.109622.
[25] Z. Hassan, S.A. Alsalhi, N. Drissi, S.A.M. Abdelmohsen, J. Sattar, P. Tehreem, H. Noor, H. Alqurashi, A.M.A. Henaish, H.M.T. Farid, Graphene nanoplatelets-polyaniline composite for the detection of cortisol, *J. Phys. Chem. Solids* 191 (2024) 112031. https://doi.org/10.1016/j.jpcs.2024.112031.
[26] S.K. Tuteja, C. Ormsby, S. Neethirajan, Noninvasive label-free detection of cortisol and lactate using graphene embedded screen-printed electrode, *Nano-Micro Lett.* 10 (2018) 41. https://doi.org/10.1007/s40820-018-0193-5.
[27] X. Cui, Z. Luo, M. Guo, J. Xu, L. Wang, G. Chen, G. Wang, C. Chang, A. Zeng, J. Zhang, Q. Fu, A new multi-identification system based on a poly(l-cysteine) sensor for simultaneous detection of multiple steroid hormones in serum, *Chem. Eng. J.* 455 (2023) 140812. https://doi.org/10.1016/J.CEJ.2022.140812.
[28] S. Tang, A. Liang, M. Liu, W. Wang, F. Zhang, A. Luo, A glassy carbon electrode modified with a composite consisting of electrodeposited chitosan and carboxylated multi-walled carbon nanotubes for simultaneous voltammetric determination of dopamine, serotonin and melatonin, *Carbon Lett.* 33 (2023) 2129–2139. https://doi.org/10.1007/s42823-023-00556-6.
[29] A. Ejaz, Y. Joo, S. Jeon, Fabrication of 1,4-bis(aminomethyl)benzene and cobalt hydroxide @ graphene oxide for selective detection of dopamine in the presence of ascorbic acid and serotonin, *Sensors Actuators B Chem.* 240 (2017) 297–307. https://doi.org/10.1016/j.snb.2016.08.171.
[30] H.M. Alsoghier, M. Abd-Elsabour, A.G. Alhamzani, M.M. Abou-Krisha, H.F. Assaf, Real samples sensitive dopamine sensor based on poly 1,3-benzothiazol-2-yl((4-carboxlicphenyl)hydrazono) acetonitrile on a glassy carbon electrode, *Sci. Rep.* 14 (2024) 16601. https://doi.org/10.1038/s41598-024-65192-0.
[31] S. Wang, P. Guo, G. Ma, J. Wei, Z. Wang, L. Cui, L. Sun, A. Wang, Three-dimensional hierarchical mesoporous carbon for regenerative electrochemical dopamine sensor, *Electrochim. Acta* 360 (2020) 137016. https://doi.org/10.1016/j.electacta.2020.137016.
[32] S. Liu, S. Gao, T. Fei, T. Zhang, Highly sensitive and selective dopamine detection utilizing nitrogen-doped mesoporous carbon prepared by a molten glucose-assisted hard-template approach, *Chempluschem* 84 (2019) 845–852. https://doi.org/10.1002/cplu.201900291.
[33] J.-W. Oh, J. Heo, T.H. Kim, An electrochemically modulated single-walled carbon nanotube network for the development of a transparent flexible sensor for dopamine, *Sensors Actuators B Chem.* 267 (2018) 438–447. https://doi.org/10.1016/j.snb.2018.04.048.
[34] C. Wang, J. Du, H. Wang, C. Zou, F. Jiang, P. Yang, Y. Du, A facile electrochemical sensor based on reduced graphene oxide and Au nanoplates modified glassy carbon electrode for simultaneous detection of ascorbic acid, dopamine and uric acid, *Sensors Actuators B Chem.* 204 (2014) 302–309. https://doi.org/10.1016/j.snb.2014.07.077.
[35] S. Wang, W. Zhang, X. Zhong, Y. Chai, R. Yuan, Simultaneous determination of dopamine, ascorbic acid and uric acid using a multi-walled carbon nanotube and reduced graphene oxide hybrid functionalized by PAMAM and Au nanoparticles, *Anal. Methods* 7 (2015) 1471–1477. https://doi.org/10.1039/C4AY02086C.
[36] X. Feng, Y. Zhang, J. Zhou, Y. Li, S. Chen, L. Zhang, Y. Ma, L. Wang, X. Yan, Three-dimensional nitrogen-doped graphene as an ultrasensitive electrochemical sensor for the detection of dopamine, *Nanoscale* 7 (2015) 2427–2432. https://doi.org/10.1039/C4NR06623E.
[37] P. Wiench, Z. González, R. Menéndez, B. Grzyb, G. Gryglewicz, Beneficial impact of oxygen on the electrochemical performance of dopamine sensors based on N-doped reduced graphene oxides, *Sensors Actuators B Chem.* 257 (2018) 143–153. https://doi.org/10.1016/j.snb.2017.10.106.
[38] F.H. Narouei, H.K. Tammandani, Y. Ghalandarzehi, N. Sabbaghi, M. Noroozifar, An electrochemical sensor based on conductive polymers/graphite paste electrode for simultaneous determination of dopamine, uric acid and tryptophan in biological samples, *Int. J. Electrochem. Sci.* 12 (2017) 7739–7753. https://doi.org/10.20964/2017.08.50.

[39] Z. Zhao, L. Zheng, Cobalt-generated reactive oxygen species promoted FeS2/CoS2 for electrochemical detection of acetaminophen with the widest linear range and simultaneous detection of dopamine, *J. Electroanal. Chem.* 956 (2024) 118101. https://doi.org/10.1016/J.JELECHEM.2024.118101.

[40] A.S. Manjunatha, P.R. Prakruthi, J. Puneetha, M. Shashank, G. Nagaraju, Cocos nucifera mediated green synthesis and characterization of BiOCl-Fe2O3 nanocomposite for photocatalytic dye degradation and electrochemical sensing of dopamine, *Sustain. Chem. Environ.* 7 (2024) 100138. https://doi.org/10.1016/J.SCENV.2024.100138.

[41] B. Suriya Devi, R. Karthikeyan, M. Anitha, S. Prakash, Electrochemical incorporation of PVA-ZnO composite on screen printed carbon electrode as dopamine sensor, *Surfaces and Interfaces* 52 (2024) 104882. https://doi.org/10.1016/J.SURFIN.2024.104882.

[42] F. Zhang, S. Li, H. Zhang, H. Li, A silver hexacyanoferrate-graphene modified glassy carbon electrode in electrochemical sensing of uric acid and dopamine, *Int. J. Electrochem. Sci.* 12 (2017) 11181–11194. https://doi.org/10.20964/2017.12.13.

[43] P.K. Kannan, S.A. Moshkalev, C.S. Rout, Highly sensitive and selective electrochemical dopamine sensing properties of multilayer graphene nanobelts, *Nanotechnology* 27 (2016) 075504. https://doi.org/10.1088/0957-4484/27/7/075504.

[44] T.P. Tsele, A.S. Adekunle, O.E. Fayemi, E.E. Ebenso, Electrochemical detection of Epinephrine using Polyaniline nanocomposite films doped with TiO 2 and RuO 2 nanoparticles on multi-walled carbon nanotube, *Electrochim. Acta* 243 (2017) 331–348. https://doi.org/10.1016/j.electacta.2017.05.031.

[45] N.G. Mphuthi, A.S. Adekunle, E.E. Ebenso, Electrocatalytic oxidation of epinephrine and norepinephrine at metal oxide doped phthalocyanine/MWCNT composite sensor, *Sci. Rep.* 6 (2016) 26938. https://doi.org/10.1038/srep26938.

[46] G. Karim-Nezhad, Z. Khorablou, Selective analysis of epinephrine in the presence of uric acid by using an amplified electrochemical sensor employing a gold nanoparticle decorated cysteic acid film, *Anal. Methods* 9 (2017) 6394–6402. https://doi.org/10.1039/C7AY02312J.

[47] M. Jahanbakhshi, Mesoporous carbon foam, synthesized via modified Pechini method, in a new dispersant of Salep as a novel substrate for electroanalytical determination of epinephrine in the presence of uric acid, Mater. *Sci. Eng.* C 70 (2017) 544–551. https://doi.org/10.1016/j.msec.2016.09.013.

[48] A. Yang, Y. Xue, H. Zhao, X. Li, Z. Yuan, One-pot synthesis of ternary hybrid nanomaterial composed of a porphyrin-functionalized graphene, tin oxide, and gold nanoparticles, and its application to the simultaneous determination of epinephrine and uric acid, *Microchim. Acta* 182 (2015) 341–349. https://doi.org/10.1007/s00604-014-1328-0.

[49] M. Taei, G. Ramazani, Simultaneous determination of norepinephrine, acetaminophen and tyrosine by differential pulse voltammetry using Au-nanoparticles/poly(2-amino-2-hydroxymethyl-propane-1,3-diol) film modified glassy carbon electrode, *Colloids Surfaces B Biointerfaces* 123 (2014) 23–32. https://doi.org/10.1016/j.colsurfb.2014.09.005.

[50] H. Devnani, S.P. Satsangee, R. Jain, A novel graphene-chitosan-Bi2O3 nanocomposite modified sensor for sensitive and selective electrochemical determination of a monoamine neurotransmitter epinephrine, *Ionics (Kiel).* 22 (2016) 943–956. https://doi.org/10.1007/s11581-015-1620-y.

[51] M. Buleandră, D.E. Popa, I.G. David, A.A. Ciucu, A simple and efficient cyclic square wave voltammetric method for simultaneous determination of epinephrine and norepinephrine using an activated pencil graphite electrode, *Microchem. J.* 160 (2021) 105621. https://doi.org/10.1016/J.MICROC.2020.105621.

[52] K. Chetankumar, B.E. Kumara Swamy, S.C. Sharma, Safranin amplified carbon paste electrode sensor for analysis of paracetamol and epinephrine in presence of folic acid and ascorbic acid, *Microchem. J.* 160 (2021) 105729. https://doi.org/10.1016/J.MICROC.2020.105729.

[53] X. Xin, S. Sun, M. Wang, R. Jia, Sensitive detection of estriol with an electrochemical sensor based on core-shell N-MWCNT/GONR-imprinted electrode, *Ionics (Kiel).* 26 (2020) 2633–2641. https://doi.org/10.1007/s11581-019-03377-8.

[54] F.H. Cincotto, T.C. Canevari, S.A.S. Machado, A. Sánchez, M.A.R. Barrio, R. Villalonga, J.M. Pingarrón, Reduced graphene oxide-Sb2O5 hybrid nanomaterial for the design of a laccase-based amperometric biosensor for estriol, *Electrochim. Acta* 174 (2015) 332–339. https://doi.org/10.1016/j.electacta.2015.06.013.

[55] L.R.G. Silva, J.G.A. Rodrigues, J.P. Franco, L.P. Santos, E. D'Elia, W. Romão, R. de Q. Ferreira, Development of a portable electroanalytical method using nickel modified screen-printed carbon electrode for ethinylestradiol determination in organic fertilizers, *Ecotoxicol. Environ. Saf.* 208 (2021) 111430. https://doi.org/10.1016/j.ecoenv.2020.111430.

[56] Y. Fan, Y. Guo, S. Shi, J. Ma, An electrochemical immunosensor based on reduced graphene oxide/multiwalled carbon nanotubes/thionine/gold nanoparticle nanocomposites for the sensitive testing of follicle-stimulating hormone, *Anal. Methods* 13 (2021) 3821–3828. https://doi.org/10.1039/D1A Y01032H.

[57] J. Luo, Z. Kong, Y. Wang, J. Xie, J. Liu, H. Jin, X. Cai, Label-free paper-based immunosensor with graphene nanocomposites for electrochemical detection of follicle-stimulating hormone, *Proc. Annu. Int. Conf. IEEE Eng. Med. Biol. Soc. EMBS 2018-July* (2018) 2901–2904. https://doi.org/10.1109/EMBC.2018.8512923.

[58] S. Pareek, U. Jain, S. Balayan, N. Chauhan, Ultra-sensitive nano- molecular imprinting polymer-based electrochemical sensor for Follicle-Stimulating Hormone (FSH) detection, *Biochem. Eng. J.* 180 (2022) 108329. https://doi.org/10.1016/J.BEJ.2021.108329.

[59] M. Roushani, A. Valipour, M. Valipour, Layer-by-layer assembly of gold nanoparticles and cysteamine on gold electrode for immunosensing of human chorionic gonadotropin at picogram levels, *Mater. Sci. Eng. C* 61 (2016) 344–350. https://doi.org/10.1016/j.msec.2015.12.088.

[60] S. Teixeira, G. Burwell, A. Castaing, D. Gonzalez, R.S. Conlan, O.J. Guy, Epitaxial graphene immunosensor for human chorionic gonadotropin, *Sensors Actuators B Chem.* 190 (2014) 723–729. https://doi.org/10.1016/j.snb.2013.09.019.

[61] F. Asadpour, M. Mazloum-Ardakani, F. Hoseynidokht, S.M. Moshtaghioun, In situ monitoring of gating approach on mesoporous silica nanoparticles thin-film generated by the EASA method for electrochemical detection of insulin, Biosens. *Bioelectron.* 180 (2021) 113124. https://doi.org/10.1016/J.BIOS.2021.113124.

[62] S. Khataee, G. Dehghan, Z. Shaghaghi, A. Khataee, M. Amini, A novel bifunctional electrochemical nanosensor for simultaneous detection of glucose and insulin based on NiO/Co3O4@CuAl LDH-MWCNT nanocomposite-modified carbon paste electrode, *Microchem. J.* 201 (2024) 110644. https://doi.org/10.1016/J.MICROC.2024.110644.

[63] N. Amini, M.B. Gholivand, M. Shamsipur, Electrocatalytic determination of traces of insulin using a novel silica nanoparticles-Nafion modified glassy carbon electrode, *J. Electroanal. Chem.* 714–715 (2014) 70–75. https://doi.org/10.1016/J.JELECHEM.2013.12.015.

[64] M. Farrokhnia, G. Amoabediny, M. Ebrahimi, M. Ganjali, M. Arjmand, Ultrasensitive early detection of insulin antibody employing novel electrochemical nano-biosensor based on controllable electro-fabrication process, *Talanta* 238 (2022) 122947. https://doi.org/10.1016/J.TALANTA.2021.122947.

[65] N.I. Wardani, T. Kangkamano, R. Wannapob, P. Kanatharana, P. Thavarungkul, W. Limbut, Electrochemical sensor based on molecularly imprinted polymer cryogel and multiwalled carbon nanotubes for direct insulin detection, *Talanta* 254 (2023) 124137. https://doi.org/10.1016/J.TALA NTA.2022.124137.

[66] S. Kouchakinejad, S. Babaee, F. Roshani, R. Kouchakinejad, N. shirmohammadi, S. Kaki, The performance of the new modified pencil graphite electrode in quantifying of insulin, *Chem. Phys. Lett.* 759 (2020) 137987. https://doi.org/10.1016/J.CPLETT.2020.137987.

[67] E. Habibi, E. Omidinia, H. Heidari, M. Fazli, Flow injection amperometric detection of insulin at cobalt hydroxide nanoparticles modified carbon ceramic electrode, Anal. *Biochem.* 495 (2016) 37–41. https://doi.org/10.1016/j.ab.2015.11.014.

[68] F. Abazar, E. Sharifi, A. Noorbakhsh, Antifouling properties of carbon quantum dots-based electrochemical sensor as a promising platform for highly sensitive detection of insulin, *Microchem. J.* 180 (2022) 107560. https://doi.org/10.1016/J.MICROC.2022.107560.

[69] F. Abazar, A. Noorbakhsh, Chitosan-carbon quantum dots as a new platform for highly sensitive insulin impedimetric aptasensor, *Sensors Actuators B Chem.* 304 (2020) 127281. https://doi.org/10.1016/J.SNB.2019.127281.

[70] A.A. Ensafi, E. Khoddami, B. Rezaei, Aptamer@Au-o-phenylenediamine modified pencil graphite electrode: A new selective electrochemical impedance biosensor for the determination of insulin, *Colloids Surfaces B Biointerfaces* 159 (2017) 47–53. https://doi.org/10.1016/j.colsu rfb.2017.07.076.

[71] N. Sebastian, W.-C. Yu, D. Balram, Q. Chen, A. Shiue, M.T. Noman, N. Amor, Porous hematite embedded C and Fe codoped graphitic carbon nitride for electrochemical detection of pineal gland hormone melatonin, *Mater. Today Chem.* (2023). https://doi.org/10.1016/j.mtchem.2023.101406.
[72] P. Sunon, K. Ngamchuea, Methylene blue molecularly imprinted polymer for melatonin determination in urine and saliva samples, *Microchim. Acta* 190 (2023) 348. https://doi.org/10.1007/s00604-023-05930-9.
[73] D. Duan, Y. Ding, L. Li, G. Ma, Rapid quantitative detection of melatonin by electrochemical sensor based on carbon nanofibers embedded with FeCo alloy nanoparticles, *J. Electroanal. Chem.* 873 (2020) 114422. https://doi.org/10.1016/j.jelechem.2020.114422.
[74] N. Tavakkoli, N. Soltani, F. Shahdost-fard, M. Ramezani, H. Salavati, M.R. Jalali, Simultaneous voltammetric sensing of acetaminophen, epinephrine and melatonin using a carbon paste electrode modified with zinc ferrite nanoparticles, *Microchim. Acta* 185 (2018) 479. https://doi.org/10.1007/s00604-018-3009-x.
[75] H. Bagheri, A. Afkhami, P. Hashemi, M. Ghanei, Simultaneous and sensitive determination of melatonin and dopamine with Fe3O4 nanoparticle-decorated reduced graphene oxide modified electrode, *RSC Adv.* 5 (2015) 21659–21669. https://doi.org/10.1039/C4RA16802J.
[76] H. Zeinali, H. Bagheri, Z. Monsef-Khoshhesab, H. Khoshsafar, A. Hajian, Nanomolar simultaneous determination of tryptophan and melatonin by a new ionic liquid carbon paste electrode modified with SnO2-Co3O4@rGO nanocomposite, *Mater. Sci. Eng.* C 71 (2017) 386–394. https://doi.org/10.1016/j.msec.2016.10.020.
[77] R.C. Freitas, L.O. Orzari, L.M.C. Ferreira, T.R.L.C. Paixão, W.K.T. Coltro, F.C. Vicentini, B.C. Janegitz, Electrochemical determination of melatonin using disposable self-adhesive inked paper electrode, *J. Electroanal. Chem.* 897 (2021) 115550. https://doi.org/10.1016/J.JELECHEM.2021.115550.
[78] X. Jin, M. Nodehi, M. Baghayeri, Y. Xu, Z. Hua, Y. Lei, M. Shao, P. Makvandi, Development of an impedimetric sensor for susceptible detection of melatonin at picomolar concentrations in diverse pharmaceutical and human specimens, *Environ. Res.* 238 (2023) 117080. https://doi.org/10.1016/j.envres.2023.117080.
[79] P. Malla, H.P. Liao, C.H. Liu, W.C. Wu, Electrochemical immunoassay for serum parathyroid hormone using screen-printed carbon electrode and magnetic beads, *J. Electroanal. Chem.* 895 (2021) 115463. https://doi.org/10.1016/J.JELECHEM.2021.115463.
[80] A. Das, M.V. Sangaranarayanan, A sensitive electrochemical detection of progesterone using tin-nanorods modified glassy carbon electrodes: Voltammetric and computational studies, *Sensors Actuators B Chem.* 256 (2018) 775–789. https://doi.org/10.1016/j.snb.2017.10.008.
[81] M. Arvand, S. Hemmati, Magnetic nanoparticles embedded with graphene quantum dots and multiwalled carbon nanotubes as a sensing platform for electrochemical detection of progesterone, *Sensors Actuators B Chem.* 238 (2017) 346–356. https://doi.org/10.1016/j.snb.2016.07.066.
[82] M. Arvand, S. Hemmati, Analytical methodology for the electro-catalytic determination of estradiol and progesterone based on graphene quantum dots and poly(sulfosalicylic acid) co-modified electrode, *Talanta* 174 (2017) 243–255. https://doi.org/10.1016/j.talanta.2017.05.083.
[83] A. Gevaerd, S.F. Blaskievicz, A.J.G. Zarbin, E.S. Orth, M.F. Bergamini, L.H. Marcolino-Junior, Nonenzymatic electrochemical sensor based on imidazole-functionalized graphene oxide for progesterone detection, *Biosens. Bioelectron.* 112 (2018) 108–113. https://doi.org/10.1016/j.bios.2018.04.044.
[84] X. Zhao, L. Zheng, Y. Yan, R. Cao, J. Zhang, An electrocatalytic active AuNPs/5-Amino-2-mercaptobenzimidazole/rGO/SPCE composite electrode for ultrasensitive detection of progesterone, *J. Electroanal. Chem.* 882 (2021) 115023. https://doi.org/10.1016/j.jelechem.2021.115023.
[85] M. Shamsipur, A. Pashabadi, A. (Arman) Taherpour, K. Bahrami, H. Sharghi, Manganese mediated oxidation of progesterone in alkaline medium: Mechanism study and quantitative determination, *Electrochim. Acta* 225 (2017) 292–302. https://doi.org/10.1016/j.electacta.2016.12.174.
[86] M. Govindasamy, B. Subramanian, S.-F. Wang, S. Chinnapaiyan, R. Jothi Ramalingam, H.A. Al-lohedan, Ultrasound-assisted synthesis of tungsten trioxide entrapped with graphene nanosheets for developing nanomolar electrochemical (hormone) sensor and enhanced sensitivity of the catalytic performance, *Ultrason. Sonochem.* 56 (2019) 134–142. https://doi.org/10.1016/j.ultsonch.2019.03.021.

[87] S. Huang, R. Feng, A new electrochemical biosensor based on graphene oxide for rapid detection of synthetic testosterone as performance-enhancing drugs in athletes, *Alexandria Eng. J.* 98 (2024) 281–289. https://doi.org/10.1016/j.aej.2024.04.054.

[88] R. Heidarimoghadam, O. Akhavan, E. Ghaderi, E. Hashemi, S.S. Mortazavi, A. Farmany, Graphene oxide for rapid determination of testosterone in the presence of cetyltrimethylammonium bromide in urine and blood plasma of athletes, *Mater. Sci. Eng.* C 61 (2016) 246–250. https://doi.org/10.1016/j.msec.2015.12.005.

[89] B. Bozdoğan, Electrochemical testosterone biosensor based on pencil graphite electrode electrodeposited with copper oxide nanoparticles, *Meas. Sci. Technol.* 34 (2023) 105106. https://doi.org/10.1088/1361-6501/ace126.

[90] A. Levent, A. Altun, Y. Yardım, Z. Şentürk, Sensitive voltammetric determination of testosterone in pharmaceuticals and human urine using a glassy carbon electrode in the presence of cationic surfactant, *Electrochim. Acta* 128 (2014) 54–60. https://doi.org/10.1016/j.electacta.2013.10.024.

[91] E. Kellens, H. Bové, T. Vandenryt, J. Lambrichts, J. Dekens, S. Drijkoningen, J. D'Haen, W. De Ceuninck, R. Thoelen, T. Junkers, K. Haenen, A. Ethirajan, Micro-patterned molecularly imprinted polymer structures on functionalized diamond-coated substrates for testosterone detection, *Biosens. Bioelectron.* 118 (2018) 58–65. https://doi.org/10.1016/j.bios.2018.07.032.

[92] G. Li, M. Zhu, L. Ma, J. Yan, X. Lu, Y. Shen, Y. Wan, Generation of small single domain nanobody binders for sensitive detection of testosterone by electrochemical impedance spectroscopy, *ACS Appl. Mater. Interfaces* 8 (2016) 13830–13839. https://doi.org/10.1021/acsami.6b04658.

[93] W. Liu, Y. Ma, G. Sun, S. Wang, J. Deng, H. Wei, Molecularly imprinted polymers on graphene oxide surface for EIS sensing of testosterone, *Biosens. Bioelectron.* 92 (2017) 305–312. https://doi.org/10.1016/J.BIOS.2016.11.007.

[94] Z. Wang, P. Wang, X. Tu, Y. Wu, G. Zhan, C. Li, A novel electrochemical sensor for estradiol based on nanoporous polymeric film bearing poly{1-butyl-3-[3-(N-pyrrole)propyl]imidazole dodecyl sulfonate} moiety, *Sensors Actuators B Chem.* 193 (2014) 190–197. https://doi.org/10.1016/J.SNB.2013.11.053.

[95] P.A. Raymundo-Pereira, N.O. Gomes, S.A.S. Machado, O.N. Oliveira, Simultaneous, ultrasensitive detection of hydroquinone, paracetamol and estradiol for quality control of tap water with a simple electrochemical method, *J. Electroanal. Chem.* 848 (2019) 113319. https://doi.org/10.1016/J.JELECHEM.2019.113319.

[96] F.C. Moraes, B. Rossi, M.C. Donatoni, K.T. de Oliveira, E.C. Pereira, Sensitive determination of 17β-estradiol in river water using a graphene based electrochemical sensor, *Anal. Chim. Acta* 881 (2015) 37–43. https://doi.org/10.1016/J.ACA.2015.04.043.

[97] X. Dong, L. He, Y. Liu, Y. Piao, Preparation of highly conductive biochar nanoparticles for rapid and sensitive detection of 17β-estradiol in water, *Electrochim. Acta* 292 (2018) 55–62. https://doi.org/10.1016/J.ELECTACTA.2018.09.129.

[98] P. Xie, Z. Liu, S. Huang, J. Chen, Y. Yan, N. Li, M. Zhang, M. Jin, L. Shui, A sensitive electrochemical sensor based on wrinkled mesoporous carbon nanomaterials for rapid and reliable assay of 17β-estradiol, *Electrochim. Acta* 408 (2022) 139960. https://doi.org/10.1016/J.ELECTACTA.2022.139960.

[99] T. Ming, Y. Wang, J. Luo, J. Liu, S. Sun, Y. Xing, G. Xiao, H. Jin, X. Cai, Folding paper-based aptasensor platform coated with novel nanoassemblies for instant and highly sensitive detection of 17β-estradiol, *ACS Sensors* 4 (2019) 3186–3194. https://doi.org/10.1021/ACSSENSORS.9B01633/SUPPL_FILE/SE9B01633_SI_001.PDF.

[100] A.A. Lahcen, A.A. Baleg, P. Baker, E. Iwuoha, A. Amine, Synthesis and electrochemical characterization of nanostructured magnetic molecularly imprinted polymers for 17-β-estradiol determination, *Sensors Actuators B Chem.* 241 (2017) 698–705. https://doi.org/10.1016/J.SNB.2016.10.132.

[101] G.C. Mauruto de Oliveira, E.P. de Palma, M.H. Kunita, R. Antigo Medeiros, R. de Matos, K.R. Francisco, B.C. Janegitz, Tapioca biofilm containing nitrogen-doped titanium dioxide nanoparticles for electrochemical detection of 17-β estradiol, *Electroanalysis* 29 (2017) 2638–2645. https://doi.org/10.1002/ELAN.201700392.

7 Sensors in Medical Diagnostics

Swati Rani, Kalawati Saini, and Dipak Maity

7.1 INTRODUCTION

Sensors are essential devices that convert physical parameters into signals that can be measured and analyzed.[1] Medical diagnostic sensors play a crucial role in contemporary healthcare, providing accurate and timely information for the diagnosis and monitoring of various health conditions.[2] These sensors leverage technological advancements to detect physiological and biochemical changes within the body, enabling early diagnosis and enhancing patient outcomes. The chemistry behind these sensors involves complex interactions between biological molecules, chemical reactions, and physical principles, leading to various applications in medical diagnostics.[3] Common types of sensors used in medical diagnostics include electrochemical sensors, biosensors, imaging sensors, optical sensors, and physical sensors. Biosensors are analytical devices that integrate a biological component, such as enzymes, antibodies, or nucleic acids, with a physicochemical detector.[4–6] The biological component engages with the target analyte, creating a signal that the detector translates into measurable data.[7] Biosensors are commonly used for detecting biomarkers, glucose levels, pathogens, and many more.[8–11] Glucose sensors are a prime example of biosensors used in medical diagnostics. These sensors are crucial for diabetic patients to track their blood glucose levels.[12]

A standard glucose biosensor employs the enzyme glucose oxidase to catalyze the oxidation of glucose into gluconic acid and hydrogen peroxide.[13] The resulting hydrogen peroxide is then detected electrochemically, generating a current that is proportional to the glucose concentration.[14] Continuous glucose monitors (CGMs), such as the FreeStyle Libre by Abbott, track glucose levels in the interstitial fluid, delivering real-time data to diabetic patients.[15,16] This technology allows for continuous monitoring without the need for frequent finger-prick tests, significantly improving patient compliance and glucose management. Electrochemical biosensors are a category of biosensors that utilize redox reactions of target molecules at the sensor's surface.[17–19] These sensors are employed to detect a range of biomolecules, such as nucleic acids, enzymes, and antibodies.[20–22] The epoc® blood analysis system uses electrochemical principles to measure blood gases, electrolytes, and metabolites.[23,24] The system's electrodes are coated with specific materials that facilitate the redox reactions of the analytes, generating a measurable current.[25] This allows for rapid and accurate analysis of critical blood parameters, crucial in emergency and critical care settings. Imaging sensors play an important role in capturing detailed images of the body's interior, aiding in diagnosing various conditions.[26] These sensors utilize different physical principles, including X-rays, magnetic fields, and sound waves. X-ray sensors are widely used to detect bone fractures, infections, and tumors.[27] Digital X-ray sensors detect X-rays that pass through the body and interact with a digital detector. The detector converts the X-rays into electrical signals, creating high-resolution images.

DOI: 10.1201/9781003500322-7

The DRX Plus by Carestream Health is an example of a digital radiography system that offers fast, high-quality imaging for diagnostic purposes.[28]

Digital X-rays reduce radiation exposure and provide immediate results, enhancing patient safety and diagnostic efficiency. Magnetic resonance imaging (MRI) sensors use MRI technology to produce detailed images of organs and tissues. MRI sensors employ powerful magnetic fields and radiofrequency pulses to produce images of soft tissues, making them invaluable for diagnosing tumors, neurological conditions, and joint disorders.[29] The process starts with hydrogen nuclei in the body's water molecules aligning with the magnetic field. Radiofrequency pulses then disturb this alignment, and as the nuclei revert to their original state, they emit signals that are detected and converted into images. The Siemens MAGNETOM Terra is an advanced MRI system that offers high-resolution imaging for brain and musculoskeletal diagnosis, providing detailed insights into complex anatomical structures.[30,31] Wearable sensors continuously monitor physiological parameters, providing valuable data for managing chronic diseases and improving overall health.[32] These sensors are integrated into wearable devices, offering convenience and real-time monitoring.[33] Heart rate monitors utilize photoplethysmography (PPG) to measure heart rate.[34] PPG sensors send light into the skin and measure how much of it is absorbed or reflected by the blood flow. The Apple Watch, for instance, uses PPG sensors to monitor heart rate and can detect irregular rhythms, alerting users to potential cardiovascular issues.[35] This ongoing monitoring makes possible the early detection of arrhythmias and other heart conditions, enabling timely medical intervention. Activity trackers use accelerometers and gyroscopes to monitor sleep patterns, physical activity, and other health metrics.[36] Accelerometers measure body movement and orientation using microfabricated structures that change capacitance or resistance with motion. Accelerometers detect changes in motion, while gyroscopes measure orientation.[37] The Fitbit Charge series tracks steps, sleep, and heart rate, helping users manage their health and wellness.[38] The Shimmer3 IMU is an inertial measurement unit that tracks motion and orientation, used in clinical research and patient monitoring.[39] These devices provide insights into daily activity levels and sleep quality, encouraging healthier lifestyle choices and improving overall well-being.

Optical sensors utilize light to detect changes in biological samples, enabling non-invasive diagnostic methods. These sensors are used in various medical applications, from monitoring vital signs to detecting specific biomolecules.[40] Pulse oximeters assess blood oxygen saturation by measuring light absorption through a finger or earlobe. Pulse oximeters emit red and infrared light, and the ratio of absorbed light at these wavelengths indicates oxygen saturation. The principle of this measurement relies on the differing absorption spectra of deoxygenated and oxygenated hemoglobin.[41] The Masimo Rad-97 is a pulse oximeter that provides continuous monitoring of oxygen saturation and pulse rate, essential for patients with respiratory conditions. This non-invasive technique permits for rapid assessment of oxygen levels, critical in clinical settings. Fluorescence-based sensors detect specific biomolecules using fluorescent markers.[42] These sensors are commonly used in laboratory diagnostics and point-of-care testing. When the target molecule binds to the sensor, it emits light at a specific wavelength, which the sensor detects.[43] This technology allows for rapid and accurate detection of biomarkers, facilitating timely medical decisions in critical situations.[44] These sensors are used in various applications, from monitoring pressure to tracking body movement. Pressure sensors monitor blood pressure, intracranial pressure, and intraocular pressure.[45] Micro-electromechanical system (MEMS) pressure sensors use piezoelectric materials to produce an electrical charge when subjected to mechanical stress.[46–48] For example, MEMS pressure sensor by STMicroelectronics measures intraocular pressure, aiding in diagnosing and managing glaucoma.[49] These sensors provide continuous and accurate pressure measurements, essential for monitoring and treating pressure-related conditions. From biosensors detecting glucose levels to advanced imaging sensors capturing detailed body images, these technologies significantly ameliorate patient outcomes. As technology continues to advance, the development of more sophisticated and accurate sensors will further enhance healthcare delivery, leading to better health outcomes and quality of life for patients. Hence, the importance of sensors in medical diagnostics cannot be overstated, as they play a crucial role in enhancing the accuracy, efficiency, and accessibility of healthcare services. Sensors can be divided into various types based on their applications, the parameters they measure, and the technology they use. Different types of sensors used in medical diagnostics are discussed in detail here.

7.2 TYPES OF SENSORS IN MEDICAL DIAGNOSTICS

Medical diagnostics rely heavily on various types of sensors to provide accurate and timely information about a patient's health. In medical diagnostics, sensors are critical tools characterized by several essential properties to ensure their effectiveness and reliability in clinical applications. Accuracy is paramount, as sensors must deliver precise measurements within specified tolerances to support accurate diagnosis and treatment decisions.[50] Precision ensures consistent and reproducible results, reducing variability in clinical outcomes.[51] Also, high sensitivity enables sensors to detect subtle changes in biological signals or analytes, facilitating early detection of diseases and monitoring of patient conditions. Moreover, specificity ensures that sensors accurately measure target analytes without interference from other substances, enhancing diagnostic accuracy.[52] In addition, rapid response times are crucial for timely monitoring and intervention in patient care. Durability and reliability are vital for sensors to maintain performance integrity over extended use. Biocompatibility is essential for sensors used directly on or inside the body, ensuring compatibility with biological systems and minimizing risks to patient health.[53] Sterilizability is necessary to prevent cross-contamination and maintain hygiene standards in clinical settings. Ease of use and integration into healthcare workflows, along with cost-effectiveness, further enhance the practical utility of medical sensors, supporting efficient diagnosis and management of health conditions. When developing sensors, we must consider all the above-mentioned factors and these can be categorized based on several criteria. Here are some common types of sensors in medical diagnostics and their categories (Figure 7.1).

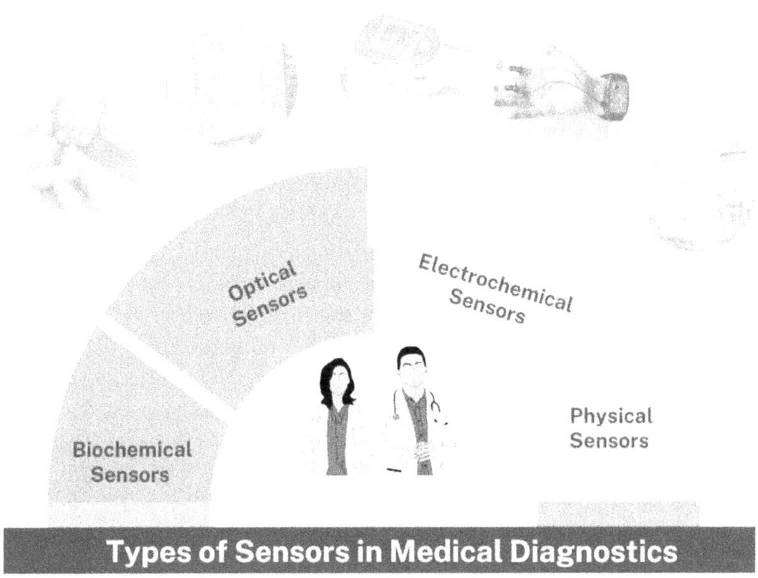

FIGURE 7.1 Figure representing types of sensors used in medical diagnostics.

7.2.1 Biochemical Sensors

Biochemical sensors play a significant role in contemporary medical diagnostics by enabling the precise and real-time detection of biological molecules and biochemical reactions.[54] These sensors are integral to monitoring biomarkers that reflect physiological states and disease conditions.[55] Biosensor technologies have advanced significantly for analyzing various physical components. A biosensor consists of several components: the Analyte (the substance being analyzed), Bioreceptor (the biological recognition element), Transducer (that converts biological signal into electrical signal), Electronics (processes the electrical signal), and Reader Display (displays results).[56] Reader devices are custom-designed to accommodate different biosensor principles, often representing a substantial portion of development costs.

The proliferation of infectious diseases such as Hendra, avian influenza, Nipah, and SARS represents a significant global health threat, necessitating substantial efforts to mitigate their spread.[57] The complexity and challenges posed by these infectious diseases highlight the urgent need for advanced diagnostic tools to prevent or reduce the likelihood of virus outbreaks.[58] Biosensors have emerged as crucial tools in the fight against COVID-19 and various other diseases, offering rapid, sensitive, and specific detection methods that are essential for early diagnosis and effective management.[59] Different types of biosensors tailored for COVID-19 and other diseases utilize diverse detection principles and technologies, each designed to target specific biomarkers or pathogens. For COVID-19, biosensors often focus on detecting viral antigens, antibodies, or nucleic acids (RNA or DNA) associated with SARS-CoV-2.[60] Antigen biosensors typically employ immunological methods, where specific antibodies bind to viral proteins, triggering a measurable signal such as fluorescence or electrochemical change.[61,62] These biosensors are valued for their speed and sensitivity, providing results within minutes and enabling rapid screening in various settings, from healthcare facilities to airports and workplaces. Nucleic acid biosensors, especially those using polymerase chain reaction (PCR) or loop-mediated isothermal amplification (LAMP) techniques, detect viral RNA sequences with high specificity.[63]

These biosensors are pivotal in confirming COVID-19 infections and monitoring viral load, critical for patient management and infection control.[64] In cancer diagnostics, biosensors can detect particular biomarkers such as circulating tumor cells or cancer-linked proteins in blood samples.[65,66] These biosensors leverage various biorecognition elements, including aptamers, antibodies, or molecularly imprinted polymers, to achieve high sensitivity and specificity. In cardiovascular diseases, biosensors measure biomarkers like cardiac troponins or B-type natriuretic peptide (BNP), aiding in the early diagnosis of myocardial infarction or heart failure.[67,68]

Recent research has demonstrated the versatile application of biosensors across diverse medical fields, demonstrating their effectiveness in disease diagnosis and monitoring. For example, researchers developed a biosensor that detects SARS-CoV-2 spike protein using graphene-based field-effect transistors.[69,70] In cancer diagnostics, a study published by Nguyen *et al.* highlighted the use of plasmonic biosensors for detecting circulating tumor DNA (ctDNA) in blood samples.[71] In a report researchers have developed a biosensor capable of detecting cardiac troponin levels in blood samples, providing early diagnosis of myocardial infarction.[72] The research conducted by Wen *et al.* highlights the biosensor's ability to detect minute concentrations of ctDNA mutations associated with various cancers, providing early and non-invasive detection capabilities that can guide personalized treatment strategies.[73] Furthermore, biosensors have shown promise in neurological disorders.[74] Research by Kujawska *et al.* focused on developing graphene-based biosensors to detect dopamine for efficient diagnostics of Parkinson's disease. This technology offers insights into neurotransmitter dynamics and their role in neurological diseases like Parkinson's disease, potentially paving the way for novel therapeutic approaches.[75] In cardiovascular medicine, a recent study published in *Chemosensors* presented a biosensor that measures multiple biomarkers simultaneously in sweat.[76] In a report by Erfkamp *et al.*, a hydrogel-based biosensor has been developed for the detection of urea (Figure 7.2). Urease catalyzes the hydrolysis of urea, leading to a rise in pH

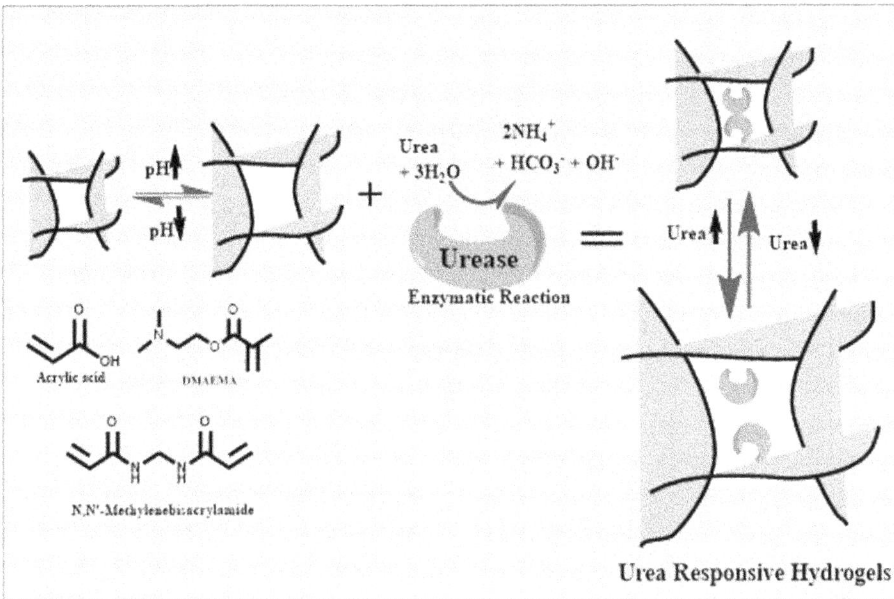

FIGURE 7.2 The schematic illustration of the hydrogel-based urea biosensor.

that is detected by a pH-sensitive poly(acrylic acid-co-dimethylaminoethyl methacrylate) hydrogel. The enzyme is physically confined within the hydrogel during its polymerization. This enzyme-hydrogel system offers high sensitivity for urea concentrations between 1 mmol/L and 20 mmol/L and remains stable for at least eight weeks. This wearable biosensor offers continuous tracking of electrolytes, metabolites, and proteins, providing valuable insights into cardiovascular health and disease progression. These examples underscore the ongoing innovation in biosensor technology, leveraging advanced materials, nanotechnology, and biorecognition elements to enhance sensitivity, specificity, and usability in clinical settings. As research continues to refine biosensor capabilities, these devices hold promise for revolutionizing disease diagnosis, personalized medicine, and improving patient outcomes across a spectrum of medical conditions.

7.2.2 Optical Sensors

Optical sensors are devices that work on the principles of optics to find and assess various physical or chemical properties of substances. These sensors depend on how light interacts with the target material to produce a detectable signal, which can then be analyzed to provide valuable information for various applications. These sensors utilize principles such as absorption, fluorescence, reflectance, and interference to provide real-time data crucial for the detection and monitoring of diseases and to give personalized treatment strategies (Figure 7.3).

7.2.2.1 Fluorescence-Based Sensors

Fluorescence sensors depend on the emission of light from fluorescent molecules that occurs when they are excited by a specific wavelength of light. This technique is highly sensitive and allows for the detection of minute quantities of target molecules. For example, Lai *et al.* developed a fluorescence aptasensor that utilizes the TLS11a aptamer, a nucleic acid-based molecule, to detect liver cancer cell lines (HepG2) and frozen cancer tissue.[77] This type of aptamer is particularly advantageous due to its high affinity and specificity for the target. Chen *et al.* designed an innovative fluorescence-based sensor for the identification of circulating tumor cells in blood samples.[78] In

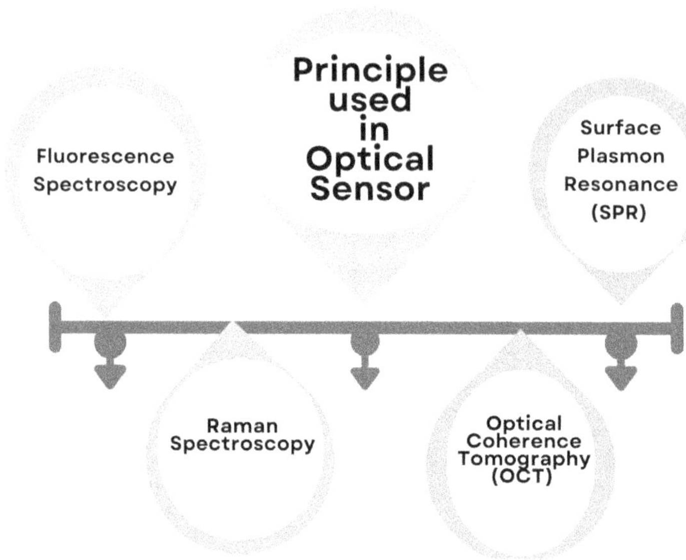

FIGURE 7.3 Principle used in optical sensor for medical diagnostics.

this work CdTe quantum dots show selective recognition of Ag+ and C-Ag+-C complexes when utilizing mucin 1 as a marker for circulating tumor cells (CTCs) and an aptamer as the recognition probe. The detection limits were 0.15 fg/mL for mucin 1 and 3 cells/mL for A549 cells, respectively. This sensor utilizes specific fluorescent markers that bind to CTCs, enabling sensitive and early detection of metastatic cancer.

7.2.2.2 Surface Plasmon Resonance (SPR) Sensors

SPR sensors identify changes in refractive index near a metal surface after the binding of target molecules. This label-free technique is widely used in biochemistry for studying biomolecular interactions, such as protein–protein interactions or antigen–antibody binding kinetics.[79] A study published in *Biosensors* demonstrated the use of a self-contained microfluidic chip with fiberoptic particle plasmon resonance aptasensor for rapid detection of SARS-CoV-2 nucleocapsid protein.[80] The SPR sensor provided rapid and accurate results, facilitating the diagnosis of COVID-19 and monitoring of immune responses.[81]

7.2.2.3 Raman Spectroscopy

Raman spectroscopy analyzes the scattering of laser light by molecular vibrations, providing a characteristic fingerprint spectrum. This technique is used in medical diagnostics to identify and quantify biomolecules in tissues or biofluids, aiding in disease diagnosis and monitoring.[82] In a research Raman spectroscopy has been applied in cancer detection and tissue characterization studies.[83,84] Riva *et al.* employed Raman spectroscopy to analyze tissue samples from patients with gliomas.[85] Another study identified characteristic Raman spectra associated with different grades of brain tumors, offering a non-invasive method for precise tumor diagnosis and surgical planning.[86]

7.2.2.4 Optical Coherence Tomography (OCT)

OCT is a non-invasive imaging technique that leverages low-coherence interferometry to generate detailed cross-sectional images of tissue structures. It is used extensively in ophthalmology for retinal imaging and glaucoma diagnosis, as well as in cardiology for imaging blood vessels and assessing cardiovascular diseases.[87,88] A study published demonstrated the use of OCT to visualize coronary artery plaques in patients with coronary artery disease.[89] The high-resolution images produced by OCT helped in assessing plaque composition and identifying vulnerable plaques prone to rupture, guiding treatment strategies.[90]

Also, researchers at the University of Southern California, California, designed a portable optical sensor for rapid diagnosis of malaria parasites in blood samples.[91] The sensor combines fluorescence imaging with microfluidic technology to achieve sensitive and specific detection of malaria infection, suitable for use in remote and resource-limited settings.[92] These examples underscore the versatility and impact of optical sensors in advancing medical diagnostics, from cancer detection and infectious disease diagnosis to tissue characterization and cardiovascular imaging.

7.2.3 Electrochemical Sensors

Electrochemical sensors are devices that convert chemical information into an electrical signal via electrochemical reactions. These sensors are commonly used for the detection and quantification of various analytes, from simple ions to complex biomolecules, because of their high selectivity, sensitivity, and effectiveness in aqueous environments.[93] Electrochemical sensors typically consist of three main components: electrolyte, electrode, and transducer.[94] Electrode is the interface where the electrochemical reaction occurs. It can be made of various materials such as platinum, gold, carbon, or conducting polymers, depending on the application and the analyte to be detected.[7] An electrolyte is the medium in which the electrochemical reaction occurs. It usually contains ions that participate in the redox (reduction-oxidation) reactions at the electrode surface.[95] A transducer converts the electrochemical signal produced by the interaction between the electrode and analyte into a measurable electrical signal, such as current or potential. The transducer amplifies and processes this signal for analysis.

Electrochemical sensors can be further classified into different types based on their mode of operation and detection principle.

7.2.3.1 Potentiometric Sensors

These sensors detect the potential difference (voltage) between two electrodes placed in an electrolyte solution. They are mainly utilized for pH measurement and ion-selective electrodes for detecting specific ions like sodium, potassium, or chloride ions.[96]

7.2.3.2 Amperometric Sensors

Amperometric sensors detect the current produced when a constant potential is applied to an electrode in the presence of the analyte. They are widely used for detecting gases (e.g., oxygen sensors in medical devices) and for enzymatic biosensors where the enzyme-catalyzed reaction produces an electroactive species.[97]

7.2.3.3 Voltammetric Sensors

Voltammetric sensors involve sweeping the applied potential to an electrode and measuring the resulting current.[98] Techniques like cyclic voltammetry (CV) and differential pulse voltammetry are employed to analyze redox-active compounds, metals, and organic molecules in complex matrices.

A study conducted by Leung *et al.* presented the use of a miniaturized electrochemical sensor for the tracking of glucose levels in diabetic patients.[99] The sensor, integrated into a wearable device, offered continuous glucose monitoring with high accuracy and minimal invasiveness, significantly improving patient management and quality of life.[93] Recent research by Alafeef *et al.* presented a portable electrochemical sensor for the detection of viral RNA, enabling rapid screening for pathogens such as SARS-CoV-2.[100] They used gold nanoparticles (AuNPs) coated with highly specific antisense oligonucleotides (ssDNA) designed to target the viral nucleocapsid phosphoprotein

(N-gene). In a report by Wang *et al.*, wearable electrochemical sensors have been utilized for monitoring neurotransmitter levels, such as dopamine, in patients with neurological disorders like Parkinson's disease, providing insights into disease progression and treatment efficacy.[101] Recent report has shown that electrochemical DNA biosensors are used for genetic testing and screening, offering rapid and cost-effective analysis of genetic mutations associated with hereditary diseases.[102]

7.2.4 Physical Sensors

Physical sensors are devices engineered to detect and respond to environmental stimuli, converting them into measurable signals such as electrical, optical, or mechanical outputs. They play a pivotal role in a wide range of applications across industries, from healthcare and automotive to environmental monitoring and consumer electronics. Here are several types of physical sensors commonly used in medical diagnostics. Temperature sensors are employed for checking body temperature in patients, detecting fever, and ensuring optimal conditions for medical equipment (Figure 7.4). Pressure sensors are used for monitoring of blood pressure, respiratory devices (e.g., ventilators), and intracranial pressure monitoring systems. Electrocardiogram (ECG) sensors track electrical activity of the heart, aiding in diagnosing arrhythmias, myocardial infarction, and other cardiac conditions. Blood Oxygen (SpO_2) sensor measures oxygen saturation levels in arterial blood, which is essential for assessing respiratory function and managing conditions like chronic obstructive pulmonary disease. Electromyography sensors detect electrical activity in muscles, helping diagnose neuromuscular disorders and evaluate muscle function. Electroencephalogram sensors measure electrical activity in the brain to diagnose epilepsy, sleep disorders, and monitor brain function during surgeries. Bioimpedance sensors assess body composition, hydration status, and fluid shifts in patients, often used in dialysis and nutritional assessment. These sensors enable healthcare providers to gather immediate data, monitor patient conditions, and make informed clinical decisions. Companies like Texas Instruments (TMP series) and Maxim Integrated (DS18B20) offer high-precision digital

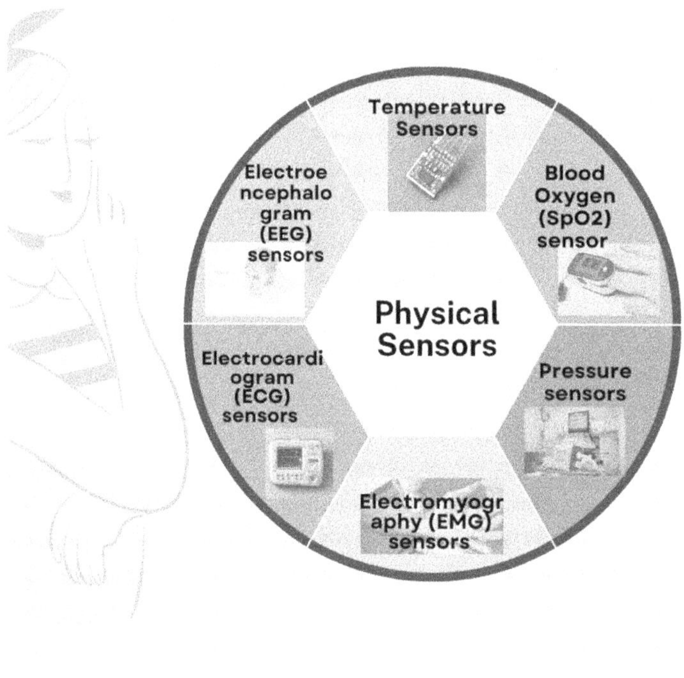

FIGURE 7.4 Different types of physical sensors used in medical diagnostics.

temperature sensors suitable for medical applications. These sensors provide accurate temperature measurements for patient monitoring, incubators, and medical equipment. Honeywell (TruStability and Sensotec series) and Freescale (MPX series) provide a range of pressure sensors designed for medical tools, such as blood pressure monitors, ventilators, and infusion pumps. These sensors offer precise pressure measurement capabilities essential for critical care settings. Also, companies like ADInstruments and Biopac Systems offer ECG sensor systems for research and clinical applications. These systems include wearable electrodes and amplifiers that capture and analyze cardiac electrical signals, aiding in the diagnosis and continuous tracking of heart conditions. In addition, Masimo (SET series) and Nonin Medical (PureLight series) manufacture pulse oximetry sensors for measuring oxygen saturation levels in arterial blood. These sensors are widely used in hospitals, emergency rooms, and home care settings for continuous monitoring of respiratory function. Dexcom and Abbott Diabetes Care offer continuous glucose monitoring (CGM) systems that include wearable sensors and transmitters. These sensors provide real-time glucose readings to diabetic patients, enhancing insulin therapy management and reducing the risk of hypoglycemia.

Liu *et al.* discussed high-sensitivity temperature sensors suitable for clinical applications, emphasizing their role in detecting fever and maintaining incubator stability.[103] Wang *et al.* developed piezoresistive pressure sensors for continuous blood pressure monitoring, highlighting their significance in critical care.[104] Patel *et al.* explored advanced ECG sensors for arrhythmia detection and cardiac rhythm analysis.[105] Zhang *et al.* reviewed the advancements and clinical applications of pulse oximetry sensors in monitoring respiratory function.[106] In a recent report Shung *et al.* discussed new advancement in ultrasound sensor technology for medical imaging applications.[107] Leung *et al.* discussed the advancements in glucose sensor technologies and their applications in diabetes management.[99] These examples demonstrate how physical sensors are integral to medical diagnostics, providing clinicians with valuable insights into patient health status, supporting clinical decision-making, and improving overall healthcare outcomes across diverse medical specialties.

7.3 APPLICATIONS OF SENSORS IN MEDICAL DIAGNOSTICS

The application of sensors in medical diagnostics has significantly transformed healthcare by enabling precise, real-time monitoring and non-invasive detection of various physiological parameters. Electrochemical sensors, biosensors, optical sensors, pressure sensors, and thermal sensors are among the key types used in medical diagnostics. These sensors facilitate the early detection and continuous monitoring of conditions such as cardiovascular diseases, diabetes, respiratory disorders, infectious diseases, and cancer. Additionally, these sensors facilitate remote patient monitoring and personalized medicine, improving patient compliance and lowering healthcare costs by helping to prevent complications and reduce hospital readmissions. In this chapter, some of the applications of sensor in medical diagnostic are discussed below.

7.3.1 Glucose Monitoring

In the past decade, advancements in sensor technology for glucose monitoring have significantly transformed diabetes management and enhanced patient outcomes. The development of CGMs has been a cornerstone of this evolution, providing real-time, continuous tracking of glucose levels and reducing the burden of frequent finger-pricking. Traditional glucose monitoring methods, which rely on intermittent blood glucose measurements through finger-pricking, have several limitations, including patient discomfort and inability to detect rapid glucose fluctuations. CGMs overcome these limitations by using minimally invasive sensors that determine glucose levels in the interstitial fluid every few minutes. Recent advancements in CGM technology have focused on improving sensor accuracy, reducing calibration frequency, and extending sensor lifespan. For instance, the development of factory-calibrated CGMs has minimized the need for frequent finger-stick calibrations, enhancing user convenience and compliance.[108] Studies have shown that factory-calibrated CGMs, such as the Dexcom G6, provide accurate glucose readings without the need for daily calibration, making them more user-friendly.[109] A breakthrough in CGM technology is its integration with insulin delivery systems, like insulin pumps, to form hybrid closed-loop systems, often referred to as "artificial pancreas" devices. These systems automatically adjust insulin delivery based on CGM data, enhancing glucose control and minimizing the risks of hypoglycemia and hyperglycemia. For example, the MiniMed 670G system by Medtronic uses an algorithm that adjusts basal insulin delivery every five minutes based on real-time CGM data.[110] Clinical trials have demonstrated that hybrid closed-loop systems significantly improve glycemic control in patients with type 1 diabetes. A study by Mcauley *et al.* showed that participants using the hybrid closed-loop system spent more time in the target glucose range and had lower HbA1c levels compared to those using standard insulin pump therapy.[111] These findings highlight the potential of sensor technology to enhance diabetes management and improve clinical outcomes.

Advancements in material science have greatly improved the performance of glucose sensors. For example, the use of nanomaterials, such as graphene, carbon nanotubes (CNTs), and AuNPss, has improved the sensitivity and specificity of glucose sensors.[112–114] These materials possess a high surface area, superior electrical conductivity, and biocompatibility, making them well-suited for glucose sensing applications. Graphene-based sensors, specifically, have demonstrated significant potential in glucose monitoring. Graphene's unique properties, including large surface-to-volume ratio and high electron mobility, enhance the detection of glucose at very low concentrations.[115] A report by Jothi *et al.* demonstrated that graphene-based electrochemical sensors could achieve rapid and accurate detection of glucose, with very low detection limit.[116] This report discusses the incorporation of nickel nanoparticles into graphene nanosheets and graphene nanoribbons (GS/GNR/Ni) composite through an in-situ chemical reduction method. The resulting GS/GNR/Ni-modified glassy carbon electrode (GCE) was subsequently employed for glucose detection. In another report, a greener electrochemical approach was carried out for the fabrication of non-enzymatic Pt nanoflower-rGO/GCE glucose sensor and the detection limit was found to be 53 μM.[117] Additionally, enzyme-based sensors using advanced materials have also shown improved performance. For instance, glucose oxidase immobilized on nanostructured materials, such as metal-organic frameworks (MOFs) and conductive polymers, has demonstrated enhanced stability and catalytic activity.[118,119] These enzyme-based sensors provide accurate and stable glucose measurements over extended periods, reducing the need for frequent sensor replacement. A recent study by Kumar *et al.* demonstrated the use of a $Ti_3C_2T_x$ MXene-based electrochemical sensor for glucose monitoring.[120] This sensor exhibited excellent sensitivity and a wide detection range due to the high surface area and conductivity of MXene. These $Ti_3C_2T_x$ MXene sensor also showed good selectivity for glucose over other common interferents, such as ascorbic acid and uric acid, indicating its potential for accurate glucose monitoring in complex biological samples. Furthermore, another study by Feng et al. explored the development of flexible $Ti_3C_2T_x$ MXene composites decorated with CuxO for non-enzymatic glucose sensing, offering a wide linear detection range.[121] This sensor integrated $Ti_3C_2T_x$ MXene with a flexible substrate, enabling continuous glucose monitoring with high accuracy and stability. Li *et al.* developed a Pt/MXene-based flexible, wearable non-enzymatic electrochemical sensor for continuous glucose monitoring in sweat (Figure 7.5).[122]

Despite the numerous advantages of technology, several challenges remain. Sensor accuracy can be affected by factors such as sensor lag, calibration errors, and interferences from medications or physiological conditions. Ongoing research aims to address these issues by developing more sophisticated algorithms and sensor materials that enhance accuracy and reliability. Another key area of interest is the development of non-invasive glucose monitoring technologies. Research in the field includes approaches such as optical sensing, transdermal patches, and saliva-based glucose sensors.[123] Although these technologies are still in the experimental stages, they hold promise for the future of glucose monitoring.

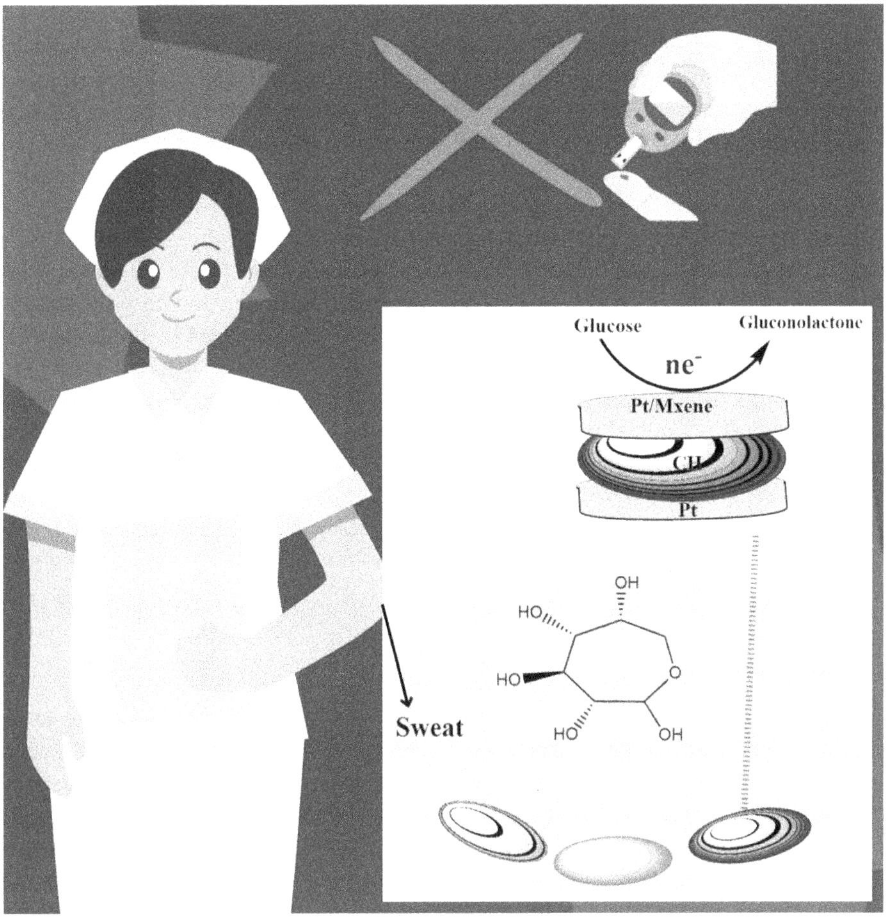

FIGURE 7.5 A schematic diagram illustrating the Pt/MXene-based flexible wearable sensor for continuous glucose monitoring in sweat.

7.3.2 Cardiovascular Health

The field of cardiovascular health monitoring has witnessed remarkable advancements due to the integration of novel materials into sensor technology.

One of the most notable advancements in sensor technology for cardiovascular health is the utilization of nanomaterials. Materials like polymers, functionalized graphitic carbon nitrides, CNTs, MOFs, graphene oxide, and AuNPs have been utilized.[124–130] Cheong *et al.* proposed inductively powered implantable blood flow sensor with bidirectional telemetry for monitoring vascular grafts.[130] Similarly, CNTs have been employed to develop highly sensitive pressure sensors capable of accurately monitoring blood pressure.[131] Flexible electronics represent another major breakthrough in cardiovascular health monitoring. Traditional rigid sensors can be uncomfortable and restrict patient movement, but flexible sensors made from materials like polyimide, polydimethylsiloxane, and flexible silicon can conform to the body's contours, providing continuous monitoring without discomfort. These sensors can be incorporated into wearable devices like smartwatches and patches to continuously monitor heart rate, blood pressure, and other vital signs in real time.[132]

Additionally, advancements in biocompatible and biodegradable materials have significantly transformed cardiovascular health sensors. These materials reduce the risk of adverse reactions and can be safely absorbed or dissolved in the body after their useful life. For example, sensors made from silk fibroin and magnesium can monitor cardiovascular parameters and then safely degrade within the body, removing the need for surgical extraction.[132–134] This technology is particularly advantageous for short-term monitoring after surgical procedures or in temporary therapeutic applications.

Moreover, hydrogel-based sensors have emerged as a promising technology for cardiovascular monitoring. Due to their high flexibility and ability to mimic the mechanical properties of biological tissues, hydrogels are ideal for integration with the human body. Recent studies have shown that conductive hydrogels can be used in wearable ECG sensors to achieve high-quality signal acquisition while ensuring excellent biocompatibility and comfort.[135] These hydrogel-based sensors could be used for continuous tracking of heart activity, providing valuable data for early detection of arrhythmias and other cardiac conditions. Recent studies have continued to push the boundaries of sensor technology for cardiovascular health. For example, a study by Han *et al.* presented the utilization of a hybrid nanomaterial composed of graphene and molybdenum disulfide (MoS_2) for high-performance ECG monitoring.[136] This hybrid material exhibited excellent electrical properties and biocompatibility, resulting in highly sensitive and accurate heart rate detection. Moreover, another promising area of research involves the advancement of smart textiles embedded with cardiovascular sensors. These textiles are capable of monitoring heart rate, blood pressure, and other vital signs while being worn as everyday clothing. A study by Zaman *et al.* examined the advancements in this area, highlighting the potential of smart textiles to provide seamless and continuous cardiovascular monitoring in daily life.[137]

7.3.3 Cancer Detection

In the past two decades, notable progress in material science has transformed sensor technology for cancer detection. These developments have resulted in the creation of highly sensitive, specific, and non-invasive diagnostic tools. The integration of novel materials such as nanomaterials, biocompatible polymers, and advanced biomolecules has enhanced the performance of biosensors, enabling early detection, real-time monitoring, and improved prognostic capabilities. Nanomaterials in cancer, including AuNPs, graphene, CNTs, and quantum dots, have played a crucial role in cancer biosensors.[138–141] AuNPs, with their distinctive optical and electronic properties, have been widely utilized in colorimetric and electrochemical sensors. For example, AuNPs conjugated with antibodies have been used to detect tumor markers like prostate-specific antigen with high sensitivity.[142] Similarly, CNTs have been employed to enhance the electrical conductivity of biosensors, allowing for the detection of low concentrations of cancer biomarkers.[143] Numerous reports have shown that graphene-based sensors can detect cancer biomarkers such as carcinoembryonic antigen at femtomolar levels, significantly enhancing early detection capabilities.[144] Quantum dots with unique optical properties have also been employed in fluorescent biosensors for cancer detection. These materials provide high signal-to-noise ratios, enabling the visualization of cancer cells and tissues with high precision. Biocompatible polymers such as polyaniline, polypyrrole, and polydopamine have been utilized to fabricate biosensors with enhanced stability and functionality.[145] These polymers offer excellent biocompatibility, making them suitable for in vivo applications.[146] For instance, polyaniline-based biosensors have been employed to detect breast cancer biomarkers in serum samples, showing high sensitivity and specificity.[147] Polydopamine, inspired by mussel adhesive proteins, has been employed to functionalize sensor surfaces, improving the immobilization of biomolecules and enhancing sensor performance.[148] Furthermore, the development of molecularly imprinted polymers (MIPs) has provided a significant boost to cancer sensor technology.[149] Molecularly imprinted polymers (MIPs) are synthetic polymers designed with specific binding sites for target molecules, replicating the function of natural antibodies. These polymers have been utilized to develop highly selective sensors for various cancer biomarkers, such as alpha-fetoprotein and human epidermal growth factor receptor 2. The use of MIPs in biosensors provides benefits including low cost, high stability, and the capability to function in harsh conditions.

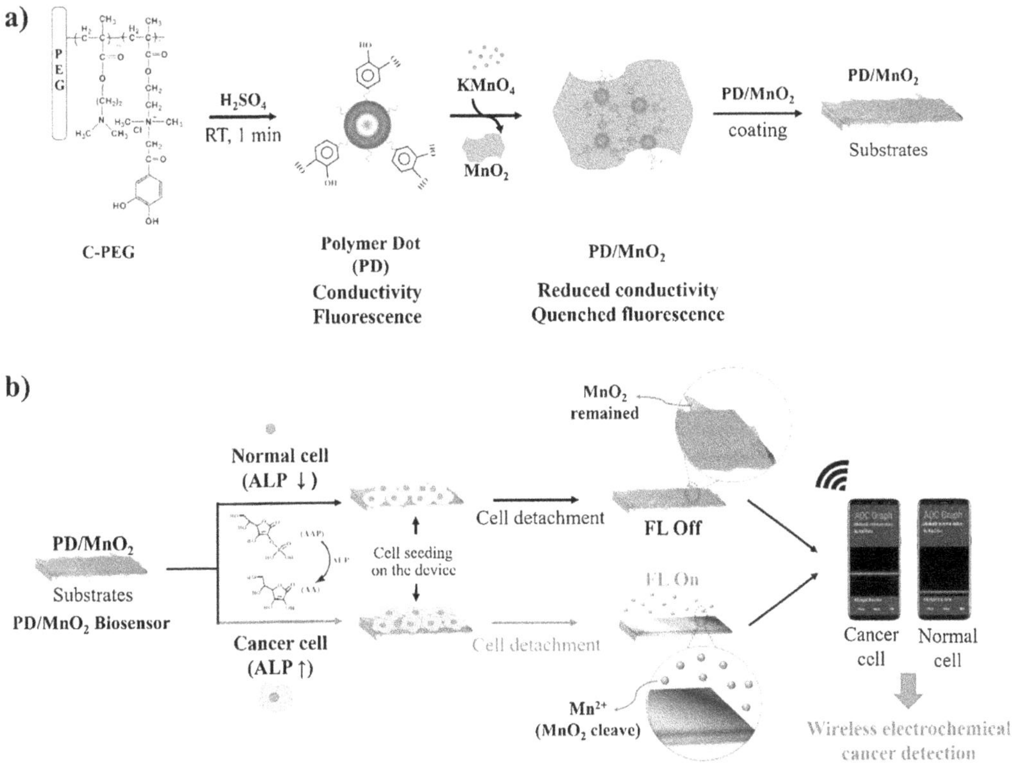

FIGURE 7.6 Schematic representation of wireless cancer detection using PD/MnO_2: (a) preparation of the PD/MnO_2 biosensor and (b) proposed mechanisms for cancer detection with PD/MnO_2 biosensors. (Reproduced with permission from reference 145.)

In a report by Won *et al.*, a groundbreaking wireless electrochemical biosensor utilizing polymer dot-manganese oxide complexes (PD/MnO_2) is introduced, featuring a dual-mode detection system (Figure 7.6).[150] The biosensor offers a dual-mode detection system, combining both fluorescence-based and electrochemical methods for identifying cancer cells on a coated surface. The electrochemical approach achieves a limit of detection (LOD) of 3.98 cells/mL, while the fluorescence method has an LOD of 1995 cells/mL. Additionally, the integration of advanced biomolecular techniques, such as aptamer-based sensors and DNA nanotechnology, has significantly enhanced cancer detection capabilities.[151] Aptamers, which are short single-stranded DNA or RNA sequences with high-affinity binding to specific targets, have been employed to create highly sensitive biosensors for cancer biomarkers. For instance, aptamer-based sensors have been engineered to detect thrombin, a protein implicated in cancer metastasis, with exceptional sensitivity.[152] DNA nanotechnology has also contributed to the development of novel cancer sensors. DNA-based sensors, or DNAzymes, leverage the catalytic properties of DNA to detect specific cancer-related molecules. These sensors offer high specificity and can be easily modified for various applications. Recent research has demonstrated the use of DNA-based sensors to detect microRNAs associated with cancer, providing a powerful tool for early diagnosis and monitoring.[153] The integration of advanced materials into point-of-care (POC) and wearable devices has made cancer detection more accessible and convenient. POC devices, which utilize biosensors for rapid diagnostics, have benefited from the advancements in material science. For instance, paper-based microfluidic devices incorporating nanomaterials have been produced for the rapid detection of cancer biomarkers in blood and urine samples.[154] These devices offer low-cost, portable, and user-friendly solutions for cancer screening, particularly in resource-limited settings. Wearable biosensors, designed to monitor cancer-related biomarkers continuously, have also emerged as a promising technology.[155] These devices can detect biomarkers such as lactate and glucose, which are indicative of metabolic changes associated with cancer, enabling continuous monitoring and early intervention.

Recent research has centered on enhancing the multiplexing capabilities of cancer biosensors, enabling the simultaneous detection of multiple biomarkers. For instance, multiplexed microfluidic devices enhanced with nanomaterials have been developed to simultaneously detect multiple cancer biomarkers, thereby enhancing diagnostic accuracy and minimizing false positives.[156] Moreover, the application of artificial intelligence (AI) and machine learning (ML) algorithms in analyzing sensor data has demonstrated significant potential for enhancing the accuracy and predictive capability of cancer diagnostics. The development of non-invasive sensors, such as breath and sweat sensors, could provide new avenues for early cancer detection. Moreover, the integration of biosensors with personalized medicine approaches, including the monitoring of individual patient responses to therapy, could significantly enhance cancer treatment outcomes. Ongoing research and development in this domain are poised to further transform cancer diagnostics, offering innovative tools and technologies to combat this formidable disease.

7.3.4 Infectious Disease Diagnostics

The integration of novel materials, microfluidics, and advanced biomolecular techniques has enhanced the performance and applicability of sensors, providing critical tools for global health.

CNTs and graphene have garnered significant attention in the realm of electrochemical sensing due to their exceptional ability to detect pathogens at ultra-low concentrations. These advanced materials have been employed in the development of highly sensitive CNT-based sensors for viral RNA detection, enabling the early diagnosis of diseases such as influenza and Zika virus.[157] Likewise, graphene-based sensors have demonstrated considerable potential in the detection of bacterial infections, such as tuberculosis, by precisely targeting specific bacterial DNA sequences.[158] Quantum dots, with distinct fluorescence characteristics, have been employed in optical biosensors for the diagnosis of infectious diseases. These materials offer high signal-to-noise ratios and enable the multiplexed detection of multiple pathogens within a single assay. Recent research has highlighted the application of quantum dots in detecting dengue virus, hepatitis B virus, and *Escherichia coli*, underscoring their potential for the simultaneous identification of diverse infectious agents.[159] The integration of microfluidics into sensor technology has resulted in the creation of lab-on-a-chip (LOC) devices, which provide substantial benefits for infectious disease diagnostics. Microfluidic devices facilitate the handling of minute fluid volumes, enabling the miniaturization and automation of intricate laboratory processes. These devices are capable of conducting multiple diagnostic tests concurrently, thereby reducing both time and cost compared to conventional diagnostic methods. LOC devices have been employed in the detection of various infectious diseases, including HIV, malaria, and COVID-19.[160] For example, a microfluidic chip developed by Mauk *et al.* is capable of detecting HIV RNA in whole blood samples with high sensitivity and specificity.[161]

This chip combines nucleic acid amplification and detection into a single platform, delivering rapid results within an hour. Similarly, microfluidic devices have been designed for detecting malaria parasites by analyzing blood samples for specific biomarkers.[162] During the COVID-19 pandemic, LOC devices were crucial for the swift diagnosis of SARS-CoV-2.[163] Researchers developed microfluidic chips that can detect viral RNA from nasal swab samples using techniques like reverse transcription loop-mediated isothermal amplification.[164] These devices delivered results within minutes, facilitating widespread testing and prompt isolation of infected individuals, which significantly aided in controlling the spread of the virus. Advancements in biomolecular techniques have markedly improved the development of highly sensitive and specific biosensors for infectious disease diagnostics. The integration of techniques such as PCR, enzyme-linked immunosorbent assay

(ELISA), and nucleic acid amplification with sensor technology has significantly enhanced diagnostic capabilities. PCR-based biosensors, in particular, are extensively employed for detecting viral and bacterial infections by amplifying specific DNA or RNA sequences of pathogens, thereby enabling the detection of minute quantities of infectious agents. For instance, a PCR-based biosensor can detect Ebola virus RNA with high sensitivity, providing a valuable tool for outbreak management. Similarly, real-time PCR (qPCR) sensors have been employed to identify a range of bacterial infections, including tuberculosis and methicillin-resistant *Staphylococcus aureus*.[165] ELISA-based biosensors have been widely utilized for detecting antigens and antibodies associated with infectious diseases. These sensors use enzyme-linked antibodies to produce a detectable signal upon binding to the target antigen or antibody. Recent advancements have enabled the creation of portable ELISA devices for POC testing. For instance, a portable ELISA sensor can rapidly and accurately detect malaria antigens in blood samples, delivering results within minutes. Additionally, nucleic acid amplification techniques, such as LAMP, have been integrated with sensor technology to develop rapid and highly sensitive diagnostic tools. LAMP-based sensors facilitate nucleic acid amplification at a constant temperature, eliminating the need for complex thermal cycling equipment. These sensors have been utilized to detect a range of pathogens, including the Zika virus, dengue virus, and SARS-CoV-2.[166] The simplicity and speed of LAMP-based sensors make them particularly well-suited for POC testing in resource-limited settings. The advancement of POC and wearable sensors has significantly enhanced the accessibility and convenience of infectious disease diagnostics. These sensors often incorporate nanomaterials, microfluidics, and biomolecular techniques to enhance their performance. For example, in a report POC sensor used AuNPs and lateral flow assay technology to detect HIV antibodies in blood samples, providing results within 20 minutes.[167] Similarly, POC sensors for malaria diagnosis use microfluidic chips to analyze blood samples for specific antigens, enabling rapid and accurate detection. A wearable sensor can continuously monitor body temperature and detect fever, providing early warning signs of infections such as influenza and COVID-19. Additionally, wearable sensors equipped with sweat analysis capabilities have been used to detect biomarkers related to bacterial and viral infections, offering a non-invasive approach to infectious disease diagnostic.

7.3.5 Wearable Health Technologies

The development of advanced sensor materials for wearable health technologies represents a significant breakthrough in modern healthcare. These materials facilitate continuous monitoring of vital signs, physiological parameters, and environmental factors, providing valuable data that can enhance personal health management and clinical decision-making. Wearable health technologies are poised to revolutionize preventive medicine, chronic disease management, and overall wellness by offering real-time, noninvasive monitoring capabilities. Numerous reports have highlighted that conductive polymers such as polyaniline (PANI), polypyrrole, and poly(3,4-ethylenedioxythiophene) are leading materials in the development of wearable sensors.[168] These polymers integrate the mechanical properties of plastics with the electrical characteristics of metals or semiconductors. The conjugated π-electron system in these polymers facilitates electrical conductivity. Doping with various agents (e.g., camphorsulfonic acid for PANI) can enhance conductivity and tailor the material's properties for specific applications. Conductive polymers are lightweight, flexible, and can be processed into thin films, making them ideal for wearable sensors. Furthermore, hydrogels can be engineered to react to various stimuli—such as pH, temperature, and glucose levels—by integrating responsive polymers or nanoparticles. Poly(N-isopropyl acrylamide) is a thermoresponsive hydrogel that exhibits a pronounced phase transition at its lower critical solution temperature, rendering it highly suitable for temperature-sensing applications.[169] Incorporating conductive materials like silver nanowires or CNTs into hydrogels can further enhance their electrical properties for biosensing applications. Moreover, polymeric hydrogels incorporating boronic acid derivatives can bind glucose

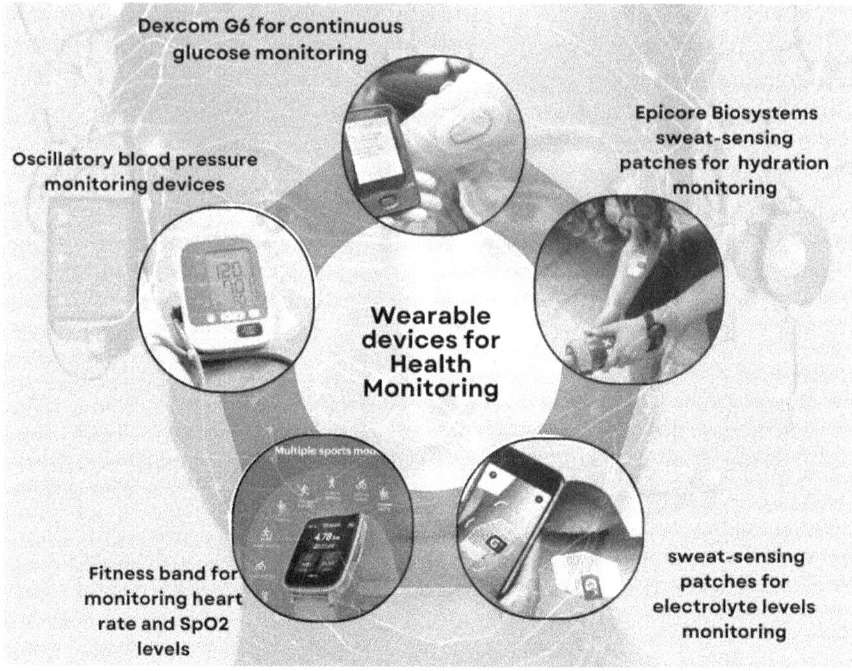

FIGURE 7.7 Different types of wearable devices for health monitoring.

molecules, changing their swelling behavior and providing a direct optical or electrical readout.[170] Companies like Dexcom and Abbott have developed commercial CGM systems, such as Dexcom G6 and FreeStyle Libre, which utilize advanced sensor materials for precise and continuous glucose monitoring (Figure 7.7).

Similarly, polyvinyl chloride membranes doped with ionophores can selectively bind specific ions, enabling accurate measurement of electrolyte levels in sweat.[171] Companies like Gatorade and Epicore Biosystems have developed sweat-sensing patches that monitor hydration and electrolyte levels, helping athletes optimize their performance and hydration strategies. Conversely, PPG is the principal technology in wearable devices for monitoring SpO_2 and heart rate. PPG is a non-invasive optical method that assesses fluctuations in blood volume within the microvascular bed of tissue. The device emits light, typically in red and infrared wavelengths, into the skin and detects the amount of light absorbed or reflected by a photodetector. These periodic variations in light absorption correspond to the cardiac cycle, providing insights into heart rate and blood oxygen levels. PPG sensors generally utilize LEDs that emit light at specific wavelengths, such as 660 nm for red light and 940 nm for infrared light.

The choice of wavelength is crucial because oxyhemoglobin and deoxyhemoglobin absorb light differently at these wavelengths. For accurate SpO_2 measurement, the ratio of absorption at these two wavelengths is calculated. Wearable devices like smartwatches and fitness bands are popular for tracking exercise intensity and recovery by monitoring heart rate and SpO_2 levels. These devices help users optimize their workouts and track their fitness progress. Devices like the Garmin Venu 2 and the Polar Vantage V2 provide detailed heart rate and SpO_2 data, allowing users to monitor their performance during various activities, including high-intensity interval training and endurance sports. A study published in *Science Advances* demonstrated a graphene-based PPG sensor integrated into a wearable patch, showing high accuracy in SpO_2 and heart rate monitoring during various physical activities.[172] Apple Watch Series 6 and 7, Fitbit Sense, and Garmin Forerunner series are examples of consumer-grade wearables with integrated PPG sensors for SpO_2 and heart rate monitoring. ML algorithms are being designed to improve the accuracy of PPG-based SpO_2 and heart rate measurements. These algorithms can filter out noise, correct for motion artifacts, and offer real-time analysis. A report in *Journal of Sensors* highlighted the use of deep learning models to process PPG signals from wearable devices, significantly improving the accuracy of heart rate and SpO_2 readings during physical activities.[173] A report published in *Frontiers in Digital Health* demonstrated that remote monitoring of COVID-19 patients using wearable devices enhanced patient outcomes and alleviated the strain on healthcare facilities.[174]

7.4 CHALLENGES IN SENSOR DEVELOPMENT

Sensors in healthcare technologies have rapidly advanced, integrating sophisticated devices for continuous monitoring of physiological parameters including heart rate, blood oxygen saturation (SpO_2), glucose levels, and many more. Despite these significant strides, ongoing challenges in sensor development affect their performance, integration, and overall user acceptance. The main challenges in sensor development for wearable health technologies include issues related to accuracy, stability, integration, power consumption, and user comfort. Some of the challenges are discussed next.

7.4.1 Accuracy and Sensitivity

Accuracy and sensitivity are two critical performance metrics for sensors used in healthcare. Achieving high accuracy ensures that the measurements are correct, while high sensitivity allows the sensor to detect small changes or low concentrations of analytes. However, numerous challenges hinder the optimization of these parameters in health sensors. Motion artifacts remain a major challenge in wearable sensor technology. When a user is active, the movement can cause significant noise in the sensor readings, particularly in optical sensors like PPG used for heart rate and SpO_2 measurements. In sensors that detect biochemical analytes in sweat, tears, or interstitial fluid, movement can cause fluctuations in the sample volume and disrupt the sensor–analyte interface, leading to inaccurate readings. The accuracy of wearable sensors can drift over time due to environmental factors, physical wear, and changes in the sensor's interface with the skin. Regular calibration is essential to maintain accuracy, but this can be cumbersome for users. Moreover, wearable sensors are frequently subjected to fluctuating environmental conditions, such as humidity, temperature, and sweat, impacting their performance and durability. Researchers have developed protective coatings and encapsulation techniques to shield sensors from environmental effects. A study in *Advanced Materials* discusses the use of hydrophobic coatings and flexible encapsulants that protect the sensors while maintaining their flexibility and sensitivity.[175]

The materials used in wearable sensors, especially conductive polymers and hydrogels, can degrade over time due to repeated mechanical stress and exposure to biological fluids. A report highlights the development of durable, biocompatible materials such as silicone elastomers and self-healing polymers that extend the lifespan of wearable sensors.[176] These materials can recover their original properties after deformation, improving the overall durability of the devices.

Addressing all these challenges requires a multidisciplinary approach, combining advances in material science, mechanical design, signal processing, and AI. Ongoing research and technological innovations continue to push the boundaries, offering promising solutions to enhance the performance of health sensors. As these technologies advance, they are poised to play a pivotal role in enhancing personal health management and clinical outcomes.

7.4.2 Specificity and Selectivity

In healthcare, the specificity and selectivity of sensors are paramount for accurate diagnosis and monitoring of various conditions. Specificity refers to a sensor's capacity to exclusively detect the target analyte amidst other substances, whereas selectivity pertains to its ability to differentiate between various analytes within a complex mixture. Achieving high specificity and selectivity is critical to avoid false positives and false negatives, ensuring reliable readings and effective patient management.

A major challenge in sensor development is cross-sensitivity, where the sensor reacts to multiple substances in addition to the target analyte. This lack of specificity can lead to erroneous readings and misdiagnoses. In complex biological environments like blood, sweat, or interstitial fluid, numerous compounds can interfere with sensor performance. These sensors, often employed for glucose monitoring, can be influenced by other electroactive substances present in the sample, such as ascorbic acid, uric acid, and acetaminophen. Photometric sensors used for measuring blood oxygen levels (SpO_2) can be influenced by factors like skin pigmentation, ambient light, and motion artifacts. Recent research has emphasized the application of multi-wavelength optical sensing and sophisticated algorithms to enhance selectivity and minimize interference from external factors. The choice of materials in sensors is pivotal for achieving specificity and selectivity. The challenge lies in engineering materials that selectively interact with the target analyte while remaining inert to other substances. A study published in *ACS Applied Materials & Interfaces* detailed the use of antibody-functionalized graphene oxide for the targeted detection of biomarkers in complex biological samples.[177]

Biological fouling, where proteins and other biomolecules non-specifically adsorb onto sensor surfaces, can severely impair specificity and selectivity. This issue is particularly prevalent in implantable sensors that must operate in vivo for extended periods of time. Developing anti-fouling coatings that prevent non-specific adsorption is a critical area of research. Research is also exploring dynamic or self-renewing coatings that can actively repel or degrade fouling agents. A publication in *Materials* discussed the use of self-cleaning surfaces and coatings that release anti-fouling agents in response to environmental triggers, thus preserving sensor specificity.[178] Sensors designed for simultaneous detection of multiple analytes encounter the challenge of differentiating signals with high specificity and selectivity. This capability is essential for applications such as wearable health monitors, which track various biomarkers to deliver a comprehensive health profile.

Utilizing sensor arrays, where each sensor element is tailored to a specific analyte, can enhance selectivity. However, ensuring that the signals from each sensor do not interfere with each other is challenging.

In addition, maintaining high specificity and selectivity over time and across different conditions is challenging. Sensors must be stable and reproducible to ensure consistent performance. The stability of functionalized surfaces and recognition elements is critical. The challenges associated with achieving high specificity and selectivity in healthcare sensors are multifaceted, involving issues related to cross-sensitivity, material properties, biological interactions, multi-analyte detection, and stability.

7.4.3 Integration with Digital Health Systems

The integration of sensors with digital health systems marks a substantial advancement in healthcare technology, enabling continuous monitoring, real-time data analysis, and enhanced patient outcomes. Nonetheless, this integration presents several challenges, including issues with interoperability, data accuracy, security, power management, and user engagement. Addressing these challenges is essential for the effective implementation of sensor-based digital health systems in both clinical and consumer settings. One of the primary challenges in integrating sensors with digital health systems is the lack of standardization across devices and platforms. Sensors from various manufacturers frequently utilize proprietary data formats and communication protocols, complicating efforts to achieve seamless interoperability. Variability in data formats can lead to incompatibilities between sensors and health information systems. A study highlighted the need for standardized data models and exchange formats to facilitate interoperability. The integration of sensors with digital health systems presents notable security and privacy challenges. Safeguarding sensitive health data from unauthorized access and breaches is crucial. Ensuring secure data transmission and storage necessitates robust encryption and authentication measures. Adherence to regulations such as general data protection regulation (GDPR) and health insurance portability and accountability (HIPAA) is essential to maintain patient privacy.

In addition, wearable sensors often rely on battery power, which can limit their operational lifespan and usability. Efficient power management is crucial for continuous monitoring applications. Energy harvesting technologies, including solar, thermal, and kinetic energy, can complement battery power. Designing low-power sensors and communication protocols can reduce energy consumption. A research report discusses the development of energy-efficient hardware and software solutions for wearable health sensors.[179]

Moreover, the design and usability of wearable sensors significantly impact user engagement and compliance. Devices that are uncomfortable or difficult to use are less likely to be worn consistently. Ensuring that sensors are comfortable and unobtrusive is critical. A study underscores the advancement of ultra-thin, flexible sensors that conform to the body, thereby improving comfort and wearability.[180] The user interface for data interaction should be intuitive and user-friendly. A report has the importance of user-centered design in the development of wearable health technologies. Integrating sensors with digital health systems presents several challenges, including issues related to interoperability, data accuracy, security, power management, and user engagement. Addressing these challenges requires a multidisciplinary approach, leveraging advances in technology, robust standards, and user-centered design principles.

7.5 FUTURE PROSPECTS

The progression of sensor technology in healthcare has already transformed how we monitor and manage health. As these technologies advance, they hold the potential for even more significant impacts, including personalized medicine, early disease detection, and improved patient care. The potential future developments and applications of sensors in healthcare, highlighting emerging trends, innovative research, and the integration of cutting-edge technologies.

7.5.1 Personalized Medicine

The future of healthcare is centered on personalized medicine, where treatments and interventions are customized according to each patient's distinct physiological and genetic profiles. Sensors are pivotal in this paradigm shift. Advances in sensor technology will facilitate the ongoing monitoring of genetic markers and specific biomarkers linked to various diseases, leading to earlier detection and tailored treatment plans. Continuous monitoring of metabolic parameters, such as glucose, cholesterol, and lactate levels, will enable the formulation of personalized dietary and lifestyle recommendations. A study explores the use of wearable sensors that track metabolic changes and provide personalized feedback for diabetes management.[181] One of the most promising future applications of sensors in healthcare is the early detection of diseases through continuous monitoring. Wearable sensors capable of detecting cancer biomarkers in sweat, blood, or interstitial fluid can enable early diagnosis and timely intervention. Research in *Nanoscale* discusses the development of nanosensors that can detect tumor markers at very low concentrations, facilitating early cancer detection.[182] Additionally, sensors that monitor brain activity and neurotransmitter levels can assist in the early diagnosis of neurological disorders like Parkinson's and Alzheimer's diseases. A study published in *Nature Biotechnology* investigates the use of implantable sensors that track alterations in brain chemistry, offering early warning signs of neurodegenerative diseases.[183]

The capability to monitor patients in real time and deliver immediate feedback is a major advantage of sensor technology in healthcare. Continuous monitoring of chronic conditions, such as heart disease, diabetes, and respiratory disorders, enhances disease management and patient outcomes. Sensors can track vital signs and recovery progress after surgery, facilitating timely interventions and reducing hospital readmissions. A study explores the use of wearable sensors for post-operative monitoring, highlighting their potential to enhance patient care and safety.

7.5.2 Advancements in Nanotechnology

Nanotechnology, which involves manipulating materials at the atomic or molecular scale, has emerged as a transformative force across various fields, especially in healthcare. The development of nanosensors is a promising area within this domain, offering unprecedented opportunities for disease diagnosis, monitoring, and treatment. One of the primary benefits of nanosensors is their enhanced sensitivity and specificity. Nanoscale materials, including nanoparticles, nanowires, and nanotubes, possess distinctive physical and chemical properties that enable highly specific interactions with biological molecules. For example, AuNPs can be functionalized with targeted antibodies to detect biomarkers at very low concentrations. This capability is crucial for early disease detection, where biomarkers may be present in minute quantities. The integration of nanotechnology into sensor development has facilitated the creation of portable and real-time monitoring devices. These devices offer instant feedback on a patient's health status, facilitating prompt medical interventions. This real-time monitoring capability is crucial for managing chronic diseases and minimizing the need for hospital visits. Nanotechnology enables the development of non-invasive or minimally invasive sensors, thereby minimizing patient discomfort and reducing infection risk. For example, nanosensors embedded in wearable devices can continuously monitor vital signs—including heart rate, respiratory rate, and body temperature—without the need for invasive interventions.[184] Furthermore, nanoscale biosensors can be incorporated into LOC devices, enabling the analysis of small blood samples collected through a simple finger prick rather than a conventional venous blood draw.

Recent scientific research has concentrated on advancing the development and surface functionalization of nanomaterials to enhance sensor performance. Research studies are being conducted on a range of nanomaterials, including carbon-based materials (such as graphene and CNTs), metal nanoparticles (like gold and silver), and quantum dots. Surface functionalization techniques, such as coating nanomaterials with specific ligands or polymers, enhance their biocompatibility and target specificity, making them more effective for medical applications.

The integration of microfluidics and LOC technologies with nanotechnology is leading to the development of compact, efficient, and cost-effective diagnostic devices. These integrated systems can process and analyze small fluid samples with high precision, enabling rapid and accurate detection of biomarkers. For instance, researchers have engineered microfluidic devices integrated with nanosensors capable of detecting cancer biomarkers from just a single drop of blood.[185] This integration streamlines the diagnostic process and has the potential to revolutionize point of care testing. Signal transduction mechanisms are critical for converting biological interactions into measurable signals. Advancements in nanotechnology have facilitated the development of innovative transduction methods, including plasmonic, electrochemical, and fluorescence-based techniques. For instance, plasmonic nanosensors utilize the resonance of surface plasmons on metal nanoparticles to detect changes in the local environment, providing highly sensitive measurements. These advancements enhance the detection capabilities of nanosensors and expand their potential applications in healthcare. Two-dimensional materials, including graphene and transition metal dichalcogenides, have proven to be promising for nanosensor applications due to their outstanding electrical, mechanical, and optical characteristics. Recent research has demonstrated that graphene-based nanosensors can detect DNA, proteins, and small molecules with remarkable sensitivity and specificity.[186] Also, researchers are exploring MOF-based nanosensors for detecting gases, volatile organic compounds, and biological molecules. For example, MOF nanosensors have been developed for detecting hydrogen sulfide, a biomarker for certain diseases, with high sensitivity and selectivity.[187]

Nanocomposites and hybrid materials leverage the benefits of various nanomaterials to achieve enhanced sensor performance. For example, integrating CNTS with metal nanoparticles can boost the electrical and catalytic properties of the resulting nanocomposite, thereby improving its sensitivity and detection limits. Recent research has investigated the application of nanocomposites for detecting glucose, cholesterol, and other biomarkers, showcasing their potential for creating multifunctional and high-performance nanosensors.

The future prospects of advancements in nanotechnology for the development of sensors in healthcare are promising. Nanosensors offer several key advantages, including enhanced sensitivity and specificity, real-time monitoring, and non-invasive techniques.

7.5.3 Artificial Intelligence and Machine Learning

AI and ML are leading the transformation of sensor development in healthcare, offering groundbreaking improvements in diagnostic accuracy, personalized patient care, and real-time health monitoring. Recent research and technological advancements have highlighted the transformative potential of AI and ML in enhancing sensor capabilities and advancing healthcare outcomes. AI's ability to interpret and analyze vast amounts of sensor data with high accuracy has been a cornerstone of recent developments. Studies have demonstrated AI's efficacy in early disease detection by identifying subtle patterns in sensor data that human analysis might overlook. For instance, AI algorithms can analyze ECG readings to predict cardiovascular events with greater precision than traditional methods, thereby enabling proactive intervention and management of cardiac conditions.[188] ML, on the other hand, leverages data-driven insights from wearable sensors to personalize healthcare interventions. By continuously monitoring patient vitals and health parameters, ML algorithms can adjust treatment plans in real time based on individualized data trends. This capability not only enhances treatment efficacy but also elevates patient satisfaction and outcomes by personalizing medical interventions to meet the unique needs of each individual.

Real-time monitoring facilitated by AI-powered sensors is another significant advancement in healthcare. These sensors can detect deviations from normal health parameters and issue alerts to healthcare providers promptly, enabling timely interventions. This is especially advantageous for patients with chronic conditions who need ongoing monitoring and proactive management to avert complications. The technological progress in AI-driven sensor development is not limited to improving accuracy and real-time monitoring but also extends to enhancing sensor capabilities. Researchers are exploring ways to improve sensor sensitivity, durability, and versatility to detect a broader range of biomarkers and physiological signals. However, alongside these technological advancements come critical ethical considerations. The integration of AI and ML into healthcare sensors introduces concerns regarding data privacy, patient consent, and the ethical application of AI in medical decision-making. Balancing innovation with ethical responsibility is essential to ensure that AI-driven healthcare technologies serve patients effectively while upholding their rights and preserving trust in healthcare systems. By harnessing the analytical power of AI and the data-driven insights of ML, healthcare providers can offer more accurate, timely, and personalized care. However, realizing these advantages necessitates continuous research, interdisciplinary collaboration, and careful consideration of ethical and regulatory frameworks to ensure the responsible deployment of AI and ML technologies in healthcare.

7.5.4 Global Health Impact

Sensors have become essential to contemporary healthcare, facilitating real-time monitoring, early disease detection, and tailored treatment strategies. As technology advances, the future prospects of sensors in global health are increasingly promising. In healthcare, sensors now encompass a broad spectrum of applications, including wearable devices that track vital signs and diagnostic tools that detect biomarkers for a variety of diseases. These technologies have already demonstrated significant benefits in improving patient outcomes, reducing healthcare costs, and enhancing the quality of care.

Sensors facilitate continuous monitoring of patient health parameters, enabling early intervention and personalized care from a distance. This capability has been particularly vital during the COVID-19 pandemic, allowing healthcare providers to oversee patients without direct contact and thereby mitigating transmission risks. Also, AI-powered sensors leverage ML algorithms to examine vast datasets, providing predictive insights into health conditions and the refinement of treatment strategies. Sensors enable remote monitoring and diagnostics, bridging gaps in healthcare access for underserved populations in remote or rural areas. This technology facilitates early detection of diseases and timely interventions, potentially reducing disparities in healthcare outcomes globally. Real-time data from sensors contribute to enhanced public health surveillance by monitoring disease outbreaks, tracking epidemiological trends, and predicting health risks. This capability is crucial for proactive health management and emergency response planning. The integration of sensors with AI and ML allows for personalized healthcare interventions based on individual health data. This approach not only boosts treatment effectiveness but also increases patient engagement and adherence to treatment plans. By enabling early detection and preventive care, sensor technologies have the potential to lower healthcare costs related to hospitalizations and advanced-stage treatments. This cost efficiency makes healthcare more sustainable and accessible on a global scale. In low-resource settings, where access to healthcare infrastructure and trained personnel is limited, sensors offer a cost-effective solution for disease monitoring and management. Mobile health initiatives leveraging sensor technologies have shown promise in improving maternal and child health outcomes, infectious disease management, and chronic disease surveillance. Sensors can contribute to achieving health equity by providing timely and accurate healthcare interventions to marginalized populations.

7.6 CONCLUSION

The advancements in sensor technologies represent a transformative leap forward in healthcare, promising to revolutionize how we diagnose, monitor, and treat medical conditions. By leveraging AI, ML, and cutting-edge sensor designs, healthcare providers can offer patients more personalized, efficient, and effective care globally. These technologies facilitate real-time monitoring of vital signs, early disease detection, and customized treatment strategies tailored to individual health data. They empower patients by delivering actionable health insights, thereby enhancing engagement and adherence to treatment plans. Moreover, in resource-limited settings, sensors offer a cost-effective solution for expanding access to healthcare and improving health outcomes. Nevertheless, incorporating sensor technologies into healthcare systems presents challenges such as data security, interoperability, and ethical concerns. Addressing these issues is critical to ensure the responsible implementation and adoption of these technologies while protecting patient privacy and upholding trust in healthcare systems. Moving forward, ongoing research and interdisciplinary collaboration will be vital for advancing sensor capabilities, improving reliability, and broadening their applications across various healthcare environments. By overcoming these challenges and leveraging the full potential of sensor technologies, we can pave the way for a future where healthcare is more proactive, precise, and patient-centered than ever before. In conclusion, the advancement of sensors in healthcare holds immense promise for improving health outcomes, reducing healthcare costs, and advancing global health equity. By embracing these technologies responsibly and innovatively, we can create a healthcare landscape that is more accessible, efficient, and equitable for all.

ABBREVIATIONS

AI	Artificial intelligence
AuNPs	Gold nanoparticles
BNP	B-type natriuretic peptides
CGMs	Continuous glucose monitors
CNTs	Carbon nanotubes
CTCs	Circulating tumor cells
ECG	Electrocardiogram
ELISA	Enzyme-linked immunosorbent assay
GCE	Glassy carbon electrode
GNR	Graphene nanoribbons
Gox	Glucose oxidase
LAMP	Loop-mediated isothermal amplification
LOD	Limit of detection
MEMS	Micro-electromechanical system
ML	Machine learning
MOFs	Metal-organic frameworks
MoS_2	Molybdenum disulfide
MRI	Magnetic resonance imaging
OCT	Optical coherence tomography
PANI	Polyaniline
PCR	Polymerase chain reaction
POC	Point-of-care
PPG	Photoplethysmography
SpO_2	Blood oxygen saturation
SPR	Surface plasmon resonance

REFERENCES

(1) Javaid, M.; Haleem, A.; Rab, S.; Singh, R. P.; Suman, R. Sensors for daily life: A review. *Sensors International* **2021**, *2*, 100121.

(2) Albahri, O. S.; Zaidan, A.; Zaidan, B.; Hashim, M.; Albahri, A. S.; Alsalem, M. Real-time remote health-monitoring Systems in a Medical Centre: A review of the provision of healthcare services-based body sensor information, open challenges and methodological aspects. *Journal of Medical Systems* **2018**, *42*, 1–47.

(3) Majumder, S.; Deen, M. J. Smartphone sensors for health monitoring and diagnosis. *Sensors* **2019**, *19* (9), 2164.

(4) Turner, A.; Karube, I.; Wilson, G. S. *Biosensors: Fundamentals and Applications*; Oxford University Press, **1987**.

(5) El-Ansary, A.; Faddah, L. M. Nanoparticles as biochemical sensors. *Nanotechnology, Science and Applications* **2010**, 65–76.

(6) Göpel, W.; Jones, T.; Kleitz, M.; Lundström, I.; Seiyama, T. *Sensors, Chemical and Biochemical Sensors*; John Wiley & Sons, **2008**.

(7) Kimmel, D. W.; LeBlanc, G.; Meschievitz, M. E.; Cliffel, D. E. Electrochemical sensors and biosensors. *Analytical Chemistry* **2012**, *84* (2), 685–707.

(8) Leva-Bueno, J.; Peyman, S. A.; Millner, P. A review on impedimetric immunosensors for pathogen and biomarker detection. *Medical Microbiology and Immunology* **2020**, *209* (3), 343–362.

(9) Qureshi, A.; Niazi, J. H. Biosensors for detecting viral and bacterial infections using host biomarkers: A review. *Analyst* **2020**, *145* (24), 7825–7848.
(10) Lino, C.; Barrias, S.; Chaves, R.; Adega, F.; Martins-Lopes, P.; Fernandes, J. Biosensors as diagnostic tools in clinical applications. *Biochimica et Biophysica Acta (BBA)-Reviews on Cancer* **2022**, *1877* (3), 188726.
(11) Oliver, N.; Toumazou, C.; Cass, A.; Johnston, D. Glucose sensors: A review of current and emerging technology. *Diabetic Medicine* **2009**, *26* (3), 197–210.
(12) Wang, H.-C.; Lee, A.-R. Recent developments in blood glucose sensors. *Journal of Food and Drug Analysis* **2015**, *23* (2), 191–200.
(13) Kornecki, J. F.; Carballares, D.; Tardioli, P. W.; Rodrigues, R. C.; Berenguer-Murcia, Á.; Alcántara, A. R.; Fernandez-Lafuente, R. Enzyme production of D-gluconic acid and glucose oxidase: Successful tales of cascade reactions. *Catalysis Science & Technology* **2020**, *10* (17), 5740–5771.
(14) Wong, C. M.; Wong, K. H.; Chen, X. D. Glucose oxidase: Natural occurrence, function, properties and industrial applications. *Applied Microbiology and Biotechnology* **2008**, *78*, 927–938.
(15) Kubihal, S.; Goyal, A.; Gupta, Y.; Khadgawat, R. Glucose measurement in body fluids: A ready reckoner for clinicians. *Diabetes & Metabolic Syndrome: Clinical Research & Reviews* **2021**, *15* (1), 45–53.
(16) Mian, Z.; Hermayer, K. L.; Jenkins, A. Continuous glucose monitoring: Review of an innovation in diabetes management. *The American Journal of the Medical Sciences* **2019**, *358* (5), 332–339.
(17) Frew, J. E.; Hill, H. A. O. Electrochemical biosensors. *Analytical Chemistry* **1987**, *59* (15), 933A–944A.
(18) Ronkainen, N. J.; Halsall, H. B.; Heineman, W. R. Electrochemical biosensors. *Chemical Society Reviews* **2010**, *39* (5), 1747–1763.
(19) Hareesha, N.; Manjunatha, J. Electro-oxidation of formoterol fumarate on the surface of novel poly (thiazole yellow-G) layered multi-walled carbon nanotube paste electrode. *Scientific Reports* **2021**, *11* (1), 12797.
(20) Raril, C.; Manjunatha, J. G.; Ravishankar, D. K.; Fattepur, S.; Siddaraju, G.; Nanjundaswamy, L. Validated electrochemical method for simultaneous resolution of tyrosine, uric acid, and ascorbic acid at polymer modified nano-composite paste electrode. *Surface Engineering and Applied Electrochemistry* **2020**, *56*, 415–426.
(21) Pushpanjali, P. A.; Manjunatha, J. G.; Hareesha, N. An overview of recent developments of carbon-based sensors for the analysis of drug molecules. *Journal of Electrochemical Science and Engineering* **2021**, *11* (3), 161–177.
(22) Tigari, G.; Manjunatha, J. Poly (glutamine) film-coated carbon nanotube paste electrode for the determination of curcumin with vanillin: An electroanalytical approach. *Monatshefte für Chemie-Chemical Monthly* **2020**, *151*, 1681–1688.
(23) Stotler, B. A.; Kratz, A. Analytical and clinical performance of the epoc blood analysis system: Experience at a large tertiary academic medical center. *American Journal of Clinical Pathology* **2013**, *140* (5), 715–720.
(24) Chen, J.; Gorman, M.; O'Reilly, B.; Chen, Y. Analytical evaluation of the epoc® point-of-care blood analysis system in cardiopulmonary bypass patients. *Clinical Biochemistry* **2016**, *49* (9), 708–712.
(25) Danielsen, J. S.; Voigt, C. T.; Koefod, H.; Jacobsen, K. K. Assessment of the analytical performance of 14 analytes using the Epoc® blood analysis system. *International Journal of Biomedical Laboratory Science* **2022**, *11* (2), 104–113.
(26) Hussain, S.; Mubeen, I.; Ullah, N.; Shah, S. S. U. D.; Khan, B. A.; Zahoor, M.; Ullah, R.; Khan, F. A.; Sultan, M. A. Modern diagnostic imaging technique applications and risk factors in the medical field: A review. *BioMed Research International* **2022**, *2022* (1), 5164970.
(27) Rockall, A. G.; Hatrick, A.; Armstrong, P.; Wastie, M. *Diagnostic Imaging*; John Wiley & Sons, **2013**.
(28) Keen, C.; Massat, M. B. RSNA 2017 technology: Innovation in abundance. *Applied Radiology* **2018**, *47* (1), 20–29.
(29) Qian, C.; Zabow, G.; Koretsky, A. Engineering novel detectors and sensors for MRI. *Journal of Magnetic Resonance* **2013**, *229*, 67–74.
(30) Schmidt, M. A.; Engelhorn, T.; Römer, F.; Mennecke, A.; Laun, F. B.; Nagel, A. M.; Uder, M.; Doerfler, A. Initial experience with the MAGNETOM Terra 7T system. *Focus on Neuroradiology and Knee Imaging. Clinical Neurology MAGNETOM Flash* **2017**, 68 (2) 38–44.

(31) Feinberg, D. A.; Beckett, A. J.; Vu, A. T.; Stockmann, J.; Huber, L.; Ma, S.; Ahn, S.; Setsompop, K.; Cao, X.; Park, S. Next-generation MRI scanner designed for ultra-high-resolution human brain imaging at 7 Tesla. *Nature Methods* **2023**, *20* (12), 2048–2057.
(32) Gao, W.; Emaminejad, S.; Nyein, H. Y. Y.; Challa, S.; Chen, K.; Peck, A.; Fahad, H. M.; Ota, H.; Shiraki, H.; Kiriya, D. Fully integrated wearable sensor arrays for multiplexed in situ perspiration analysis. *Nature* **2016**, *529* (7587), 509–514.
(33) Bandodkar, A. J.; Jeerapan, I.; Wang, J. Wearable chemical sensors: Present challenges and future prospects. *Acs Sensors* **2016**, *1* (5), 464–482.
(34) Parak, J.; Korhonen, I. Evaluation of wearable consumer heart rate monitors based on photopletysmography. In *2014 36th annual international conference of the IEEE engineering in medicine and biology society*; IEEE, **2014**: pp 3670–3673.
(35) Isakadze, N.; Martin, S. S. How useful is the smartwatch ECG? *Trends in Cardiovascular Medicine* **2020**, *30* (7), 442–448.
(36) Henriksen, A.; Haugen Mikalsen, M.; Woldaregay, A. Z.; Muzny, M.; Hartvigsen, G.; Hopstock, L. A.; Grimsgaard, S. Using fitness trackers and smartwatches to measure physical activity in research: Analysis of consumer wrist-worn wearables. *Journal of Medical Internet Research* **2018**, *20* (3), e110.
(37) Spilz, A.; Munz, M. Synchronisation of wearable inertial measurement units based on magnetometer data. *Biomedical Engineering/Biomedizinische Technik* **2023**, *68* (3), 263–273.
(38) Leininger, L. J.; Cook, B. J.; Adams, K. J. Validation and accuracy of Fitbit Charge: A pilot study in a university worksite walking program. *Journal of Fitness Research* **2016**, *5*, 3–9.
(39) Machireddy, A.; Van Santen, J.; Wilson, J. L.; Myers, J.; Hadders-Algra, M.; Song, X. A video/IMU hybrid system for movement estimation in infants. In *2017 39th Annual International Conference of the IEEE Engineering in Medicine and Biology Society (EMBC)*; IEEE, **2017**: pp 730–733.
(40) Long, F.; Zhu, A.; Shi, H. Recent advances in optical biosensors for environmental monitoring and early warning. *Sensors* **2013**, *13* (10), 13928–13948.
(41) Alexander, C. M.; Teller, L. E.; Gross, J. B. Principles of pulse oximetry: Theoretical and practical considerations. *Anesthesia & Analgesia* **1989**, *68* (3), 368–376.
(42) Sargazi, S.; Fatima, I.; Kiani, M. H.; Mohammadzadeh, V.; Arshad, R.; Bilal, M.; Rahdar, A.; Díez-Pascual, A. M.; Behzadmehr, R. Fluorescent-based nanosensors for selective detection of a wide range of biological macromolecules: A comprehensive review. *International Journal of Biological Macromolecules* **2022**, *206*, 115–147.
(43) Goryacheva, O. A.; Ponomaryova, T. D.; Drozd, D. D.; Kokorina, A. A.; Rusanova, T. Y.; Mishra, P. K.; Goryacheva, I. Y. Heart failure biomarkers BNP and NT-proBNP detection using optical labels. *TrAC Trends in Analytical Chemistry* **2022**, *146*, 116477.
(44) Luppa, P. B.; Bietenbeck, A.; Beaudoin, C.; Giannetti, A. Clinically relevant analytical techniques, organizational concepts for application and future perspectives of point-of-care testing. *Biotechnology Advances* **2016**, *34* (3), 139–160.
(45) Raboel, P.; Bartek Jr, J.; Andresen, M.; Bellander, B.; Romner, B. Intracranial pressure monitoring: Invasive versus non-invasive methods—a review. *Critical Care Research and Practice* **2012**, *2012* (1), 950393.
(46) Sonetha, V.; Agarwal, P.; Doshi, S.; Kumar, R.; Mehta, B. Microelectromechanical systems in medicine. *Journal of Medical and Biological Engineering* **2017**, *37*, 580–601.
(47) Judy, J. W. Microelectromechanical systems (MEMS): Fabrication, design and applications. *Smart Materials and Structures* **2001**, *10* (6), 1115.
(48) Romig Jr, A.; Dugger, M. T.; McWhorter, P. J. Materials issues in microelectromechanical devices: Science, engineering, manufacturability and reliability. *Acta Materialia* **2003**, *51* (19), 5837–5866.
(49) Upadhyaya, A. M.; Hasan, M. K.; Abdel-Khalek, S.; Hassan, R.; Srivastava, M. C.; Sharan, P.; Islam, S.; Saad, A. M. E.; Vo, N. A comprehensive review on the optical micro-electromechanical sensors for the biomedical application. *Frontiers in Public Health* **2021**, *9*, 759032.
(50) Yin, H.; Mukadam, B.; Dai, X.; Jha, N. K. DiabDeep: Pervasive diabetes diagnosis based on wearable medical sensors and efficient neural networks. *IEEE Transactions on Emerging Topics in Computing* **2019**, *9* (3), 1139–1150.

(51) Zhan, X.; Long, H.; Gou, F.; Duan, X.; Kong, G.; Wu, J. A convolutional neural network-based intelligent medical system with sensors for assistive diagnosis and decision-making in non-small cell lung cancer. *Sensors* **2021**, *21* (23), 7996.
(52) Haick, H.; Tang, N. Artificial intelligence in medical sensors for clinical decisions. *ACS Nano* **2021**, *15* (3), 3557–3567.
(53) Kozitsina, A. N.; Svalova, T. S.; Malysheva, N. N.; Okhokhonin, A. V.; Vidrevich, M. B.; Brainina, K. Z. Sensors based on bio and biomimetic receptors in medical diagnostic, environment, and food analysis. *Biosensors* **2018**, *8* (2), 35.
(54) Metkar, S. K.; Girigoswami, K. Diagnostic biosensors in medicine–a review. *Biocatalysis and Agricultural Biotechnology* **2019**, *17*, 271–283.
(55) Mehrotra, P. Biosensors and their applications–A review. *Journal of Oral Biology and Craniofacial Research* **2016**, *6* (2), 153–159.
(56) Karunakaran, C.; Rajkumar, R.; Bhargava, K. Introduction to biosensors. In *Biosensors and bioelectronics*, Elsevier, **2015**: pp 1–68.
(57) Fronczek, C. F.; Yoon, J.-Y. Biosensors for monitoring airborne pathogens. *Journal of Laboratory Automation* **2015**, *20* (4), 390–410.
(58) Vidic, J.; Manzano, M.; Chang, C.-M.; Jaffrezic-Renault, N. Advanced biosensors for detection of pathogens related to livestock and poultry. *Veterinary Research* **2017**, *48*, 1–22.
(59) Samson, R.; Navale, G. R.; Dharne, M. S. Biosensors: Frontiers in rapid detection of COVID-19. *3 Biotech* **2020**, *10* (9), 385.
(60) Xu, L.; Li, D.; Ramadan, S.; Li, Y.; Klein, N. Facile biosensors for rapid detection of COVID-19. *Biosensors and Bioelectronics* **2020**, *170*, 112673.
(61) Van Regenmortel, M. H.; Altschuh, D.; Chatellier, J.; Bruyère, N. R.; Sécordel, P. R.; Saunal, H. Uses of biosensors in the study of viral antigens. *Immunological Investigations* **1997**, *26* (1–2), 67–82.
(62) Caygill, R. L.; Blair, G. E.; Millner, P. A. A review on viral biosensors to detect human pathogens. *Analytica Chimica Acta* **2010**, *681* (1-2), 8–15.
(63) Becherer, L.; Borst, N.; Bakheit, M.; Frischmann, S.; Zengerle, R.; von Stetten, F. Loop-mediated isothermal amplification (LAMP)–review and classification of methods for sequence-specific detection. *Analytical Methods* **2020**, *12* (6), 717–746.
(64) Islam, M. M.; Koirala, D. Toward a next-generation diagnostic tool: A review on emerging isothermal nucleic acid amplification techniques for the detection of SARS-CoV-2 and other infectious viruses. *Analytica Chimica Acta* **2022**, *1209*, 339338.
(65) Hussain, S. H.; Huertas, C. S.; Mitchell, A.; Deman, A.-L.; Laurenceau, E. Biosensors for circulating tumor cells (CTCs)-biomarker detection in lung and prostate cancer: Trends and prospects. *Biosensors and Bioelectronics* **2022**, *197*, 113770.
(66) Ferhan, A. R.; Jackman, J. A.; Park, J. H.; Cho, N.-J.; Kim, D.-H. Nanoplasmonic sensors for detecting circulating cancer biomarkers. *Advanced Drug Delivery Reviews* **2018**, *125*, 48–77.
(67) Alawieh, H.; El Chemaly, T.; Alam, S.; Khraiche, M. Towards point-of-care heart failure diagnostic platforms: BNP and NT-proBNP biosensors. *Sensors* **2019**, *19* (22), 5003.
(68) Shinde, R.; Juwarwala, I.; Modi, V.; Chandarana, C. V. Utility of cardiac biomarkers and biosensors for diagnosis of acute myocardial infarction. *Global Translational Medicine* **2023**, *2* (2), 0403.
(69) Park, I.; Lim, J.; You, S.; Hwang, M. T.; Kwon, J.; Koprowski, K.; Kim, S.; Heredia, J.; Stewart de Ramirez, S. A.; Valera, E. Detection of SARS-CoV-2 virus amplification using a crumpled graphene field-effect transistor biosensor. *ACS Sensors* **2021**, *6* (12), 4461–4470.
(70) Nguyen, N. H. L.; Kim, S.; Lindemann, G.; Berry, V. COVID-19 spike protein induced phononic modification in antibody-coupled graphene for viral detection application. *ACS Nano* **2021**, *15* (7), 11743–11752.
(71) Nguyen, A. H.; Sim, S. J. Nanoplasmonic biosensor: Detection and amplification of dual biosignatures of circulating tumor DNA. *Biosensors and Bioelectronics* **2015**, *67*, 443–449.
(72) Rasmi, Y.; Mosa, O. F.; Alipour, S.; Heidari, N.; Javanmard, F.; Golchin, A.; Gholizadeh-Ghaleh Aziz, S. Significance of cardiac troponins as an identification tool in COVID-19 patients using biosensors: An update. *Frontiers in Molecular Biosciences* **2022**, *9*, 821155.
(73) Wen, X.; Pu, H.; Liu, Q.; Guo, Z.; Luo, D. Circulating Tumor DNA—A Novel Biomarker of Tumor Progression and Its Favorable Detection Techniques. *Cancers* **2022**, *14* (24), 6025.

(74) Karaboğa, M. N. S.; Sezgintürk, M. K. Biosensor approaches on the diagnosis of neurodegenerative diseases: Sensing the past to the future. *Journal of Pharmaceutical and Biomedical Analysis* **2022**, *209*, 114479.
(75) Kujawska, M.; Bhardwaj, S. K.; Mishra, Y. K.; Kaushik, A. Using graphene-based biosensors to detect dopamine for efficient parkinson's disease diagnostics. *Biosensors* **2021**, *11* (11), 433.
(76) Qiao, Y.; Qiao, L.; Chen, Z.; Liu, B.; Gao, L.; Zhang, L. Wearable sensor for continuous sweat biomarker monitoring. *Chemosensors* **2022**, *10* (7), 273.
(77) Lai, Z.; Tan, J.; Wan, R.; Tan, J.; Zhang, Z.; Hu, Z.; Li, J.; Yang, W.; Wang, Y.; Jiang, Y. An 'activatable'aptamer-based fluorescence probe for the detection of HepG2 cells. *Oncology Reports* **2017**, *37* (5), 2688–2694.
(78) Chen, P.; Wang, Y.; He, Y.; Huang, K.; Wang, X.; Zhou, R.; Liu, T.; Qu, R.; Zhou, J.; Peng, W. Homogeneous visual and fluorescence detection of circulating tumor cells in clinical samples via selective recognition reaction and enzyme-free amplification. *ACS Nano* **2021**, *15* (7), 11634--11643.
(79) Boozer, C.; Kim, G.; Cong, S.; Guan, H.; Londergan, T. Looking towards label-free biomolecular interaction analysis in a high-throughput format: A review of new surface plasmon resonance technologies. *Current Opinion in Biotechnology* **2006**, *17* (4), 400–405.
(80) Chang, T.-C.; Sun, A. Y.; Huang, Y.-C.; Wang, C.-H.; Wang, S.-C.; Chau, L.-K. Integration of power-free and self-contained microfluidic chip with fiber optic particle plasmon resonance aptasensor for rapid detection of SARS-CoV-2 nucleocapsid protein. *Biosensors* **2022**, *12* (10), 785.
(81) Lim, W. Y.; Lan, B. L.; Ramakrishnan, N. Emerging biosensors to detect severe acute respiratory syndrome coronavirus 2 (SARS-CoV-2): A review. *Biosensors* **2021**, *11* (11), 434.
(82) Kong, K.; Kendall, C.; Stone, N.; Notingher, I. Raman spectroscopy for medical diagnostics—From in-vitro biofluid assays to in-vivo cancer detection. *Advanced Drug Delivery Reviews* **2015**, *89*, 121--134.
(83) Kalkanis, S. N.; Kast, R. E.; Rosenblum, M. L.; Mikkelsen, T.; Yurgelevic, S. M.; Nelson, K. M.; Raghunathan, A.; Poisson, L. M.; Auner, G. W. Raman spectroscopy to distinguish grey matter, necrosis, and glioblastoma multiforme in frozen tissue sections. *Journal of Neuro-oncology* **2014**, *116*, 477–485.
(84) Aguiar, R. P.; Falcão, E. T.; Pasqualucci, C. A.; Silveira Jr, L. Use of Raman spectroscopy to evaluate the biochemical composition of normal and tumoral human brain tissues for diagnosis. *Lasers in Medical Science* **2022**, *37* (1), 121–133.
(85) Riva, M.; Sciortino, T.; Secoli, R.; D'Amico, E.; Moccia, S.; Fernandes, B.; Conti Nibali, M.; Gay, L.; Rossi, M.; De Momi, E. Glioma biopsies classification using Raman spectroscopy and machine learning models on fresh tissue samples. *Cancers* **2021**, *13* (5), 1073.
(86) Galli, R.; Meinhardt, M.; Koch, E.; Schackert, G.; Steiner, G.; Kirsch, M.; Uckermann, O. Rapid label-free analysis of brain tumor biopsies by near infrared Raman and fluorescence spectroscopy—a study of 209 patients. *Frontiers in Oncology* **2019**, *9*, 1165.
(87) Budoff, M. J.; Achenbach, S.; Blumenthal, R. S.; Carr, J. J.; Goldin, J. G.; Greenland, P.; Guerci, A. D.; Lima, J. A.; Rader, D. J.; Rubin, G. D. Assessment of coronary artery disease by cardiac computed tomography: A scientific statement from the American Heart Association Committee on Cardiovascular Imaging and Intervention, Council on Cardiovascular Radiology and Intervention, and Committee on Cardiac Imaging, Council on Clinical Cardiology. *Circulation* **2006**, *114* (16), 1761–1791.
(88) Prakash, A.; Powell, A. J.; Geva, T. Multimodality noninvasive imaging for assessment of congenital heart disease. *Circulation: Cardiovascular Imaging* **2010**, *3* (1), 112–125.
(89) Jang, I.-K.; Tearney, G. J.; MacNeill, B.; Takano, M.; Moselewski, F.; Iftima, N.; Shishkov, M.; Houser, S.; Aretz, H. T.; Halpern, E. F. In vivo characterization of coronary atherosclerotic plaque by use of optical coherence tomography. *Circulation* **2005**, *111* (12), 1551–1555.
(90) Jang, I.-K.; Bouma, B. E.; Kang, D.-H.; Park, S.-J.; Park, S.-W.; Seung, K.-B.; Choi, K.-B.; Shishkov, M.; Schlendorf, K.; Pomerantsev, E. Visualization of coronary atherosclerotic plaques in patients using optical coherence tomography: Comparison with intravascular ultrasound. *Journal of the American College of Cardiology* **2002**, *39* (4), 604–609.
(91) McBirney, S. E.; Chen, D.; Scholtz, A.; Ameri, H.; Armani, A. M. Rapid diagnostic for point-of-care malaria screening. *ACS Sensors* **2018**, *3* (7), 1264–1270.

(92) Armani, A.; Chen, D.; McBirney, S.; Kaypaghian, K.; Huber, H.; Ameri, H. Portable optical diagnostics for early malaria detection. In *Frontiers in Biological Detection: From Nanosensors to Systems XI*; SPIE, **2019**: Vol. 10895, pp 29–33.
(93) Manjunatha, J.; Swamy, K.; Mamatha, G.; Gilbert, O.; Sherigara, B. Poly (maleic acid) modified carbon paste electrode for simultaneous detection of dopamine in the presence of uric acid: A cyclic voltammetric study. *Analytical and Bioanalytical Electrochemistry* **2011**, *3* (2), 146–159.
(94) Privett, B. J.; Shin, J. H.; Schoenfisch, M. H. Electrochemical sensors. *Analytical Chemistry* **2010**, *82* (12), 4723–4741.
(95) Saputra, H. A. Electrochemical sensors: Basic principles, engineering, and state of the art. *Monatshefte für Chemie-Chemical Monthly* **2023**, *154* (10), 1083–1100.
(96) Zdrachek, E.; Bakker, E. Potentiometric sensing. *Analytical Chemistry* **2020**, *93* (1), 72–102.
(97) Baranwal, J.; Barse, B.; Gatto, G.; Broncova, G.; Kumar, A. Electrochemical sensors and their applications: A review. *Chemosensors* **2022**, *10* (9), 363.
(98) Pushpanjali, P.; Manjunatha, J.; Hareesha, N.; D'Souza, E. S.; Charithra, M.; Prinith, N. Voltammetric analysis of antihistamine drug cetirizine and paracetamol at poly (L-Leucine) layered carbon nanotube paste electrode. *Surfaces and Interfaces* **2021**, *24*, 101154.
(99) Leung, H. M. C.; Forlenza, G. P.; Prioleau, T. O.; Zhou, X. Noninvasive glucose sensing in vivo. *Sensors* **2023**, *23* (16), 7057.
(100) Alafeef, M.; Dighe, K.; Moitra, P.; Pan, D. Rapid, ultrasensitive, and quantitative detection of SARS-CoV-2 using antisense oligonucleotides directed electrochemical biosensor chip. *ACS Nano* **2020**, *14* (12), 17028–17045.
(101) Wang, C.; Zhang, Y.; Liu, Y.; Zeng, X.; Jin, C.; Huo, D.; Hou, J.; Hou, C. A wearable flexible electrochemical biosensor with CuNi-MOF@ rGO modification for simultaneous detection of uric acid and dopamine in sweat. *Analytica Chimica Acta* **2024**, *1299*, 342441.
(102) Saha, U.; Todi, K.; Malhotra, B. D. Emerging DNA-based multifunctional nano-biomaterials towards electrochemical sensing applications. *Nanoscale* **2021**, *13* (23), 10305–10319.
(103) Liu, M.; Sun, Y.; Teh, D. B. L.; Zhang, Y.; Cao, D.; Mei, Q. Nanothermometry for cellular temperature monitoring and disease diagnostics. *Interdisciplinary Medicine* **2024**, *2* (2), e20230059.
(104) Wang, T.-W.; Lin, S.-F. Wearable piezoelectric-based system for continuous beat-to-beat blood pressure measurement. *Sensors* **2020**, *20* (3), 851.
(105) Patel, S.; Wang, M.; Guo, J.; Smith, G.; Chen, C. A study of RR interval transition matrix features for machine learning algorithms in AFib detection. *Sensors* **2023**, *23* (7), 3700.
(106) Zhang, X.; Wang, C.; Zheng, T.; Wu, H.; Wu, Q.; Wang, Y. Wearable optical fiber sensors in medical monitoring applications: A review. *Sensors* **2023**, *23* (15), 6671.
(107) Shung, K. K.; Cannata, J.; Zhou, Q. Piezoelectric materials for high frequency medical imaging applications: A review. *Journal of Electroceramics* **2007**, *19*, 141–147.
(108) Bailey, T.; Bode, B. W.; Christiansen, M. P.; Klaff, L. J.; Alva, S. The performance and usability of a factory-calibrated flash glucose monitoring system. *Diabetes Technology & Therapeutics* **2015**, *17* (11), 787–794.
(109) Shah, V. N.; Laffel, L. M.; Wadwa, R. P.; Garg, S. K. Performance of a factory-calibrated real-time continuous glucose monitoring system utilizing an automated sensor applicator. *Diabetes Technology & Therapeutics* **2018**, *20* (6), 428–433.
(110) Lausch, M. J. Technology advancement: Hybrid closed-loop insulin pump. *AADE in Practice* **2018**, *6* (5), 10–15.
(111) McAuley, S. A.; De Bock, M. I.; Sundararajan, V.; Lee, M. H.; Paldus, B.; Ambler, G. R.; Bach, L. A.; Burt, M. G.; Cameron, F. J.; Clarke, P. M. Effect of 6 months of hybrid closed-loop insulin delivery in adults with type 1 diabetes: A randomised controlled trial protocol. *BMJ Open* **2018**, *8* (6), e020274.
(112) Zeng, X.; Zhang, Y.; Du, X.; Li, Y.; Tang, W. A highly sensitive glucose sensor based on a gold nanoparticles/polyaniline/multi-walled carbon nanotubes composite modified glassy carbon electrode. *New Journal of Chemistry* **2018**, *42* (14), 11944–11953.
(113) Jeong, H.; Nguyen, D. M.; Lee, M. S.; Kim, H. G.; Ko, S. C.; Kwac, L. K. N-doped graphene-carbon nanotube hybrid networks attaching with gold nanoparticles for glucose non-enzymatic sensor. *Materials Science and Engineering: C* **2018**, *90*, 38–45.

(114) Peña-Bahamonde, J.; Nguyen, H. N.; Fanourakis, S. K.; Rodrigues, D. F. Recent advances in graphene-based biosensor technology with applications in life sciences. *Journal of Nanobiotechnology* **2018**, *16* (1), 75.
(115) Vilian, A. E.; Dinesh, B.; Rethinasabapathy, M.; Hwang, S.-K.; Jin, C.-S.; Huh, Y. S.; Han, Y.-K. Hexagonal Co_3O_4 anchored reduced graphene oxide sheets for high-performance supercapacitors and non-enzymatic glucose sensing. *Journal of Materials Chemistry A* **2018**, *6* (29), 14367–14379.
(116) Jothi, L.; Jayakumar, N.; Jaganathan, S.; Nageswaran, G. Ultrasensitive and selective non-enzymatic electrochemical glucose sensor based on hybrid material of graphene nanosheets/graphene nanoribbons/nickel nanoparticle. *Materials Research Bulletin* **2018**, *98*, 300–307.
(117) Gengan, S.; Gnanamuthu, R.; Sankaranarayanan, S.; Reddy, V. M.; Marepally, B. C.; Biroju, R. K. Electrochemical modified Pt nanoflower@ rGO for non-enzymatic electrochemical sensing of glucose. *Sensors and Actuators A: Physical* **2023**, *353*, 114232.
(118) Adeel, M.; Asif, K.; Rahman, M. M.; Daniele, S.; Canzonieri, V.; Rizzolio, F. Glucose detection devices and methods based on metal–organic frameworks and related materials. *Advanced Functional Materials* **2021**, *31* (52), 2106023.
(119) Xu, W.; Jiao, L.; Yan, H.; Wu, Y.; Chen, L.; Gu, W.; Du, D.; Lin, Y.; Zhu, C. Glucose oxidase-integrated metal–organic framework hybrids as biomimetic cascade nanozymes for ultrasensitive glucose biosensing. *ACS Applied Materials & Interfaces* **2019**, *11* (25), 22096–22101.
(120) Kumar, R.; Singh, L. $Ti_3C_2T_x$ MXene as electrocatalyst for designing robust glucose biosensors. *Advanced Materials Technologies* **2022**, *7* (12), 2200151.
(121) Feng, L.; Chen, J.; Yang, M.; Wang, J.; Yin, S.; Zhang, D.; Qin, W.; Song, J. CuxO decorated $Ti_3C_2T_x$ MXene composites for non-enzymatic glucose sensing with large linear ranges. *Microchimica Acta* **2024**, *191* (8), 1–10.
(122) Li, Q.-F.; Chen, X.; Wang, H.; Liu, M.; Peng, H.-L. Pt/MXene-based flexible wearable non-enzymatic electrochemical sensor for continuous glucose detection in sweat. *ACS Applied Materials & Interfaces* **2023**, *15* (10), 13290–13298.
(123) Kim, J.; Campbell, A. S.; de Ávila, B. E.-F.; Wang, J. Wearable biosensors for healthcare monitoring. *Nature Biotechnology* **2019**, *37* (4), 389–406.
(124) Arshad, A.; Irfan, H.; Iftikhar, S.; Yameen, B. Conductive polymers for cardiovascular applications. In *Nanoscale Engineering of Biomaterials: Properties and Applications*, Springer, **2022**: pp 319–347.
(125) Gupta, S.; Sharma, A.; Verma, R. S. Polymers in biosensor devices for cardiovascular applications. *Current Opinion in Biomedical Engineering* **2020**, *13*, 69–75.
(126) Khushaim, W.; Peramaiah, K.; Beduk, T.; Vijjapu, M. T.; de Oliveira Filho, J. I.; Huang, K.-W.; Mani, V.; Salama, K. N. Porous graphitic carbon nitrides integrated biosensor for sensitive detection of cardiac troponin I. *Biosensors and Bioelectronics: X* **2022**, *12*, 100234.
(127) Satyanarayana, M.; Goud, K. Y.; Reddy, K. K.; Kumar, V. S.; Gobi, K. V. Silver nanoparticles impregnated chitosan layered carbon nanotube as sensor interface for electrochemical detection of clopidogrel in-vitro. *Materials Science and Engineering: C* **2019**, *101*, 103–110.
(128) Rajesh, R.; Sharma, V.; Puri, N. K.; Singh, R. K.; Biradar, A. M.; Mulchanadani, A. Label-free detection of cardiac troponin-I using gold nanoparticles functionalized single-walled carbon nanotubes based chemiresistive biosensor. *Applied Physics Letters* **2013**, *103* (20).
(129) Liu, Y.; Meng, F.; Zhou, Y.; Mugo, S. M.; Zhang, Q. Graphene oxide films prepared using gelatin nanofibers as wearable sensors for monitoring cardiovascular health. *Advanced Materials Technologies* **2019**, *4* (11), 1900540.
(130) Palanisamy, S.; Senthil Raja, D.; Subramani, B.; Wu, T.-H.; Wang, Y.-M. Label-free bimetallic in situ-grown 3D nickel-foam-supported NH2-MIL-88B (Fe2Co)-MOF-based impedimetric immunosensor for the detection of cardiac troponin I. *ACS Applied Materials & Interfaces* **2020**, *12* (29), 32468–32476.
(131) Pang, C.; Lee, C.; Suh, K. Y. Recent advances in flexible sensors for wearable and implantable devices. *Journal of Applied Polymer Science* **2013**, *130* (3), 1429–1441.
(132) Chen, S.; Qi, J.; Fan, S.; Qiao, Z.; Yeo, J. C.; Lim, C. T. Flexible wearable sensors for cardiovascular health monitoring. *Advanced Healthcare Materials* **2021**, *10* (17), 2100116.
(133) Prakash, N. J.; Mane, P. P.; George, S. M.; Kandasubramanian, B. Silk fibroin as an immobilization matrix for sensing applications. *ACS Biomaterials Science & Engineering* **2021**, *7* (6), 2015–2042.

(134) Zhang, Y.; Aniruddha, P.; Hou, C.; Lu, D.; Qiu, W.; Kong, L.; Wu, R.; Ma, L.; Yu, R.; Yu, W. Reconstructed silk fibroin mediated smart wristband for physiological signal detection. *Chemical Engineering Journal* **2022**, *428*, 132362.
(135) Lyu, Q.; Gong, S.; Yin, J.; Dyson, J. M.; Cheng, W. Soft wearable healthcare materials and devices. *Advanced Healthcare Materials* **2021**, *10* (17), 2100577.
(136) Han, Y.; Su, X.; Fan, L.; Liu, Z.; Guo, Y. Electrochemical aptasensor for sensitive detection of Cardiac troponin I based on CuNWs/MoS2/rGO nanocomposite. *Microchemical Journal* **2021**, *169*, 106598.
(137) Zaman, S. u.; Tao, X.; Cochrane, C.; Koncar, V. Smart E-textile systems: A review for healthcare applications. *Electronics* **2021**, *11* (1), 99.
(138) Medici, S.; Peana, M.; Coradduzza, D.; Zoroddu, M. A. Gold nanoparticles and cancer: Detection, diagnosis and therapy. In *Seminars in cancer biology*; Elsevier, **2021**: Vol. 76, pp 27–37.
(139) Sheikhpour, M.; Golbabaie, A.; Kasaeian, A. Carbon nanotubes: A review of novel strategies for cancer diagnosis and treatment. *Materials Science and Engineering: C* **2017**, *76*, 1289–1304.
(140) Pourmadadi, M.; Soleimani Dinani, H.; Saeidi Tabar, F.; Khassi, K.; Janfaza, S.; Tasnim, N.; Hoorfar, M. Properties and applications of graphene and its derivatives in biosensors for cancer detection: A comprehensive review. *Biosensors* **2022**, *12* (5), 269.
(141) Liang, Z.; Khawar, M. B.; Liang, J.; Sun, H. Bio-conjugated quantum dots for cancer research: Detection and imaging. *Frontiers in Oncology* **2021**, *11*, 749970.
(142) Jazayeri, M.; Amani, H.; Pourfatollah, A.; Avan, A.; Ferns, G.; Pazoki-Toroudi, H. Enhanced detection sensitivity of prostate-specific antigen via PSA-conjugated gold nanoparticles based on localized surface plasmon resonance: GNP-coated anti-PSA/LSPR as a novel approach for the identification of prostate anomalies. *Cancer Gene Therapy* **2016**, *23* (10), 365–369.
(143) Materon, E. M.; Gómez, F. R.; Joshi, N.; Dalmaschio, C. J.; Carrilho, E.; Oliveira Jr, O. N. Smart materials for electrochemical flexible nanosensors: Advances and applications. *Nanosensors for Smart Manufacturing* 2021, 347–371.
(144) Hasanzadeh, M.; Shadjou, N.; Lin, Y.; de la Guardia, M. Nanomaterials for use in immunosensing of carcinoembryonic antigen (CEA): Recent advances. *TrAC Trends in Analytical Chemistry* **2017**, *86*, 185–205.
(145) Zare, E. N.; Agarwal, T.; Zarepour, A.; Pinelli, F.; Zarrabi, A.; Rossi, F.; Ashrafizadeh, M.; Maleki, A.; Shahbazi, M.-A.; Maiti, T. K. Electroconductive multi-functional polypyrrole composites for biomedical applications. *Applied Materials Today* **2021**, *24*, 101117.
(146) Bavatharani, C.; Muthusankar, E.; Wabaidur, S. M.; Alothman, Z. A.; Alsheetan, K. M.; mana AL-Anazy, M.; Ragupathy, D. Electrospinning technique for production of polyaniline nanocomposites/nanofibres for multi-functional applications: A review. *Synthetic Metals* **2021**, *271*, 116609.
(147) Kazemi, F.; Naghib, S. M.; Zare, Y.; Rhee, K. Y. Biosensing applications of polyaniline (PANI)-based nanocomposites: A review. *Polymer Reviews* **2021**, *61* (3), 553–597.
(148) Rocha, J. F.; Hasimoto, L. H.; Santhiago, M. Recent progress and future perspectives of polydopamine nanofilms toward functional electrochemical sensors. *Analytical and Bioanalytical Chemistry* **2023**, *415* (18), 3799–3816.
(149) Bhakta, S.; Mishra, P. Molecularly imprinted polymer-based sensors for cancer biomarker detection. *Sensors and Actuators Reports* **2021**, *3*, 100061.
(150) Won, H. J.; Robby, A. I.; Jhon, H. S.; In, I.; Ryu, J. H.; Park, S. Y. Wireless label-free electrochemical detection of cancer cells by MnO2-Decorated polymer dots. *Sensors and Actuators B: Chemical* **2020**, *320*, 128391.
(151) Meng, H.-M.; Liu, H.; Kuai, H.; Peng, R.; Mo, L.; Zhang, X.-B. Aptamer-integrated DNA nanostructures for biosensing, bioimaging and cancer therapy. *Chemical Society Reviews* **2016**, *45* (9), 2583–2602.
(152) Huang, Z.; Wang, D.; Zhang, Q.; Zhang, Y.; Peng, R.; Tan, W. Leveraging aptamer-based DNA nanotechnology for bioanalysis and cancer therapeutics. *Accounts of Materials Research* **2024**, *5* (4), 438–452.
(153) Chandrasekaran, A. R.; Punnoose, J. A.; Zhou, L.; Dey, P.; Dey, B. K.; Halvorsen, K. DNA nanotechnology approaches for microRNA detection and diagnosis. *Nucleic Acids Research* **2019**, *47* (20), 10489–10505.

(154) Asci Erkocyigit, B.; Ozufuklar, O.; Yardim, A.; Guler Celik, E.; Timur, S. Biomarker detection in early diagnosis of cancer: Recent achievements in point-of-care devices based on paper microfluidics. *Biosensors* **2023**, *13* (3), 387.

(155) Yu, Q.; Chen, J.; Fu, W.; Muhammad, K. G.; Li, Y.; Liu, W.; Xu, L.; Dong, H.; Wang, D.; Liu, J. Smartphone-based platforms for clinical detections in lung-cancer-related exhaled breath biomarkers: A review. *Biosensors* **2022**, *12* (4), 223.

(156) Yonet-Tanyeri, N.; Ahlmark, B. Z.; Little, S. R. Advances in multiplexed paper-based analytical devices for cancer diagnosis: A review of technological developments. *Advanced Materials Technologies* **2021**, *6* (8), 2001138.

(157) Ehtesabi, H. Application of carbon nanomaterials in human virus detection. *Journal of Science: Advanced Materials and Devices* **2020**, *5* (4), 436–450.

(158) Fahmy, H. M.; Abu Serea, E. S.; Salah-Eldin, R. E.; Al-Hafiry, S. A.; Ali, M. K.; Shalan, A. E.; Lanceros-Méndez, S. Recent progress in graphene-and related carbon-nanomaterial-based electrochemical biosensors for early disease detection. *ACS Biomaterials Science & Engineering* **2022**, *8* (3), 964–1000.

(159) Mousavi, S. M.; Hashemi, S. A.; Yari Kalashgrani, M.; Omidifar, N.; Lai, C. W.; Vijayakameswara Rao, N.; Gholami, A.; Chiang, W.-H. The pivotal role of quantum dots-based biomarkers integrated with ultra-sensitive probes for multiplex detection of human viral infections. *Pharmaceuticals* **2022**, *15* (7), 880.

(160) Kumar, S.; Rathee, G.; Bartwal, G.; Solanki, P. R. Biosensors for point-of-care (POC) applications: The flag bearer of the modern medicinal technology to tackle infectious diseases. *Point-of-Care Biosensors for Infectious Diseases* **2023**, 69–86.

(161) Mauk, M.; Song, J.; Bau, H. H.; Gross, R.; Bushman, F. D.; Collman, R. G.; Liu, C. Miniaturized devices for point of care molecular detection of HIV. *Lab on a Chip* **2017**, *17* (3), 382–394.

(162) Fraser, L. A.; Kinghorn, A. B.; Dirkzwager, R. M.; Liang, S.; Cheung, Y.-W.; Lim, B.; Shiu, S. C.-C.; Tang, M. S.; Andrew, D.; Manitta, J. A portable microfluidic Aptamer-Tethered Enzyme Capture (APTEC) biosensor for malaria diagnosis. *Biosensors and Bioelectronics* **2018**, *100*, 591–596.

(163) Rodriguez-Manzano, J.; Malpartida-Cardenas, K.; Moser, N.; Pennisi, I.; Cavuto, M.; Miglietta, L.; Moniri, A.; Penn, R.; Satta, G.; Randell, P. Handheld point-of-care system for rapid detection of SARS-CoV-2 extracted RNA in under 20 min. *ACS Central Science* **2021**, *7* (2), 307–317.

(164) Tarim, E. A.; Karakuzu, B.; Oksuz, C.; Sarigil, O.; Kizilkaya, M.; Al-Ruweidi, M. K. A.; Yalcin, H. C.; Ozcivici, E.; Tekin, H. C. Microfluidic-based virus detection methods for respiratory diseases. *Emergent Materials* **2021**, *4*, 143–168.

(165) Corrigan, D.; Schulze, H.; Henihan, G.; Hardie, A.; Ciani, I.; Giraud, G.; Terry, J.; Walton, A.; Pethig, R.; Ghazal, P. Development of a PCR-free electrochemical point of care test for clinical detection of methicillin resistant *Staphylococcus aureus* (MRSA). *Analyst* **2013**, *138* (22), 6997–7005.

(166) Das, D.; Lin, C.-W.; Chuang, H.-S. LAMP-based point-of-care biosensors for rapid pathogen detection. *Biosensors* **2022**, *12* (12), 1068.

(167) Hu, J.; Wang, L.; Li, F.; Han, Y. L.; Lin, M.; Lu, T. J.; Xu, F. Oligonucleotide-linked gold nanoparticle aggregates for enhanced sensitivity in lateral flow assays. *Lab on a Chip* **2013**, *13* (22), 4352–4357.

(168) Raman, S.; Ravi Sankar, A. Intrinsically conducting polymers in flexible and stretchable resistive strain sensors: A review. *Journal of Materials Science* **2022**, *57* (28), 13152–13178.

(169) Ansari, M. J.; Rajendran, R. R.; Mohanto, S.; Agarwal, U.; Panda, K.; Dhotre, K.; Manne, R.; Deepak, A.; Zafar, A.; Yasir, M. Poly (N-isopropylacrylamide)-based hydrogels for biomedical applications: A review of the state-of-the-art. *Gels* **2022**, *8* (7), 454.

(170) Horkay, F.; Cho, S.; Tathireddy, P.; Rieth, L.; Solzbacher, F.; Magda, J. Thermodynamic analysis of the selectivity enhancement obtained by using smart hydrogels that are zwitterionic when detecting glucose with boronic acid moieties. *Sensors and Actuators B: Chemical* **2011**, *160* (1), 1363–1371.

(171) Brothers, M. C.; DeBrosse, M.; Grigsby, C. C.; Naik, R. R.; Hussain, S. M.; Heikenfeld, J.; Kim, S. S. Achievements and challenges for real-time sensing of analytes in sweat within wearable platforms. *Accounts of Chemical Research* **2019**, *52* (2), 297–306.

(172) Polat, E. O.; Mercier, G.; Nikitskiy, I.; Puma, E.; Galan, T.; Gupta, S.; Montagut, M.; Piqueras, J. J.; Bouwens, M.; Durduran, T. Flexible graphene photodetectors for wearable fitness monitoring. *Science Advances* **2019**, *5* (9), eaaw7846.

(173) Zambrana-Vinaroz, D.; Vicente-Samper, J. M.; Manrique-Cordoba, J.; Sabater-Navarro, J. M. Wearable epileptic seizure prediction system based on machine learning techniques using ECG, PPG and EEG signals. *Sensors* **2022**, *22* (23), 9372.
(174) Santos, M. D.; Roman, C.; Pimentel, M. A.; Vollam, S.; Areia, C.; Young, L.; Watkinson, P.; Tarassenko, L. A real-time wearable system for monitoring vital signs of COVID-19 patients in a hospital setting. *Frontiers in Digital Health* **2021**, *3*, 630273.
(175) Hassan, M.; Abbas, G.; Li, N.; Afzal, A.; Haider, Z.; Ahmed, S.; Xu, X.; Pan, C.; Peng, Z. Significance of flexible substrates for wearable and implantable devices: Recent advances and perspectives. *Advanced Materials Technologies* **2022**, *7* (3), 2100773.
(176) Ma, Z.; Li, H.; Jing, X.; Liu, Y.; Mi, H.-Y. Recent advancements in self-healing composite elastomers for flexible strain sensors: Materials, healing systems, and features. *Sensors and Actuators A: Physical* **2021**, *329*, 112800.
(177) Ren, X.; Ma, H.; Zhang, T.; Zhang, Y.; Yan, T.; Du, B.; Wei, Q. Sulfur-doped graphene-based immunological biosensing platform for multianalysis of cancer biomarkers. *ACS Applied Materials & Interfaces* **2017**, *9* (43), 37637–37644.
(178) Ielo, I.; Giacobello, F.; Sfameni, S.; Rando, G.; Galletta, M.; Trovato, V.; Rosace, G.; Plutino, M. R. Nanostructured surface finishing and coatings: Functional properties and applications. *Materials* **2021**, *14* (11), 2733.
(179) Linh, V.T.N.; Han, S.; Koh, E.; Kim, S.; Jung, H.S.; Koo, J. Advances in wearable electronics for monitoring human organs: bridging external and internal health assessments. *Biomaterials* **2025**, *314*, 122865.
(180) Wattamwar, R.V.; Dhotre, D.; Madachane, S.S.; Mulani, N.T.; Kumbhar, S.A.; Kamble, R.S. Human-centric design for wearable technology: bridging aesthetic and functional aspects in smart devices. *Panamerican Mathematical Journal*, **2025**, *35* (2s) 116–131.
(181) Rodriguez-León, C.; Villalonga, C.; Munoz-Torres, M.; Ruiz, J. R.; Banos, O. Mobile and wearable technology for the monitoring of diabetes-related parameters: Systematic review. *JMIR mHealth and uHealth* **2021**, *9* (6), e25138.
(182) Fu, Y.; Ma, Q. Recent developments in electrochemiluminescence nanosensors for cancer diagnosis applications. *Nanoscale* **2020**, *12* (26), 13879–13898.
(183) Patel, S. R.; Lieber, C. M. Precision electronic medicine in the brain. *Nature Biotechnology* **2019**, *37* (9), 1007–1012.
(184) Ramasamy, M.; Kumar, P. S.; Varadan, V. K. Wearable nanosensor systems and their applications in healthcare. In *Nanosensors, Biosensors, Info-Tech Sensors and 3D Systems 2017*; SPIE, **2017**: Vol. 10167, p 1016703.
(185) Swierczewska, M.; Liu, G.; Lee, S.; Chen, X. High-sensitivity nanosensors for biomarker detection. *Chemical Society Reviews* **2012**, *41* (7), 2641–2655.
(186) Zheng, X. T.; Tan, Y. N. Recent development of nucleic acid nanosensors to detect sequence-specific binding interactions: From metal ions, small molecules to proteins and pathogens. *Sensors International* **2020**, *1*, 100034.
(187) Velusamy, P.; Su, C.-H.; Ramasamy, P.; Arun, V.; Rajnish, N.; Raman, P.; Baskaralingam, V.; Senthil Kumar, S. M.; Gopinath, S. C. Volatile organic compounds as potential biomarkers for noninvasive disease detection by nanosensors: A comprehensive review. *Critical Reviews in Analytical Chemistry* **2023**, *53* (8), 1828–1839.
(188) Stamate, E.; Piraianu, A.-I.; Ciobotaru, O. R.; Crassas, R.; Duca, O.; Fulga, A.; Grigore, I.; Vintila, V.; Fulga, I.; Ciobotaru, O. C. Revolutionizing cardiology through artificial intelligence—Big Data from proactive prevention to precise diagnostics and cutting-edge treatment—A comprehensive review of the past 5 years. *Diagnostics* **2024**, *14* (11), 1103.

8 Applications of Sensors in Environmental Monitoring

Sevinc Kurbanoglu and Seda Nur Topkaya Cetin

8.1 ELECTROCHEMICAL SENSORS

Sensors are integral components in modern technology, playing a vital role in the seamless interaction of devices and systems across various industries. These devices are engineered to detect and respond to various physical, chemical, or biological stimuli, converting these signals into interpretable data for evaluation and decision-making (Ensafi 2019; Iftikhar et al. 2019). In industrial automation, sensors oversee processes and machinery, ensuring both efficiency and safety. Environmental sensors track changes in climate, pollution levels, and natural disasters, providing essential data for risk mitigation and ecosystem preservation (Beitollahi et al. 2020; Qian et al. 2021; Ho et al. 2005). In the healthcare sector, biosensors are employed in diagnostics and patient monitoring, delivering real-time insights into vital signs and disease indicators. In today's consumer electronics, such as smartphones and wearable devices, sensors enhance user experiences and enable fitness tracking. Additionally, ongoing developments in sensor technology involving portability with miniaturization, increased sensitivity, and energy efficiency are paving the way for innovative applications across diverse fields (Chung, Fortunato, and Radacsi 2019; Z. Zhang et al. 2019; Lam et al. 2009; Yarman, Kurbanoglu, and Scheller 2020; Pankratov et al. 2016; Tseng et al. 2018; Angelov et al. 2019; Figure 8.1).

Electrochemical sensors are instruments that convert chemical information into a quantifiable electrical signal. At the surface sensor's electrode, converted meaningful transformation occurs through a chemical reaction, where the target analyte engages with the sensor, leading to a measurable electrical response. Typically, electrochemical sensors are composed of three major parts: the working electrode, the reference electrode, and the auxiliary electrode (Kimmel et al. 2012; Iftikhar et al. 2019; Erkmen et al. 2022; Beitollahi et al. 2020; Windmiller and Wang 2013). The working electrode is where the analyte interacts, the reference provides a consistent potential, and the counter completes the circuit. There are several kinds of electrochemical sensors, including potentiometric, conductometric, and amperometric sensors, each of which measures a different type of electrical property. Potentiometric sensors measure the potential changings (voltage), amperometric sensors measure the current provided by redox reactions, and conductometric sensors measure the changing in conductivity caused by the presence of a target (Y. Wang et al. 2008; Wring and Hart 1992; J. Wang 2002; Tallman and Petersen 1990; Baranwal et al. 2022).

In electrochemical sensors, transducer systems are pivotal in transforming chemical information into measurable electrical signals, essential for detecting and quantifying analytes with high sensitivity and precision. The primary types of transducers used in these sensors include voltammetric, amperometric, potentiometric, and conductometric transducers (Vamvakaki and Chaniotakis 2007; Švorc et al. 1997; Singh et al. 2021). Each type offers distinct advantages, enabling the creation

DOI: 10.1201/9781003500322-8

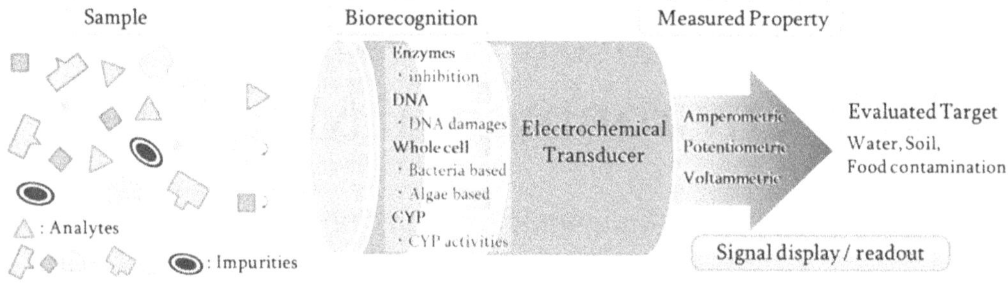

FIGURE 8.1 Schematic demonstration of electrochemical sensors and biosensors. (Reprinted with permission from Islam et al. 2023.)

of versatile and efficient electrochemical sensors for utilization in medical diagnostics, point-of-care (POC) devices, food safety, environmental monitoring, and industry processes. There are various types of electrochemical methods available and users need to choose different techniques depending on what to sense (Pushpanjali and Manjunatha 2019; Manjunatha 2019; Manjunatha et al. 2021; Hareesha and Manjunatha 2021; Hareesha et al. 2019). Cyclic voltammetry (CV) is a generally employed electrochemical approach that measures the current output of a redox-active substance as a linearly varied potential is applied. The principle behind CV includes scanning a potential across an electrochemical cell while recording the corresponding current response. The potential is assorted linearly over time, typically in a triangular waveform, and the current is recorded as the potential is cycled back and forth. This generates a cyclic voltammogram, a plot of current versus potential, which provides detailed insights into the electrochemical characteristics of the analyte, such as its redox potentials, reaction kinetics, and diffusion coefficients. The main strength of cyclic voltammetry lies in its ability to offer a comprehensive electrochemical profile of a substance. It is particularly effective for identifying redox processes, analyzing reaction mechanisms, and assessing the stability of intermediates. CV is also sensitive to changes at the electrode surface, making it ideal for studying surface modifications and interactions at the electrode interface (Rafiee et al. 2024; Uslu and Ozkan 2011; Kissinger, Heineman, and Achterberg 1996; Kissinger and Heineman 1983).

Differential pulse voltammetry (DPV) is an advanced electrochemical technique employed to measure the current response of an analyte to a sequence of potential pulses superimposed on a linear potential sweep. In DPV, the potential is gradually raised in small pulses, and the current is recorded just before and after every single pulse. The variation in current is recorded, resulting in a differential current plot against the potential. This technique improves the resolution of the electrochemical signal by reducing background currents and emphasizing the faradaic current associated with the analyte's redox process. The primary advantage of DPV is its heightened sensitivity and specificity, enabling the detection of low analyte concentrations. The differential measurement technique significantly minimizes the effects of capacitive currents and noise, resulting in clearer and more distinct peak signals. This makes DPV particularly useful for trace analysis and detecting species with closely spaced redox potentials. Additionally, DPV provides detailed information about redox behavior, including peak potentials and current magnitudes, which can be used for identifying and quantifying specific analytes (Ángela Molina and Joaquín González 2016; Romani et al. 2000; Kissinger, Heineman, and Achterberg 1996; Agrawal and Prajapati 2012; Kurbanoglu et al. 2019; Kurbanoglu and Ozkan 2018).

Square wave voltammetry (SWV) is an alternative electrochemical technique that applies a square wave potential modulation on top of a staircase potential. In SWV, the current is measured at the end of each forward and reverse pulse, and the change between these recordings is plotted toward the applied potential (Chen and Shah 2013). This technique enhances sensitivity and resolution by reducing background currents and boosting faradaic current responses. SWV is especially effective for rapid and quantitative analysis of electroactive species, making it ideal for trace detection in complex samples (Ramaley and Krause 1969; Temerk et al. 2016; Mirceski, Skrzypek, and Stojanov 2018; Mirceski et al. 2013).

Electrochemical impedance spectroscopy (EIS) is a powerful method utilized to assess a system's impedance over a spectrum of frequencies. In EIS, a small alternating current signal is applied to an electrochemical cell, and the subsequent current signal is recorded (S. Wang et al. 2021; Chang and Park 2010; Lazanas and Prodromidis 2023). The impedance, defined as the ratio of voltage-to-current, is then examined as a function of frequency. This technique offers comprehensive information about electrochemical activities occurring at the electrode interface, for example, charge double-layer capacitance, transfer resistance, and diffusion phenomena. The primary advantage of EIS is its ability to dissect complex electrochemical reactions into individual components, providing insights into the kinetics and mechanisms of these reactions. EIS is extensively sensitive to changes in the electrode surface and the features of the electrolyte, making it ideal for studying corrosion, battery performance, and sensor interfaces. It can detect subtle changes in system properties, allowing for the monitoring of surface modifications and the detection of low-concentration analytes (Prodromidis 2010; Ciucci 2019; Bahadır et al. 2016). Amperometry is an electrochemical technique that assesses the current generated from the oxidation or reduction of an analyte at a fixed potential. In an amperometric sensor, a constant potential is maintained on an electrode, and the resulting current, which correlates directly with the analytes' concentration, is recorded. The primary benefits of amperometry include its simplicity, high sensitivity, and fast response time. The ability to continuously measure allows for real-time monitoring, making it ideal for applications requiring fast and accurate analysis (Amine and Mohammadi 2019; Patel 2020; Kissinger, Heineman, and Achterberg 1996). Potentiometric sensors(e.g., ion-selective electrodes), are extensively utilized for measuring pH, ion concentrations (e.g., sodium, potassium, and chloride), and other analytes (Pohanka and Skládal 2008; Baranwal et al. 2022). All these electrochemical methods are widely used in environmental monitoring studies, even to confirm each other, and depending on the type of analytes, researchers need to choose and move forward (Islam et al. 2023).

8.2 IMPORTANCE OF ELECTROCHEMICAL SENSORS IN ENVIRONMENTAL MONITORING

One of the main factors that make electrochemical sensors vital for environmental monitoring is their outstanding selectivity and sensitivity. Environmental contaminants, including heavy metals, pesticides, and organic contaminants, often exist in trace amounts. Detecting these trace levels requires sensors that can respond to very low concentrations of analytes (Suzuki et al. 2020; Badihi-Mossberg, Buchner, and Rishpon 2007; Pierce and Zhao 2010; Pallavi Singh Chauhan et al. 2022). Electrochemical sensors are capable of detecting pollutants at nanomolar or even picomolar levels, making them ideal for monitoring low concentrations of toxic substances in the environment. Additionally, electrochemical sensors can be designed to be significantly selective for targets. This is achieved through modifying the electrode surface with selective recognition elements, such as enzymes, molecularly imprinted polymers, or antibodies, that interact specifically with the target pollutant (Taillefert and Rozan 2002; Arduini et al. 2017; Meredith et al. 2016). This selectivity is crucial for accurately monitoring specific pollutants in complex environmental matrices, for example, water, soil, and air. The ability to perform real-time monitoring is another significant advantage of electrochemical sensors in environmental monitoring (Barhoum et al. 2023; Rassaei et al. 2011). Conventional techniques of environmental analysis, such as mass spectrometry and gas chromatography, often require sample collection and transportation to a laboratory for analysis. This procedure is time-consuming and may not provide prompt information about the environmental conditions. Electrochemical sensors, on the other hand, can provide instant feedback on pollutant levels, allowing for timely decision-making and intervention. This real-time monitoring capability is particularly important in situations where rapid detection of pollutants is necessary, such as in the case of accidental spills, industrial emissions, or natural disasters (Hanrahan, Patil, and Wang 2004; Sadik and Van Emon 1996).

Moreover, electrochemical sensors are portable and can be used for on-site analysis. This portability allows for the continuous monitoring of environmental pollutants in remote or hard-to-reach areas without the need for complex laboratory equipment (Hanrahan, Patil, and Wang 2004). The ability to conduct on-site analysis also reduces the risk of sample contamination or degradation during transportation. Furthermore, electrochemical sensors are relatively inexpensive compared to other analytical techniques. The materials used in their construction, such as carbon, gold, or platinum electrodes, are cost-effective, and the fabrication processes are straightforward. Additionally, electrochemical sensors have low power requirements, making them suitable for deployment in resource-constrained settings or remote locations where power availability is limited. The cost-effectiveness of electrochemical sensors makes them accessible to a large scale of users, including government agencies, industries, and environmental organizations, facilitating widespread environmental monitoring efforts (Taillefert and Rozan 2002; Arduini et al. 2017; Meredith et al. 2016; Hanrahan, Patil, and Wang 2004; Sadik and Van Emon 1996; Barhoum et al. 2023; Rassaei et al. 2011).

Electrochemical sensors are highly versatile and can be regulated to determine a large range of environmental pollutants. By modifying the sensor surface with different recognition elements or by employing different detection strategies, electrochemical sensors can be tailored to determine specific pollutants such as heavy metals, nitrates, phosphates, organic solvents, and even biological contaminants like bacteria and viruses. This adaptability is crucial for comprehensive environmental monitoring, where multiple pollutants may need to be detected simultaneously or in different environmental compartments (air, water, and soil). Electrochemical sensors can be integrated into multi-sensor arrays or networks, enabling the simultaneous determination of numerous pollutants in real time (Barhoum et al. 2023; Rassaei et al. 2011; Rishpon 2002).

In addition to their role in monitoring, electrochemical sensors also contribute to environmental sustainability. The development of sensors that use green and sustainable materials, such as biodegradable polymers or renewable electrode materials, is gaining attention. These environmentally friendly sensors decrease the environmental effect of the monitoring procedure itself, aligning with the broader goals of sustainability. Moreover, electrochemical sensors can help reduce the use of hazardous chemicals in environmental analysis. Traditional analytical methods often require large quantities of reagents and solvents, which can be harmful to the environment. Electrochemical sensors, in contrast, require minimal or no reagents, making them a greener alternative for environmental monitoring (Barhoum et al. 2023; Rassaei et al. 2011; Rishpon 2002; Taillefert and Rozan 2002; Arduini et al. 2017; Meredith et al. 2016; Hanrahan, Patil, and Wang 2004; Sadik and Van Emon 1996; Figure 8.2).

8.3 POLLUTANTS AND MONITORING TECHNIQUES

Water, soil, and air are vital resources for all forms of life, and their quality is paramount for human health, agriculture, and ecosystems. Monitoring their quality involves assessing physical, chemical, and biological parameters to detect contaminants. Rising human activities have caused significant contamination of water and soil with heavy metals (Pallavi Singh Chauhan et al. 2022; Fleet and Gunasingham 1992). Water pollutants can generally be separated into two main classes: organic contaminants and inorganic contaminants. In this chapter, we emphasize key examples of each, specifically discussing organic pollutants such as pesticides and antibiotics alongside inorganic pollutants like arsenic (As), lead (Pb), chromium (Cr), and cadmium (Cd), ions chosen for their

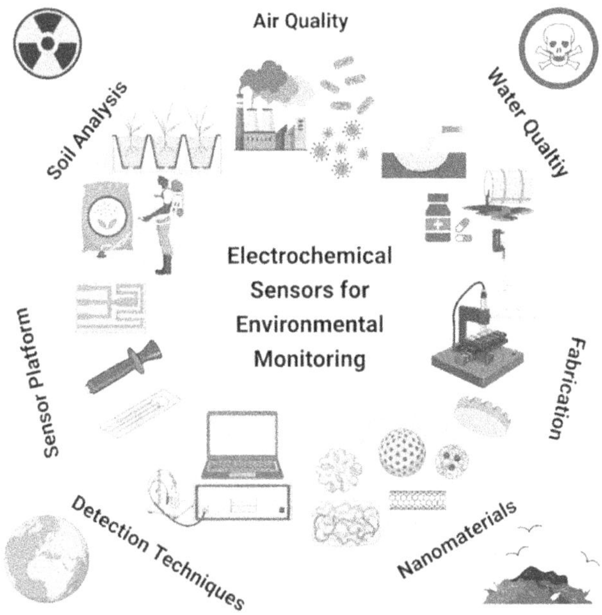

FIGURE 8.2 Utilizing electrochemical sensors for monitoring environmental factors. (Reprinted with permission from Barhoum et al. 2023.)

prevalent utilization and significant toxicity. *Soil pollutants* consist of heavy metals, toxic organic chemicals like biological pathogens, pesticides, and plastic refuse. While air pollution is the most apparent and thoroughly examined type of contamination, often depicted through visuals of smoke billowing from train engines and fumes from vehicle tailpipes, soil pollution is much less conspicuous. Consequently, the harmful impacts of soil pollution on public health are not as well understood or thoroughly quantified (Mostert, Ayoko, and Kokot 2010; Hirano and Tamae 2011; Brookes 1995; Münzel et al. 2023). *Common air pollutants* involve nitrogen oxides, particulate matter, sulfur dioxide, volatile organic compounds, and pesticides. Monitoring these contaminants involves using advanced technologies such as gas chromatography, mass spectrometry, and remote sensing. Monitoring air quality helps in identifying pollution sources, understanding pollutant dispersion patterns, and assessing the effectiveness of air quality management strategies. It also provides valuable data for developing policies and regulations to guard public health and the environment (Yang and Omaye 2009; Suh et al. 2000; Brunekreef and Holgate 2002).

The principle behind using electrochemical sensors (or electrodes) for detecting pollutants is grounded on the adsorption of these pollutants at the electrode interface, which is then followed by a redox reaction driven by electro-catalysis on the electrode's surface. These redox reactions, which entail electron transfer, allow the quantification of electron exchange by measuring the current at a macroscopic level. This establishes a measurable correlation between the concentration of pollutants and the resulting electrical signal. Additionally, the electrode can distinguish qualitatively and quantitatively various contaminants depending on variations in their oxidation or reduction peaks. Contaminants differ in both concentration and type across various aquatic environments, and changes in chemical conditions (such as chemical composition, pH levels, and interfering factors) can greatly influence the sensitivity, interference resistance, stability, and durability of the electrode. These environmental differences must be considered to ensure optimal electrode function and durability (Liu et al. 2022; Fleet and Gunasingham 1992; Hanrahan, Patil, and Wang 2004).

8.3.1 Detection of Antibiotics and Pesticides with Electrochemical Sensors

Herbicides and pesticides are chemicals employed to control unwanted plants and pests. While they contribute significantly to agricultural productivity, their extensive use poses serious environmental risks. These chemicals can persist in the environmental conditions for extended periods of time, accumulate in the food chain, and cause harmful impacts on non-target species, including humans. The persistence of these chemicals in the environment corresponds to their chemical features, environmental conditions, and biological factors (Sidhu et al. 2019; C. Li, Begum, and Xue 2020; Abdollahi et al. 2004; Du, Xie, and Wang 2021). In many countries, the registration and use of herbicides and pesticides are governed by regulation organizations such as the European Chemicals Agency for the European Union, the Environmental Protection Agency for the United States, and similar bodies in other regions. These agencies evaluate the safety and effectiveness of chemical products before approving them for use. They adjust maximum residue limits (MRLs) for pesticides in water and food, establish guidelines for safe application, and monitor compliance with regulations. Thus, the precise assessment of herbicides and pesticides in real-world samples within MRLs is of paramount importance. Electrochemical-based sensors offer an efficient solution for this critical task (Mohamed and Fouad 2020; Bilitewski and Turner 2000). To illustrate the latest studies on electrochemical-based sensors for environmental monitoring, MXene has been extensively examined for its remarkable structure and physical properties, including magnificent electrical conductivity, strong hydrophilicity, diverse surface chemistry, superior adsorption capacity, and preparation simplicity via solution processing (Riu, Maroto, and Rius 2006; Abdel-Karim, Reda, and Abdel-Fattah 2020). A novel electrochemical investigation is demonstrated by Zhang et al. (2024)for the ultrasensitive assessment of the organophosphate pesticide (OP), fenitrothion, utilizing a Ti_3C_2MXene/CoAl-LDH nanocomposite for electrode's modification. The synthesis of this nanocomposite involves the direct growth of CoAl-LDH on MXene nanosheets. The integration of these two ultrathin 2D materials offers additional active spots, a wider surface area, enhanced electrical conductivity, and superior adsorption properties. This enables rapid transfer of electrons and mass between the electrode and the analytes when used as a material for electrochemical measuring. Additionally, the binding of phosphate groups with Ti defect sites in MXene facilitates the attachment of OP onto the electrode. As a result, the fabricated electrode provides a broad dynamic range of 0.03–120 μM for the assessment of fenitrothion and limit of detection (LOD) of 5.8 nM. The technique ensures excellent selectivity, reproducibility, durability, and feasibility for real sample applications. This approach serves as an exemplary application for the electrochemical sensing of trace amounts of OP residues using MXene/LDH-based nanocomposites. The outstanding electroanalytical performance is primarily attributed to the comprehensive enhancement of the nanocomposite's electrochemical features, such as broad surface area, exceptional electrical conductivity, better adsorption capabilities, and accessible interlayer space. Consequently, the presented technique illustrates an effective application of MXene/LDH-based nanocomposites in the ultra-sensitive electrochemical monitoring of OP residue (H. Zhang et al. 2024). Developing highly efficient artificial nanozymes for pesticide determination is crucial for human health and ecological systems (Erkmen, Kurbanoglu, and Uslu 2021; Ballesteros et al. 2021; X. Li et al. 2019). In another study, by Zhang et al. in 2024, the authors offered a two-phase method to synthesize TAPB-DMTP-COF (denoted as COF-OMe, with TAPB being 1,3,5-tris(4-aminophenyl) benzene and DMTP being 2,5-dimethoxyterephaldehyde) integrated with cerium oxide nanozymes (CeO_2) exhibiting phosphatase-like behaviors. These nanozymes are able to break down phosphate bonds (P—O) in OPs to produce electroactive p-nitrophenol (p-NP). Specifically, Ce (IV)/Ce (III) species act as active areas that polarize and hydrolyze P—O bonds in organophosphates, while the outstanding adsorption properties of the porous COF-OMe facilitate the hydrolysis process. Additionally, the effective transfer of charge at the covalent interface of COF-OMe@Valine-CeO_2 improves the efficiency of this catalytic process. Leveraging these benefits, the COF-OMe@Valine-CeO_2 sensing

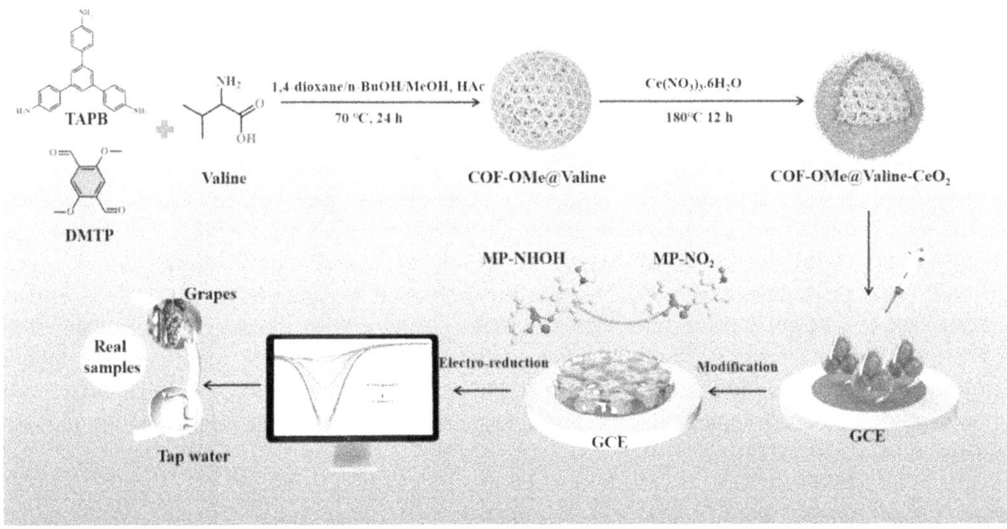

FIGURE 8.3 The representation scheme of the synthesis of COF-OMe@Valine-CeO_2 nanozymes for MP determination. (Reprinted with permission from X. Zhang et al. 2024.)

platform was effectively used to detect methyl-paraoxon. The obtained electrochemical sensors demonstrated a broad linear range of 0.034–76 μM and LOD of 0.011 μM with good sensitivity for methyl-paraoxon. This investigation not only introduces a technique for creating COFs with exceptional nanozyme activity but also offers new approaches for designing highly effective nanozymes for the timely identification of OPs (X. Zhang et al. 2024; Figure 8.3).

In another study, an enzyme-based biosensor was fabricated for the indirect determination of OPs by leveraging inhibition of acetylcholinesterase (AChE). This biosensor was designed to enhance performance using a screen-printed carbon electrode (SPCE) that was designed with copper nanowires (CuNWs) combined and reduced graphene oxide (rGO). The oxidation current, which results from the enzymatic reaction between AChE and its substrate, acetylthiocholine (ATCh), was measured using the CV method. The biosensing activity is characterized by an increase in signal due to acetylcholinesterase inhibition when an organophosphate inhibitor is present. Testing revealed that the nanocomposite significantly increased the signal current and lowered the oxidation potential, enabling effective electrochemical determination of the chlorpyrifos. The sensor demonstrated a broad linear range of 10 μg/L to 200 μg/L, with an LOD of 3.1 μg/L and a quantification limit of 12.5 μg/L. The offered sensor proved efficient for analyzing chlorpyrifos in orange juice and drinking water, demonstrating good recovery values and no interference effects. The authors stated that the fabricated enzyme-based electrochemical biosensor with high catalytic activity for thiocholine oxidation was developed utilizing a CuNWs/rGO modification onto an SPCE. This modification improved the electrochemical outputs by eight times and reduced the overpotential by 330 mV relative to an unmodified SPCE. Chlorpyrifos caused irreversible inhibition mechanism of AChE, leading to a decrease in the recorded output (Suwannachat et al. 2024).

The prevalent use and improper handling of antibiotics have resulted in water sources becoming contaminated with antibiotic residues, posing risks to the environment, public health, and the economy. Accordingly, it is crucial to improve a detection method with high sensitivity and selectivity for identifying these residues in real water samples (X. X. Zhang, Zhang, and Fang 2009; Baquero, Martínez, and Cantón 2008). In their study, Adane et al. (2024) fabricated an ultrasensitive electrochemical sensor through integrating gold–silver alloy nanocoral clusters (Au–AgANCCs) with functionalized multi-walled carbon nanotube (*f*-MWCNT) onto a carbon paste electrode (CPE) and choline chloride (ChCl) nanocomposites. This sensor was designed to simultaneously detect residues of the antimicrobial drugs norfloxacin (NFX) and rifampicin (RAMP) in water samples. The fabricated sensor (Au–Ag-ANCCs/f-MWCNTs-CPE/ChCl) was thoroughly examined utilizing various analytical techniques such as ultraviolet visible (UV–Vis) spectroscopy, X-ray diffraction, energy dispersive X-ray, and scanning electron microscopy, and Fourier transform infrared spectroscopy (FT-IR) and electrochemical methods such as CV, EIS, and SWV). It demonstrated excellent performance across a broad linear range of 0.9 nM–200 μM for NFX, and 14 pM-115 μM for RAMP, with detection and quantification limits of 0.14 nM and 0.47 nM for NFX, and 2.7 pM and 8.85 pM for RAMP, respectively. Moreover, the sensor demonstrated outstanding stability, reproducibility, and interference resistance. It was effectively used to detect NFX and RAMP residues in river samples, hospital wastewater, and tap water samples, with recovery rates ranging from 96.8% to 103% and relative standard deviations under 5%. Overall, the developed sensor exhibited exceptional capability in determining the target analytes, positioning it as an optimal device, and an innovative approach to combating the global issue of antibiotic residue contamination in water sources (Adane et al. 2024).

8.3.2 Detection of Heavy Metals with Electrochemical Sensors

Heavy metals refer to a class of metals and metalloids distinguished by their high atomic mass and density, which are at least five times greater than that of water (Ty et al. 2024; Zaynab et al. 2022). These elements include Cd, As, Cr, mercury (Hg), and Pb, among others. They are poisonous or toxic at very low concentrations and cause significant health issues, for example, neurological damage, kidney or bone disorders, cancer, and cardiovascular diseases (Cui, Wu, and Ju 2015; Zaynab et al. 2022). These metals persist in the environment due to their inability to degrade easily, leading to bioaccumulation in living organisms, and biomagnification by the food chain. Upon entering the organism, metal ions may be converted into more hazardous types or directly disrupt metabolic operations, resulting in oxidative stress or even a range of disorders (Meng et al. 2023; Ding et al. 2021).

Various analytical strategies are employed to detect heavy metals in water sources, each offering distinct advantages and facing specific limitations. Inductively coupled plasma mass spectrometry (ICP-MS) method ionizes the sample with an inductively coupled plasma and then utilizes a mass spectrometer to detect the ions. It is known for its remarkably low detection limits, often in the parts-per-trillion range, and can analyze multiple metals simultaneously. Despite its high sensitivity, ICP-MS is expensive, requires extensive sample preparation, and is not portable, making it more suitable for laboratory use. X-Ray fluorescence (XRF) and atomic absorption spectroscopy (AAS) are also widely used for their accuracy and reliability. AAS measures the absorption of light by metal ions in a sample, providing precise assessment with good sensitivity, though it typically requires separate analysis for each metal. XRF, on the other hand, is a non-destructive technique that detects metals based on the characteristic X-rays emitted from the sample upon excitation. It is useful for rapid screening but generally has higher detection limits compared to ICP-MS (Koester and Moulik 2005; Ballesteros-Gómez and Rubio 2011; Clement, Koester, and Eiceman 1993). Moreover, emerging techniques such as biosensors and colorimetric assays are increasingly being recognized for their potential in heavy metal detection. Biosensors leverage biological components to detect heavy metals with high specificity and sensitivity, offering potential for in-field applications. Colorimetric assays use color changes to indicate the presence of metals, providing a simple and cost-effective detection method, although they are generally less sensitive than other techniques. Each of these methods offers distinct benefits and drawbacks, with the selection of the appropriate technique frequently determined by specific factors such as sensitivity, cost, portability, and the nature of the sample being analyzed (Ribeiro et al. 2014; Koester and Moulik 2005; Ballesteros-Gómez and Rubio 2011; Clement, Koester, and Eiceman 1393).

In contemporary environmental monitoring systems, sensors have become indispensable tools, delivering critical data on heavy metal pollution and supporting sustainable resource regulation and pollution control strategies. Sensors for detecting heavy metals often use electrochemical methods, where the metal ions are reduced or oxidized at an electrode surface, producing a measurable current proportional to the concentration. Considering that heavy metal concentrations are frequently at very low levels (typically in the nanomolar range), it is crucial to further improve electrode sensitivity and detection limits (Meredith et al. 2016; Riu, Maroto, and Rius 2006). Thin film technology enables efficient heavy metal sensing by providing a large wide area for target ion interaction, thereby enhancing the sensor's sensitivity and responsiveness to trace amounts. Additionally, thin film membrane sensors provide portability and adaptability, making them ideal for incorporation into handheld detectors or wearable devices. This allows for continuous and on-site monitoring of heavy metal quantities. Their user-friendliness and quick output times are critical for field applications, where prompt determination is vital for mitigating possible risks. Thus, thin film sensors are valuable tools for environmental monitoring and pollution control (Arida, Kloock, and Schöning 2006; Kumar, Raghuwanshi, and Kumar 2022; Nikoleli et al. 2018). For instance, the porous Al_2O_3 membrane, created through anodization, offers precise control over pore size and shape, making it ideal for fabricating nanomaterials with tailored properties. It features high chemical stability, corrosion resistance, mechanical strength, thermal stability, and optical transparency, suitable for harsh environments and biomedical applications. By integrating TiON/TiO_2 onto the Al_2O_3 membrane, enhanced reactivity and catalytic capabilities are achieved, enabling effective Pb^{2+} ion detection in aqueous solutions using cyclic voltammetry (Elsayed et al. 2024). Another thin film-based biosensor was developed with high sensitivity, selectivity, and reusability and food (II) (Hg^{2+}) ions in the food and environment. The biosensor utilizes a three-dimensional macroporous dealloyed (3DOMD) Au-Ag thin-film electrode with a multifunctional encoded Au@Pt nanocage for signal amplification. The 3DOMD Au-Ag thin film, produced using a dealloying technique, offers a surface area that is 4.8 times greater than that of conventional 3D macroporous gold films. This enhanced surface area and electron transport capacity enable the biosensor to detect Hg^{2+} ions with a linear range of 0.01–1000 nM and a highly low LOD of 2 pM (Jiang et al. 2023).

The incorporation of novel materials, particularly nanomaterials, has greatly improved sensors for detecting heavy metals. Notable examples include graphene, carbon nanotubes, rGO, and carbon dots. Additionally, polymers like Polyaniline, Nafion, Polyethylene glycol, and Polyvinyl alcohol are commonly used. Metallic nanoparticles such as silver, gold, and copper, along with metal oxides like CuO, ZnO, Fe_3O_4, MnO_2, MgO, TiO_2, SnO_2, and ZrO_2, enhance sensor performance. These materials contribute to making sensors more efficient, selective, and sensitive, leading to better environmental monitoring and public health protection (Wu et al. 2024; Filippidou and Chatzandroulis 2023).

A MWCNT/ZnO nanocomposite electrode was fabricated for sensitive voltammetric determination of dissolved heavy metal ions. Combining the heightened conductivity, catalytic activity, and large surface area of MWCNTs with the selective binding properties of ZnO nanoparticles, the sensor demonstrated enhanced metal ion accumulation and resolution, with low detection limits. The optimized MWCNT/ZnO ratio and ink casting process achieved a robust, reproducible sensor with a wide linear range and low limit of detections. Notably, it enabled synchronous determination of Cu(II), Pb(II), and Hg(II) even in complex water matrices, eliminating the need for separation steps (Zhou 2024).

Combined methods such as microfluidic and electrochemical or electrochemical/optical methods are preferred for dual or multi-detection of heavy metals. In particular, combined microfluidic systems allow the consolidation of numerous laboratory bench processes into a single device, thereby meeting the requirements of POC systems by using smaller, smarter, and more efficient electronic systems (Sequeira et al. 2002; Yew et al. 2019; Maguire et al. 2018; Dhar and Lee 2018;

Pol et al. 2017; Jokerst, Emory, and Henry 2011; Medina-Sánchez et al. 2016). For instance, a polydimethylsiloxane microfluidic tool incorporating graphene oxide/aptamer sensors for the determination of Hg^{2+} and Pb^{2+} ions was developed (Hao et al. 2024). This device achieved detection limits of approximately 0.70 ppb for Hg^{2+} and 0.53 ppb for Pb^{2+}, both of which are below the thresholds set by the World Health Organization. An advanced autonomous robotic system has been fabricated for the detection of Pb^{2+} in surface water. This cutting-edge system combines a microfluidic tool with an electrochemical sensor utilizing SPCEs. It can conduct 39 measurements daily, with LOD for Pb^{2+} of 4 μg/L. This integrated system provides a highly effective and efficient method for continuous and precise water quality monitoring (De Vito-Francesco et al. 2022). The device, integrated with a microfluidic channel, featured three electrodes deposited onto a paper substrate via a sputtering process. This advanced setup facilitated the voltammetric detection of Cd^{2+} and Zn^{2+} using anodic stripping voltammetry. The LOD was calculated to be 0.9 ppb for Cd^{2+} and 1.1 ppb for Zn^{2+} (Kokkinos, Economou, and Giokas 2018).

Combined electrochemical-microfluidic sensors also have several drawbacks. They are complex to fabricate and require advanced external components, limiting their ease of use and autonomy. Their reproducibility is often low, with limited testing on real water samples or in industrial environments. Challenges such as electrode fouling and interference from complex matrices must still be overcome to ensure accurate and reliable quantification of heavy metals in real samples. To address these issues, better fabrication methods, improved detection schemes, and extensive real-sample testing are needed (Sequeira et al. 2002; Yew et al. 2019; Maguire et al. 2018; Dhar and Lee 2018; Pol et al. 2017; Jokerst, Emory, and Henry 2011; Medina-Sánchez et al. 2016).

Paper-based electrochemical sensors for determining heavy metals represent a cutting-edge solution that merges the advantages of paper substrates with the precision of electrochemical analysis. These sensors consist of a simple paper platform integrated with conductive electrodes, typically made from materials like carbon, gold, or silver. The paper acts not only as a structural support but also as a fluidic channel, facilitating the transport of environmental samples such as water or soil extracts to the detection zone. The electrodes, which are often screen-printed or drop-cast onto the paper, serve as the active sensing elements and can be modified with nanomaterials like carbon nanotubes or gold nanoparticles to advance sensitivity and selectivity. The electrochemical detection methods utilized, such as DPV or anodic stripping voltammetry, are particularly effective for identifying and quantifying trace amounts of heavy metals like lead Pb^{2+}, Hg^{2+}, Cd^{2+}, and As^{3+}. These techniques measure differences in current or potential as the metal ions interact with the electrode surface, allowing for precise quantification even at low concentrations (Pankaj Singh Chauhan, Pandey, and Bhattacharya 2019; Kummari et al. 2023; Sun et al. 2022; Mazurkiewicz et al. 2020; Deroco, Wachholz Junior, and Kubota 2023).

The most leading strengths of paper-based electrochemical sensors are their miniaturization and ease of use, making them ideal for in-situ testing in remote or resource-limited settings. They are also highly cost-effective, as the paper substrate is inexpensive and the fabrication process is straightforward, often requiring minimal laboratory infrastructure. This makes them a valuable tool for environmental monitoring, especially in developing regions where heavy metal pollution is a significant concern. Paper-based electrochemical sensors for heavy metal determination, despite their advantages, face several limitations, including limited stability and durability due to the vulnerability of paper substrates to environmental conditions like humidity and temperature fluctuations. Sensitivity and selectivity can be compromised by interference from other substances in complex real-world samples, leading to potential cross-reactivity and false positives. Reproducibility issues arise from variability in fabrication processes and inconsistent paper quality. Additionally, these sensors often have a limited detection range, struggling with very low or high levels of heavy metals. Storage and handling constraints, due to their fragile nature, further limit their robustness, and there are challenges in integrating them effectively with digital systems for accurate field quantification

(Pankaj Singh Chauhan, Pandey, and Bhattacharya 2019; Kummari et al. 2023; Sun et al. 2022; Mazurkiewicz et al. 2020; Deroco, Wachholz Junior, and Kubota 2023).

A major component of paper, cellulose, contains many carboxyl and hydroxyl groups, providing negatively charged adsorption sites that can attract heavy metals (Ding et al. 2021; Meng et al. 2023). This characteristic gives filter papers a notable sorption potential for heavy metals, with the adsorption procedure following first-order reversible kinetics and adhering to both Freundlich and Langmuir models. The ability of adsorption to increase with higher ion concentrations is accompanied by an exothermic process. However, this adsorption can alter the initial levels of heavy metals in samples, potentially leading to measurement errors. The presence of competing electrolytes or high acid concentrations can significantly reduce the adsorption of divalent metal ions. This adsorption property can cause systematic errors for the assessment of heavy metal ion concentrations, particularly at low levels and varying pH levels, where ion-exchange mechanisms become more significant. A paper-based study introduces a cost-effective, paper-based heavy metal sensor for POC determination. The sensor utilizes typical three-electrode system screen-printed on PET, with a carbon paste working electrode containing antimony precursors, for reference electrode an Ag/AgCl, and a carbon paste auxiliary electrode. Designed for easy substance delivery via a paper channel, the sensor showed high sensitivity for Cd detection, particularly with 5 wt/wt% antimony, enhancing sensitivity by 2.18 times compared to standard electrodes. It detects Cd within a 20–2000 μg/L range, with strong linearity ($R^2 = 0.988$) between 20 and 500 μg/L, rendering it ideal for various heavy metal determination applications (Han et al. 2022). A multi-channel, paper-based electrochemical chip was fabricated for the fast and highly sensitive determination of various heavy metals, employing an aptamer competition approach. The chip's microfluidic patterns were created via automatic screen printing, enabling self-driven liquid flow by capillary action without the need for external pumps. When target ions are present, methylene blue-modified aptamers decompose from their complementary sequences on the chip, resulting in a difference in the electrochemical output. This multi-channel platform exhibited a broad linear detection range (10 pM to 1000 nM) and low LODs for Pb^{2+}, Cd^{2+}, AsO_2^-, and Hg^{2+}, with limits of 1.55, 0.84, 1.18, and 1.63 pM, respectively. The sensor also successfully detected these ions in crab meat, achieving recovery rates between 84.79% and 101.55%. Paper has proven to be highly effective for real sample testing, particularly in separating liquids from solids, as samples are wicked through its pores. This natural filtration capability allows for efficient separation, making paper ideal for handling samples with particulate matter (Guan and Sun 2020).

8.4 CONCLUSION AND FUTURE PERSPECTIVES

Environmental monitoring programs are crucial for assessing the presence and levels of pollutants in soil, water, and air. Over the previous decade, advancements in micro- and nanotechnology have driven the portabilization of conventional bench-top instruments, paving the way for the advancement of affordable, straightforward, lab-on-a-chip sensing platforms. These innovations make possible fast on-site assessments. Electrochemical sensors have already established themselves as indispensable tools in environmental monitoring, providing precise, real-time data on pollutants and other environmental parameters. However, the future holds even more promise for these technologies as advancements in materials science, nanotechnology, and data analytics drive further innovation. The evolution of electrochemical sensors is expected to enhance their sensitivity, selectivity, and versatility, making them even more effective in addressing the complex challenges of environmental monitoring. The incorporation of electrochemical sensors with the Internet of Things (IoT) and smart systems is poised to revolutionize environmental monitoring. By embedding sensors into IoT networks, it will be possible to create extensive, interconnected sensor arrays capable of continuously monitoring environmental parameters across vast geographic areas. This connectivity will enable real-time data collection, transmission, and analysis, facilitating rapid responses to environmental hazards. Moreover, these networks can be used to monitor long-term environmental trends, providing valuable insights into climate change, pollution patterns, and ecosystem health. Moreover, smart nanomaterials will also take part in the future. Nanotechnology is projected to have a crucial impact on advancing next-generation electrochemical sensors. The utilization of nanomaterials, for example, graphene, carbon nanotubes, and metallic nanoparticles, can greatly increase the performance of sensors by enhancing their surface area, improving electron transfer rates, and enabling the determination of pollutants at ultra-low concentrations. These advancements will make electrochemical sensors more selective and sensitive, allowing for the detection of emerging contaminants that are currently challenging to monitor with existing technologies. The trend toward miniaturization will continue to impact the design and application of electrochemical sensors. Smaller, more portable sensors will make it easier to conduct on-site environmental monitoring in remote or hard-to-reach areas. Portable electrochemical sensors can be integrated into wearable devices or deployed as part of autonomous drones or robots, enabling environmental monitoring in locations that were previously inaccessible. This increased portability will also facilitate the widespread adoption of electrochemical sensors in citizen science initiatives, where individuals can contribute to environmental data collection.

In addition, the advancements in paper-based electrochemical sensors for heavy metal determination demonstrate significant progress in environmental monitoring technology. These sensors offer a highly effective, low-cost, and portable solution for real-time detection of multiple heavy metals, leveraging innovations such as nanocomposites and aptamer-based strategies to enhance sensitivity and selectivity. The ability to fabricate these sensors using simple techniques like screen printing and their operation without the need for external pumps underscores their practicality for on-site applications. Moving forward, continued improvements in sensor design, particularly in terms of stability, robustness, and digital integration, will be crucial in expanding their application and ensuring that they remain a vital tool in environmental protection and public health monitoring. The extensive data produced by electrochemical sensors can be utilized through sophisticated data analytics and machine learning techniques. These tools can help identify patterns and trends in environmental data, predict potential environmental hazards, and optimize sensor performance. Machine learning algorithms can also be used to enhance the accuracy and reliability of sensor data by compensating for interference from other environmental factors. This will result in more robust and actionable insights, aiding in environmental management and policymaking.

Moreover, as environmental monitoring aims to preserve and protect the environment, there is growing interest in developing sustainable and eco-friendly electrochemical sensors. Researchers are exploring the utilization of biodegradable materials, renewable energy sources, and green manufacturing processes to reduce the environmental footprint of sensor production and deployment. These efforts will align the improvement of electrochemical sensors with the broader goals of sustainability and environmental stewardship. Electrochemical sensors are set to play a progressively critical role in environmental monitoring as they evolve to meet the demands of a changing world. Their capability to offer real-time, accurate, and sensitive detection of environmental pollutants makes them indispensable in efforts to preserve human well-being and the environmental quality. The future of electrochemical sensors lies in their integration with advanced technologies—for example, IoT, nanotechnology, and machine learning—which will enhance their capabilities and broaden their applications. As the global community continues to grapple with environmental challenges such as climate change, contamination, and resource depletion, the importance of reliable and efficient monitoring tools cannot be overstated. Electrochemical sensors, with their versatility and adaptability, are well-positioned to address these challenges and contribute to a sustainable future. The continued research and advancement in this field will ensure that electrochemical sensors remain at the forefront of environmental monitoring, providing the data necessary to safeguard our planet for generations to come.

ACKNOWLEDGEMENTS

S. Kurbanoglu acknowledges Turkish Academy of Sciences-The Young Scientists Award program (TUBA-GEBIP) and The Science Academy, Turkiye for Young Scientist Award Program (BAGEP).

REFERENCES

Abdel-Karim, Randa., Yasser Reda, and Aliaa Abdel-Fattah. 2020. "Review — Nanostructured Materials-Based Nanosensors." *Journal of the Electrochemical Society* 167 (3): 037554. https://doi.org/10.1149/1945-7111/ab67aa.

Abdollahi, Mohammad, Akram Ranjbar, Shahin Shadnia, Shekoufeh Nikfar, and Ali Rezaie. 2004. "Pesticides and Oxidative Stress: A Review." *Medical Science Monitor* 10 (6).

Adane, Wondimeneh Dubale, Bhagwan Singh Chandravanshi, Negash Getachew, and Merid Tessema. 2024. "A Cutting-Edge Electrochemical Sensing Platform for the Simultaneous Determination of the Residues of Antimicrobial Drugs, Rifampicin and Norfloxacin, in Water Samples." *Analytica Chimica Acta* 1312 (July): 342746. https://doi.org/10.1016/J.ACA.2024.342746.

Agrawal, Surendra, and Mahendra Prajapati. 2012. "Nanosensors and Their Pharmaceutical Applications: A Review." *International Journal of Pharmaceutical Science and Nanotechnology* 4 (4): 1528–36. www.researchgate.net/profile/Surendra_Agrawal3/publication/259493886_Nanosensors_and_their_Pharmaceutical_Applications_A_Review/links/54b613710cf2318f0f9a080d.pdf.

Amine, Aziz, and Hasna Mohammadi. 2019. "Amperometry." *Encyclopedia of Analytical Science*, January, 85–98. https://doi.org/10.1016/B978-0-12-409547-2.14204-0.

Ángela Molina, and Joaquín González. 2016. "Pulse Voltammetry in Physical Electrochemistry and Electroanalysis: Theory and Application." *Monographs in Electrochemistry*, 671. https://doi.org/10.1007/978-3-319-21251-7.

Angelov, George Vasilev, Dimitar Petrov Nikolakov, Ivelina Nikolaeva Ruskova, Elitsa Emilova Gieva, and Maria Liubomirova Spasova. 2019. "Healthcare Sensing and Monitoring." In *Lecture Notes in Computer Science (Including Subseries Lecture Notes in Artificial Intelligence and Lecture Notes in Bioinformatics)*, 11369 LNCS:226–62. Springer Verlag. https://doi.org/10.1007/978-3-030-10752-9_10.

Arduini, Fabiana, Stefano Cinti, Viviana Scognamiglio, Danila Moscone, and Giuseppe Palleschi. 2017. "How Cutting-Edge Technologies Impact the Design of Electrochemical (Bio)Sensors for Environmental Analysis. A Review." *Analytica Chimica Acta* 959 (March): 15–42.

Arida, Hassan A., Joachim P. Kloock, and Michael J. Schöning. 2006. "Novel Organic Membrane-Based Thin-Film Microsensors for the Determination of Heavy Metal Cations." *Sensors* 6 (4): 435–44. https://doi.org/10.3390/S6040435.

Badihi-Mossberg, Michal, Virginia Buchner, and Judith Rishpon. 2007. "Electrochemical Biosensors for Pollutants in the Environment." *Electroanalysis* 19 (19–20): 2015–28. https://doi.org/10.1002/ELAN.200703946.

Bahadır, Elif Burcu, Mustafa Kemal Sezgintürk. 2016. "A Review on Impedimetric Biosensors." *Artificial Cells, Nanomedicine and Biotechnology* 44 (1): 248–62. https://doi.org/10.3109/21691401.2014.942456.

Ballesteros-Gómez, Ana, and Soledad Rubio. 2011. "Recent Advances in Environmental Analysis." *Analytical Chemistry* 83 (12): 4579–4613. https://doi.org/10.1021/AC200921J/ASSET/IMAGES/LARGE/AC-2011-00921J_0011.JPEG.

Ballesteros, Camilo A.S., Luiza A. Mercante, Augusto D. Alvarenga, Murilo H.M. Facure, Rodrigo Schneider, and Daniel S. Correa. 2021. "Recent Trends in Nanozymes Design: From Materials and Structures to Environmental Applications." *Materials Chemistry Frontiers* 5 (20): 7419–51. https://doi.org/10.1039/d1qm00947h.

Baquero, Fernando, José Luis Martínez, and Rafael Cantón. 2008. "Antibiotics and Antibiotic Resistance in Water Environments." *Current Opinion in Biotechnology* 19 (3): 260–65. https://doi.org/10.1016/J.COPBIO.2008.05.006.

Baranwal, Jaya, Brajesh Barse, Gianluca Gatto, Gabriela Broncova, and Amit Kumar. 2022. "Electrochemical Sensors and Their Applications: A Review." *Chemosensors* 10 (9): 363. https://doi.org/10.3390/chemosensors10090363.

Barhoum, Ahmed, Selma Hamimed, Hamda Slimi, Amina Othmani, Fatehy M. Abdel-Haleem, and Mikhael Bechelany. 2023. "Modern Designs of Electrochemical Sensor Platforms for Environmental

Analyses: Principles, Nanofabrication Opportunities, and Challenges." *Trends in Environmental Analytical Chemistry* 38 (June): e00199. https://doi.org/10.1016/J.TEAC.2023.E00199.

Beitollahi, Hadi, Sayed Zia Mohammadi, Mohadeseh Safaei, and Somayeh Tajik. 2020. "Applications of Electrochemical Sensors and Biosensors Based on Modified Screen-Printed Electrodes: A Review." *Analytical Methods* 12 (12): 1547–60. https://doi.org/10.1039/c9ay02598g.

Bilitewski, Ursula, and Anthony P. F. Turner. 2000. *Biosensors for Environmental Monitoring*. Harwood Academic Publishers.

Brookes, Philip Charles. 1995. "The Use of Microbial Parameters in Monitoring Soil Pollution by Heavy Metals." *Biology and Fertility of Soils* 19 (4): 269–79. https://doi.org/10.1007/BF00336094/METRICS.

Brunekreef, Bert, and Stephen T. Holgate. 2002. "Air Pollution and Health." *The Lancet* 360 (9341): 1233–42. https://doi.org/10.1016/S0140-6736(02)11274-8.

Chang, Byoung Yong, and Su Moon Park. 2010. "Electrochemical Impedance Spectroscopy." *Annual Review of Analytical Chemistry* 3 (1): 207–29. https://doi.org/10.1146/ANNUREV.ANCHEM.012809.102211/CITE/REFWORKS.

Chauhan, Pallavi Singh, Neha Sharma, Asha Singh, and Rajesh Singh Tomar. 2022. "Recent Trends in Rapid Environmental Monitoring of Toxicants Using Nanobiosensors." *Nanobiosensors for Environmental Monitoring: Fundamentals and Application*, January, 393–405. https://doi.org/10.1007/978-3-031-16106-3_19/FIGURES/2.

Chauhan, Pankaj Singh, Mohit Pandey, and Shantanu Bhattacharya. 2019. "Paper Based Sensors for Environmental Monitoring." In *Paper Microfluidics. Advanced Functional Materials and Sensors*, edited by S. Bhattacharya, S. Kumar, and A. Agarwal, 165–81. Springer. https://doi.org/10.1007/978-981-15-0489-1_10.

Chen, Aicheng, and Badal Shah. 2013. "Electrochemical Sensing and Biosensing Based on Square Wave Voltammetry." *Analytical Methods* 5 (9): 2158–73. https://doi.org/10.1039/c3ay40155c.

Chung, Michael, Giuseppino Fortunato, and Norbert Radacsi. 2019. "Wearable Flexible Sweat Sensors for Healthcare Monitoring: A Review." *Journal of the Royal Society Interface*, 16 (159): 1–15. https://doi.org/10.1098/rsif.2019.0217.

Ciucci, Francesco. 2019. "Modeling Electrochemical Impedance Spectroscopy." *Current Opinion in Electrochemistry* 13 (13): 132–39. https://doi.org/10.1016/j.coelec.2018.12.003.

Clement, Ray E, Carolyn J Koester, and Gary A Eiceman. 1393. "Environmental Analysis." *Analytical Chemistry* 65: 85–116. https://pubs.acs.org/sharingguidelines.

Cui, Lin, Jie Wu, and Huangxian Ju. 2015. "Electrochemical Sensing of Heavy Metal Ions with Inorganic, Organic and Bio-Materials." *Biosensors and Bioelectronics* 63 (January): 276–86. https://doi.org/10.1016/J.BIOS.2014.07.052.

Deroco, Patricia Batista, Dagwin Wachholz Junior, and Lauro Tatsuo Kubota. 2023. "Paper-Based Wearable Electrochemical Sensors: A New Generation of Analytical Devices." *Electroanalysis* 35 (1): e202200177. https://doi.org/10.1002/ELAN.202200177.

Dhar, Bidhan Chandra, and Nae Yoon Lee. 2018. "Lab-on-a-Chip Technology for Environmental Monitoring of Microorganisms." *Biochip Journal* 12 (3): 173–83. https://doi.org/10.1007/S13206-018-2301-5/METRICS.

Ding, Ruiyu, Yi Heng Cheong, Ashiq Ahamed, and Grzegorz Lisak. 2021. "Heavy Metals Detection with Paper-Based Electrochemical Sensors." *Analytical Chemistry* 93 (4): 1880–88. https://doi.org/10.1021/ACS.ANALCHEM.0C04247/ASSET/IMAGES/LARGE/AC0C04247_0004.JPEG.

Du, Hao, Yuqun Xie, and Jun Wang. 2021. "Nanomaterial-Sensors for Herbicides Detection Using Electrochemical Techniques and Prospect Applications." *TrAC Trends in Analytical Chemistry* 135 (February): 116178. https://doi.org/10.1016/J.TRAC.2020.116178.

Elsayed, Asmaa M., Ashour M. Ahmed, M. T. Tammam, M. F. Eissa, and Arafa H. Aly. 2024. "Sensing of Heavy Metal Pb^{2+} Ions in Water Utilizing the Photonic Structure of Highly Controlled Hexagonal TiON/TiO_2 Nanotubes." *Scientific Reports* 14 (1): 1–9. https://doi.org/10.1038/s41598-023-50428-2.

Ensafi, Ali A. 2019. "An Introduction to Sensors and Biosensors." In *Electrochemical Biosensors*, 1–10. Elsevier. https://doi.org/10.1016/B978-0-12-816491-4.00001-2.

Erkmen, Cem, Sevinc Kurbanoglu, and Bengi Uslu. 2021. "Nanozymes in Electrochemical Analysis." *Nanozymes*, November, 159–98. https://doi.org/10.1201/9781003109228-9.

Erkmen, Cem, Didem N Unal, Sevinc Kurbanoglu, and Bengi Uslu. 2022. "Basics of Electrochemical Sensors BT – Organic Electrodes: Fundamental to Advanced Emerging Applications." In *Organic Electrodes*, edited by Ram K Gupta, 81–99. Springer International Publishing. https://doi.org/10.1007/978-3-030-98021-4_5.

Filippidou, Myrto Kyriaki, and Stavros Chatzandroulis. 2023. "Microfluidic Devices for Heavy Metal Ions Detection: A Review." *Micromachines*. https://doi.org/10.3390/mi14081520.

Fleet, Bernard, and Hari Gunasingham. 1992. "Electrochemical Sensors for Monitoring Environmental Pollutants." *Talanta* 39 (11): 1449–57. https://doi.org/10.1016/0039-9140(92)80125-W.

Guan, Yanfang, and Baichuan Sun. 2020. "Detection and Extraction of Heavy Metal Ions Using Paper-Based Analytical Devices Fabricated via Atom Stamp Printing." *Microsystems & Nanoengineering* 6 (1): 1–12. https://doi.org/10.1038/s41378-019-0123-9.

Han, Ji Hoon, Jinkyeong Kim, Joon Hyung Jin, and Joon Hyub Kim. 2022. "Electrochemical Stripping Detection of Cadmium with Paper-Based Channels for Point-of-Care Detection." *Microchemical Journal* 183 (December): 108111. https://doi.org/10.1016/J.MICROC.2022.108111.

Hanrahan, Grady, Deepa G. Patil, and Joseph Wang. 2004. "Electrochemical Sensors for Environmental Monitoring: Design, Development and Applications." *Journal of Environmental Monitoring* 6 (8): 657–64. https://doi.org/10.1039/B403975K.

Hao, Xuanxuan, Zhimin Liu, Tongtong Zheng, Yunfeng Fan, and Leqian Hu. 2024. "Label-Free Electrochemical Aptasensor for Ultrasensitive Lead Ion Detection Based on Flower-like AuNPs@MoS2 and Core–Shell Pt@Pd Bimetallic Nanoparticles." *Microchimica Acta* 191 (6): 1–11. https://doi.org/10.1007/S00604-024-06414-0/TABLES/2.

Hareesha, Nagarajappa, and Jamballi Gangadharappa Gowda Manjunatha. 2021. "Electro-Oxidation of Formoterol Fumarate on the Surface of Novel Poly(Thiazole Yellow-G) Layered Multi-Walled Carbon Nanotube Paste Electrode." *Scientific Reports* 11 (1): 1–14. https://doi.org/10.1038/s41598-021-92099-x.

Hareesha, Nagarajappa, Jamballi Gangadharappa Gowda Manjunatha, Chenthattil Raril, and Girish Tigari. 2019. "Design of Novel Surfactant Modified Carbon Nanotube Paste Electrochemical Sensor for the Sensitive Investigation of Tyrosine as a Pharmaceutical Drug." *Advanced Pharmaceutical Bulletin* 9 (1): 132–37. https://doi.org/10.15171/apb.2019.016.

Hirano, Takeshi, and Kazuyoshi Tamae. 2011. "Earthworms and Soil Pollutants." *Sensors* 11 (12): 11157–67. https://doi.org/10.3390/S111211157.

Ho, Clifford K., Alex Robinson, David R. Miller, and Mary J. Davis. 2005. "Overview of Sensors and Needs for Environmental Monitoring." *Sensors* 5 (1–2): 4–37. https://doi.org/10.3390/s5010004.

Iftikhar, Faiza Jan, Afzal Shah, Mohammad Salim Akhter, Sevinc Kurbanoglu, and Sibel A. Ozkan. 2019. "Introduction to Nanosensors." In *New Developments in Nanosensors for Pharmaceutical Analysis*, 1–46. Elsevier. https://doi.org/10.1016/B978-0-12-816144-9.00001-8.

Islam, Md Saiful, Kazuto Sazawa, Kazuharu Sugawara, and Hideki Kuramitz. 2023. "Electrochemical Biosensor for Evaluation of Environmental Pollutants Toxicity." *Environments* 10 (4): 63. https://doi.org/10.3390/ENVIRONMENTS10040063.

Jiang, Nan, Chengzhou Zhang, Lingna Ge, Shan Huang, and Xiaojun Chen. 2023. "Unique Three-Dimensional Ordered Macroporous Dealloyed Gold-Silver Electrochemical Sensing Platforms for Ultrasensitive Mercury(Ii) Monitoring." *Analyst* 149 (4): 1141–50. https://doi.org/10.1039/d3an02075d.

Jokerst, Jana C., Jason M. Emory, and Charles S. Henry. 2011. "Advances in Microfluidics for Environmental Analysis." *Analyst* 137 (1): 24–34. https://doi.org/10.1039/C1AN15368D.

Kimmel, Danielle W., Gabriel LeBlanc, Mika E. Meschievitz, and David E. Cliffel. 2012. "Electrochemical Sensors and Biosensors." *Analytical Chemistry* 84 (2): 685–707. https://doi.org/10.1021/ac202878q.

Kissinger, Peter T., William R. Heineman, and Eric P. Achterberg. 1996. "Laboratory Techniques in Electroanalytical Chemistry." *TrAC Trends in Analytical Chemistry* 15 (10): 550. https://doi.org/10.1016/S0165-9936(96)80740-0.

Kissinger, Peter T., and William R. Heineman. 1983. "Cyclic Voltammetry." *Journal of Chemical Education* 60 (9): 702–6. https://doi.org/10.1021/ed060p702.

Koester, Carolyn J., and Amal Moulik. 2005. "Trends in Environmental Analysis." *Analytical Chemistry* 77 (12): 3737–54. https://doi.org/10.1021/AC0505674.

Kokkinos, Christos, Anastasios Economou, and Dimosthenis Giokas. 2018. "Paper-Based Device with a Sputtered Tin-Film Electrode for the Voltammetric Determination of Cd(II) and Zn(II)" 260 (May): 223–26.

Kumar, Vikash, Sanjeev Kumar Raghuwanshi, and Santosh Kumar. 2022. "Advances in Nanocomposite Thin-Film-Based Optical Fiber Sensors for Environmental Health Monitoring – A Review." *IEEE Sensors Journal* 22 (15): 14696–707. https://doi.org/10.1109/JSEN.2022.3185004.

Kummari, Shekher, Lakshmi R. Panicker, Jagadeeswara Rao Bommi, Sampath Karingula, Venisheety Sunil Kumar, Kuldeep Mahato, and Kotagiri Yugender Goud. 2023. "Trends in Paper-Based Sensing Devices for Clinical and Environmental Monitoring." *Biosensors* 13 (4): 420. https://doi.org/10.3390/BIOS13040420.

Kurbanoglu, Sevinc, Nurgul K. Bakirhan, Afzal Shah, and Sibel A. Ozkan. 2019. "Chemical Nanosensors in Pharmaceutical Analysis." *New Developments in Nanosensors for Pharmaceutical Analysis*, January, 141–70. https://doi.org/10.1016/B978-0-12-816144-9.00005-5.

Kurbanoglu, Sevinc, and Sibel A. Ozkan. 2018. "Electrochemical Carbon Based Nanosensors: A Promising Tool in Pharmaceutical and Biomedical Analysis." *Journal of Pharmaceutical and Biomedical Analysis* 147 (January): 439–57. https://doi.org/10.1016/j.jpba.2017.06.062.

Lam, Stephen Chiu Kwok, Kai Lap Wong, Kwok On Wong, Wenxiu Wong, and Wai Ho Mow. 2009. "A Smartphone-Centric Platform for Personal Health Monitoring Using Wireless Wearable Biosensors." In *ICICS 2009 – Conference Proceedings of the 7th International Conference on Information, Communications and Signal Processing*, 1–7. IEEE. https://doi.org/10.1109/ICICS.2009.5397628.

Lazanas, Alexandros Ch, and Mamas I. Prodromidis. 2023. "Electrochemical Impedance Spectroscopy—A Tutorial." *ACS Measurement Science Au* 3 (3): 162–93. https://doi.org/10.1021/ACSMEASURESCIAU.2C00070/ASSET/IMAGES/LARGE/TG2C00070_0032.JPEG.

Li, Chao, Afruza Begum, and Jinkai Xue. 2020. "Analytical Methods to Analyze Pesticides and Herbicides." *Water Environment Research* 92 (10): 1770–85. https://doi.org/10.1002/WER.1431.

Li, Xin, Linjie Wang, Dan Du, Liang Ni, Jianming Pan, and Xiangheng Niu. 2019. "Emerging Applications of Nanozymes in Environmental Analysis: Opportunities and Trends." *TrAC – Trends in Analytical Chemistry* 120: 1–13. https://doi.org/10.1016/j.trac.2019.115653.

Liu, Yao, Qiang Xue, Chunwen Chang, Rong Wang, Zeyu Liu, and Lin He. 2022. "Recent Progress Regarding Electrochemical Sensors for the Detection of Typical Pollutants in Water Environments." *Analytical Sciences* 38 (1): 55–70. https://doi.org/10.2116/ANALSCI.21SAR12/TABLES/1.

Maguire, Ivan, Richard O'Kennedy, Jens Ducrée, and Fiona Regan. 2018. "A Review of Centrifugal Microfluidics in Environmental Monitoring." *Analytical Methods* 10 (13): 1497–1515. https://doi.org/10.1039/C8AY00361K.

Manjunatha, Jamballi Gangadharappa Gowda. 2019. "A Surfactant Enhanced Graphene Paste Electrode as an Effective Electrochemical Sensor for the Sensitive and Simultaneous Determination of Catechol and Resorcinol." *Chemical Data Collections*, 100331. https://doi.org/10.1016/j.cdc.2019.100331.

Manjunatha, Jamballi Gangadharappa Gowda, Madikeri M. Charithra, Nagaraja Sreeharsha, Syed Mohammed, and Basheeruddin Asdaq. 2021. "Polymerized Carbon Nanotube Paste Electrode as a Sensing Material for the Detection of Adrenaline with Folic Acid." *Monatshefte Für Chemie – Chemical Monthly* 152 (4): 411–20. https://doi.org/10.1007/s00706-021-02756-0.

Mazurkiewicz, Wojciech, Marta Podrażka, Elżbieta Jarosińska, Karthika Kappalakandy Valapil, Magdalena Wiloch, Martin Jönsson-Niedziółka, and Emilia Witkowska Nery. 2020. "Paper-Based Electrochemical Sensors and How to Make Them (Work)." *ChemElectroChem* 7 (14): 2939–56. https://doi.org/10.1002/CELC.202000512.

Medina-Sánchez, Mariana, Carmen C. Mayorga-Martinez, Takeshi Watanabe, Tribidasari A. Ivandini, Yuki Honda, Flavio Pino, Azuya Nakata, Akira Fujishima, Yasuaki Einaga, and Arben Merkoçi. 2016. "Microfluidic Platform for Environmental Contaminants Sensing and Degradation Based on Boron-Doped Diamond Electrodes." *Biosensors and Bioelectronics* 75 (January): 365–74. https://doi.org/10.1016/J.BIOS.2015.08.058.

Meng, Ruonan, Qiujin Zhu, Tingyu Long, Xuli He, Zewei Luo, Ronghui Gu, Wenzhong Wang, and Ping Xiang. 2023. "The Innovative and Accurate Detection of Heavy Metals in Foods: A Critical Review on Electrochemical Sensors." *Food Control* 150 (August): 109743. https://doi.org/10.1016/J.FOODCONT.2023.109743.

Meredith, Nathan A., Casey Quinn, David M. Cate, Thomas H. Reilly, John Volckens, and Charles S. Henry. 2016. "Paper-Based Analytical Devices for Environmental Analysis." *Analyst* 141 (6): 1874–87. https://doi.org/10.1039/C5AN02572A.

Mirceski, Valentin, Rubin Gulaboski, Milivoj Lovric, Ivan Bogeski, Reinhard Kappl, and Markus Hoth. 2013. "Square-Wave Voltammetry: A Review on the Recent Progress." *Electroanalysis* 25 (11): 2411–22. https://doi.org/10.1002/ELAN.201300369.

Mirceski, Valentin, Sławomira Skrzypek, and Leon Stojanov. 2018. "Square-Wave Voltammetry." *ChemTexts* 4 (4): 1–14. https://doi.org/10.1007/S40828-018-0073-0/FIGURES/9.

Mohamed, Dalia, and Manal M. Fouad. 2020. "Application of NEMI, Analytical Eco-Scale and GAPI Tools for Greenness Assessment of Three Developed Chromatographic Methods for Quantification of Sulfadiazine and Trimethoprim in Bovine Meat and Chicken Muscles: Comparison to Greenness Profile of Reporte." *Microchemical Journal* 157 (February): 104873. https://doi.org/10.1016/j.microc.2020.104873.

Mostert, Maria M.R., Godwin A. Ayoko, and Serge Kokot. 2010. "Application of Chemometrics to Analysis of Soil Pollutants." *TrAC Trends in Analytical Chemistry* 29 (5): 430–45. https://doi.org/10.1016/J.TRAC.2010.02.009.

Münzel, Thomas, Omar Hahad, Andreas Daiber, and Philip J. Landrigan. 2023. "Soil and Water Pollution and Human Health: What Should Cardiologists Worry About?" *Cardiovascular Research* 119 (2): 440–49. https://doi.org/10.1093/CVR/CVAC082.

Nikoleli, Georgia Paraskevi, Dimitrios Nikolelis, Christina G. Siontorou, and Stephanos Karapetis. 2018. "Lipid Membrane Nanosensors for Environmental Monitoring: The Art, the Opportunities, and the Challenges." *Sensors* 18 (1): 284. https://doi.org/10.3390/S18010284.

Pankratov, Dmitry, Elena González-Arribas, Zoltan Blum, and Sergey Shleev. 2016. "Tear Based Bioelectronics." In *Electroanalysis*. Wiley-VCH Verlag. https://doi.org/10.1002/elan.201501116.

Patel, Bhavik A. 2020. "Chapter 2. Amperometry and Potential Step Techniques." In *Electrochemistry for Bioanalysis*, 9–26. Elsevier. https://doi.org/10.1016/c2019-0-03108-1.

Pierce, David T., and Julia Xiaojun Zhao. 2010. *Trace Analysis with Nanomaterials*. Wiley. https://doi.org/10.1002/9783527632015.

Pohanka, Miroslav, and Petr Skládal. 2008. "Electrochemical Biosensors – Principles and Applications." *Journal of Applied Biomedicine* 6 (2): 57–64. https://doi.org/10.32725/jab.2008.008.

Pol, Roberto, Francisco Céspedes, David Gabriel, and Mireia Baeza. 2017. "Microfluidic Lab-on-a-Chip Platforms for Environmental Monitoring." *TrAC Trends in Analytical Chemistry* 95 (October): 62–68. https://doi.org/10.1016/J.TRAC.2017.08.001.

Prodromidis, Mamas I. MI. 2010. "Impedimetric Immunosensors—A Review." *Electrochimica Acta* 55 (14): 4227–33. https://doi.org/10.1016/J.ELECTACTA.2009.01.081.

Pushpanjali, P. Amrutha, and Jamballi Gangadharappa Gowda Manjunatha. 2019. "A Sensitive Novel Approach towards the Detection of 8-Hydroxyquinoline at Anionic Surfactant Modified Carbon Nanotube Based Biosensor: A Voltammetric Study." *Physical Chemistry Research* 7 (4): 813–22. https://doi.org/10.22036/pcr.2019.198397.1663.

Qian, Lanting, Sharmila Durairaj, Scott Prins, and Aicheng Chen. 2021. "Nanomaterial-Based Electrochemical Sensors and Biosensors for the Detection of Pharmaceutical Compounds." *Biosensors and Bioelectronics* 175: 112836. https://doi.org/https://doi.org/10.1016/j.bios.2020.112836.

Rafiee, Mohammad, Dylan J. Abrams, Luana Cardinale, Zachary Goss, Antonio Romero-Arenas, and Shannon S. Stahl. 2024. "Cyclic Voltammetry and Chronoamperometry: Mechanistic Tools for Organic Electrosynthesis." *Chemical Society Reviews* 53 (2): 566–85. https://doi.org/10.1039/d2cs00706a.

Ramaley, Louis, and Matthew S. Krause. 1969. "Theory of Square Wave Voltammetry." *Analytical Chemistry* 41 (11): 1362–65. https://doi.org/10.1021/ac60280a005.

Rassaei, Liza, Frank Marken, Mika Sillanpää, Mandana Amiri, Ciprian Mihai Cirtiu, and Markus Sillanpää. 2011. "Nanoparticles in Electrochemical Sensors for Environmental Monitoring." *TrAC Trends in Analytical Chemistry* 30 (11): 1704–15. https://doi.org/10.1016/J.TRAC.2011.05.009.

Ribeiro, Cláudia, Ana Rita Ribeiro, Alexandra S. Maia, Virgínia M.F. Gonçalves, and Maria Elizabeth Tiritan. 2014. "New Trends in Sample Preparation Techniques for Environmental Analysis." *Critical Reviews in Analytical Chemistry* 44 (2): 142–85. https://doi.org/10.1080/10408347.2013.833850.

Rishpon, Judith. 2002. "Electrochemical Biosensors for Environmental Monitoring." *Reviews on Environmental Health* 17 (3): 219–47. https://doi.org/10.1515/REVEH.2002.17.3.219.

Riu, Jordi, Alicia Maroto, and F. Xavier Rius. 2006. "Nanosensors in Environmental Analysis." *Talanta* 69 (2): 288–301.

Romani, Annalisa, Maria Minunni, Nadia Mulinacci, Patrizia Pinelli, Franco F. Vincieri, M. Del Carlo, and Marco Mascini. 2000. "Comparison among Differential Pulse Voltammetry, Amperometric Biosensor, and HPLC/DAD Analysis for Polyphenol Determination." *Journal of Agricultural and Food Chemistry* 48 (4). https://doi.org/10.1021/jf990767e.

Sadik, Omowunmi A., and Jeanette M. Van Emon. 1996. *Applications of Electrochemical Immunosensors to Environmental Monitoring. Biosensors and Bioelectronics*. Vol. 11. Elsevier. https://doi.org/10.1016/0956-5663(96)85936-7.

Sequeira, Margaret, Michaela Bowden, Edel Minogue, and Dermot Diamond. 2002. "Towards Autonomous Environmental Monitoring Systems." *Talanta* 56 (2): 355–63. https://doi.org/10.1016/S0039-9140(01)00601-4.

Sidhu, Gurpreet Kaur, Simranjeet Singh, Vijay Kumar, Daljeet Singh Dhanjal, Shivika Datta, and Joginder Singh. 2019. "Toxicity, Monitoring and Biodegradation of Organophosphate Pesticides: A Review." *Critical Reviews in Environmental Science and Technology* 49 (13). https://doi.org/10.1080/10643389.2019.1565554.

Singh, Anoop, Asha Sharma, Aamir Ahmed, Ashok K. Sundramoorthy, Hidemitsu Furukawa, Sandeep Arya, and Ajit Khosla. 2021. "Recent Advances in Electrochemical Biosensors: Applications, Challenges, and Future Scope." *Biosensors* 11 (9): 336. https://doi.org/10.3390/BIOS11090336.

Suh, H. Helen, Tina Bahadori, Jose Vallarino, and John. D. Spengler. 2000. "Criteria Air Pollutants and Toxic Air Pollutants." *Environmental Health Perspectives* 108 (SUPPL. 4): 625–33. https://doi.org/10.1289/EHP.00108S4625.

Sun, Yang, Qiao Yan Jiang, Feng Chen, and Yue Cao. 2022. "Paper-Based Electrochemical Sensor." *Electrochemical Science Advances* 2 (4): e2100057. https://doi.org/10.1002/ELSA.202100057.

Suwannachat, Jakkrit, Apichart Saenchoopa, Wonn Shweyi Thet Tun, Rina Patramanon, Sakda Daduang, Jureerat Daduang, and Sirinan Kulchat. 2024. "An Electrochemical AChE-Based Biosensor for Organophosphate Pesticides Using a Modified CuNWs/RGO Nanocomposite on a Screen-Printed Carbon Electrode." *Food Chemistry* 434 (February): 137431.

Suzuki, Takafumi, Takanori Hidaka, Yoshito Kumagai, and Masayuki Yamamoto. 2020. "Environmental Pollutants and the Immune Response." *Nature Immunology* 21 (12): 1486–95. https://doi.org/10.1038/s41590-020-0802-6.

Švorc, Jozef, Jaroslav Miertuš, Miroslav Stred'ansky´, and Miroslav Stred'ansky´. 1997. "Composite Transducers for Amperometric Biosensors. The Glucose Sensor." *Analytical Chemistry* 69 (12): 2086–90. https://doi.org/10.1021/AC9609485/ASSET/IMAGES/LARGE/AC9609485F00003.JPEG.

Taillefert, Martial, and Tim F. Rozan. 2002. "Electrochemical Methods for the Environmental Analysis of Trace Elements Biogeochemistry." *ACS Symposium Series* 811: 2–14. https://doi.org/10.1021/BK-2002-0811.CH001.

Tallman, Dennis E., and Steven L. Petersen. 1990. "Composite Electrodes for Electroanalysis: Principles and Applications." *Electroanalysis* 2 (7): 499–510. https://doi.org/10.1002/ELAN.1140020702.

Temerk, Yassien, Mohamed Ibrahim, Hossieny Ibrahim, and Mohamed Kotb. 2016. "Adsorptive Stripping Voltammetric Determination of Anticancer Drug Lomustine in Biological Fluids Using in Situ Mercury Film Coated Graphite Pencil Electrode." *Journal of Electroanalytical Chemistry* 760 (January): 135–42. https://doi.org/10.1016/J.JELECHEM.2015.11.026.

Tseng, Ryan Chang, Ching-Chuen Chen, Sheng-Min Hsu, and Han-Sheng Chuang. 2018. "Contact-Lens Biosensors." *Sensors (Basel)*, 18 (8): 1–24. https://doi.org/10.3390/s18082651.

Ty, Boreborey, Chanvorleak Phat, Kuok Fidero, Eden M. Andrews, Winarto Kurniawan, and Hirofumi Hinode. 2024. "Heavy Metals." *Water and Life in Tonle Sap Lake,* February, 379–86. https://doi.org/10.1007/978-981-16-6632-2_36.

Uslu, Bengi, and Sibel A. Ozkan. 2011. "Electroanalytical Methods for the Determination of Pharmaceuticals: A Review of Recent Trends and Developments" 44 (16): 2644–2702. https://doi.org/10.1080/00032719.2011.553010.

Vamvakaki, Vicky, and Nikos A. Chaniotakis. 2007. "Carbon Nanostructures as Transducers in Biosensors." *Sensors and Actuators B: Chemical* 126 (1): 193–97. https://doi.org/https://doi.org/10.1016/j.snb.2006.11.042.

Vito-Francesco, Elisabetta De, Alessandro Farinelli, Qiuyue Yang, Bhawna Nagar, Ruslan Álvarez, Arben Merkoçi, Thorsten Knutz, et al. 2022. "An Innovative Autonomous Robotic System for On-Site Detection of Heavy Metal Pollution Plumes in Surface Water." *Environmental Monitoring and Assessment* 194 (2): 1–19. https://doi.org/10.1007/S10661-021-09738-Z/FIGURES/9.

Wang, Joseph. 2002. "Electrochemical Detection for Microscale Analytical Systems: A Review." *Talanta* 56 (2): 223–31. https://doi.org/10.1016/S0039-9140(01)00592-6.

Wang, Shangshang, Jianbo Zhang, Oumaïma Gharbi, Vincent Vivier, Ming Gao, and Mark E. Orazem. 2021. "Electrochemical Impedance Spectroscopy." *Nature Reviews Methods Primers* 1 (1): 1–21. https://doi.org/10.1038/s43586-021-00039-w.

Wang, You, Hui Xu, Jianming Zhang, and Guang Li. 2008. "Electrochemical Sensors for Clinic Analysis." *Sensors* 8 (4): 2043–81. https://doi.org/10.3390/s8042043.

Windmiller, Joshua Ray, and Joseph Wang. 2013. "Wearable Electrochemical Sensors and Biosensors: A Review." *Electroanalysis* 25 (1): 29–46. https://doi.org/10.1002/elan.201200349.

Wring, Stephen A., and John P. Hart. 1992. "Chemically Modified, Carbon-Based Electrodes and Their Application as Electrochemical Sensors for the Analysis of Biologically Important Compounds. A Review." *Analyst* 117 (8). https://pubs.rsc.org/en/content/articlehtml/1992/an/an9921701215.

Wu, Guangfu, Huijie Li, Jialong Chen, King Wai Chiu Lai, Lize Xiong, and Runhuai Yang. 2024. "Microfluidics Enhanced Biosensor Technology for Advancing Food Safety Monitoring." *Trends in Food Science & Technology* 149 (July): 104556. https://doi.org/10.1016/J.TIFS.2024.104556.

Yang, Wei, and Stanley T. Omaye. 2009. "Air Pollutants, Oxidative Stress and Human Health." *Mutation Research/Genetic Toxicology and Environmental Mutagenesis* 674 (1–2): 45–54. https://doi.org/10.1016/J.MRGENTOX.2008.10.005.

Yarman, Aysu, Sevinc Kurbanoglu, and Frieder W. Scheller. 2020. "Noninvasive Biosensors for Diagnostic Biomarkers." In *Commercial Biosensors and Their Applications: Clinical, Food, and Beyond*, 167. Elsevier. https://doi.org/10.1016/B978-0-12-818592-6.00007-4.

Yew, Maxine, Yong Ren, Kai Seng Koh, Chenggong Sun, and Colin Snape. 2019. "A Review of State-of-the-Art Microfluidic Technologies for Environmental Applications: Detection and Remediation." *Global Challenges* 3 (1): 1800060. https://doi.org/10.1002/GCH2.201800060.

Zaynab, Madiha, Rashid Al-Yahyai, Ayesha Ameen, Yasir Sharif, Liaqat Ali, Mahpara Fatima, Khalid Ali Khan, and Shuangfei Li. 2022. "Health and Environmental Effects of Heavy Metals." *Journal of King Saud University – Science* 34 (1): 101653. https://doi.org/10.1016/J.JKSUS.2021.101653.

Zhang, Hui, Jun Wang, Kaili Li, Ruixue Yang, Shifeng Cai, Yang Li, and Tianrong Zhan. 2024. "Highly Conductive Ti_3C_2 MXene-Supported CoAl-Layered Double Hydroxide Nanosheets for Ultrasensitive Electrochemical Detection of Organophosphate Pesticide Fenitrothion." *Microchimica Acta* 191 (8): 1–14. https://doi.org/10.1007/S00604-024-06549-0/TABLES/3.

Zhang, Xu Xiang, Tong Zhang, and Herbert H.P. Fang. 2009. "Antibiotic Resistance Genes in Water Environment." *Applied Microbiology and Biotechnology* 82 (3): 397–414. https://doi.org/10.1007/S00253-008-1829-Z.

Zhang, Xuan, Nan Hao, Shucheng Liu, Kai Wei, Changchang Ma, Jianming Pan, and Sheng Feng. 2024. "Construction of Phosphatase-like COF-OMe@Valine-CeO_2 Nanozymes for Ultrasensitive Electrochemical Detection of Organophosphorus Pesticides." *Sensors and Actuators B: Chemical* 417 (October): 136068. https://doi.org/10.1016/J.SNB.2024.136068.

Zhang, Zhong, Morteza Azizi, Michelle Lee, Philip Davidowsky, Peter Lawrence, and Alireza Abbaspourrad. 2019. "A Versatile, Cost-Effective, and Flexible Wearable Biosensor for: In Situ and Ex Situ Sweat Analysis, and Personalized Nutrition Assessment." *Lab on a Chip* 19 (20): 3448–60. https://doi.org/10.1039/c9lc00734b.

Zhou, Xiang. 2024. "Electrochemical Detection of Heavy Metal Ions in Water Using MWCNT/ZnO Nanocomposite." *International Journal of Electrochemical Science* 19 (5): 100559. https://doi.org/10.1016/J.IJOES.2024.100559.

9 Applications of Sensors in Food Quality Control

Nazife Aslan, Melike Akan, Cigdem Kanbes-Dindar, and Bengi Uslu

9.1 INTRODUCTION

In recent years, the rise of foodborne diseases and food safety issues has led to shifts in consumer preferences. Consequently, food safety is evolving into a public health concern and a vital tool for market development.

Problems including pesticide residues, illicit additives, allergens, pathogens, and other hazards not only endanger human health but also restrict the rapid growth of the food industry. One of the most important aspects of food safety research is the development of analytical techniques or tools to identify different harmful compounds in food (Fu et al., 2024). Owing to the complex nature of food samples and the existence of trace hazardous materials, there is an increasing need for convenient analytical methods or equipment that are high-throughput, inexpensive, precise, sensitive, and easily portable (Ali et al., 2024). The disadvantages of expensive and popular instrumental procedures can be addressed by sensors that combine an identification element and a signal transducer. Such sensors are suitable for on-site and real-time qualitative and quantitative analysis of harmful substances in food, offering a broader scope for research and development (Figure 9.1). This is an important field of research in food safety since recent studies have focused on creating distinct chemical and biological sensing devices based on diverse principles to identify harmful compounds in food.

9.2 IMPORTANCE OF FOOD QUALITY CONTROL

Health is the fundamental need and right of every individual. Food products significantly impact our health. Consequently, ensuring the safety of food products is a paramount requirement for consumers.

Food that has been considered safe for ingestion by eliminating all possible sources of contamination and spoilage, posing no risk to health, and keeping all of its nutritional content is considered safe or trustworthy (Chhetri, 2024). Throughout its shelf life, safe food should not present any physical, chemical, or biological risks. If it does, it cannot be deemed suitable for public consumption. Properly identifying safe food requires recognizing these potential issues first. Physical risks involve the presence of foreign substances such as stones, soil, metal, glass, bones, wood, hair, and other materials that are undesirable in foodstuffs. Chemical risks include veterinary and pesticide residues misused in animal husbandry and crop production, detergent residues, heavy metals, harmful compounds formed during food processing, chemical contamination from packaging materials, and improper use of additives (Chhetri, 2024; T. Li et al., 2022). Microbiological risk refers to the group of risks posed by microorganisms such as bacteria, molds, viruses, parasites, and the toxic substances they may leave behind in foods (Varilla et al., 2020).

 DOI: 10.1201/9781003500322-9

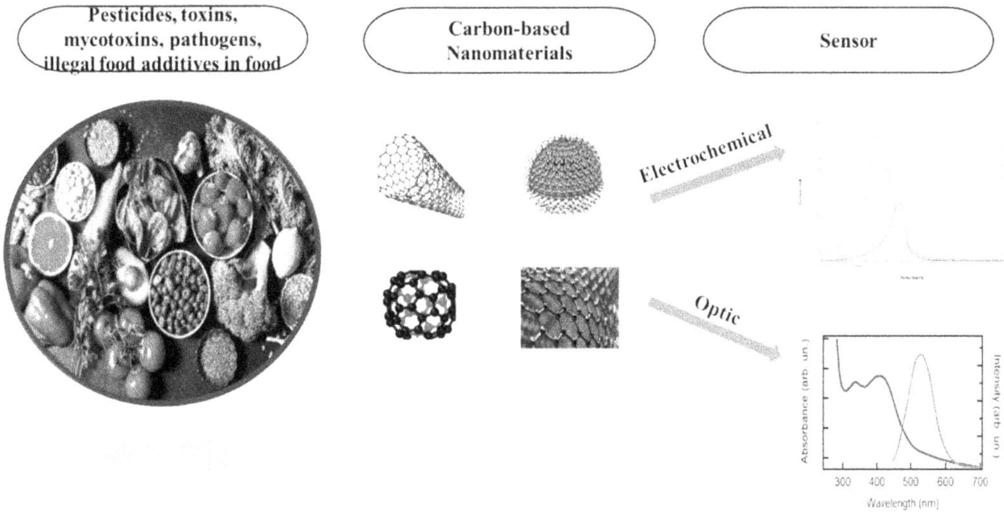

FIGURE 9.1 Schematic representation of the combination of carbon-based materials in food quality analyses.

Foods are classified as "safe" if they do not contain any of the three major health hazards and are considered safe for consumption. The term "food safety" refers to a scientific system that details how food should be processed, prepared, stored, and served to consumers. Its goal is to reduce the risk of microbial, chemical, and physical factors leading to foodborne illnesses and spoilage. "Field to fork" conceptualizes "food safety" as a chain that encompasses the entire journey of food. Primary producers, such as farmers or animal husbandry owners, represent the first link in this system by supplying the raw ingredients (V. Kumar et al., 2024; T. Li et al., 2022; Shruti et al., 2024). After initial production, items move into the food industry where they are processed, packaged, and prepared for consumption. Processed food then goes through several phases, such as distribution and retail, before it is finally consumed by the consumer. Food safety relies on the collaborative and effective functioning of all stages in the chain, as any weak link can disrupt it. The important point here is that the consumer is aware that he/she is a part of this chain (Chhetri, 2024; V. Kumar et al., 2024).

Food-borne diseases not only impact the health of individuals but also have extensive effects on the social and economic structures of families, communities, industries, and even entire nations. Worldwide, food trade enhances the industry's value and contributes to national income by providing foreign exchange for food-exporting nations. However, despite these benefits, foodborne illnesses can significantly impact public health across all levels of society, decrease workplace productivity, and lead to financial losses (Luo et al., 2024).

The globalization of the food trade has broadened food diversity, making safe, high-quality, and affordable food products that meet consumer demands more effectively. However, prioritizing health is crucial for the production and consumption of safe food. Consequently, new systems such as Hazard Analysis and Critical Control Point, Good Manufacturing Practices, Good Hygiene Practices, and Good Agricultural Practices need to be established in response to these changes (Mercer, 2021).

Despite advancements in many countries to improve food safety, consuming contaminated food still causes foodborne illnesses in millions of people each year. The advent of innovations like genetic engineering, irradiation, packaging technology, food additives, pesticide and veterinary drug applications (and their residues), and mycotoxins highlights the increasing importance of addressing food safety risks (Zhou, 2022). The advancement of new technologies not only enhances food production and diversity but also raises concerns about product safety and consumer acceptance. To address these concerns effectively, it is essential to evaluate them using internationally recognized methods in a collaborative and transparent manner.

9.3 APPLICATIONS OF NANOTECHNOLOGY IN FOOD INDUSTRY

Nanotechnology is an applied science and technology field that allows the manipulation to understand the scale of 1 nm originates from the Greek word for "dwarf" (Sekhon, 2010). To understand the scale of 1 nm (10^{-9} m), which is one-millionth of a meter, consider that an ordinary sheet of paper is about 100,000 nm thick, a red blood cell is 2,000–5,000 nm in diameter, and DNA is approximately 2.5 nm wide (Dingman, 2008). Nanotechnology focuses on the production, characterization, and targeted modification of naturally existing or synthetically produced materials at the atomic, molecular, or colloidal levels (Mir & Shah, 2017). It enables the control of individual atoms and molecules, allowing for the creation and design of new materials, nanomachines, and nanodevices (Mohammad et al., 2012). Many new products that we come across in our daily lives, such as touch screens, sun cream, cosmetic products, tennis rackets, bicycles, computer memory, etc., are actually nanotechnological products (Mousavi et al., 2011). Nanomaterials demonstrate distinct physical and chemical properties compared to their normal-sized equivalents (Duncan, 2011). They are manufactured in diverse forms, encompassing one-dimensional structures like films and coatings, two-dimensional structures such as nanofibers and nanotubes, and three-dimensional structures like nanoparticles.

The term "nano food" refers to the use of nanotechnology in the production, processing, and packaging of foods. It's important to note that this term does not imply that foods are produced or modified using nanomachines. In basic terms, food nanotechnology involves applying nanoscience to the food industry. Specifically, it focuses on developing new advancements and applications that improve the structure, texture, and quality of foods using nanoscience and engineering (Singh & Sengar, 2020).

There is definitely some hesitation and caution among consumers' attitudes toward nano foods.

While public perception of nanotechnology, in general, may lean toward neutrality or slight positivity, the application of nanotechnology in the food industry often generates a more cautious response from consumers. This caution primarily stems from concerns about potential health risks, safety implications, and ethical considerations related to nano-enabled food products, often referred to as "nano foods." Issues such as the unknown long-term effects of ingesting nanomaterials, regulatory oversight, and transparency in labeling and communication further contribute to public apprehension. Therefore, while nanotechnology holds promise for enhancing food quality, safety, and sustainability, addressing these concerns through clear communication, rigorous safety assessments, and transparent regulatory frameworks is crucial for gaining consumer trust and acceptance.

The applications of nanotechnology in the food industry mainly focus on improving food processing techniques, developing functional food products, creating advanced packaging systems to maintain food quality and safety, enhancing traceability through biosensors (Hareesha & Manjunatha, 2021; Siangproh et al., 2011), and developing active and smart packaging systems for detecting bacteria (Mir & Shah, 2017; Valdés et al., 2009).

Nanotechnology is employed in producing functional foods that can fulfill the body's nutritional needs more effectively by enhancing nutrient utilization (Singh & Sengar, 2020). Furthermore, nanotechnology techniques are employed to specifically design substances such as food flavors and antioxidants, which are crucial in functional foods. The objective is to improve the functionality and efficiency of these components within food products. Functional food compounds such as nanoparticulate lycopene and carotenoids are now being commercially produced. These nanotechnology-enhanced versions reportedly offer higher bioavailability and better dispersibility compared to their conventionally produced counterparts (Weiss et al., 2006).

Nanoemulsions represent one of the most prevalent applications of nanotechnology in the food industry. They are colloidal systems where tiny droplets of oil or water, typically ranging from 20 to 200 nm in size, are dispersed within another immiscible phase. These nano-sized droplets are stabilized using surfactants or emulsifiers to prevent aggregation and coalescence. Nanoemulsions have several important characteristics and applications in food technology. For example, they are more stable than conventional emulsions, resisting gravitational separation, creaming, and flocculation. This stability enhances the shelf life and consistency of food products. Recent studies have demonstrated that various lipophilic bioactive compounds such as vitamins, antioxidants, and phytochemicals can be encapsulated into nanoemulsions to enhance their water dispersibility, stability, and bioavailability (Lee et al., 2024). Nanoemulsions are indeed utilized to deliver functional ingredients such as flavors, colors, nutrients, and preservatives into food products. They improve the solubility, dispersion, and efficacy of these ingredients, enhancing their overall performance in food applications.

Nanoemulsions exhibit distinct physical properties compared to microemulsions. For instance, microemulsions scatter light in multiple directions and appear white and opaque. In contrast, nanoemulsions have droplet sizes smaller than visible wavelengths, making them optically transparent. This transparency is particularly advantageous for using nanoemulsions as carrier components in beverages.

Nanoemulsions serve as effective carrier mediums for water-insoluble food components like fish oil and lipophilic vitamins, facilitating their absorption in the gastrointestinal tract and allowing them to exhibit their bioactivities. Nanoemulsions can exhibit higher viscosity even at very low concentrations, which enables the production of less fatty foods without compromising texture or mouthfeel (Gökmen et al., 2012). In a study conducted at Georgia Tech in the USA, modified carbon nanotubes (CNTs) were utilized as biosensors for detecting microorganisms, toxic substances, and spoilage in food and beverages (Alfadul & Elneshw, 2010; Qureshi et al., 2012). It has been reported that nanotubes, which can also be used in packaging materials, exhibit strong antimicrobial effects, and direct contact with nanotubes damages *Escherichia coli*. In recent years, there have been developments in self-assembled nanotubes formed through the hydrolysis of the milk protein lactalbumin. These nanotubes can serve as natural carriers for encapsulating nutrients, additives, and drugs at the nanoscale (Chaudhry et al., 2008).

Foods are packaged to maintain their quality and ensure that they are kept under hygienic conditions until consumption. Today, increasingly educated and aware consumers prefer ready-to-eat meals with long shelf lives. They favor foodstuffs that can be prepared for consumption with minimal processing techniques, maintain natural quality, and have guaranteed safety. By using the latest technological developments, food packaging manufacturers are making an intensive effort to develop and produce innovative packaging that can provide the highest level of competition in the sector, these advancements aim to not only protect and distribute products during transport but also minimize associated risks and adequately inform consumers about the products (Duncan, 2011; Sharma et al., 2017).

Nanomaterials with enhanced mechanical and thermal characteristics are being developed to provide more robust protection for foodstuffs from external mechanical, thermal, chemical, or microbiological influences. Nanocomposites, which are nanoparticles bonded in polymers, offer materials with enhanced properties, including a reduction in weight, improved recyclability, and an increased resistance to spoilage and flavor issues.

The utilization of nanosensors for the detection of pathogens and contaminants can serve to enhance the security of food production, processing, and shipment. Silver nanoparticle-based food containers are already in use, offering the advantage of maintaining high-quality foods for extended periods of time, thereby reducing food waste (Handford et al., 2014). Incorporating silver nanoparticles into food packages was reported to reduce bacterial growth by over 98% within a 24-hour period (Zorraquín-Peña et al., 2020; Alfadul & Elneshwy, 2010).

Nanotechnology provides new applications in detecting counterfeiting and adulteration, which are increasingly prevalent with technological advancements, as well as in identifying substances harmful to health. In 2008, reports were published showing that thousands of people, especially children, became ill due to melamine contamination in dairy products in China. Following this incident, in 2010, Wei and her colleagues at the University of Miami quickly and effectively detected melamine in milk using gold nanoparticles (Wei et al., 2010). When gold nanoparticles are added to milk, the color of the mixture changes from pink to blue in the presence of melamine. In another study conducted at Clemson University, nanoparticles were added to chicken feed as an alternative to antibiotics for controlling pathogens in chickens (Tzeng et al., 2021). These nanoparticles, functioning as artificial cells, are incorporated into chicken feed and consumed by the chickens. Pathogens, perceiving these artificial cells as real cells, adhere to them and are subsequently eliminated from the chicken's digestive system. This process helps enhance chicken health and ensures safer poultry for human consumption (Mousavi et al., 2011; Sadr et al., 2023).

Predicting the long-term effects of a new technology like nanotechnology is very challenging (Shruti et al., 2024). Nanotechnology offers numerous contributions, such as enhancing food quality and safety, reducing agricultural inputs, improving food processing, and boosting nutrition. However, alongside these benefits, the nano-sizing of materials leads to human exposure to some insoluble and potentially biodegradable nanoparticles through food and beverage consumption (Chaudhry et al., 2008).

According to published data, these compounds are retained by various cells, including nerve cells, pulmonary epithelium (lung tissue), intestinal epithelium, alveolar macrophages (a type of lymphoid cell), and endothelial cells (blood vessel cells). Additionally, it has been proposed that increased immune system illnesses and gastrointestinal tract inflammations, such as Crohn's disease, may be related to nanomaterials (Momin et al., 2013).

However, concerns about the safety of nanofoods persist due to the limited number of in vitro and in vivo studies on nanofoods and the insufficient information about the interaction of nanoparticles with the human body. To address these concerns, more in vitro and in vivo studies are necessary.

9.4 CARBON NANOMATERIALS

The technological advancement and potential applications of nanomaterials in developing sensitive electrochemical sensors are inevitable in the food industry for safety and quality control monitoring. Carbon-based nanomaterials, in particular, have garnered significant scientific interest since their discovery. They can be roughly categorized into many types based on their spatial dimensions, such as graphene – GR (two-dimensional), fullerenes (zero-dimensional), CNTs (one-dimensional), and GR coils (multidimensional) (Nasir et al., 2018; Mauter & Elimelech, 2008). Extensive research on the synthesis, modification, or use of carbon-based nanomaterials has been actively published in the past few years and successfully applied into various research areas (Cha et al., 2013; Szunerits & Boukherroub, 2018; Z. Wang et al., 2016).

The heterocyclic state of the C-C bonds in carbon-based nanomaterials determines their unique spatial structure, leading to remarkable chemical and electronic properties. These materials are characterized by features such as small size, interface properties, surface characteristics, dielectric confinement, and macroscopic quantum tunneling effects (Armenta & Esteve-Turrillas, 2019; Scida et al., 2011). Their benefits include ease of preparation, stability, and high heat and electronic conductivity. These advantages make carbon-based nanomaterials widely applicable in various fields, including environmental monitoring, energy storage, and life sciences (Marmisollé & Azzaroni, 2016; Paquin et al., 2015).

Carbon-based nanomaterials have been utilized in the development of high-performance sensing devices for food safety inspection. These materials are used to produce, identify, and enhance sensing signals. In particular, in-depth studies of new carbon materials, such as GR and carbon dots (CDs), have expanded the potential of carbon-based sensors. These advancements improve the development of food safety inspection devices, which are characterized by high precision, high protection from interference, and convenience (Arduini et al., 2016; Dridi et al., 2016).

9.4.1 Carbon Nanotubes

CNTs were first introduced to the world of science by Lijima in 1991. In its simplest form, they are created by cylindrically wrapping a honeycomb-shaped sheet of carbon atoms. In uniform CNT structures, the atoms are connected in an sp^2 configuration (as in a graphite sheet), forming a hexagonal geometry with each atom bonded to three neighbors.

Tubes formed by rolling a single graphite sheet are called single-walled carbon nanotubes (SWCNTs). Multiple carbon cylinders formed by the coaxial interweaving of nanotubes are known as multi-walled carbon nanotubes (MWCNTs). Their diameters range from a few nanometers to 10–20 nm, with lengths around several microns. Due to their larger radii, MWCNTs are less flexible than SWCNTs. The most significant advantage of MWCNTs is their cost-effectiveness in production.

9.4.2 Graphene and Graphene Oxide

Graphene, which has gained significant excitement in the scientific world, is one of the newest nanoparticles widely used in various fields of nanotechnology. It is a two-dimensional carbon allotrope, just one atom thick, formed by arranging carbon atoms in a hexagonal structure within a single plane. Graphite, commonly found in pencils, is composed of overlapping layers of GR (Javed et al., 2014). This structure gives GR its remarkable mechanical, electrical, and thermal properties, contributing to its wide-ranging applications in various scientific and technological fields. GR exhibits impressive properties: it has a conductivity of 15,000 $cm^2V^{-1}s^{-1}$ at room temperature (Karteri et al., 2015), a thermal conductivity of 5000 $Wm^{-1}K^{-1}$ (Choi et al., 2012; Hsiao et al., 2010; Nair et al., 2008), and a large surface area of approximately 2630 m^2g^{-1} (Choi et al., 2012; Hsiao et al., 2011; Wolf, 2014). These characteristics make GR highly valuable for applications in electronics, energy storage, and various other advanced materials. GR finds applications in a wide range of fields, including:

- *Polymer supports:* Enhancing mechanical and electrical properties.
- *Composite materials:* Improving strength, conductivity, and lightweight characteristics.
- *Gas sensors:* Offering high sensitivity and selectivity.
- *Biological sensors:* Facilitating precise detection in biomedical applications.
- *Field-effect transistors:* Enabling faster and more efficient electronic devices.
- *Hydrogen storage devices:* Enhancing storage capacity and efficiency.
- *Transparent touch screens:* Providing durability and conductivity.
- *Light panels:* Improving efficiency and flexibility in lighting technology.

With the development of functionalized GR materials, their versatility and applications in composites are rapidly expanding, promising significant advancements across various fields. Functionalized GR, which involves modifying GR's surface with different functional groups or incorporating it into hybrid structures, enhances its compatibility with different matrices like polymers and metals. This capability allows for improvements in mechanical strength, electrical conductivity, thermal stability, and barrier properties in composite materials. As a result, functionalized GR is increasingly being explored for applications in aerospace, automotive, electronics, energy storage, and biomedical devices, among others. The ongoing research and development in this area are expected to lead to innovative solutions and new possibilities in material science and engineering (Hsiao et al., 2011; Ryu & Shanmugharaj, 2014; S. Wang et al., 2013; B. Yuan et al., 2014). The exceptional flexibility and conductivity properties of GR make it highly suitable for applications in flexible wearable electronic textiles, including fibers, yarns, and fabrics (Javed et al., 2014; Shanmugharaj et al., 2013; Y. J. Yun et al., 2013). GR's single-atom thickness and robust mechanical strength enable it to be integrated into textiles without compromising comfort or durability. This integration enhances the textiles with functionalities such as sensing, heating, and energy storage, making them ideal for smart clothing, wearable sensors, and other innovative electronic applications.

Graphene oxide (GO), a derivative of GR, incorporates oxygen-containing groups into its structure through an oxidation process. These functional groups include hydroxyl, epoxy, and carboxyl groups, which enhance GO's dispersibility in water and other solvents. This improved dispersibility makes GO more versatile for applications such as sensors, catalysis, and drug delivery systems. While GO retains many of GR's beneficial properties like mechanical strength and conductivity, the presence of oxygen groups also allows for additional chemical modifications. This flexibility expands the potential uses of GO in fields ranging from materials science and electronics to biomedical applications, where tailored properties are advantageous.

9.4.3 Quantum Dots

Quantum dots are indeed a fascinating example of nanostructures that have gained significant attention in recent years. Due to their unique optical, chemical, and electrical properties, quantum dots find applications in various fields such as medicine, food, environmental monitoring, and computing (Algar et al., 2010). Quantum dots emit fluorescence in different colors depending on their size; smaller quantum dots emit blue fluorescence, while larger ones emit fluorescence shifting toward red (Pisanic et al., 2014). These fluorescence properties make quantum dots valuable as biosensors. Examples of quantum dots include lead sulphide (PbS), lead selenium (PbSe), cadmium selenium (CdSe), and cadmium telluride (CdTe) (Drbohlavova et al., 2009). Their ability to emit light of precise wavelengths and their tunable properties make them versatile tools in various technological and scientific applications.

Carbon quantum dots (CQDs) or fluorescent carbon nanoparticles represent a novel class of carbon nanomaterials that have recently emerged and have gained considerable interest as potential competitors to conventional semiconductor quantum dots (J. H. Qu et al., 2018; Lim et al., 2015). They are composed of carbon atoms with dimensions less than 10 nm, and exhibit distinct properties compared to other quantum dots. Unlike quantum dots that contain toxic heavy metals such as cadmium and lead, CDs are non-toxic because they do not contain these heavy metals. Moreover, CDs are water-soluble, produced using environmentally friendly methods, and exhibit photostability. This characteristic makes them favorable alternatives for further development in biomedical imaging, environmental monitoring, and food applications (Lim, Shen, & Gao, 2015; Dinç, 2016; Dinç et al., 2017; Shen & Liu, 2016). It has been demonstrated by numerous researchers that CDs can serve as biosensors for analyzing a wide range of substances, from metals to riboflavin, thiamine to glucose (She et al., 2016; W. Shi et al., 2011; Tan et al., 2015).

9.4.4 Fullerenes

Fullerenes are carbon molecules with closed cage structures composed of pentagonal and hexagonal rings, described by the formula C_{20+m} where m is an integer (Mauter & Elimelech, 2008; Ebert, 1997). The first fullerene, C_{60}, was synthesized in 1985 by Kroto et al. through spectrometric measurements, and its structure was subsequently confirmed in the laboratory (Smalley, 1985). C_{60} consists of 60 carbon atoms arranged in 12 pentagons and 20 hexagons, resembling the shape of a football. It was named "Buckminsterfullerene" in tribute to architect Buckminster Fuller. Depending on the number of hexagons, fullerenes can vary in size. Beyond C_{60} and C_{70}, fullerene molecules include a wide range of isomers such as C_{240}, C_{540}, and C_{720}, forming a series of similar structures

(Calaminici et al., 2012). Various fullerene structures, including cone and tube shapes, have been discovered in addition to the originally presumed spherical shape observed in C_{60} molecules.

Physicists, chemists, materials scientists, and engineers have expressed that these new spherical carbon structures have extraordinary potential for use in superconducting materials, new alloy sources, self-assembled nanostructures, and various optical devices (Ebert, 1997).

Fullerenes have a wide range of applications due to their unique properties. They are utilized in medical fields for diagnostics, super drugs, anti-HIV and anti-cancer therapies, nuclear magnetic resonance systems, and cosmetic additives. Additionally, fullerenes are employed in fuel cells, solar cells, transistors, inductors, diodes, electronic cameras, nonlinear optical materials, and fluorescent imaging tubes in electronic and optical applications.

In the industrial sector, fullerenes are used in the production of wear-resistant materials, lubricants, flame retardant materials, high-performance membranes, polymer additives, artificial diamonds, hard alloys, catalysts, electrically conductive fluids, high-performance coatings, ink filters, fire retardant coatings, and embedded molecular and bioactive materials. They also find environmental applications in gas adsorption and gas storage.

9.5 POLYMERIC NANOMATERIALS

Polymeric nanomaterials are a class of nanostructured materials composed of polymers that have gained significant attention due to their diverse applications across various fields. There are three main types of polymeric nanomaterials available:

1. ***Natural substance-based polymeric nanomaterials***
   These nanomaterials utilize renewable resources and are ecologically benign as they can decompose into water, carbon dioxide, and inorganic fragments. Examples include starch, chitosan, cellulose, alginate, and hyaluronic acid. They can be synthesized using physical or chemical methods, often incorporating various functional groups or modifications through advancements in nanoscience.
2. ***Biosynthesized polymer-based nanomaterials***
   These nanomaterials are produced through enzyme hydrolysis using microbial enzymes to create substances such as microbial polyesters and polysaccharides. Examples include poly-hydroxybutyrate (PHB), poly (3-hydroxybutyrate-co-3-hydroxyvalerate), and biofiber bundles.
3. ***Synthetic polymer-based nanomaterials***
   These are chemically synthesized polymers that include materials like poly(lactic-co-glycolic acid), polyurethane, poly(methyl methacrylate) (PMMA), polylactic acid), polyester, silicone rubber, polyvinylpyrrolidone, and polyvinyl alcohol.

These categories highlight the diversity and versatility of polymeric nanomaterials, each with unique properties and applications across various industries including food packaging, medicine, and materials science.

When nanomaterials are added to polymers, they enhance the mechanical strength, electrical conductivity, and thermal stability of the resulting material. However, the structure of nanomaterials is entirely dependent on their composition and the processing conditions used during their synthesis. The choice of materials and the specific methods employed play crucial roles in determining the final structure, which impacts their properties and potential applications in various fields, including food packaging, medicine, and environmental technologies.

The application of polymer nanotechnology in food packaging aims to enhance the fundamental properties of traditional packaging systems. These improvements include better preservation (ease of transport and distribution), increased convenience (consumer-friendliness), enhanced protection and preservation (preventing leakage or breakage, protecting against microbial contaminants, and offering longer shelf life), and improved marketing and communication (providing real-time information about the quality of packaged food, as well as nutritional components and preparation guidelines). By addressing these aspects, polymer nanotechnology significantly elevates the overall functionality and efficiency of food packaging systems (Mihindukulasuriya & Lim, 2014). These advancements lead to more robust and effective packaging solutions, capable of extending the shelf life of food products and maintaining their quality and safety during storage and transport.

9.6 APPLICATIONS OF CARBON-BASED SENSOR FOR DETERMINATION OF FOOD QUALITY CONTROL

Carbon-based optical and electrochemical sensors have found numerous applications in the field of food quality control due to their high sensitivity, selectivity, and versatility. These sensors utilize the unique properties of carbon nanomaterials such as GR, CNTs, and CDs to detect and analyze various harmful substances in food. Optical sensors, including those based on fluorescence and surface plasmon resonance (SPR), can rapidly identify contaminants like allergens, pesticide residues, and illegal food additives with excellent precision. Similarly, electrochemical sensors employ carbon nanomaterials to enhance electron transfer and signal amplification, making them effective for monitoring veterinary drugs and other toxic substances. The integration of enzymes, antibodies, and aptamers further improves the performance of these sensors, enabling real-time and accurate detection. The development of these advanced sensor technologies represents a significant step forward in ensuring food safety and protecting public health (Arduini et al., 2016; Dridi et al., 2016).

In this chapter, we will include current studies on the use of carbon-based optical and electrochemical sensors for analyzing and detecting some allergens, veterinary drugs, pesticide residues, illegal food additives, and other significant toxic and harmful substances.

9.6.1 Applications of Carbon Based-Optic Sensor for Determination of Food Quality Control

Optical biosensors operate on principles such as absorption, internal reflection, fluorescence, SPR, luminescence, or light scattering spectroscopy during biochemical reactions (Korotkaya, 2014; Meshram et al., 2018; Thakur & Ragavan, 2013). These biosensors, also known as optodes, resemble electrodes but function using light-based systems (Thakur & Ragavan, 2013). In the food industry, SPR is a predominant technique for optical biosensing. Similar to electrochemical biosensors, optodes integrate with enzymes, antigens, antibodies, microbial cells, DNA, or aptamers to enhance their efficiency in detecting food contaminants (De Marcos et al., 2014; Ko & Grant, 2006; Bilir et al., 2016; J. Shi et al., 2015). Nanoparticles are frequently used for enhancement of optosensor sensitivity. In addition to the widespread use of metal-based nanoparticles, carbon-based nanomaterials are also preferred. For example, A quantum dot-based immunochromatographic assay (QICA) and a fluorescence quenching immunochromatographic assay (FQICA) were developed to visually detect bisphenol A (BPA). The proposed QICA and FQICA are based on conventional immunoassay and fluorescence resonance energy transfer (FRET) from fluorescent donors (quantum dots) to fluorescent acceptors (colloidal gold nanoparticles), respectively. The visual limits of detection (LOD) of the proposed QICA and FQICA were 10 and 4 $\mu g\ kg^{-1}$ ($\mu g\ L^{-1}$) in samples, respectively. These two proposed fluorescence immunochromatographic assays can be applied on a large scale to rapidly, sensitively, and accurately detect BPA in distilled spirits (42%, v/v), cabbage, grass carp, and river water samples (Sheng et al., 2018).

In 2017, Sun et al. developed new optic sensor method for thiram detection in apple juice by using Raman spectroscopy. For this aim, they first prepared a single layer of GR grown by CVD on Cu foil, which was then transferred onto a silicon wafer using a PMMA coating layer as the transfer medium. Then, a very thin Ag film was deposited on the GR by electron deposition method, and then a transparent and Raman inactive PMMA film was spin-coated on the Ag/1LG composite structure and cured at 180 °C for 20 min. Finally, the resulting silicon wafer was etched with hydrofluoric acid. The surface-enhanced Raman scattering (SERS) film was easily separated from the hard substrate and was washed with distilled water. When using prepared Raman SERS surface, the minimum detectable concentration was shown to reach 0.24 ppm. This is below the maximum residue limit of 7 ppm prescribed by the US Environmental Protection Agency in fruit. Also, reusability demonstrates the potential of GR-enhanced SERS film to detect even trace amounts of pesticide residues in water and agricultural products (H. Sun et al., 2017).

Moreover, CDs can be used as fluorescent probes to detect various important food contaminant analytes. The core principle of fluorescence sensor analysis involves the interaction between the recognition component and the target analyte, which modifies the fluorescence properties of CDs. This fluorescence change is quantitatively linked to the structure or concentration of the target analyte and operates on one of four common principles (Zu et al., 2017):

1. Direct fluorescence quenching (turn-off)
2. Direct fluorescence enhancement (turn-on)
3. Fluorescence recovery (off-on)
4. Fluorescence quenching followed by recovery (off-on-off).

Fluorescent methods for pesticide detection have been developed based on the inhibitory effects of pesticides on enzyme activity. These methods utilize enzymes whose activity is reduced or altered by the presence of pesticides, leading to detectable changes in fluorescence (H. Li et al., 2016). Haitao Li et al. (2016) proposed a novel sensitive method for the detection of carbaryl using CDs as fluorescent probes. In the study, novel CQDs doped with nitrogen (N) and sulfur (S) have been synthesized from an ionic liquid source. This doping significantly improves the photoluminescence (PL) lifetime of the CQDs. These doped CQDs have been utilized in a two-enzyme coupled reaction system for pesticide detection and this method was satisfactorily applied to the analysis of carbaryl in apple samples.

J. Liu et al. (2016) developed a CQD-MnO_2 sensing platform that utilizes fluorescent recovery to measure ascorbic acid (AA). This platform was successfully applied to detect AA in fresh fruits, vegetables, and commercial fruit juices. Ahmed et al. (2015) utilized CDs for the direct measurement of tannic acid (TA) in red and white wine samples. The method achieved a limit of detection (LOD) of 0.018 mg/L, as TA effectively quenches the fluorescence of the CDs. Ma and his colleagues developed a fluorescence turn-off-on probe for detecting copper(II) ions (Cu^{2+}), AA, and acid phosphatase (Ahmed et al., 2015). In this system, Cu^{2+} causes fluorescence quenching of CDs. The LOD for Cu^{2+} is 2.4 μM. When AA is introduced into the CDs–Cu^{2+} solution, it reduces Cu^{2+} to Cu^{+}, leading to the recovery of fluorescence in the CDs. The fluorescence intensity exhibits a linear correlation with the concentration of AA in the range of 100–2800 μM, with an LOD of 60 μM. Additionally, this probe shows potential for detecting AA in real samples, including vitamin C tablets, orange juice, and fresh oranges.

CDs are increasingly used for detecting microbial growth and the toxins produced by microbes, such as aflatoxin B1 and *Salmonella typhimurium* (R. Wang et al., 2015). Additionally, CDs have proven to be effective in identifying toxins in food products. For example, L. Wang et al. (2016) utilized the fluorescence on-off-on principle by combining CDs with gold nanoparticles (AuNPs) and an aptamer to specifically detect aflatoxin B1 (AFB1). This method achieved a detection limit of 5 pg/mL (16 pM).

Recently, researchers have started investigating the potential applications of CDs in food preservation. These studies focus on developing new technologies to improve the freshness and shelf life of food products, designing tools and methods for rapid on-site analysis, and creating environmentally friendly and functional food packaging. Other sensing platforms for detecting nutrient substances, pesticides, banned or limited additives, pathogenic microorganisms, microbial toxins, and applications in food packaging using CDs are summarized in Table 9.1.

9.6.2 Applications of Carbon Based-Electrochemical Sensor for Determination of Food Quality Control

CNTs, known for their high surface area and excellent electrical properties, are often utilized in sensor applications to enhance sensitivity and selectivity, making them valuable tools in environmental monitoring and food safety assessment. The diverse electrochemical properties of SWCNTs and MWCNTs, including catalytic activity (Türk et al., 2018; X. Xu et al., 2010), stability (Chiou et al., 2013; Stamatin et al., 2015), electrical conductivity (Amekpewu et al., 2015; Mashkour et al., 2018), and biocompatibility (M. Khan et al., 2017; Prencipe et al., 2009), play crucial roles in developing chemical and biological sensors for food safety (Barsan et al., 2015; T. H. V. Kumar et al., 2018). These properties are pivotal in enhancing sensor performance and reliability in detecting food contaminants and ensuring safety standards. The use of CNT-based nanomaterials in creating electrochemical sensors for food safety is extensive and varied. These sensors are capable of detecting pesticide residues, veterinary drugs, pathogenic microorganisms, and prohibited or restricted additives.

Numerous studies have demonstrated that SWNTs exhibit high electrical conductivity, significantly enhancing the signal-to-noise ratio. This property has facilitated the development of ultrasensitive electrochemical sensors, capable of detecting various food analytes with high precision. In a study published by Yamada et al. (2016), an SWCNT-based multi-junction sensor was designed to detect *E. coli* with a detection time of less than 5 min.

As indicated in Figure 9.2, Bhardwaj and coauthors developed a convenient, low-cost paper-based electrochemical immunosensor for label-free detection of *Staphylococcus aureus* using Ab-SWCNT bioconjugates (Bhardwaj et al., 2017). The anti-*S. aureus* antibodies (Abs) were covalently attached to SWCNTs and immobilized on the working electrode surface to recognize the analyte, which caused changes in peak current. This remarkable sensor demonstrated good linearity ($R^2 = 0.976$) between the increase in peak current and the logarithm of *S. aureus* concentration ($10–10^7$ CFU mL^{-1}) within a short time (30 min) and had a limit of detection of 15 CFU mL^{-1} in milk.

The combination of carbon-based nanomaterials represents another strategy to enhance the performance of food safety sensors. Lin et al. developed a hybrid electrode modified with SWCNTs and MWCNTs for the simultaneous determination of toxic compounds like ractopamine and salbutamol in pork samples (Lin et al., 2013). Both SWCNT and MWCNT primarily function to provide a large active surface area, which significantly influences the current response in electrochemical systems. Additionally, these active materials notably reduce overpotential, especially in SMWCNT hybrid composites compared to bare electrodes. The interaction between the toxic compounds and the CNTs (SWCNT or MWCNT) alters the electrochemical behavior of the ractopamine or salbutamol reactants, resulting in a shift of the reactants' voltammetric peaks toward a lower potential.

Zhang et al. used carboxylic short-chain multi-wall carbon nanotubes (short-MWNTs-COOH) and silver nanoparticles as electrode modifiers (P. Zhang et al., 2019). They immobilized Short-MWNTs-COOH onto a glassy carbon electrode (GCE) and electrodeposited Ag nanodendrites onto the modified electrodes. This platform allowed for accurate detection of chloramphenicol in real samples, including honey, milk powder, and eye drops.

Chen et al. designed an ultrasensitive biosensor based on MWCNTs and acetylcholinesterase enzyme for the detection of carbofuran and dichlorvos (DDV) pesticides in spiked beverages (H. Chen et al., 2008). MWCNTs were coated onto the Prussian blue (PB) prefunctionalized glassy carbon electrode surface and the enzyme was immobilized onto the polymer film due to crosslinking by glutaraldehyde. The use of MWCNTs as an electrode modifier both increased the surface area

TABLE 9.1
Selected application of optosensors based on carbon dots in food analysis

ANALYTES	SENSORS	METHOD	LINEAR WORKING RANGE	DETECTION LİMİT	SAMPLE	REF.
			PESTICIDES			
Carbaryl	N,S-CDs/AChE/ChOx/AChE	Inhibition on AChE	6.3 ng/L–0.63 mg/L	5.4×10^{-9} g/L	Apple	(H. Li et al., 2016)
Dichlorvos	CDs/Cu2+/ATChCl/AChE	Quenching effect of Cu^{2+}, inhibition on AChE	6–60 nM	3.8 nM	Cabbage,fruit juice	(Hou et al., 2016)
Glyphosate	CDs/AgNPs	IFE, aggregation of AgNPs by glyphosate	0.025–2.5 μg/mL	12 ng/mL	Spiked cereal	(L. Wang et al., 2016)
Paraoxon	N-CDs/AuNPs/ATChCl/AChE	FRET, inhibition on BChE	10^{-4}–10^{-9} g/L	0.0036 nM	Apple juice	(Gong et al., 2016)
			VETERINARY DRUGS AND ANTIBIOTICS			
Tetracycline (TC)	CDs/Al3+	Combination of TC and Al^{3+}	0.001–30 μM	0.52 nM	Spiked milk, fish, pork	(F. Qu et al., 2016)
Clenbuterol	CDs/AuNPs	Competitive absorption	8–200 nm	3 nM	Pork	(Y. Liu et al., 2017)
Doxycycline	N,S-CDs	Static quenching	0.1–100 μM	25 nM	Raw milk	(Song et al., 2017)
Paraoxon	CDs/DNBA/ATChCl/AChE	Inhibition on AChE	0.001–1.0 μg/mL	0.4 ng/mL	Water, rice, cabbage	(H. Li et al., 2018)
			ARTIFICAL ADDITIVE			
Tannic acid	PEGA-C-dots and DAP-C-dots	Spectrofluorimetry	0.1–10 mg L^{-1}	0.018 mg L^{-1}	Wine	(Ahmed et al., 2015)
Vitamin C	CDs/Fe^{3+}	Reduction from Fe^{3+} to Fe^{2+} by Vitamin C	24–40 μg/mL	-	Vitamin C soluble tablet	(Fong et al., 2016)
Phyticacid	CDs/Fe^{3+}	Combination of phytic acid and Fe3+	0.68–18.69 μM	0.36 μM	Corn grain	(Gao et al., 2015)
Caffeic acid	CDs	Static quenching	0.79–100 μM	0.24 μM	Redwine	(Q. Sun et al., 2018)
Curcumin	CDs	FRET	0.78–50 μM	40 nM	Turmeric powder	(Baig & Chen, 2017)
Sunsetyellow	CDs	FRET	0.3–8 μM	79.6 nM	Soft drinks	(Y. Yuan et al., 2016)
Melamine	AuNPs-CDs	FRET	0.05–0.5 μM	0.036 μM	Raw milk, milk powder	(Dai et al., 2014)

(continued)

TABLE 9.1 (Continued)
Selected application of optosensors based on carbon dots in food analysis

ANALYTES	SENSORS	METHOD	LINEAR WORKING RANGE	DETECTION LİMİT	SAMPLE	REF.
			PATHOGENS			
Salmonella typhimurium	CDs-Aptamer	Salmonella typhimurium and aptamer combination	10^3–10^5 CFU mL^{-1}	50 CFU mL^{-1}	Egg, tap water	(R. Wang et al., 2015)
Escherichia coli (E. coli)	Man–CQDs	Manose and E.coli combination	-	100 CFU mL^{-1}	Drinking water, apple juice,	(Lai et al., 2016)
Staphylococcus aureus	Vancomycin-CDs	Vancomycin and Staphylococcus aureus combination	3.18×10^5 –1.59×10^8 CFU mL^{-1}	9.40×10^4 CFU mL^{-1}	Orange juice	(Zhong et al., 2015)
Escherichia coli O157:H7	LEFA	Fluorescent aptasensing	10^2–10^8 CFU mL^{-1}	1 CFU mL^{-1}	Spiked milk and tap water	(S. Chen et al., 2018)
Escherichia coli	CDs@amikacin		3.904×10^5 –7.625×10^2 CFU mL^{-1}	552 CFU mL^{-1}	Apple, pineapple and orange	(Chandra et al., 2016)
			MYCOTOXINS			
Ochratoxin A (OTA)	MPC and NGQDs-Apt	Static quenching	10–5000 nM	2.28 nM	Corn, wheat	(Chandra et al., 2016)
Ochratoxin A (OTA)	GQDs	Fluorescence sensing process	0–1 ng mL^{-1}	13 pg mL^{-1}	Red wine	(Y. Zhang et al., 2017)

Note: PEGA-C-dots – polyethyleneglycol bis(3-aminopropyl; DAP – 1,2-aminopropane; MPC – Iron-doped porous carbon; NGQDs-Apt – Aptamer-functionalized nitrogen-doped graphene quantum dots; FRET – Förster resonance energy transfer; Man-CQDs – Mannose-functionalized carbon quantum dots; LEFA – Label-free and enzyme-free fluorescent aptasensor; GQDs – Graphene quantum dots.

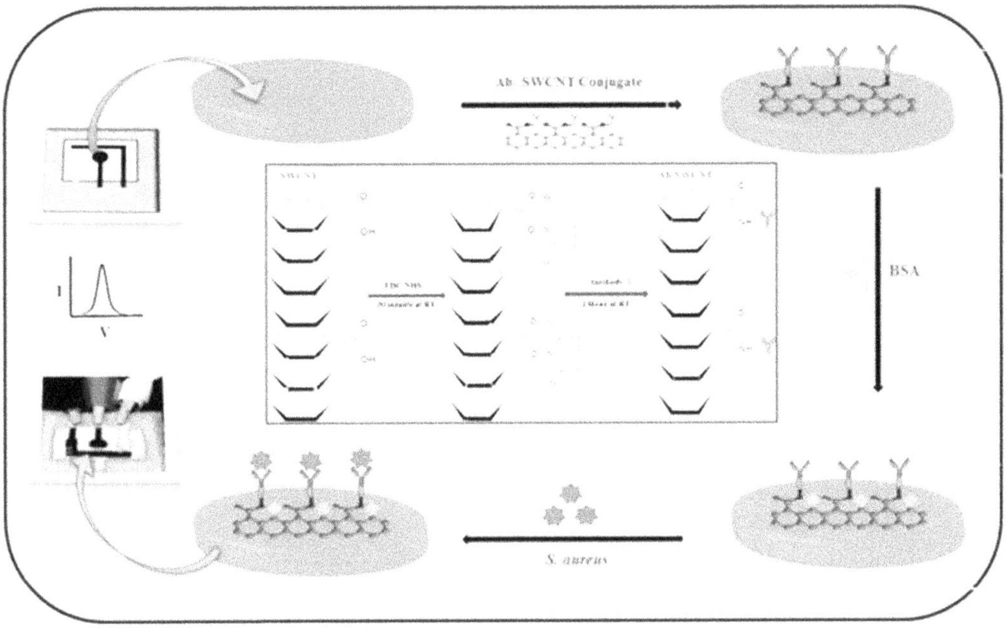

FIGURE 9.2 Schematic of the Ab-SWCNTs bio-conjugate: attachment of antibodies to SWCNTs using EDC/NHS coupling reagent, electrochemical biosensor fabrication, and assay procedure. (Reproduced from Bhardwaj et al., 2017 with permission from Elsevier.)

of electrodes and enhanced the enzymatic activity of AChE, significantly improving the sensor performance.

Poly(allylamine hydro-chloride) was also exploited for MWCNTs functionalization to immobilize AChE for monocrotophos detection. The modification of a carbon paste electrode with poly(allylamine hydro-chloride) functionalized MWCNTs allowed to get a current peak, due to enzymatic byproduct (thiocholine) oxidation, at about 0.7 V (vs. Ag/AgCl). Due to the presence of poly(allylamine hydro-chloride)/ MWCNTs, monocrotophos was successfully detected down to 0.88 pg mL^{-1} in cabbage, broccoli, and apple samples naturally contaminated (Çevik et al., 2013).

Sun et al. developed an amperometric immunosensor using a nanocomposite consisting of MWCNTs, thionine, and chitosan for the detection of chlorpyrifos residues (X. Sun et al., 2012). This sensor likely utilized the unique properties of MWCNTs to enhance sensitivity and stability, while thionine and chitosan likely contributed to the immobilization and signal amplification of the immunosensor. To evaluate the feasibility of the proposed immunosensor for real sample analysis, cabbage, lettuce, and Chinese chive samples were spiked with chlorpyrifos solutions at different concentrations. The recoveries obtained from these spiked samples ranged from 85.2% to 104.3%. This demonstrates the effectiveness of the immunosensor in accurately detecting chlorpyrifos residues in complex food matrices, highlighting its potential for practical applications in food safety and quality control.

MWCNTs have also been used for many other applications in food science. In this regard, it is worth mentioning the excellent work performed by Creville´n and co-author in the determination of flavonoids and antioxidant profiles (Crevillén et al., 2007) and the work from Kachoosangi's group for the determination of capsaicin, the chemical responsible for the hot taste of chilli peppers (Kachoosangi et al., 2008). The other examples of CNT-based electrochemical sensors for food safety analysis in the past 10 years have been summarized in Table 9.2.

TABLE 9.2
Application of electrochemical sensors based on carbon nanotubes (CNTs) in food analysis

ANALYTES	SENSORS	METHOD	LINEAR WORKING RANGE	DETECTION LİMİT	SAMPLE	REF.
			PESTICIDES			
Methyl parathion	f-SWNTs–β-CD	DPV	0.002–17.5 μg mL^{-1}	0.4 ng mL^{-1}	Vegetable	Zeng Y et al., 2016
Carbendazim	$Nd_2Mo_3O_9$/MWCNTs/GCE	DPV	0.5–9.0 μM	0.0167 nM	Paddy water	Xie et al., 2019
Carbofuran	MWNT-AChE/PB/MWNT	Amperometry	_	0.1 μg mL^{-1} carbofuran	Spiked beverages	Chen et al., 2008
Dichlorvos (DDV)			_	0.04 μg mL^{-1} DDV		
Monocrotophos	PAH/MWCNT/AChE-CPE	CV	1–25 pg/mL	0.88 pg/mL	Tap water, mineralized water, tomato and grape	Çevik et al., 2013
Chlorpyrifos	MWCNTs-THI-CHIT/GCE	DPV	0.1–1.0 × 10^5 ng/mL	0.046 ng/mL	Vegetables and fruits	Sun et al., 2012
			VETERINARY DRUGS AND ANTIBIOTICS			
Ractopamine salbutamol	SWNTs/MWNTs	DPV	0.1–5 μM	0.05 μM	Pork	K. C. Lin et al., 2013
Chloramphenicol	MWNTs@MIP	DPV	5.0 nM–0.5 μM	0.1 nM	Milk, honey	Wu et al., 2016
Amoxicillin	MWNTs@MIP	DPV	1.0 nM–1.0 μM	0.89 nM	Milk, honey	Yang & Zhao, 2015
			ARTIFICAL ADDITIVE			
Capsaicin	MWCNT-BPPGE	AdsSV	0.5–15 μM	0.31 μM	hot pepper sauces	Kachoosangi et al, 2008
Neotame	CuNPs–APDC–MWNTs–β-CD	DPV	0.05–2.0 mM	0.013 mM	Food	Bathinapatla et al., 2015
Sunset yellow	GO/MWNTs	LSV	0.09–8.0 μM	0.025 μM	Orange juice	Qiu et al., 2016
TBHQ	Au-SnO_2/GNs-SWNTs	DPV	0.5 nM–2300 μM	0.58 nM	Food	Gao, 2014
Rebaudioside A	Cytc/AuNPs-GO/MWNTs	DPV	0.075–1.25 mM	0.264 μM	Food	Bathinapatla et al., 2016
Amaranth	CNT/GO-[BMIM][PF6]	SWV	0.0005–4.0 μM	0.1 nM	Drinks	Wang et al. 2015
Chrysoidine	DNA/MWNTs-PDDA/PGE	DPV	0.05–15.00 g mL^{-1}	0.03 g mL^{-1}	Food	Ensafi et al., 2014
Quinoline yellow	Polypyrrole/SWNTs	SWV	0.8–100 μM	0.08 μM	Soft drinks	Gao et al., 2014

			PATHOGENS			
S. aureus	Ab-SWCNT/CPE	DPV	10–10^7 CFU mL^{-1}	15 CFU mL^{-1}	Milk	Bhardwaj et al. 2017
			INORGANIC IONS			
Sulfite	CNTs–PDDA–AuNPs	FIA	2–200 mg L^{-1}	0.03 mg L^{-1}	Fruit juice	Amatatongchai et al. 2015
Nitrite	AuNPs/MWCPE	SWV	0.05–250.0 μM	0.01 μM	Food stuff	Afkhami et al., 2014
	Polythionine/CNTs	Amperometry	2.0 μM–1.0 mM	1.4 μM	Milk	Deng et al., 2012
	ZnO/Pt/MWNTs	DPV	0.4 μM–0.2 mM	0.82 nM	Pickled food	M. Le Zhang et al., 2015
	IL–SWNTs	Amperometry	1.0 μM–12 mM	0.1 μM	Milk	Zhou et al., 2013
	Cu-Co PBA/MWCNTs	DPV	10–400 μM and 400–2100 μM	0.5 μM	Ham sausage and mustard samples	H. Y. Liu, 2019
			ORGANICS			
Bisphenol A	MWCNT-NH2	DPV	0.1–10 μM	0.02 μM	Plastic products	Lin et al., 2014
Bisphenol A	AuNPs/rGO/CNTs	DPV	1.45–1490 nM	800 pM	Food	Y. C. Wang et al., 2015
Polyphenols	SWNTs-polytyrosine	Amperometry	0.5–170 μM	8.8 nM	Tea	Eguílaz et al., 2016
Acrylamide	COOH-SWNTs	CV	10–200 μM	0.03 μM	Food	González-Fuentes et al., 2014
Bisphenol F	MWNTs–COOH	DPV	0.12–6.01 μg mL^{-1}	0.11 μg mL^{-1}	Food	(X. Wang et al., 2014)
Hydroquinone	rGO–MWNTs	DPV	8.0–391 μM	2.6 μM	River water	Hu et al., 2012
Hydroquinone	Fe_2O_3/CNTs	DPV	10–80 μM	0.5 μM	Tap water	Wang et al., 2014
Aflatoxin B_1	Chitosan/MWCNT	CV	1–30 ng mL^{-1}	0.62 ng mL^{-1}	Maize flour	Migliorini et al., 2020

Note: CV – cyclic voltammetry; DPV – differential pulse voltammetry; LSV – linear sweep voltammetry; SWV – square-wave voltammetry; AdsSV – adsorptive stripping voltammetry; PAH – poly(allylamine hydrochloride); CPE – carbon paste electrode; BPPGE – basal plane pyrolytic graphite electrode; FIA – fluoroimmunoassay; MIP – molecularly imprinted polymer; β-CD – β-cyclodextrin; Cyt c – cytochrome *c*; GNs – graphene nanosheets; GO – graphene oxide; PB – Prussian blue; AChE – acetylcholinesterase; MWCPE – multi walled carbon nanotube/carbon paste electrode; f-SWNTs-carboxylic acid-functionalized SWNTs; THI – Thionine; CHIT – chitosan; GCE – glassy carbon electrode; PEI – Polyethylenimine; [BMIM] [PF6] – 1-butyl-3-methylimidazolium hexafluorophosphat; PDDA – poly(diallyldimethylammo-nium chloride); PGA – polyglutamate acid; rGO – reduced graphene oxide.

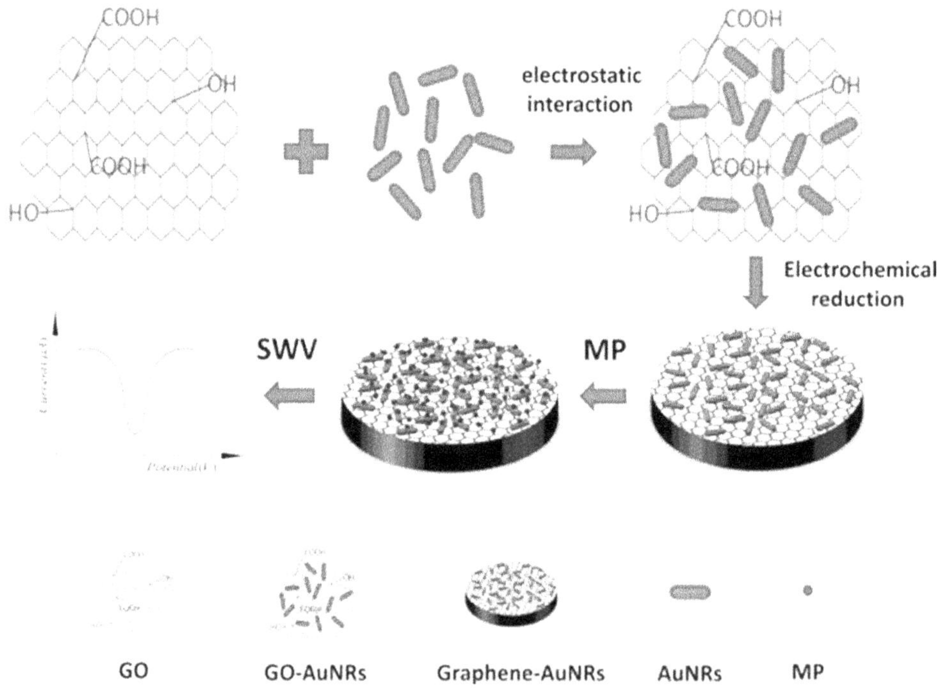

FIGURE 9.3 Schematic illustration for the facile green synthesis of GN-AuNRs/GCE and its application for the extraction and electrochemical detection of MP. (Reproduced from Zhu et al., 2014 with permission from Elsevier.)

On the other hand, GR materials have a relatively complete structure and stable surface, which results in poor dispersibility and solubility. Moreover, the strong Van der Waals force between GR layers often causes agglomeration, thereby limiting its widespread application (Cho et al., 2016). To overcome these issues, researchers have investigated the use of various inorganic and organic materials or polymers to enhance the properties of GR in sensing applications. These GR composite materials exhibit diverse properties and play crucial roles in developing new sensors for food safety. Table 9.3 shows the applications of GR-based nanomaterials in the development of electrochemical sensors for food safety.

In Figure 9.3, based on the electrostatic interaction between GR and gold nanorods (AuNRs), Zhu et al. developed a facile green electrochemical approach to synthesize a GRAuNRs nanocomposite on a GCE for the electrocatalytic analysis of methyl parathion (MP) (Zhu et al., 2014). This method utilizes the synergistic properties of GR and gold nanorods, resulting in a nanocomposite that enhances the sensor's sensitivity and selectivity for detecting methyl parathion, a common pesticide. This advancement highlights the potential of GR-based nanomaterials in developing highly effective electrochemical sensors for food safety applications.

In another study, Ma and Chen utilized cyclic voltammetry to reduce $HAuCl_4$ to AuNPs on a GR-modified GCE (Ma et al., 2013). The GR/AuNPs/GCE sensor showed excellent catalytic efficiency in oxidizing diethylstilbestrol (DES), a synthetic estrogen commonly utilized as a growth promoter in livestock and for medical treatments addressing estrogen-related deficiencies in animals (Ma et al., 2013). Its notable selectivity and stability in detecting DES residues within food samples highlight its potential role in safeguarding food safety and enhancing quality control measures.

Electrochemical sensors incorporating CDs are attracting significant attention in food quality control due to their distinctive properties (Mansuriya & Altintas, 2020). CDs offer a large surface area

TABLE 9.3
Application of electrochemical sensors based on graphene and graphene oxide in food analysis

ANALYTES	SENSORS	METHOD	LINEAR WORKING RANGE	DETECTION LİMİT	SAMPLE	REF.
			PESTICIDES			
Imidacloprid	PoPD/rGO	LSV	0.75–70 μM	0.4 μM	Spiked pear	(Kong et al., 2013)
Malathion	FTO/PA6/PPy/ERGO	DPV	0.5–20.0 μgmL^{-1}	0.8 ng mL^{-1}	Tap water and river water	(Migliorini et al., 2020)
Methyl parathion	GR-AuNRs	SWV	10–500 ng mL^{-1}	0.82 ng mL^{-1}	Kiwi	(Zhu et al., 2014)
Napropamide	Sulfonated GR	ECL	0.1–10,000 mg L^{-1}	1.0 mg L^{-1}	Spiked water	(L. Wang et al., 2014)
Isoproturon and carbendazim	GR	SWSV	0.02–10.0 mg L^{-1}, 0.50–10.0 mg L^{-1}	0.02 mg L^{-1}, 0.11 mg L^{-1}	Water, vegetable	(Noyrod et al., 2014)
			VETERINARY DRUGS AND ANTIBIOTICS			
Diethylstilbestrol	Gr/Nano-Au/GCE	CV	1.20×10^{-8}–1.20×10^{-5} M	9.80×10^{-9} M	Beef, milk powder, fish	(Ma & Chen, 2015)
Chlorampenicol	RGO@Cu2S NC/SPCE	Amperometry	60 nM–1954.5 μM	15.3 nM	Fresh milk, powder milk, ice cream	(Govindasamy et al., 2019)
Clenbuterol	isopropanol/nafion/PSS/GR	LSV	0.075–25 μM	22 nM	Pork	(L. Wang et al., 2014)
Metronidazole	AgNPs/SF-GR	DPSV	0.1–20.0 μM	0.05 μM	Shrimp	(Zhai et al., 2015)
			ARTIFICAL ADDITIVE			
Sunset yellow	AuNPs/rGO	DPV	0.002–109.14 μM	2 nM	Soft drinks	(J. Wang et al., 2015)
Sunset yellow	AuNPs/magnetic GR	DPV	5.0–2000 nM	2 nM	Drinks	(J. Li et al., 2016)
Sunset yellow	rGO/CPE	DPV	0.05–10 μM	27 nM	Fanta, Miranda, Snack, and Chitoz	(Vatandost et al., 2020)
Ferulic acid	rGO	DPV	0.849 nM–38.9 μM	0.206 nM	Sinensis	(L. Liu et al., 2014)
Orange II	rGO-ATDT-Pt	CV	10–600 nM	0.34 nM	Ketchup and chili	(M. Yun et al., 2015)
Tyramine	ITO/APTES/ErGO	DPV	1–100 nM, and 1–100 μM	0.1 nM	Human urine and commercial milk, beer samples	(M. Z. H. Khan et al., 2018)
Sudan I	GR-ATDT	CV	0.075–7.50 μM	40 nM	Ketchup	(Ma et al., 2013)
Sudan I	AuNPs/rGO	LSV	0.01–70 μM	1.0 nM	Chili powder	(J. Li et al., 2016)
Tartrazine	TiO_2–ErGO–GCE	DPV	2.0 × 10^{-8}–2.0 × 10^{-5} mol/L	8.0 × 10^{-9} mol/L	Carbonated beverage	(He et al., 2018)
Vanillin	Ag-PdNPs/GO	DPV	0.02–45 nM	5 nM	Baked goods	(J. Li et al., 2015)
Rhodamine B	MnO_2NRs-ERGO/GCE	SDLSV	0.02–1.0 μM	6 nM	Commercial chili-containing products (powder, sauce, oil)	(Sebastian et al., 2022)

(continued)

TABLE 9.3 (Continued)
Application of electrochemical sensors based on graphene and graphene oxide in food analysis

ANALYTES	SENSORS	METHOD	LINEAR WORKING RANGE	DETECTION LIMIT	SAMPLE	REF.
Caffeic acid	rGO	Amperometry	0.01–800 μM	2 nM	Wine	(Ezhil Vilian et al., 2014)
Maltol	GO-ACS	Amperometry	0.1–0.8 mM	24 nM	Wine	(Gan et al., 2016)
			PATHOGENS			
S. aureus	ssDNA/GNDs-Zeo	DPV	10^{-4}–10^{-1} μM	0.1 nM	Fruit juice	(Mathur et al., 2018)
Escherichia coli O157:H7	ssDNA/GO/CS/GCE	EIS	1.0×10^{-14}–1.0×10^{-18} M	3.5840×10^{-15} M	_	(S. Xu et al., 2017)
Salmonella	PPy-rGO/GCE	DPV	9.6–9.6×10^4 CFU mL^{-1}	8.07 CFU mL^{-1}	_	(Ye et al., 2019)
			ORGANIC CONTAMINANTS			
Bisphenol A	TA/NGR	DPV	0.05–13 mM	4.0 nM	Food packaging	(Jiao et al., 2014)
Bisphenol A	Au-Cu@BSA-GNRs/GCE	SWASVs	0.01–2.0 and 2.0–70 μM,	0.004 μM	Tap water, bottled water, baby bottle and food storage container	(Mahmoudi et al., 2019)
Anthracene	GR-PANI	SWV	0.012–1000 μM	0.0044 μM	Water	(Tovide et al., 2014)
Caffeic acid	Au–PEDOT/rGO	DPV	0.01–46 μM	0.004 μM	Red wine	(Z. Liu et al., 2016)
Bisphenol A	MNPs-rGO	DPV	0.6 nM–11 μM	0.17 nM	Plastic samples	(Y. Zhang et al., 2013)
Thiols	NGR/CoPc	Amperometry	1 μM–16 mM	1 μM	Water	(H. Xu et al., 2015)
			MYCOTOXINS			
Aflatoxin B_1	GR/AuNPs	CV	3.2 fM–0.32 pM	1 fM	Spiked food	(Linting et al., 2012)
Cadaverine	β-CD/GO-SO_3H	DPV	50–500 nM	20 nM	Fish	(Saghatforoush et al., 2014)
Theophylline	AuNPs/l-cysteine/GR	DPV	0.004–60 M	0.0004 M	Pu'er tea	(Zi et al., 2012)
A2704-12 gene sequence	ssDNA/rGO	DPV	0.001 nM–4μM	0.29 nM	Transgenic soybean	(W. Sun et al., 2013)

Note: SWV – square-wave voltammetry; SWSV – square-wave stripping voltammetry; DPSV – differential pulse stripping voltammetry; CV – cyclic voltammetry; DPV – differential pulse voltammetry; LSV – linear sweep voltammetry; SWASV– square wave anodic stripping voltammetry; EIS – Electrochemical impedance spectroscopy: GR – graphene; GNs – graphene nanosheets; NGR – nitrogen doped graphene; SF-GR – sulfonated functionalized graphene; GO – graphene oxide; rGO – reduced graphene oxide; PAR-EGR – poly(alizarin red)/electrodeposited graphene; CFG – carboxyl functionalized graphene; GQDs – graphene quantum dots; AuNRs – Au nanorod; AuNPs – Au nanoparticles; Eu-MGO – Europium doped magnetic graphene oxide MNPs – magnetic nano- particles; PANI – polyaniline; IL – ionic liquid; DPSPP – 1-[2,4-Dihydroxy-5-(phenylazo-4-sulphonicacid)phenyl]-1-phenylmethanon; CCLP – calcium ions cross linked pectin film; PA6 – polyamide6: PPy – polypyrrole: FTO –fluorinedopedtinoxide:PEDOT – poly(3,4-ethylenedioxythiophene); CoPc – cobalt phthalocyanine; GS-TH – graphene sheet and thionine nanocomposite; PoPD – poly(*o*-phenylenediamine); P3MT – poly(3- methylthiophene); PSS – poly(sodium 4-styrenesulfonate); ATDT – 5-amino-1,3,4-thiadiazole-2-thiol; ECL – electrochemiluminescence; ACS – amino-functionalized carbon sphere; ABSA – aminobenzene sulfonic acid; TA – tannic acid; P4VP – poly(4–vinylpyridine); ssDNA/GNDs-Zeo –graphene nano dots (GNDs) and zeolite (Zeo) specific DNA probe; CS – Chitosan.

to volume ratio, providing more active sites for interactions, and exhibit strong photoluminescence, which enhances detection when combined with electrochemical techniques. Additionally, their non-toxic and environmentally friendly nature makes them ideal for food-related application.

Another example of electrochemical sensors in food safety is carbon nanomaterial-based molecularly imprinted sensors. These have emerged as powerful tools in food analysis due to their exceptional properties, including high surface area, excellent conductivity, and robust mechanical strength. By integrating molecularly imprinted polymers (MIPs) with carbon nanomaterials, such as GR, CNTs, and CDs, these sensors achieve heightened sensitivity and selectivity for detecting various contaminants and residues in food products. The high surface area of carbon nanomaterials allows for more active sites, enhancing the loading capacity of the MIP and increasing the number of recognition sites available for target analytes. Their excellent electrical conductivity improves signal transduction efficiency, leading to higher sensitivity in electrochemical sensing. Additionally, the robust mechanical properties of carbon nanomaterials contribute to the durability and long-term stability of these sensors, making them ideal for ensuring food safety and quality. In recent years, the integration of CNTs with MIPs has gained significant attention for their use in detecting food analysis (Table 9.4).

GR nanomaterial-based MIPs have the capacity to selectively detect extremely low levels of pesticides in a range of food matrices, exhibiting both high sensitivity and low detection limits (Motaharian et al., 2016; Perçin et al., 2017; X. Shi et al., 2017). Shi et al. (2017), for example, created and synthesized a metal-organic framework (MIL@MIP/Fe_3O_4/rGO/CS/GCE) film with a core-shell MIP coating that is capable of selectively identifying the insecticides omethoate and metamorphose in spiked cucumber and kidney bean samples.

CNT-based MIP sensors are also used to detect residues of veterinary drugs, such as antibiotics and growth promoters, in meat and dairy products. They have been employed for the sensitive detection of carbofuran, semicarbazide and sulfadimidine (Lieberzeit, 2018; Wei et al., 2019; Yu et al., 2020). A team of researchers has developed an electrochemical sensor for the detection of amoxicillin in milk and honey samples. The researchers initially attached an ionic liquid to MWCNTs, which acted as monomers, and then proceeded to polymerize them in order to create MWCNTs@MIP (Yang & Zhao, 2015).

MIP sensors integrated with carbon nanomaterials are effective in detecting illegal additives and contaminants. For example, carbon dot-based MIP sensors have been utilized to detect melamine in milk and dairy products (S. Xu et al., 2018; Yin et al., 2018). Mycotoxins, such as patulin and ochratoxins, can be efficiently detected using carbon nanomaterial-based MIP sensors (Hatamluyi et al., 2020; Pacheco et al., 2015).

In the study of Hatamluyi et al. (2020) GO-MIP composites have shown high sensitivity and selectivity for patulin detection in apple juice (Figure 9.4). The combination of N-GQDs and Au@Cu-MOF with the MIP technique demonstrated remarkable synergistic effects, resulting in enhanced selectivity, sensitivity, stability, and reproducibility. The concentration of patulin in apple juice, as determined by the MIP sensor, exhibited a high degree of consistency with the HPLC results (Hatamluyi et al., 2020).

Carbon nanomaterial-based MIP sensors are at the same time capable of detecting bacterial pathogens like Bacillus cereus and Salmonella in food products (Ait Lahcen et al., 2018; Perçin et al., 2017). These sensors combine the selective recognition of MIPs with the high sensitivity of carbon nanomaterials, providing rapid and accurate pathogen detection. In summary, molecularly imprinted sensors are valuable tools for ensuring food safety and quality. Their ability to selectively recognize and bind specific analytes makes them ideal for detecting contaminants, residues, and pathogens in various food products. As technology advances, the applications and effectiveness of MIP-based sensors in food analysis are expected to grow.

TABLE 9.4
Application of CNTs composite-based MIP nanosensors in food analysis

ANALYTES	SENSORS	METHOD	LINEAR WORKING RANGE	DETECTION LİMİT	SAMPLE	REF.
			PESTICIDES			
Methamidophos	MIL@MIP/Fe_3O_4/rGO/CS/GCE	DPV	1.0×10^{-7}–1.0×10^{-12} M	2.67×10^{-13} M	Cucumber and kidney bean	(X. Shi et al., 2017)
Omethoate	MIL@MIP/Fe_3O_4/rGO/CS/GCE	DPV	1.0×10^{-7}–1.0×10^{-13} M	2.05×10^{-14} M	Cucumber and kidney bean	(X. Shi et al., 2017)
Diazinon	nano-MIP-CPE	SWV	2.5×10^{-9}–1.0×10^{-7} M	7.9×10^{-10} M	Well water and apple fruit	(Motaharian et al., 2016)
			VETERINARY DRUGS AND ANTIBIOTICS			
Carbofuran	MIP-CNTs-Fe_3O_4@Au/CPE	Amperometry	0.1 to 100 µM	3.8 nM	Cabbage, celery, chili, onion, and peppermint	(Lieberzeit, 2018)
Amoxicillin	MWCNTs-IL@MIP	DPV	1.0×10^{-9}–6.0×10^{-6} M	8.9×10^{-10}	Milk and honey	(Yang & Zhao, 2015)
Ampicillin	MIP/MWCNTs/AuNPs/ Pt		1.0×10^{-8}–5.0×10^{-6} M	1.0×10^{-9}	Milk, feed egg	(H. Liu et al., 2018)
Semicarbazide	MIP/SWNTs-COOH/CS	DPV	0.04–7.6 ng mL^{-1}	0.025 ng mL^{-1}	Honey, pork, shrimp, sheep casings	(Yu et al., 2020)
Sulfadimidine	MIP/NiCo2O4/3D graphene	DPV	0.2–1000 ng/mL	0.169 ng/mL	Milk	(Wei et al., 2019)
Sulfamethoxazole	MIP/H-GNs	DPV	5 µg/kg–1 mg/g	1.2 5 µg/kg	Milk, honey	(Guo et al., 2019)
			ARTIFICAL ADDITIVE			
Sunset yellow	GCE/MWCNT@MIP-PDA	DPV	2.2 nM–4.64 µM	1.4 nM	Jelly, fruit drink, chocolate, ice cream, candy	(Yin et al., 2018)
Melamin	GCE/MIP-MWCNTs	DPSV	1.0×10^{-12}–1.0×10^{-6} M	5.6×10^{-13} M	Milk	(S. Xu et al., 2018)
			NATURAL TOXINS			
Patulin	MIP/Au@Cu-MOF/N-GQDs/GCE	DPV	0.001–70.0 ng mL^{-1}	0.0007 ng mL^{-1}	Apple juice	(Hatamluyi et al., 2020)
Ochratoxin A	MIP/MWCNT/GCE MIP: Polypyrole (PPy)	DPV	0.050–1.0 µM	0.0041 µM (1.7 µg/L)	Beer	(H. Liu et al., 2018)
Histamine	MIP-apt/AuNPs/cCNTs/GCE	DPV	0.46–35 nM	0.15 nM	Canned tuna	(Mahmoud et al., 2020)
			MYCOTOXINS			
Bacillus cereus	CPE/SIP	CV	10^2–10^5 CFU mL^{-1}	–	–	(Ait Lahcen et al., 2018)
Salmonella paratyphi	Salmonella paratyphi-Imprinted SPR Chips		2.5×10^6–15×10^6 CFU mL^{-1}	1.4×10^6 CFU mL^{-1}	Apple juice	(Perçin et al., 2017)

Note: PDA – Polydopamine: PAVE – poly(acrylicacidco-(7-(4-vinylbenzyloxy)-4-methyl coumarin)-co-ethylhexyl acrylate) (poly(AAco-VMc-co-EHA); N-GQDs – Nitrogen doped graphene quantum dots; Au@Cu-MOF – AuNPs-functionalized Cu-metal organic framework; DPSV – Differential pulse stripping voltammetry; PPy – Polypyrole; CS – Chitosan; MIL – Chrom (III)-teraphtalic acid; H-GNs – hemin/graphene hybrid nanosheets.

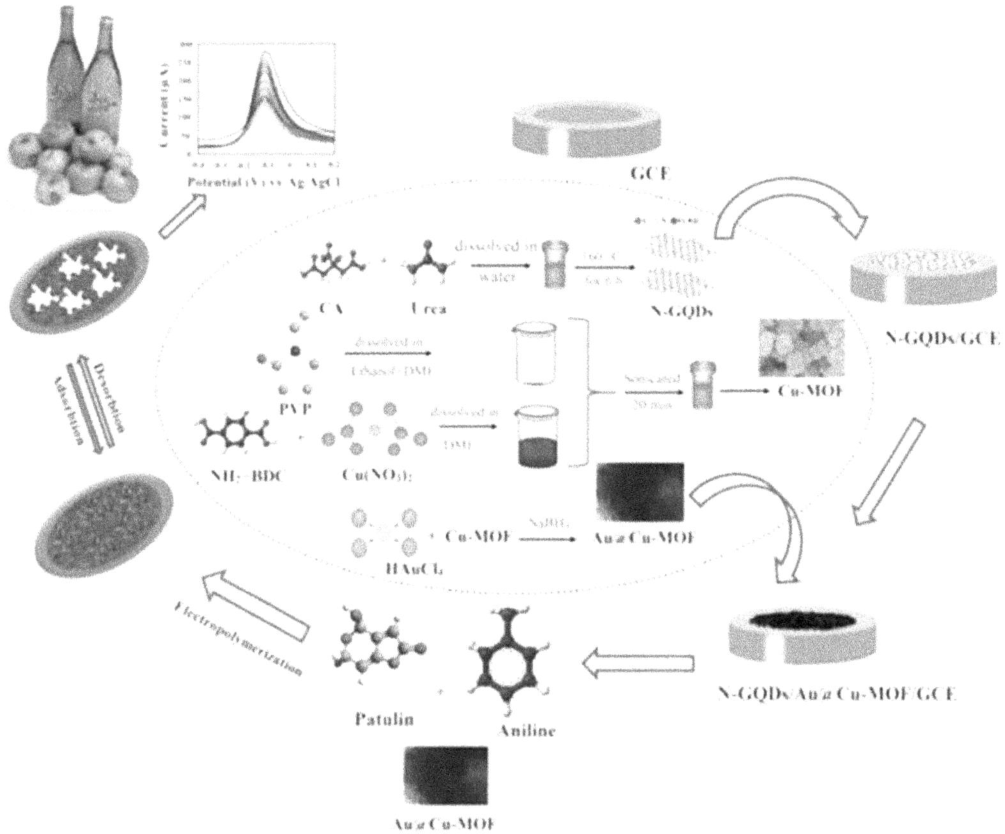

FIGURE 9.4 A schematic preparation of MIP/Au@Cu-MOF/N-DGQs/GCE and electrochemical detection of Patulin. (Reproduced from Hatamluyi et al., 2020 with permission from Elsevier.)

9.7 CONCLUSION AND FUTURE PERSPECTIVES

In conclusion, quality control represents a fundamental priority within the food industry at each stage of the production and supply chain, from initial production and storage to the point of consumption. The presence of bacteria, oxygen, ambient temperature, and humidity are all factors that play a significant role in determining the freshness of foodstuffs. Nanotechnology has the potential to provide innovative solutions for the preservation of food safety and quality. The advent of nanomaterials and nanosensors has enabled more efficacious monitoring and control of these factors, thereby ensuring superior standards of food quality and safety. The development of efficient sensors facilitates the quality monitoring process and is cost-effective. Such sensors facilitate rapid and precise identification of contaminants and quality parameters, thereby guaranteeing that food products comply with safety standards and retain consumer confidence.

While nanomaterials are adequate to fulfill the technological requirements of food safety and measurement, their utilization in food packaging and as food ingredients remains a topic of contention. The general public displays a decidedly negative attitude toward the use of nanomaterials as food additives. The prevailing view among governmental organizations and the general public is one of ambivalence and neutrality. The majority of people express a preference for natural foods over those containing nanomaterial additives. Despite the fact that nanotechnology has facilitated sustainable development in the food sector, there is still a considerable distance to be traversed. The perception of and attitudes toward certain nanomaterials are still evolving. The nanoscale dimensions of nanomaterials render them suitable for a multitude of intended applications. However, numerous studies have documented a range of adverse effects associated with their use. From the aforementioned evidence, it can be concluded that nanotechnology has the potential to enhance various aspects of the food industry, including quality, processing, preservation, freshness, and packaging, as well as the use of additives (Falguera et al., 2012). However, concerns regarding trust in this technology, both from the public and governmental organizations, present a significant challenge to its acceptance. Nevertheless, it is imperative that we continue to research and develop sustainable solutions for the advancement of the food industry.

REFERENCES

Ahmed, G. H. G., Laíño, R. B., Calzón, J. A. G., & García, M. E. D. (2015). Fluorescent carbon nanodots for sensitive and selective detection of tannic acid in wines. *Talanta, 132*, 252–257. https://doi.org/10.1016/j.talanta.2014.09.028

Ait Lahcen, A., Arduini, F., Lista, F., & Amine, A. (2018). Label-free electrochemical sensor based on spore-imprinted polymer for Bacillus cereus spore detection. *Sensors and Actuators, B: Chemical, 276*(June), 114–120. https://doi.org/10.1016/j.snb.2018.08.031

Alfadul S. M., & Elneshwy, A. A. (2010). Use of nanotechnology in food processing, packaging and safety – review. *African Journal of Food, Agriculture, Nutrition and Development, 10*(6), 2719–2739.

Algar, W. R., Tavares, A. J., & Krull, U. J. (2010). Beyond labels: A review of the application of quantum dots as integrated components of assays, bioprobes, and biosensors utilizing optical transduction. *Analytica Chimica Acta, 673*(1), 1–25. https://doi.org/10.1016/j.aca.2010.05.026

Ali, D. S., Hassan, R. O., Othman, H. O., Taha, H. T., Mousavi Khaneghah, A., & Smaoui, S. (2024). Revolutionizing detection: Smartphone-powered colorimetry for the drugs and food analysis. *Microchemical Journal, 205*. https://doi.org/10.1016/j.microc.2024.111228

Amekpewu, M., Abukari, S. S., Adu, K. W., Mensah, S. Y., & Mensah, N. G. (2015). Effect of hot electrons on the electrical conductivity of carbon nanotubes under the influence of applied dc field. *European Physical Journal B, 88*(2), 1–6. https://doi.org/10.1140/epjb/e2014-50493-5

Arduini, F., Cinti, S., Scognamiglio, V., & Moscone, D. (2016). Nanomaterials in electrochemical biosensors for pesticide detection: Advances and challenges in food analysis. *Microchimica Acta, 183*(7), 2063–2083. https://doi.org/10.1007/s00604-016-1858-8

Armenta, S., & Esteve-Turrillas, F. A. (2019). Carbon-Based Nanomaterials in Analytical Chemistry. In *Handbook of Smart Materials in Analytical Chemistry*. https://doi.org/10.1002/9781119422587.ch11

Baig, M. M. F., & Chen, Y. C. (2017). Bright carbon dots as fluorescence sensing agents for bacteria and curcumin. *Journal of Colloid and Interface Science, 501*, 341–349. https://doi.org/10.1016/j.jcis.2017.04.045

Barsan, M. M., Ghica, M. E., & Brett, C. M. A. (2015). Electrochemical sensors and biosensors based on redox polymer/carbon nanotube modified electrodes: A review. *Analytica Chimica Acta, 881*, 1–23. https://doi.org/10.1016/j.aca.2015.02.059

Bhardwaj, J., Devarakonda, S., Kumar, S., & Jang, J. (2017). Development of a paper-based electrochemical immunosensor using an antibody-single walled carbon nanotubes bio-conjugate modified electrode for label-free detection of foodborne pathogens. *Sensors and Actuators, B: Chemical, 253*, 115–123. https://doi.org/10.1016/j.snb.2017.06.108

Bilir, K., Weil, M. T., Lochead, J., Kök, F. N., & Werner, T. (2016). Construction of an oxygen detection-based optic laccase biosensor for polyphenolic compound detection. *Turkish Journal of Biology, 40*(6), 1303–1310. https://doi.org/10.3906/biy-1602-40

Calaminici, P., Carmona-Espindola, J., Geudtner, G., & Köster, A. M. (2012). Static and dynamic polarizability of C 540 fullerene. *International Journal of Quantum Chemistry, 112*(19), 3252–3255. https://doi.org/10.1002/qua.24176

Çevik, S., Timur, S., & Anik, Ü. (2013). Poly(allylamine hydrochloride) functionalized multiwalled carbon nanotube modified carbon paste electrode as acetylcholinesterase biosensor transducer. *Electroanalysis, 25*(10), 2377–2383. https://doi.org/10.1002/elan.201300260

Cha, C., Shin, S. R., Annabi, N., & Dokmeci, M. R. (2013). Carbon-based nanomaterials: Multifunctional materials for biomedical engineering. *ACS Nano, 4*, 2891–2897.

Chandra, S., Chowdhuri, A. R., Mahto, T. K., Samui, A., & Sahu, S. K. (2016). One-step synthesis of amikacin modified fluorescent carbon dots for the detection of Gram-negative bacteria like: Escherichia coli. *RSC Advances*, *6*(76), 72471–72478. https://doi.org/10.1039/c6ra15778e

Chaudhry, Q., Scotter, M., Blackburn, J., Ross, B., Boxall, A., Castle, L., Aitken, R., & Watkins, R. (2008). Applications and implications of nanotechnologies for the food sector. *Food Additives and Contaminants – Part A*, *25*(3), 241–258. https://doi.org/10.1080/02652030701744538

Chen, H., Zuo, X., Su, S., Tang, Z., Wu, A., Song, S., Zhang, D., & Fan, C. (2008). An electrochemical sensor for pesticide assays based on carbon nanotube-enhanced acetycholinesterase activity. *Analyst*, *133*(9), 1182–1186. https://doi.org/10.1039/b805334k

Chen, S., Li, Y., Fu, Z., Zeng, Y., He, L., Zhou, K., Ao, X., Liu, S., & Zou, L. (2018). Label-free and enzyme-free sensitive fluorescent method for detection of viable Escherichia coli O157:H7. *Analytical Biochemistry*, *556*(July), 145–151. https://doi.org/10.1016/j.ab.2018.07.003

Chhetri, K. B. (2024). Applications of artificial intelligence and machine learning in food quality control and safety assessment. *Food Engineering Reviews*, *16*(1), 1–21. https://doi.org/10.1007/s12393-023-09363-1

Chiou, Y. D., Tsai, D. S., Lam, H. H., Chang, C. H., Lee, K. Y., & Huang, Y. S. (2013). Cycle stability of the electrochemical capacitors patterned with vertically aligned carbon nanotubes in an LiPF6-based electrolyte. *Nanoscale*, *5*(17), 8122–8129. https://doi.org/10.1039/c3nr01980b

Cho, E. C., Huang, J. H., Li, C. P., Chang-Jian, C. W., Lee, K. C., Hsiao, Y. S., & Huang, J. H. (2016). Graphene-based thermoplastic composites and their application for LED thermal management. *Carbon, 102*, 66–73. https://doi.org/10.1016/j.carbon.2016.01.097

Choi, H. J., Jung, S. M., Seo, J. M., Chang, D. W., Dai, L., & Baek, J. B. (2012). Graphene for energy conversion and storage in fuel cells and supercapacitors. *Nano Energy*, *1*(4), 534–551. https://doi.org/10.1016/j.nanoen.2012.05.001

Crevillén, A. G., Ávila, M., Pumera, M., González, M. C., & Escarpa, A. (2007). Food analysis on microfluidic devices using ultrasensitive carbon nanotubes detectors. *Analytical Chemistry*, *79*(19), 7408–7415. https://doi.org/10.1021/ac071247i

Dai, H., Shi, Y., Wang, Y., Sun, Y., Hu, J., Ni, P., & Li, Z. (2014). A carbon dot based biosensor for melamine detection by fluorescence resonance energy transfer. *Sensors and Actuators, B: Chemical, 202*, 201–208. https://doi.org/10.1016/j.snb.2014.05.058

De Marcos, S., Callizo, E., Mateos, E., & Galbán, J. (2014). An optical sensor for pesticide determination based on the autoindicating optical properties of peroxidase. *Talanta, 122*, 251–256. https://doi.org/10.1016/j.talanta.2014.01.011

Dinç, S. (2016). A simple and green extraction of carbon dots from sugar beet molasses: Biosensor applications. *Sugar Industry, August*, 560–564. https://doi.org/10.36961/si17741

Dinç, S., Kara, M., Demirel Kars, M., Aykül, F., Çiçekci, H., & Akkuş, M. (2017). Biocompatible yogurt carbon dots: Evaluation of utilization for medical applications. *Applied Physics A: Materials Science and Processing*, *123*(9), 1–7. https://doi.org/10.1007/s00339-017-1184-y

Dingman, J. (2008). Nanotechnology: Its impact on food safety. *Journal of Environmental Health*, *70*(6), 47–50.

Drbohlavova, J., Adam, V., Kizek, R., & Hubalek, J. (2009). Quantum dots – characterization, preparation and usage in biological systems. *International Journal of Molecular Sciences*, *10*(2), 656–673. https://doi.org/10.3390/ijms10020656

Dridi, F., Marrakchi, M., Gargouri, M., Saulnier, J., Jaffrezic-Renault, N., & Lagarde, F. (2016). Nanomaterial-based electrochemical biosensors for food safety and quality assessment. *Nanobiosensors, 781*, 167–204. https://doi.org/10.1016/B978-0-12-804301-1.00005-9

Duncan, T. V. (2011). Applications of nanotechnology in food packaging and food safety: Barrier materials, antimicrobials and sensors. *Journal of Colloid and Interface Science*, *363*(1), 1–24. https://doi.org/10.1016/j.jcis.2011.07.017

Ebert, L. B. (1997). Science of fullerenes and carbon nanotubes. *Carbon*, *35*(3), 437–438. https://doi.org/10.1016/s0008-6223(97)89618-2

Ezhil Vilian, A. T., Chen, S. M., Chen, Y. H., Ajmal Ali, M., & Al-Hemaid, F. M. A. (2014). An electrocatalytic oxidation and voltammetric method using a chemically reduced graphene oxide film for the determination of caffeic acid. *Journal of Colloid and Interface Science, 423*, 33–40. https://doi.org/10.1016/j.jcis.2014.02.016

Falguera, V., Aliguer, N., & Falguera, M. (2012). An integrated approach to current trends in food consumption: Moving toward functional and organic products? *Food Control*, *26*(2), 274–281. https://doi.org/10.1016/j.foodcont.2012.01.051

Fong, J. F. Y., Chin, S. F., & Ng, S. M. (2016). A unique "turn-on" fluorescence signalling strategy for highly specific detection of ascorbic acid using carbon dots as sensing probe. *Biosensors and Bioelectronics, 85*, 844–852. https://doi.org/10.1016/j.bios.2016.05.087

Fu, W., Fu, X., Li, Z., Liu, Z., & Li, X. (2024). Advances in smartphone assisted sensors for on-site detection of food safety based on fluorescence on-off-on mode: A review. *Chemical Engineering Journal, 489*. https://doi.org/10.1016/j.cej.2024.151225

Gan, T., Lv, Z., Liu, N., Sun, J., Shi, Z., & Zhao, A. (2016). Ultrasensitive electrochemical sensor for maltol in wines using graphene oxide-wrapped amino-functionalized carbon sphere as sensing electrode materials. *Electroanalysis*, *28*(1), 103–110. https://doi.org/10.1002/elan.201500476

Gao, Z., Wang, L., Su, R., Huang, R., Qi, W., & He, Z. (2015). A carbon dot-based "off-on" fluorescent probe for highly selective and sensitive detection of phytic acid. *Biosensors and Bioelectronics, 70*, 232–238. https://doi.org/10.1016/j.bios.2015.03.043

Gökmen, S., Palamutoğlu, R., & Sarıçoban, C. (2012). Gıda Endüstrisinde Enkapsülasyon Uygulamaları [Applications of encapsulation in food industry]. *Teknolojik Araştırmalar*, *7*(1), 36–50.

Gong, N. C., Li, Y. Le, Jiang, X., Zheng, X. F., Wang, Y. Y., & Huan, S. Y. (2016). Fluorescence resonance energy transfer-based biosensor composed of nitrogen-doped carbon dots and gold nanoparticles for the highly sensitive detection of organophosphorus pesticides. *Analytical Sciences*, *32*(9), 951–956. https://doi.org/10.2116/analsci.32.951

Govindasamy, M., Wang, S. F., Kumaravel, S., Ramalingam, R. J., & Al-lohedan, H. A. (2019). Facile synthesis of copper sulfide decorated reduced graphene oxide nanocomposite for high sensitive detection of toxic antibiotic in milk. *Ultrasonics Sonochemistry*, *52*(December 2018), 382–390. https://doi.org/10.1016/j.ultsonch.2018.12.015

Guo, S. fei, Chen, X. yu, Wang, P., Chen, C., Pan, R. hua, Ling, Y. tao, & Tang, Y. zhu. (2019). Preparation of molecularly imprinted composites initiated by hemin/graphene hybrid nanosheets and its application in detection of sulfamethoxazole. *Current Medical Science*, *39*(1), 159–165. https://doi.org/10.1007/s11596-019-2014-6

Handford, C. E., Dean, M., Henchion, M., Spence, M., Elliott, C. T., & Campbell, K. (2014). Implications of nanotechnology for the agri-food industry: Opportunities, benefits and risks. *Trends in Food Science and Technology*, *40*(2), 226–241. https://doi.org/10.1016/j.tifs.2014.09.007

Hareesha, N., & Manjunatha, J. G. (2021). Electro-oxidation of formoterol fumarate on the surface of novel poly(thiazole yellow-G) layered multi-walled carbon nanotube paste electrode. *Scientific Reports*, *11*(1). https://doi.org/10.1038/s41598-021-92099-x

Hatamluyi, B., Rezayi, M., Beheshti, H. R., & Boroushaki, M. T. (2020). Ultra-sensitive molecularly imprinted electrochemical sensor for patulin detection based on a novel assembling strategy using Au@Cu-MOF/N-GQDs. *Sensors and Actuators, B: Chemical*, *318*(April). https://doi.org/10.1016/j.snb.2020.128219

He, Q., Liu, J., Liu, X., Li, G., Deng, P., Liang, J., & Chen, D. (2018). Sensitive and selective detection of tartrazine based on TiO2-electrochemically reduced graphene oxide composite-modified electrodes. *Sensors (Switzerland)*, *18*(6). https://doi.org/10.3390/s18061911

Hou, J., Dong, G., Tian, Z., Lu, J., Wang, Q., Ai, S., & Wang, M. (2016). A sensitive fluorescent sensor for selective determination of dichlorvos based on the recovered fluorescence of carbon dots-Cu(II) system. *Food Chemistry, 202*, 81–87. https://doi.org/10.1016/j.foodchem.2015.11.134

Hsiao, M. C., Liao, S. H., Lin, Y. F., Wang, C. A., Pu, N. W., Tsai, H. M., & Ma, C. C. M. (2011). Preparation and characterization of polypropylene-graft-thermally reduced graphite oxide with an improved compatibility with polypropylene-based nanocomposite. *Nanoscale*, *3*(4), 1516–1522. https://doi.org/10.1039/c0nr00981d

Hsiao, M. C., Liao, S. H., Yen, M. Y., Liu, P. I., Pu, N. W., Wang, C. A., & Ma, C. C. M. (2010). Preparation of covalently functionalized graphene using residual oxygen-containing functional groups. *ACS Applied Materials and Interfaces*, *2*(11), 3092–3099. https://doi.org/10.1021/am100597d

Javed, K., Galib, C. M. A., Yang, F., Chen, C. M., & Wang, C. (2014). A new approach to fabricate graphene electro-conductive networks on natural fibers by ultraviolet curing method. *Synthetic Metals, 193*, 41–47. https://doi.org/10.1016/j.synthmet.2014.03.028

Jiao, S., Jin, J., & Wang, L. (2014). Tannic acid functionalized N-doped graphene modified glassy carbon electrode for the determination of bisphenol A in food package. *Talanta, 122*, 140–144. https://doi.org/10.1016/j.talanta.2014.01.063

Kachoosangi, R. T., Wildgoose, G. G., & Compton, R. G. (2008). Carbon nanotube-based electrochemical sensors for quantifying the "heat" of chilli peppers: The adsorptive stripping voltammetric determination of capsaicin. *Analyst, 133*(7), 888–895. https://doi.org/10.1039/b803588a

Karteri, I., Karataş, Ş., Al-Ghamdi, A. A., & Yakuphanoılu, F. (2015). The electrical characteristics of thin film transistors with graphene oxide and organic insulators. *Synthetic Metals, 199*, 241–245. https://doi.org/10.1016/j.synthmet.2014.11.036

Khan, M., Husain, Q., & Bushra, R. (2017). Immobilization of β-galactosidase on surface modified cobalt/multiwalled carbon nanotube nanocomposite improves enzyme stability and resistance to inhibitor. *International Journal of Biological Macromolecules, 105*, 693–701. https://doi.org/10.1016/j.ijbiomac.2017.07.088

Khan, M. Z. H., Liu, X., Zhu, J., Ma, F., Hu, W., & Liu, X. (2018). Electrochemical detection of tyramine with ITO/APTES/ErGO electrode and its application in real sample analysis. *Biosensors and Bioelectronics, 108*(January), 76–81. https://doi.org/10.1016/j.bios.2018.02.042

Ko, S., & Grant, S. A. (2006). A novel FRET-based optical fiber biosensor for rapid detection of Salmonella typhimurium. *Biosensors and Bioelectronics, 21*(7), 1283–1290. https://doi.org/10.1016/j.bios.2005.05.017

Kong, L., Jiang, X., Zeng, Y., Zhou, T., & Shi, G. (2013). Molecularly imprinted sensor based on electropolmerized poly(o-phenylenediamine) membranes at reduced graphene oxide modified electrode for imidacloprid determination. *Sensors and Actuators, B: Chemical, 185*, 424–431. https://doi.org/10.1016/j.snb.2013.05.033

Korotkaya, E. V. (2014). Biosensors: Design, classification, and applications in the food industry. *Foods and Raw Materials, 2*(2), 161–171. https://doi.org/10.12737/5476

Kumar, V., Ahire, J. J., & Taneja, N. K. (2024). Advancing microbial food safety and hazard analysis through predictive mathematical modeling. *The Microbe, 2*, 100049. https://doi.org/10.1016/j.microb.2024.100049

Kumar, T. H. V., Yadav, S. K., & Sundramoorthy, A. K. (2018). Review—Electrochemical synthesis of 2D layered materials and their potential application in pesticide detection. *Journal of The Electrochemical Society, 165*(16), B848–B861. https://doi.org/10.1149/2.0641816jes

Lai, I. P. J., Harroun, S. G., Chen, S. Y., Unnikrishnan, B., Li, Y. J., & Huang, C. C. (2016). Solid-state synthesis of self-functional carbon quantum dots for detection of bacteria and tumor cells. *Sensors and Actuators, B: Chemical, 228*, 465–470. https://doi.org/10.1016/j.snb.2016.01.062

Lee, M. H., Do Kim, H., & Jang, Y. J. (2024). Delivery systems designed to enhance stability and suitability of lipophilic bioactive compounds in food processing: A review. *Food Chemistry, 437*, 137910. https://doi.org/10.1016/j.foodchem.2023.137910

Li, H., Sun, C., Vijayaraghavan, R., Zhou, F., Zhang, X., & MacFarlane, D. R. (2016). Long lifetime photoluminescence in N, S co-doped carbon quantum dots from an ionic liquid and their applications in ultrasensitive detection of pesticides. *Carbon, 104*, 33–39. https://doi.org/10.1016/j.carbon.2016.03.040

Li, H., Yan, X., Lu, G., & Su, X. (2018). Carbon dot-based bioplatform for dual colorimetric and fluorometric sensing of organophosphate pesticides. *Sensors and Actuators, B: Chemical, 260*, 563–570. https://doi.org/10.1016/j.snb.2017.12.170

Li, J., Feng, H., Lee, J., Jiang, J., Feng, Y., He, L., & Qian, D. (2015). Bimetallic Ag-Pd nanoparticles-decorated graphene oxide: A fascinating three-dimensional nanohybrid as an efficient electrochemical sensing platform for vanillin determination. *Electrochimica Acta, 176*, 827–835. https://doi.org/10.1016/j.electacta.2015.07.091

Li, T., Shang, D., Gao, S., Wang, B., Kong, H., Yang, G., Shu, W., Xu, P., & Wei, G. (2022). Sensors / Biosensors for Food Safety and Biomolecular Detection. *Biosensors, 12*(314), 1–32.

Li, J., Wang, X., Duan, H., Wang, Y., Bu, Y., & Luo, C. (2016). Based on magnetic graphene oxide highly sensitive and selective imprinted sensor for determination of sunset yellow. *Talanta, 147*, 169–176. https://doi.org/10.1016/j.talanta.2015.09.056

Lieberzeit, P. A. (2018). Selective amperometric flow-injection analysis of carbofuran using a molecularly-imprinted polymer and gold-coated-magnetite modified carbon. *Talanta 179*(November 2017), 700–709.

Lim, S. Y., Shen, W., & Gao, Z. (2015). Carbon quantum dots and their applications. *Chemical Society Reviews, 44*(1), 362–381. https://doi.org/10.1039/c4cs00269e

Lin, K. C., Hong, C. P., & Chen, S. M. (2013). Simultaneous determination for toxic ractopamine and salbutamol in pork sample using hybrid carbon nanotubes. *Sensors and Actuators, B: Chemical, 177*, 428–436. https://doi.org/10.1016/j.snb.2012.11.052

Linting, Z., Ruiyi, L., Zaijun, L., Qianfang, X., Yinjun, F., & Junkang, L. (2012). An immunosensor for ultrasensitive detection of aflatoxin B 1 with an enhanced electrochemical performance based on graphene/conducting polymer/gold nanoparticles/the ionic liquid composite film on modified gold electrode with electrodeposition. *Sensors and Actuators, B: Chemical, 174*, 359–365. https://doi.org/10.1016/j.snb.2012.06.051

Liu, J., Chen, Y., Wang, W., Feng, J., Liang, M., Ma, S., & Chen, X. (2016). "Switch-on" fluorescent sensing of ascorbic acid in food samples based on carbon quantum dots-MnO2 probe. *Journal of Agricultural and Food Chemistry, 64*(1), 371–380. https://doi.org/10.1021/acs.jafc.5b05726

Liu, L., Gou, Y., Gao, X., Zhang, P., Chen, W., Feng, S., Hu, F., & Li, Y. (2014). Electrochemically reduced graphene oxide-based electrochemical sensor for the sensitive determination of ferulic acid in *A. sinensis* and biological samples. *Materials Science and Engineering C, 42*, 227–233. https://doi.org/10.1016/j.msec.2014.05.045

Liu, Y., Lu, Q., Hu, X., Wang, H., Li, H., Zhang, Y., & Yao, S. (2017). A nanosensor based on carbon dots for recovered fluorescence detection clenbuterol in pork samples. *Journal of Fluorescence, 27*(5), 1847–1853. https://doi.org/10.1007/s10895-017-2122-2

Liu, Z., Xu, J., Yue, R., Yang, T., & Gao, L. (2016). Facile one-pot synthesis of Au-PEDOT/rGO nanocomposite for highly sensitive detection of caffeic acid in red wine sample. *Electrochimica Acta, 196*, 1–12. https://doi.org/10.1016/j.electacta.2016.02.178

Liu, H., Zhou, Y., Qi, Y., Sun, Z., & Gong, B. (2018). Preparation of thiamphenicol magnetic surface molecularly imprinted polymers for its selective recognition of thiamphenicol in milk samples. *Journal of Liquid Chromatography and Related Technologies, 41*(13–14), 868–879. https://doi.org/10.1080/10826076.2018.1531294

Luo, N., Xu, D., Xing, B., Yang, X., & Sun, C. (2024). Principles and applications of convolutional neural network for spectral analysis in food quality evaluation: A review. *Journal of Food Composition and Analysis, 128*. https://doi.org/10.1016/j.jfca.2024.105996

Ma, X., & Chen, M. (2015). Electrochemical sensor based on graphene doped gold nanoparticles modified electrode for detection of diethylstilboestrol. *Sensors and Actuators, B: Chemical, 215*, 445–450. https://doi.org/10.1016/j.snb.2015.04.016

Ma, X., Chen, M., & Chao, M. (2013). Voltammetric determination of Sudan II in food samples at graphene modified glassy carbon electrode based on the enhancement effect of sodium dodecyl sulfate. *Journal of the Chemical Society of Pakistan, 35*(1), 30–36.

Mahmoud, A. M., Alkahtani, S. A., Alyami, B. A., & El-Wekil, M. M. (2020). Dual-recognition molecularly imprinted aptasensor based on gold nanoparticles decorated carboxylated carbon nanotubes for highly selective and sensitive determination of histamine in different matrices. *Analytica Chimica Acta, 1133*, 58–65. https://doi.org/10.1016/j.aca.2020.08.001

Mahmoudi, E., Hajian, A., Rezaei, M., Afkhami, A., Amine, A., & Bagheri, H. (2019). A novel platform based on graphene nanoribbons/protein capped Au-Cu bimetallic nanoclusters: Application to the sensitive electrochemical determination of bisphenol A. *Microchemical Journal, 145*(July 2018), 242–251. https://doi.org/10.1016/j.microc.2018.10.044

Mansuriya, B. D., & Altintas, Z. (2020). Applications of graphene quantum dots in biomedical sensors. *Sensors (Switzerland), 20*(4). https://doi.org/10.3390/s20041072

Marmisollé, W. A., & Azzaroni, O. (2016). Recent developments in the layer-by-layer assembly of polyaniline and carbon nanomaterials for energy storage and sensing applications. from synthetic aspects to structural and functional characterization. *Nanoscale, 8*(19), 9890–9918. https://doi.org/10.1039/c5nr08326e

Mashkour, M., Sharifinia, M., Yousefi, H., & Afra, E. (2018). MWCNT-coated cellulose nanopapers: Droplet-coating, process factors, and electrical conductivity performance. *Carbohydrate Polymers, 202*(July), 504–512. https://doi.org/10.1016/j.carbpol.2018.09.006

Mathur, A., Gupta, R., Kondal, S., Wadhwa, S., Pudake, R. N., Shivani, Kansal, R., Pundir, C. S., & Narang, J. (2018). A new tactics for the detection of *S. aureus* via paper based geno-interface incorporated with graphene nano dots and zeolites. *International Journal of Biological Macromolecules, 112*, 364–370. https://doi.org/10.1016/j.ijbiomac.2018.01.143

Mauter, M. S., & Elimelech, M. (2008). Environmental applications of carbon-based nanomaterials: Critical review. *American Chemical Society*, *42*(16), 5843–5859.

Mercer, D. G. (2021). Concepts of Small-scale Food Processing. In D. G. Mercer (Ed.), *Concepts of Small-scale Food Processing*. The Royal Society of Chemistry. https://doi.org/10.1039/BK9781788018401-00406

Meshram, B. D., Agrawal, A. K., Adil, S., Ranvir, S., & Sande, K. K. (2018). Biosensor and its application in food and dairy industry: A review. *International Journal of Current Microbiology and Applied Sciences*, *7*(2), 3305–3324. https://doi.org/10.20546/ijcmas.2018.702.397

Migliorini, F. L., Sanfelice, R. C., Mercante, L. A., Facure, M. H. M., & Correa, D. S. (2020). Electrochemical sensor based on polyamide 6/polypyrrole electrospun nanofibers coated with reduced graphene oxide for malathion pesticide detection. *Materials Research Express*, *7*(1). https://doi.org/10.1088/2053-1591/ab5744

Mihindukulasuriya, S. D. F., & Lim, L. T. (2014). Nanotechnology development in food packaging: A review. *Trends in Food Science and Technology*, *40*(2), 149–167. https://doi.org/10.1016/j.tifs.2014.09.009

Mir, S. A., & Shah, M. A. (2017). Nanotechnology in the food industry. *Materials Science and Engineering: Concepts, Methodologies, Tools, and Applications*, *2–3*(4), 1165–1181. https://doi.org/10.4018/978-1-5225-1798-6.ch045

Mohammad, A. W., Lau, C. H., Zaharim, A., & Omar, M. Z. (2012). Elements of nanotechnology education in engineering curriculum worldwide. *Procedia – Social and Behavioral Sciences, 60*, 405–412. https://doi.org/10.1016/j.sbspro.2012.09.398

Momin, J. K., Jayakumar, C., & Prajapati, J. B. (2013). Potential of nanotechnology in functional foods. *Emirates Journal of Food and Agriculture*, *25*(1), 10–19. https://doi.org/10.9755/ejfa.v25i1.9368

Motaharian, A., Motaharian, F., Abnous, K., Hosseini, M. R. M., & Hassanzadeh-Khayyat, M. (2016). Molecularly imprinted polymer nanoparticles-based electrochemical sensor for determination of diazinon pesticide in well water and apple fruit samples. *Analytical and Bioanalytical Chemistry*, *408*(24), 6769–6779. https://doi.org/10.1007/s00216-016-9802-7

Mousavi, S. R., & Rezaei, M. (2011). Nanotechnology in agriculture and food production. *J Appl Environ Biol Sci*, *1*(10), 414-419

Nair, R. R., Blake, P., Grigorenko, A. N., Novoselov, K. S., Booth, T. J., Stauber, T., Peres, N. M. R., & Geim, A. K. (2008). Fine structure constant defines visual transparency of graphene. *Science*, *320*(5881), 1308. https://doi.org/10.1126/science.1156965

Nasir, S., Hussein, M. Z., Zainal, Z., & Yusof, N. A. (2018). Carbon-based nanomaterials/allotropes: A glimpse of their synthesis, properties and some applications. *Materials*, *11*(2), 1–24. https://doi.org/10.3390/ma11020295

Noyrod, P., Chailapakul, O., Wonsawat, W., & Chuanuwatanakul, S. (2014). The simultaneous determination of isoproturon and carbendazim pesticides by single drop analysis using a graphene-based electrochemical sensor. *Journal of Electroanalytical Chemistry, 719*, 54–59. https://doi.org/10.1016/j.jelechem.2014.02.001

Pacheco, J. G., Castro, M., Machado, S., Barroso, M. F., Nouws, H. P. A., & Delerue-Matos, C. (2015). Molecularly imprinted electrochemical sensor for ochratoxin A detection in food samples. *Sensors and Actuators, B: Chemical, 215*, 107–112. https://doi.org/10.1016/j.snb.2015.03.046

Paquin, F., Rivnay, J., Salleo, A., Stingelin, N., & Silva, C. (2015). Multi-phase semicrystalline microstructures drive exciton dissociation in neat plastic semiconductors. *Journal of Materials Chemistry C, 3*, 10715–10722. https://doi.org/10.1039/b000000x

Perçin, I., Idil, N., Bakhshpour, M., Yılmaz, E., Mattiasson, B., & Denizli, A. (2017). Microcontact imprinted plasmonic nanosensors: Powerful tools in the detection of salmonella paratyphi. *Sensors (Switzerland), 17*(6). https://doi.org/10.3390/s17061375

Pisanic, T. R., Zhang, Y., & Wang, T. H. (2014). Quantum dots in diagnostics and detection: Principles and paradigms. *Analyst*, *139*(12), 2968–2981. https://doi.org/10.1039/c4an00294f

Prencipe, G., Tabakman, S. M., Welsher, K., Liu, Z., Goodwin, A. P., Zhang, L., Henry, J., & Dai, H. (2009). PEG branched polymer for functionalization of nanomaterials with ultralong blood circulation. *Journal of the American Chemical Society*, *131*(13), 4783–4787. https://doi.org/10.1021/ja809086q

Qu, F., Sun, Z., Liu, D., Zhao, X., & You, J. (2016). Direct and indirect fluorescent detection of tetracyclines using dually emitting carbon dots. *Microchimica Acta*, *183*(9), 2547–2553. https://doi.org/10.1007/s00604-016-1901-9

Qu, J. H., Wei, Q., & Sun, D. W. (2018). Carbon dots: Principles and their applications in food quality and safety detection. *Critical Reviews in Food Science and Nutrition, 58*(14), 2466–2475. https://doi.org/10.1080/10408398.2018.1437712

Qureshi, M. A., Karthikeyan, S., Karthikeyan, P., Khan, P. A., Uprit, S., & Mishra, U. K. (2012). Application of nanotechnology in food and dairy processing: An overview. *Pakistan Journal of Food Science, 22*(1), 23–31.

Ryu, S. H., & Shanmugharaj, A. M. (2014). Influence of long-chain alkylamine-modified graphene oxide on the crystallization, mechanical and electrical properties of isotactic polypropylene nanocomposites. *Chemical Engineering Journal, 244*, 552–560. https://doi.org/10.1016/j.cej.2014.01.101

Sadr, S., Lotfalizadeh, N., Ghafouri, S. A., Delrobaei, M., Komeili, N., & Hajjafari, A. (2023). Nanotechnology innovations for increasing the productivity of poultry and the prospective of nanobiosensors. *Veterinary Medicine and Science, 9*(5), 2118–2131. https://doi.org/10.1002/vms3.1193

Saghatforoush, L., Hasanzadeh, M., & Shadjou, N. (2014). ß-Cyclodextrin/graphene oxide grafted sulfonic acid: Application for electro-oxidation and determination of cadaverine in fish samples. *Journal of Electroanalytical Chemistry, 714–715*, 79–84. https://doi.org/10.1016/j.jelechem.2013.12.021

Scida, K., Stege, P. W., Haby, G., Messina, G. A., & García, C. D. (2011). Recent applications of carbon-based nanomaterials in analytical chemistry: Critical review. *Analytica Chimica Acta, 691*(1–2), 6–17. https://doi.org/10.1016/j.aca.2011.02.025

Sebastian, N., Yu, W. C., & Balram, D. (2022). Ultrasensitive electrochemical detection and plasmon-enhanced photocatalytic degradation of rhodamine B based on dual-functional, 3D, hierarchical Ag/ZnO nanoflowers. *Sensors, 22*(13), 5049. https://doi.org/10.3390/s22135049

Sekhon, B. S. (2010). Food nanotechnology – an overview. *Nanotechnology, Science and Applications, 3*(1), 1–15. https://doi.org/10.2147/nsa.s12187498

Shanmugharaj, A. M., Yoon, J. H., Yang, W. J., & Ryu, S. H. (2013). Synthesis, characterization, and surface wettability properties of amine functionalized graphene oxide films with varying amine chain lengths. *Journal of Colloid and Interface Science, 401*, 148–154. https://doi.org/10.1016/j.jcis.2013.02.054

Sharma, C., Dhiman, R., Rokana, N., & Panwar, H. (2017). Nanotechnology: An untapped resource for food packaging. *Frontiers in Microbiology, 8*(SEP). https://doi.org/10.3389/fmicb.2017.01735

She, P., Chu, Y., Liu, C., Guo, X., Zhao, K., Li, J., Du, H., Zhang, X., Wang, H., & Deng, A. (2016). A competitive immunoassay for ultrasensitive detection of Hg^{2+} in water, human serum and urine samples using immunochromatographic test based on surface-enhanced Raman scattering. *Analytica Chimica Acta, 906*, 139–147. https://doi.org/10.1016/j.aca.2015.12.021

Shen, L., & Liu, J. (2016). New development in carbon quantum dots technical applications. *Talanta, 157*, 245–256.

Sheng, W., Liu, Y., Li, S., Lu, Y., Chang, Q., Zhang, Y., & Wang, S. (2018). Lateral flow quantum-dot-based immunochromatographic assay and fluorescence quenching immunochromatographic assay with quantum dots as fluorescence donors to visually detect bisphenol a in food and water samples. *Food Analytical Methods, 11*(3), 675–685. https://doi.org/10.1007/s12161-017-1039-3

Shi, J., Guo, J., Bai, G., Chan, C., Liu, X., Ye, W., Hao, J., Chen, S., & Yang, M. (2015). A graphene oxide based fluorescence resonance energy transfer (FRET) biosensor for ultrasensitive detection of botulinum neurotoxin A (BoNT/A) enzymatic activity. *Biosensors and Bioelectronics, 65*, 238–244. https://doi.org/10.1016/j.bios.2014.10.050

Shi, W., Wang, Q., Long, Y., Cheng, Z., Chen, S., Zheng, H., & Huang, Y. (2011). Carbon nanodots as peroxidase mimetics and their applications to glucose detection. *Chemical Communications, 47*(23), 6695–6697. https://doi.org/10.1039/c1cc11943e

Shi, X., Lu, J., Yin, H., Qiao, X., & Xu, Z. (2017). A biomimetic sensor with signal enhancement of ferriferrous oxide-reduced graphene oxide nanocomposites for ultratrace levels quantification of methamidophos or omethoate in vegetables. *Food Analytical Methods, 10*(4), 910–920. https://doi.org/10.1007/s12161-016-0641-0

Shruti, A., Bage, N., & Kar, P. (2024). Nanomaterials based sensors for analysis of food safety. *Food Chemistry, 433*. https://doi.org/10.1016/j.foodchem.2023.137284

Siangproh, W., Dungchai, W., Rattanarat, P., & Chailapakul, O. (2011). Nanoparticle-based electrochemical detection in conventional and miniaturized systems and their bioanalytical applications: A review. *Analytica Chimica Acta, 690*(1), 10–25. https://doi.org/10.1016/j.aca.2011.01.054

Singh, J., & Sengar, R. S. (2020). Nanotechnology in agriculture and food. *Annals of Horticulture, 13*(1), 14-24.

Smalley, R. E. (1985). *C 60: Buckminsterfullerene*. 62–63.
Song, J., Li, J., Guo, Z., Liu, W., Ma, Q., Feng, F., & Dong, C. (2017). A novel fluorescent sensor based on sulfur and nitrogen co-doped carbon dots with excellent stability for selective detection of doxycycline in raw milk. *RSC Advances*, *7*(21), 12827–12834. https://doi.org/10.1039/c7ra01074e
Stamatin, S. N., Borghei, M., Dhiman, R., Andersen, S. M., Ruiz, V., Kauppinen, E., & Skou, E. M. (2015). Activity and stability studies of platinized multi-walled carbon nanotubes as fuel cell electrocatalysts. *Applied Catalysis B: Environmental, 162*, 289–299. https://doi.org/10.1016/j.apcatb.2014.07.005
Sun, H., Liu, H., & Wu, Y. (2017). A green, reusable SERS film with high sensitivity for in-situ detection of thiram in apple juice. *Applied Surface Science, 416*, 704–709. https://doi.org/10.1016/j.apsusc.2017.04.159
Sun, Q., Long, Y., Li, H., Pan, S., Yang, J., Liu, S., & Hu, X. (2018). Fluorescent carbon dots as cost-effective and facile probes for caffeic acid sensing via a fluorescence quenching process. *Journal of Fluorescence*, *28*(2), 523–531. https://doi.org/10.1007/s10895-018-2213-8
Sun, X., Cao, Y., Gong, Z., Wang, X., Zhang, Y., & Gao, J. (2012). An amperometric immunosensor based on multi-walled carbon nanotubes-thionine-chitosan nanocomposite film for chlorpyrifos detection. *Sensors (Switzerland), 12*(12), 17247–17261. https://doi.org/10.3390/s121217247
Sun, W., Zhang, Y., Hu, A., Lu, Y., Shi, F., Lei, B., & Sun, Z. (2013). Electrochemical DNA biosensor based on partially reduced graphene oxide modified carbon ionic liquid electrode for the detection of transgenic soybean A2704-12 gene sequence. *Electroanalysis*, *25*(6), 1417–1424. https://doi.org/10.1002/elan.201300069
Szunerits, S., & Boukherroub, R. (2018). Graphene-based nanomaterials in innovative electrochemistry. *Current Opinion in Electrochemistry, 10*, 24–30. https://doi.org/10.1016/j.coelec.2018.03.016
Tan, H., Li, Q., Zhou, Z., Ma, C., Song, Y., Xu, F., & Wang, L. (2015). A sensitive fluorescent assay for thiamine based on metal-organic frameworks with intrinsic peroxidase-like activity. *Analytica Chimica Acta, 856*, 90–95. https://doi.org/10.1016/j.aca.2014.11.026
Thakur, M. S., & Ragavan, K. V. (2013). Biosensors in food processing. *Journal of Food Science and Technology*, *50*(4), 625–641. https://doi.org/10.1007/s13197-012-0783-z
Tovide, O., Jahed, N., Sunday, C. E., Pokpas, K., Ajayi, R. F., Makelane, H. R., Molapo, K. M., John, S. V., Baker, P. G., & Iwuoha, E. I. (2014). Electro-oxidation of anthracene on polyanilino-graphene composite electrode. *Sensors and Actuators, B: Chemical, 205*, 184–192. https://doi.org/10.1016/j.snb.2014.07.116
Türk, K. K., Kruusenberg, I., Kibena-Põldsepp, E., Bhowmick, G. D., Kook, M., Tammeveski, K., Matisen, L., Merisalu, M., Sammelselg, V., Ghangrekar, M. M., Mitra, A., & Banerjee, R. (2018). Novel multi walled carbon nanotube based nitrogen impregnated Co and Fe cathode catalysts for improved microbial fuel cell performance. *International Journal of Hydrogen Energy*, *43*(51), 23027–23035. https://doi.org/10.1016/j.ijhydene.2018.10.143
Tzeng, T. R. J., Liu, T. Y., Lin, C. W., Chang, P. E., Liao, P. X., Yang, W. Y., ... & Lin, Y. Y. (2021). Effects of dietary inclusion of dry Hydrastis canadensis on laying performance, egg quality, serum biochemical parameters and cecal microbiota in laying hens. *Animals, 11*(5), 1381.
Valdés, M. G., González, A. C. V., Calzón, J. A. G., & Díaz-García, M. E. (2009). Analytical nanotechnology for food analysis. *Microchimica Acta*, *166*(1–2), 1–19. https://doi.org/10.1007/s00604-009-0165-z
Varilla, C., Marcone, M., & Annor, G. A. (2020). Potential of cold plasma technology in ensuring the safety of foods and agricultural produce: A review. *Foods*, *9*(10), 1–17. https://doi.org/10.3390/foods9101435
Vatandost, E., Ghorbani-HasanSaraei, A., Chekin, F., Naghizadeh Raeisi, S., & Shahidi, S. A. (2020). Green tea extract assisted green synthesis of reduced graphene oxide: Application for highly sensitive electrochemical detection of sunset yellow in food products. *Food Chemistry: X*, *6*(July 2019), 0–5. https://doi.org/10.1016/j.fochx.2020.100085
Wang, L., Bi, Y., Hou, J., Li, H., Xu, Y., Wang, B., Ding, H., & Ding, L. (2016). Facile, green and clean one-step synthesis of carbon dots from wool: Application as a sensor for glyphosate detection based on the inner filter effect. *Talanta, 160*, 268–275. https://doi.org/10.1016/j.talanta.2016.07.020
Wang, S., Sun, H., Ang, H. M., & Tadé, M. O. (2013). Adsorptive remediation of environmental pollutants using novel graphene-based nanomaterials. *Chemical Engineering Journal, 226*, 336–347. https://doi.org/10.1016/j.cej.2013.04.070
Wang, R., Xu, Y., Zhang, T., & Jiang, Y. (2015). Rapid and sensitive detection of Salmonella typhimurium using aptamer-conjugated carbon dots as fluorescence probe. *Analytical Methods*, *7*(5), 1701–1706. https://doi.org/10.1039/c4ay02880e

Wang, L., Yang, R., Chen, J., Li, J., Qu, L., & De B. Harrington, P. (2014). Sensitive voltammetric sensor based on isopropanol-Nafion-PSS-GR nanocomposite modified glassy carbon electrode for determination of clenbuterol in pork. *Food Chemistry, 164*, 113–118. https://doi.org/10.1016/j.foodchem.2014.04.052

Wang, J., Yang, B., Wang, H., Yang, P., & Du, Y. (2015). Highly sensitive electrochemical determination of Sunset Yellow based on gold nanoparticles/graphene electrode. *Analytica Chimica Acta, 893*, 41–48. https://doi.org/10.1016/j.aca.2015.08.042

Wang, X., Yang, L., Jin, X., & Zhang, L. (2014). Electrochemical determination of estrogenic compound bisphenol F in food packaging using carboxyl functionalized multi-walled carbon nanotubes modified glassy carbon electrode. *Food Chemistry, 157*, 464–469. https://doi.org/10.1016/j.foodchem.2014.02.079

Wang, Y. C., Cokeliler, D., & Gunasekaran, S. (2015). Reduced graphene oxide/carbon nanotube/gold nanoparticles nanocomposite functionalized screen-printed electrode for sensitive electrochemical detection of endocrine disruptor bisphenol A. *Electroanalysis, 27*(11), 2527–2536.

Wang, Z., Yu, J., Gui, R., Jin, H., & Xia, Y. (2016). Biosensors and Bioelectronics Carbon nanomaterials-based electrochemical aptasensors. *Biosensors and Bioelectronic, 79*, 136–149.

Wei, X., Zhang, Z., Zhang, L., & Xu, X. (2019). Synthesis of molecularly imprinted polymers/$NiCo_2O_4$ nanoneedle arrays on 3D graphene electrode for determination of sulfadimidine residue in food. *Journal of Materials Science*, *54*(3), 2066–2078. https://doi.org/10.1007/s10853-018-2975-z

Weiss, J., Takhistov, P., & McClements, D. J. (2006). Functional materials in food nanotechnology. *Journal of Food Science*, *71*(9). https://doi.org/10.1111/j.1750-3841.2006.00195.x

Wolf, E. L. (2014). Applications of Graphene: An Overview. In *Springer Briefs in Materials*. Springer.

Xu, X., Jiang, S., Hu, Z., & Liu, S. (2010). Nitrogen-doped carbon nanotubes: High oxidation of hydrogen peroxide and its. *Acsnano*, *4*(7).

Xu, S., Lin, G., Zhao, W., Wu, Q., Luo, J., Wei, W., Liu, X., & Zhu, Y. (2018). Necklace-like molecularly imprinted nanohybrids based on polymeric nanoparticles decorated multiwalled carbon nanotubes for highly sensitive and selective melamine detection. *ACS Applied Materials and Interfaces*, *10*(29), 24850–24859. https://doi.org/10.1021/acsami.8b08558

Xu, H., Xiao, J., Liu, B., Griveau, S., & Bedioui, F. (2015). Enhanced electrochemical sensing of thiols based on cobalt phthalocyanine immobilized on nitrogen-doped graphene. *Biosensors and Bioelectronics, 66*, 438–444. https://doi.org/10.1016/j.bios.2014.12.011

Xu, S., Zhang, Y., Dong, K., Wen, J., Zheng, C., & Zhao, S. (2017). Electrochemical DNA biosensor based on graphene oxide-chitosan hybrid nanocomposites for detection of Escherichia coli O157:H7. *International Journal of Electrochemical Science*, *12*(4), 3443–3458. https://doi.org/10.20964/2017.04.16

Yamada, K., Choi, W., Lee, I., Cho, B. K., & Jun, S. (2016). Rapid detection of multiple foodborne pathogens using a nanoparticle-functionalized multi-junction biosensor. *Biosensors and Bioelectronics, 77*, 137–143. https://doi.org/10.1016/j.bios.2015.09.030

Yang, G., & Zhao, F. (2015). Molecularly imprinted polymer grown on multiwalled carbon nanotube surface for the sensitive electrochemical determination of amoxicillin. *Electrochimica Acta*, *174*(1), 33–40. https://doi.org/10.1016/j.electacta.2015.05.156

Ye, Y., Yan, W., Liu, Y., He, S., Cao, X., Xu, X., Zheng, H., & Gunasekaran, S. (2019). Electrochemical detection of Salmonella using an invA genosensor on polypyrrole-reduced graphene oxide modified glassy carbon electrode and AuNPs-horseradish peroxidase-streptavidin as nanotag. *Analytica Chimica Acta, 1074*, 80–88. https://doi.org/10.1016/j.aca.2019.05.012

Yin, Z. Z., Cheng, S. W., Xu, L. Bin, Liu, H. Y., Huang, K., Li, L., Zhai, Y. Y., Zeng, Y. B., Liu, H. Q., Shao, Y., Zhang, Z. L., & Lu, Y. X. (2018). Highly sensitive and selective sensor for sunset yellow based on molecularly imprinted polydopamine-coated multi-walled carbon nanotubes. *Biosensors and Bioelectronics*, *100*(June 2017), 565–570. https://doi.org/10.1016/j.bios.2017.10.010

Yu, W., Tang, Y., Sang, Y., Liu, W., Wang, S., & Wang, X. (2020). Preparation of a carboxylated single-walled carbon-nanotube-chitosan functional layer and its application to a molecularly imprinted electrochemical sensor to quantify semicarbazide. *Food Chemistry*, *333*(June). https://doi.org/10.1016/j.foodchem.2020.127524

Yuan, B., Bao, C., Song, L., Hong, N., Liew, K. M., & Hu, Y. (2014). Preparation of functionalized graphene oxide/polypropylene nanocomposite with significantly improved thermal stability and studies on the crystallization behavior and mechanical properties. *Chemical Engineering Journal, 237*, 411–420. https://doi.org/10.1016/j.cej.2013.10.030

Yuan, Y., Zhao, X., Qiao, M., Zhu, J., Liu, S., Yang, J., & Hu, X. (2016). Determination of sunset yellow in soft drinks based on fluorescence quenching of carbon dots. *Spectrochimica Acta – Part A: Molecular and Biomolecular Spectroscopy, 167*, 106–110. https://doi.org/10.1016/j.saa.2016.05.038

Yun, M., Choe, J. E., You, J. M., Ahmed, M. S., Lee, K., Üstündağ, Z., & Jeon, S. (2015). High catalytic activity of electrochemically reduced graphene composite toward electrochemical sensing of Orange II. *Food Chemistry, 169*, 114–119. https://doi.org/10.1016/j.foodchem.2014.07.143

Yun, Y. J., Hong, W. G., Kim, W. J., Jun, Y., & Kim, B. H. (2013). A novel method for applying reduced graphene oxide directly to electronic textiles from yarns to fabrics. *Advanced Materials, 25*(40), 5701–5705. https://doi.org/10.1002/adma.201303225

Zeng, Y., Zhu, Z., Du, D., & Lin, Y. (2016). Nanomaterial-based electrochemical biosensors for food safety. *Journal of Electroanalytical Chemistry, 781*, 147–154. https://doi.org/10.1016/j.jelechem.2016.10.030.

Zhai, H., Liang, Z., Chen, Z., Wang, H., Liu, Z., Su, Z., & Zhou, Q. (2015). Simultaneous detection of metronidazole and chloramphenicol by differential pulse stripping voltammetry using a silver nanoparticles/sulfonate functionalized graphene modified glassy carbon electrode. *Electrochimica Acta, 171*, 105–113. https://doi.org/10.1016/j.electacta.2015.03.140

Zhang, Y., Cheng, Y., Zhou, Y., Li, B., Gu, W., Shi, X., & Xian, Y. (2013). Electrochemical sensor for bisphenol A based on magnetic nanoparticles decorated reduced graphene oxide. *Talanta, 107*, 211–218. https://doi.org/10.1016/j.talanta.2013.01.012

Zhang, Y., Wang, S., Zhang, Y., Pang, G., & Guo, S. (2017). Tuning the aggregation/disaggregation behavior of graphene quantum dots by structure-switching aptamer for high-sensitivity fluorescent ochratoxin A sensor. *Analytical Chemistry, 89*(3), 1704–1709. https://doi.org/10.1021/acs.analchem.6b03913

Zhang, P., Zhang, N., Jing, L., Hu, B., Yang, X., & Ma, X. (2019). Silver nanoparticles/carboxylic short-chain multi-wall carbon nanotubes as electrochemical sensor for ultrasensitive detection of chloramphenicol in food. *International Journal of Electrochemical Science, 14*(9), 9337–9346. https://doi.org/10.20964/2019.09.69

Zhong, D., Zhuo, Y., Feng, Y., & Yang, X. (2015). Employing carbon dots modified with vancomycin for assaying Gram-positive bacteria like Staphylococcus aureus. *Biosensors and Bioelectronics, 74*, 546–553. https://doi.org/10.1016/j.bios.2015.07.015

Zhou, S., Liu, C., Lin, J., Zhu, Z., Hu, B., & Wu, L. (2022). Towards development of molecularly imprinted electrochemical sensors for food and drug safety: Progress and trends. *Biosensors, 12*. https://doi.org/10.3390/bios12060369

Zhu, W., Liu, W., Li, T., Yue, X., Liu, T., Zhang, W., Yu, S., Zhang, D., & Wang, J. (2014). Facile green synthesis of graphene-Au nanorod nanoassembly for on-line extraction and sensitive stripping analysis of methyl parathion. *Electrochimica Acta, 146*, 419–428. https://doi.org/10.1016/j.electacta.2014.09.085

Zi, L., Li, J., Mao, Y., Yang, R., & Qu, L. (2012). High sensitive determination of theophylline based on gold nanoparticles/l-cysteine/Graphene/Nafion modified electrode. *Electrochimica Acta, 78*, 434–439. https://doi.org/10.1016/j.electacta.2012.06.018

Zorraquín-Peña, I., Cueva, C., Bartolomé, B., & Moreno-Arribas, M. V. (2020). Silver nanoparticles against foodborne bacteria. Effects at intestinal level and health limitations. *Microorganisms, 8*(1), 132.

Zu, F., Yan, F., Bai, Z., Xu, J., Wang, Y., Huang, Y., & Zhou, X. (2017). The quenching of the fluorescence of carbon dots: A review on mechanisms and applications. *Microchimica Acta, 184*(7), 1899–1914. https://doi.org/10.1007/s00604-017-2318-9

10 Applications of Sensors in Energy and Industrial Processes

Mohamed Bendany, Khalid Ait Ben Brahim, Youssra El Hamdouni, Najoua Labjar, Hamid Nasrellah, Meryem Bensemlali, Halima Mortadi, and Souad El Hajjaji

10.1 INTRODUCTION

Sensors play a pivotal role in modern energy and industrial processes, serving as essential tools for monitoring, control, and optimization [1]. Their importance spans various applications, from ensuring operational efficiency and safety to enabling advanced technological innovations [2]. This chapter provides a comprehensive overview of the diverse applications of sensors in these critical domains, focusing on recent advancements and emerging technologies that are reshaping the landscape of industrial monitoring [3].

The introduction of sensors into industrial processes has revolutionized how industries operate, offering real-time insights and precise measurements that drive efficiency and reliability [4]. In energy sectors, sensors are crucial for monitoring parameters such as temperature, pressure, and gas concentrations, which are vital for the safe and efficient operation of power plants, both conventional and renewable [5]. For instance, electrochemical sensors are used to monitor and manage chemical processes in thermal power plants, while optical sensors play a key role in renewable energy systems like solar panels by measuring light intensity and optimizing panel orientation [6].

Industrial processes also benefit significantly from sensor technologies. Sensors are integral to machine condition monitoring, where they help track the health of equipment through parameters such as vibration and temperature [7]. This capability supports condition-based maintenance strategies, which can reduce downtime and extend the lifespan of machinery. Additionally, sensors for process control, including flow, pressure, and gas sensors, enable precise regulation of industrial operations, enhancing product quality and safety [3].

The chapter delves into the technical aspects of these sensors, highlighting the challenges and innovations that define their performance. Calibration, sensitivity, and reliability are key factors that impact sensor effectiveness, and managing these challenges is crucial for maintaining accurate and dependable measurements [8]. Moreover, the integration of emerging technologies, such as nanotechnology-based sensors and wireless sensor networks, offers new opportunities for enhancing sensor capabilities. Nanotechnology enables the development of highly sensitive and miniaturized sensors, while wireless networks facilitate seamless data collection and integration into smart industrial systems [9, 10].

By examining these aspects, this chapter underscores the transformative impact of sensors on industrial practices. It demonstrates how advancements in sensor technology not only improve operational efficiency and safety but also contribute to more sustainable and environmentally friendly

DOI: 10.1201/9781003500322-10

practices. The continuous evolution of sensor technologies promises to drive further innovation, making it essential for industries to stay abreast of these developments to fully harness their potential.

In summary, this chapter provides an in-depth look at the critical role of sensors in energy and industrial processes, exploring their applications, challenges, and the future directions of sensor technology. This examination reveals the significant contributions of sensors to industrial progress and highlights the ongoing need for innovation in this dynamic field.

10.2 SENSORS IN ENERGY PRODUCTION PROCESSES

10.2.1 Sensors in Power Plants

In thermal power plants, sensors play a crucial role in monitoring and controlling various operational parameters to ensure efficiency, safety, and environmental compliance. Temperature sensors, such as thermocouples and resistance temperature detectors (RTDs), are employed to track the temperature of critical components like boilers, turbines, and cooling systems [1]. These sensors help maintain optimal temperature ranges, preventing overheating and damage to equipment. Pressure sensors, including piezoelectric and capacitive types, monitor steam and water pressure within the system. They are vital for ensuring that pressure levels are kept within safe limits, thereby avoiding potential equipment failures or safety hazards [11].

Flow sensors, such as magnetic and ultrasonic flow meters, measure the rate of fluid flow through pipes and turbines. Accurate flow measurement is essential for maintaining proper fluid dynamics and optimizing the energy conversion process. Additionally, gas sensors are particularly important in coal-fired power plants for emission detection [4]. Electrochemical sensors are used to detect specific pollutants such as carbon dioxide (CO_2), sulfur dioxide (SO_2), and nitrogen oxides (NOx) [12], [13]. These sensors work by measuring the electrical changes that occur when a gas interacts with a reactive electrode [14]. This information is crucial for adjusting combustion processes to minimize emissions and comply with environmental regulations [15]. By integrating these various sensors into power plant control systems, operators can enhance the efficiency of energy production, ensure the safety of plant operations, and meet environmental standards.

10.2.2 Sensors in Renewable Energy

In renewable energy systems, particularly in solar energy, electrochemical sensors have a specialized role in optimizing performance and ensuring efficient operation [16]. For solar energy applications, while light intensity and temperature are typically measured using optical and thermal sensors, electrochemical sensors are instrumental in monitoring and managing the chemical conditions affecting solar panel efficiency [6]. For instance, electrochemical sensors can be used to detect the presence of corrosive substances or contaminants that may affect the integrity of photovoltaic panels over time [17]. These sensors work by detecting changes in electrical properties as chemical reactions occur at the sensor's electrodes, providing insights into potential degradation or damage to the panels. Additionally, electrochemical sensors can play a role in advanced solar energy systems that include energy storage components, such as batteries [18]. They monitor the state of charge, health, and chemical composition of battery systems, ensuring optimal performance and longevity of energy storage solutions. While electrochemical sensors are not typically used for direct measurements of light intensity or temperature, their ability to monitor chemical conditions and battery health contributes significantly to the overall efficiency and reliability of renewable energy systems [19, 20].

10.2.3 Sensors for Energy Efficiency Optimization

Sensors play a crucial role in optimizing energy efficiency in various industrial and energy production settings. By providing real-time data on key parameters such as temperature, pressure, flow rate, and equipment performance, sensors enable precise monitoring and control of processes, leading to reduced energy consumption and enhanced operational efficiency [21]. For instance, in heating, ventilation, and air conditioning systems, temperature and humidity sensors can adjust operations dynamically to maintain optimal environmental conditions while minimizing energy use [22]. In industrial processes, flow sensors can help regulate the supply of gases or liquids to maintain efficiency and reduce waste. Smart sensors, integrated with advanced data analytics and machine learning algorithms, can predict equipment failures or identify energy loss points, allowing for proactive maintenance and optimization [23]. Additionally, sensors that monitor the efficiency of electrical systems, such as voltage and current sensors, help ensure that energy is distributed and used effectively, preventing losses. By leveraging these capabilities, industries can achieve significant cost savings, reduce their carbon footprint, and comply with energy efficiency standards and regulations, thus contributing to more sustainable operations [24].

10.3 SENSORS IN INDUSTRIAL PROCESSES

10.3.1 Sensors for Machine Condition Monitoring

In the realm of machine condition monitoring, electrochemical sensors are increasingly recognized for their ability to offer insights into the chemical environment that can impact machinery health and performance. Unlike traditional sensors that monitor physical and thermal parameters, electrochemical sensors provide critical data on the chemical composition and interactions within the machine's environment [25, 26]. These sensors are particularly valuable in environments where corrosive chemicals, reactive gases, or other chemical agents are present, as they can detect changes in chemical concentrations that may signal potential issues such as corrosion, contamination, or degradation of materials [27].

For instance, electrochemical sensors can be used to monitor the presence of corrosive gases like hydrogen sulfide (H_2S) or chlorine (Cl_2), which can cause significant wear and tear on machinery over time [28]. By providing real-time data on these chemical agents, electrochemical sensors help predict and prevent equipment failures before they occur, enabling proactive maintenance [29, 30]. They also play a crucial role in detecting chemical changes in lubricants or cooling fluids, which can affect the performance and longevity of machinery components. This is particularly important in high-stress environments where chemical interactions can lead to accelerated wear [23].

Moreover, electrochemical sensors can complement other types of condition monitoring sensors, such as vibration and temperature sensors, by offering a comprehensive view of the machine's health [31, 32]. While vibration sensors detect mechanical wear and imbalances, and temperature sensors track thermal conditions, electrochemical sensors provide additional data on the chemical factors that might influence these parameters [20, 33]. This integrated approach ensures a more thorough understanding of the machinery's condition, allowing for more precise and effective maintenance strategies. By addressing chemical factors alongside physical and thermal conditions, electrochemical sensors enhance the overall reliability and efficiency of machine condition monitoring systems [32].

10.3.2 Sensors for Process Control

10.3.2.1 Flow and Level Sensors

In process control, electrochemical sensors play a crucial role in monitoring flow and level parameters, particularly in environments where chemical interactions influence the process. For flow monitoring, electrochemical sensors can be used to detect changes in the chemical composition of the fluid, which may indicate variations in flow rates or potential blockages within pipelines [34, 35]. For instance, sensors that measure changes in electrical conductivity can provide real-time data on the concentration of dissolved substances in fluids, indirectly inferring flow characteristics [36]. In transformation processes, such as chemical manufacturing or refining, these sensors help ensure that the chemical composition remains consistent, which is essential for maintaining product quality and process efficiency [34]. Additionally, electrochemical sensors can be used to monitor levels in tanks or reactors by detecting the presence of specific chemicals at various concentrations, ensuring that the levels are within the required operational ranges to prevent overflows or underfills [37].

10.3.2.2 Electrochemical Sensors for Pressure Monitoring

In chemical reactors and transport systems, electrochemical sensors provide a nuanced approach to pressure monitoring by focusing on the chemical environment that can influence pressure conditions [6, 38]. These sensors are integral for maintaining process stability and ensuring safety in environments where chemical reactions or interactions significantly affect pressure dynamics. Unlike traditional pressure sensors that measure physical pressure directly, electrochemical sensors can detect changes in the chemical composition of the reactor or pipeline contents, which can be indicative of pressure fluctuations [39–41].

In chemical reactors, electrochemical sensors are used to monitor the concentration of reactants and products, which can influence gas production and pressure changes. For example, in a reaction where gases are evolved, the electrochemical sensor detects changes in the concentration of specific reactants or byproducts [38, 42]. This information can be used to infer pressure changes and adjust operational conditions accordingly, preventing overpressure situations that could lead to equipment failure or safety hazards. By continuously monitoring these chemical indicators, operators can ensure that the reactor maintains optimal pressure levels, enhancing both process efficiency and safety [43, 44].

Similarly, in transport systems, such as pipelines that carry reactive or corrosive fluids, electrochemical sensors play a crucial role in assessing the chemical composition of the transported material. These sensors can detect changes in the chemical makeup, which might indicate issues like leaks, chemical reactions, or contamination [45]. Such insights help in maintaining proper pressure levels within the pipeline, ensuring that the system operates within safe limits and preventing potential pressure-induced failures [6, 46]. Overall, electrochemical sensors provide valuable data that complements traditional pressure measurement techniques, offering a comprehensive view of the factors influencing pressure in chemical reactors and transport systems [26].

10.3.2.3 Electrochemical Sensors for Gas and Particle Monitoring

Electrochemical sensors are vital for monitoring gas emissions and particulate matter in industrial processes, offering precise and real-time data essential for environmental management and regulatory compliance [47]. These sensors operate by detecting the presence and concentration of specific gases or particles through electrochemical reactions at their electrodes. In industrial settings, electrochemical sensors are commonly used to monitor emissions of gases such as CO_2, SO_2, nitrogen oxides (NOx), and ozone (O_3) [48]. For instance, electrochemical sensors can measure the electrical current generated when these gases interact with the sensor's electrode materials, providing accurate concentration readings. This capability is crucial for managing and reducing harmful emissions, ensuring that industrial operations comply with environmental regulations and standards [48, 49].

Additionally, electrochemical sensors play a significant role in air quality control by monitoring particulate matter, which includes pollutants like dust, smoke, and fumes. Although traditional particulate matter sensors often use optical or gravimetric methods, electrochemical sensors can be adapted to detect chemical changes associated with particle accumulation. For example, these sensors can identify changes in the chemical composition of particles or detect the presence of reactive species in the air, which may indicate high levels of pollutants [50]. By integrating electrochemical sensors into air quality management systems, industries can achieve a more comprehensive understanding of emission sources and their impact on the environment. This real-time monitoring enables proactive measures to reduce pollution, improve air quality, and enhance overall environmental sustainability [51].

10.3.3 Sensors for Safety and Environmental Protection

Electrochemical sensors are pivotal in ensuring safety and environmental protection by detecting hazardous gases and fine particles, as well as aiding in gas leak prevention and fire detection. These sensors are designed to identify and quantify toxic gases such as carbon monoxide (CO), NOx, and sulfur oxides (SOx) through electrochemical reactions at their electrodes [52]. For instance, CO sensors detect the presence of this colorless and odorless gas by measuring the electrical current generated during its reaction with a sensing electrode. Similarly, NOx and SOx sensors provide real-time data on the concentration of nitrogen and sulfur oxides, which are critical for monitoring and controlling industrial emissions. The precise detection of these hazardous gases is essential for protecting workers and the environment from pollution and health risks [48, 53].

In addition to detecting hazardous gases, electrochemical sensors are integral to gas leak prevention and fire detection systems. These sensors continuously monitor the presence of combustible gases and vapors, triggering alarms or safety protocols when gas concentrations exceed safe levels [47, 54]. For example, sensors that detect methane (CH_4) or propane (C_3H_8) can alert operators to potential leaks in facilities where these gases are used or stored. Fire detection systems can also utilize electrochemical sensors to identify the early presence of smoke or combustion products, such as CO_2 and CO, which are indicative of a fire [54]. By integrating these sensors into safety systems, industries can implement timely interventions to prevent accidents, mitigate hazards, and ensure a safer working environment. Overall, electrochemical sensors enhance safety and environmental protection by providing reliable, real-time monitoring of dangerous conditions, enabling prompt and effective responses to potential threats [39].

10.4 EMERGING TECHNOLOGIES AND INNOVATIONS IN SENSORS

10.4.1 IoT Sensors and Smart Sensors

Emerging technologies in sensors, particularly the integration of sensors into the Internet of Things (IoT) and the development of smart sensors, are revolutionizing the way data is collected, analyzed, and utilized across various industries [55]. IoT sensors are designed to seamlessly connect to networked systems, enabling the real-time transmission of data from multiple sensors to centralized platforms. This integration allows for extensive monitoring and control across diverse applications, ranging from industrial processes to smart cities [56]. By leveraging IoT technology, sensors can provide continuous, remote monitoring of parameters such as temperature, pressure, and chemical composition, facilitating more responsive and adaptive systems [57]. For instance, in industrial settings, IoT-enabled sensors can collect data on equipment performance, environmental conditions, and process parameters, which is then analyzed to optimize operations, predict maintenance needs, and enhance overall efficiency [58].

Smart sensors represent another significant advancement, characterized by their ability to not only collect data but also perform real-time analysis and decision-making. These sensors are equipped with embedded processing capabilities, allowing them to interpret data locally and make immediate adjustments without relying solely on external systems [55]. For example, a smart temperature sensor in a manufacturing plant can detect deviations from set parameters and automatically adjust cooling systems to maintain optimal conditions. This capability enhances operational efficiency, reduces latency, and minimizes the need for extensive data processing infrastructure. Smart sensors also often incorporate advanced features such as self-calibration, diagnostics, and communication capabilities, further improving their reliability and functionality [56]. Together, IoT and smart sensors are driving the development of more intelligent, interconnected, and responsive systems, transforming industries by enabling more efficient operations, improved safety, and enhanced decision-making [59].

10.4.2 Emerging Technologies and Innovations in Sensors: Nanotechnology-Based Sensors

10.4.2.1 Applications of Nanosensors for Chemical and Biomedical Detection

Nanotechnology-based sensors, or nanosensors, represent a cutting-edge advancement in the field of detection, offering unparalleled capabilities for both chemical and biomedical applications [60, 61]. In chemical detection, nanosensors utilize the unique properties of nanomaterials, such as nanoparticles, nanowires, and nanotubes, to achieve high sensitivity and selectivity [62, 63]. These sensors can detect low concentrations of chemical substances, including environmental pollutants, toxic gases, and industrial contaminants, with exceptional accuracy. For instance, gold nanoparticles are employed in colorimetric assays to detect specific analytes through color changes, while carbon nanotubes and graphene-based sensors leverage their high electrical conductivity to identify trace gases. This sensitivity is particularly valuable in applications requiring early detection of hazardous materials or real-time monitoring of chemical processes [64, 65].

In the biomedical field, nanosensors have transformed diagnostics by enabling the detection of biomarkers at incredibly low concentrations, which is crucial for early disease detection and personalized medicine [66]. Nanosensors can be engineered to target specific biomolecules associated with diseases, such as cancer or cardiovascular conditions, through the use of quantum dots, fluorescent nanoparticles, or magnetic nanoparticles. For example, nanosensors incorporating antibodies or aptamers can selectively bind to specific proteins or DNA sequences, facilitating the diagnosis of diseases from small samples of blood or other bodily fluids [67]. This capability not only enhances diagnostic accuracy but also allows for non-invasive and real-time monitoring of health conditions, improving patient outcomes and treatment efficacy [68].

10.4.2.2 Advantages of Miniaturized and High-Sensitivity Sensors

The miniaturization and high sensitivity of nanosensors provide several significant advantages, enhancing their utility across various applications. Miniaturized sensors, due to their small size, offer increased surface-to-volume ratios, which enhances their interaction with target molecules and improves detection sensitivity [69]. This high sensitivity allows nanosensors to detect analytes at very low concentrations, making them ideal for applications requiring precise measurements, such as environmental monitoring and early disease detection [68, 69].

Additionally, the compact size of nanosensors enables their integration into diverse platforms, including portable devices and wearable technology, without sacrificing performance. This miniaturization contributes to the development of on-site diagnostic tools and real-time monitoring systems that are both practical and cost-effective [70]. The rapid response time of nanosensors, coupled with their high sensitivity, also means that they can provide immediate feedback, which is crucial for timely decision-making in both industrial and medical contexts. Furthermore, the ability to produce nanosensors at a lower cost due to their small size and the efficiency of nanomaterials leads to reduced overall expenses for detection applications [71]. Overall, the advancements in nanotechnology have significantly enhanced sensor capabilities, making nanosensors a powerful tool for precise and efficient detection across a wide range of fields [72].

10.4.3 Emerging Technologies and Innovations in Sensors: Optical and Fiber-Optic Sensors

10.4.3.1 Applications of Optical Sensors in Detecting Temperature, Pressure, and Gas Concentration

Optical sensors have become increasingly important in various applications due to their ability to provide precise measurements of temperature, pressure, and gas concentration. These sensors operate by detecting changes in light properties, such as intensity, wavelength, or polarization, in response to physical or chemical changes in their environment. For temperature measurement, optical sensors utilize techniques such as infrared thermometry or fiber-optic temperature probes, which detect changes in the emitted or transmitted light as temperature varies [73]. This method is highly accurate and allows for non-contact measurements, making it suitable for monitoring high-temperature processes or delicate materials [74].

Pressure sensors based on optical principles often use the interference of light waves or the modulation of light intensity to measure changes in pressure [75]. For instance, fiber Bragg grating (FBG) sensors embed a grating structure within an optical fiber that reflects specific wavelengths of light. Changes in pressure cause shifts in these wavelengths, providing accurate pressure readings. These optical pressure sensors are valued for their high sensitivity, resistance to electromagnetic interference, and ability to operate in harsh environments [76].

In the realm of gas concentration detection, optical sensors leverage techniques such as absorption spectroscopy, where light of specific wavelengths is absorbed by target gases. The amount of absorbed light correlates with the concentration of the gas, allowing for precise and real-time monitoring of gas levels [77]. Optical sensors are widely used in environmental monitoring and industrial processes to detect gases like CO_2, methane (CH_4), and nitrogen dioxide (NO_2). Their high sensitivity and selectivity make them ideal for applications requiring accurate and continuous gas analysis [78].

10.4.3.2 Fiber-Optic Sensors for Structural Health Monitoring and Fault Detection

Fiber-optic sensors have emerged as a leading technology for structural health monitoring and fault detection due to their ability to provide detailed, real-time data over long distances and in challenging environments. These sensors utilize the properties of optical fibers to detect changes in physical parameters such as strain, temperature, and pressure within structures [79]. FBG sensors, for example, use a periodic variation in the refractive index of the optical fiber to measure strain and temperature. When the structure experiences stress or temperature changes, the resulting strain causes shifts in the reflected light wavelength, providing valuable information about the structural integrity [80].

In structural health monitoring, fiber-optic sensors are employed to assess the condition of critical infrastructure such as bridges, dams, and buildings. They can detect early signs of damage, deformation, or stress, enabling timely maintenance and reducing the risk of catastrophic failures. The distributed sensing capabilities of optical fibers allow for continuous monitoring along the entire length of a structure, offering a comprehensive view of its health [81, 82].

For fault detection, fiber-optic sensors are used to identify and locate issues such as leaks, cracks, or abnormal vibrations. The high spatial resolution and sensitivity of these sensors enable precise detection of small changes in the structure, which can be critical for preventing further damage and ensuring safety [83]. Their resistance to electromagnetic interference and ability to operate in harsh conditions further enhance their applicability in various industrial and civil engineering projects. Overall, fiber-optic sensors represent a powerful tool for maintaining the safety and reliability of infrastructure and industrial systems through advanced monitoring and fault detection capabilities [84].

10.5 CURRENT CHALLENGES AND FUTURE PERSPECTIVES

10.5.1 Technical Challenges in Sensor Utilization

The deployment and utilization of sensors, particularly in advanced applications, come with several technical challenges that impact their performance and effectiveness. Key issues include calibration, sensitivity, and reliability, as well as managing interference and operating under extreme conditions.

10.5.1.1 Calibration, Sensitivity, and Reliability

Calibration is crucial for ensuring that sensors provide accurate and consistent measurements. However, maintaining calibration over time can be challenging due to factors such as drift, wear, and environmental changes. For instance, sensors may experience shifts in their baseline measurements due to aging components or exposure to harsh conditions, leading to decreased accuracy and necessitating frequent recalibration. This challenge is particularly pronounced in high-precision applications, where even minor deviations can have significant impacts [85].

Sensitivity is another critical factor, especially for sensors designed to detect low concentrations of substances or subtle changes in physical parameters. Achieving high sensitivity while maintaining stability and avoiding false positives or negatives requires advanced materials and technologies. For example, nanosensors and optical sensors need to balance their high sensitivity with robust performance to ensure reliable data under varying conditions [86].

Reliability encompasses the sensor's ability to perform consistently over time and under different operating conditions. Sensors used in industrial or environmental monitoring must withstand harsh conditions such as extreme temperatures, high humidity, and corrosive environments. Ensuring long-term reliability often involves incorporating protective measures and designing sensors with robust materials, but these solutions can sometimes add complexity or cost [87, 88].

10.5.1.2 Managing Interference and Extreme Conditions

Interference from external factors, such as electromagnetic fields, vibrations, or chemical contaminants, can significantly affect sensor performance. For instance, in environments with high electromagnetic interference, sensors may experience signal noise or reduced accuracy. Similarly, sensors used in chemical or industrial processes must be designed to handle potential interference from other substances that could alter their readings. Effective strategies to mitigate interference include shielding, advanced signal processing algorithms, and using sensor materials resistant to environmental contaminants [89, 90].

Operating under extreme conditions poses additional challenges for sensor functionality. Sensors deployed in extreme temperatures, high-pressure environments, or corrosive atmospheres need to be designed to withstand these conditions without degradation. This may involve selecting specialized materials, employing protective coatings, or utilizing calibration techniques that account for environmental variations. Additionally, sensors in such conditions may require more frequent maintenance or replacement to ensure continued accuracy and reliability [91].

10.5.2 Future Perspectives

Addressing these challenges requires ongoing research and development in sensor technology. Innovations in materials science, signal processing, and sensor design are essential for improving calibration stability, sensitivity, and reliability. Advances in self-calibration techniques, adaptive algorithms, and ruggedized sensor designs are likely to enhance sensor performance and broaden their applicability.

Future perspectives also include the integration of sensors with emerging technologies such as the IoT and artificial intelligence (AI). IoT integration can facilitate real-time data collection and analysis, while AI can enhance sensor calibration, error correction, and predictive maintenance. Furthermore, the development of new sensing materials and methods, such as nanotechnology and advanced optical techniques, promises to overcome current limitations and expand the capabilities of sensors across various applications.

Overall, overcoming the technical challenges of sensor utilization will pave the way for more accurate, reliable, and versatile sensing solutions, driving advancements in fields ranging from industrial monitoring to environmental protection and healthcare.

10.5.3 Prospects for Improvement and Development

10.5.3.1 Advances in Sensor Materials and Detection Technology

The continuous advancement in sensor materials and detection technology is crucial for enhancing the performance, accuracy, and versatility of sensors across various applications. Recent developments in sensor materials include the exploration of nanomaterials, advanced polymers, and novel composites. Nanomaterials, such as graphene, carbon nanotubes, and quantum dots, offer exceptional properties like high surface area, electrical conductivity, and unique optical characteristics, which contribute to improved sensitivity and selectivity in sensors. These materials enable the detection of lower concentrations of target analytes and provide faster response times, which are essential for applications ranging from environmental monitoring to biomedical diagnostics.

Additionally, advances in detection technologies, such as improved electrochemical, optical, and piezoelectric methods, are expanding the capabilities of sensors. For example, innovations in electrochemical sensors include the development of new electrode materials and sensing interfaces that enhance signal-to-noise ratios and reduce interference. In optical sensing, advancements in photonic technologies, such as fiber optics and laser-based techniques, are enabling more precise and remote measurements. Furthermore, emerging technologies like lab-on-a-chip systems and microelectromechanical systems (MEMS) are facilitating miniaturized, high-performance sensors that can be integrated into portable and wearable devices.

10.5.3.2 The Future of Wireless Sensors and Their Integration into Smart Industrial Networks

The future of wireless sensors holds significant promise for revolutionizing industrial monitoring and control through seamless integration into smart industrial networks. Wireless sensors offer the advantage of eliminating the need for cumbersome cabling, allowing for more flexible and scalable deployment in complex industrial environments. These sensors can transmit data in real time to centralized systems, enabling remote monitoring, diagnostics, and control of industrial processes.

The integration of wireless sensors into smart industrial networks is a key development that leverages technologies such as the IoT and edge computing. IoT enables sensors to connect and communicate over the internet, facilitating the collection and analysis of large volumes of data from diverse sources. This connectivity supports the creation of intelligent systems that can perform predictive maintenance, optimize operational efficiency, and enhance safety. Edge computing, which involves processing data closer to the sensor's location, reduces latency and bandwidth usage, allowing for faster decision-making and more responsive control.

Future advancements in wireless sensor technology will focus on improving power efficiency, data security, and network reliability. Innovations such as energy-harvesting sensors, which can generate power from environmental sources, and secure communication protocols will address current limitations and enhance the functionality of wireless sensor networks. Additionally, advancements in sensor fusion and data analytics will enable more comprehensive insights and actionable intelligence, further driving the evolution of smart industrial networks.

Overall, the prospects for improvement and development in sensor materials, detection technologies, and wireless integration are set to transform industries by providing more accurate, efficient, and interconnected sensing solutions. These advancements will enable more precise monitoring, better decision-making, and greater operational efficiency across various sectors, including manufacturing, energy, healthcare, and environmental management.

CONCLUSION

In this chapter, we explored the transformative role of sensors in energy and industrial processes, focusing on various sensor types including electrochemical, optical, and fiber-optic sensors. The applications of sensors in monitoring critical parameters such as temperature, pressure, and gas concentration were examined, along with the challenges related to calibration, sensitivity, and reliability. The impact of emerging technologies, such as nanotechnology-based sensors and advancements in wireless sensing, was highlighted, promising to enhance sensor performance and integration. The future importance of sensors lies in their ability to optimize processes, improve safety, and support sustainable practices. To fully harness their potential, continued innovation and the adoption of next-generation sensors are essential. Industries must embrace these advancements to achieve greater accuracy, efficiency, and operational excellence, positioning themselves at the forefront of technological progress and meeting future challenges effectively.

REFERENCES

[1] W. Buratto, R. Muniz, A. Nied, C. Barros, R. Cardoso, and G. Villarrubia, "A Review of automation and sensors: Parameter control of thermal treatments for electrical power generation," *Sensors*, vol. 24, p. 967, Feb. 2024, doi: 10.3390/s24030967.

[2] M. Javaid, A. Haleem, R. Singh, S. Rab, and R. Suman, "Significance of sensors for Industry 4.0: Roles, capabilities, and applications," *Sensors Int.*, vol. 2, p. 100110, Jun. 2021, doi: 10.1016/j.sintl.2021.100110.

[3] Y. Jiang, S. Yin, J. Dong, and O. Kaynak, "A review on soft sensors for monitoring, control and optimization of industrial processes," *IEEE Sens. J.*, p. 1, Oct. 2020, doi: 10.1109/JSEN.2020.3033153.

[4] R. Mukundan and F. Garzon, "Electrochemical sensors for energy and transportation," *Electrochem. Soc. Interf.*, vol. 13, Jun. 2004, doi: 10.1149/2.F05042IF.

[5] D. Scott, "Industrial process sensors," 2007. [Online]. Available: https://api.semanticscholar.org/CorpusID:114306474

[6] U. Guth, W. Vonau, and J. Zosel, "TOPICAL REVIEW: Recent developments in electrochemical sensor application and technology---a review," *Meas. Sci. Technol.*, vol. 20, Apr. 2009, doi: 10.1088/0957-0233/20/4/042002.

[7] A. Al-Habaibeh, R. Cai, and R. M. Parkin, "Modern developments in sensor technology and their applications in condition monitoring of manufacturing processes," 2004. [Online]. Available: https://api.semanticscholar.org/CorpusID:13433856

[8] V. C. Gungor and G. P. Hancke, "Industrial wireless sensor networks: Challenges, design principles, and technical approaches," *IEEE Trans. Ind. Electron.*, vol. 56, pp. 4258–4265, 2009, [Online]. Available: https://api.semanticscholar.org/CorpusID:6820964

[9] T. R. Fadel *et al.*, "Toward the responsible development and commercialization of sensor nanotechnologies," *ACS sensors*, vol. 1, no. 3, pp. 207–216, 2016, [Online]. Available: https://api.semanticscholar.org/CorpusID:41530463

[10] H. Hu, N. Wang, J. Liao, and F. J. Tovar-Lopez, "Recent progress in micro- and nanotechnology-enabled sensors for biomedical and environmental challenges," *Sensors (Basel).*, vol. 23, 2023, [Online]. Available: https://api.semanticscholar.org/CorpusID:259143671

[11] A. Zrelli, B. Mohamed, and T. Ezzedine, "Monitoring of temperature in distributed optical sensors: Raman and Brillouin spectrum," *Opt. – Int. J. Light Electron Opt.*, vol. 127, Jan. 2016, doi: 10.1016/j.ijleo.2016.01.073.

[12] J. Fergus, "ChemInform abstract: A review of electrolyte and electrode materials for high temperature electrochemical CO_2 and SO_2 gas sensors," *Sensors Actuators B Chem.*, vol. 134, pp. 1034–1041, Sep. 2008, doi: 10.1016/j.snb.2008.07.005.

[13] S. Halley, K. Ramaiyan, L. Tsui, and F. Garzon, "A review of Zirconia oxygen, NOx, and mixed potential gas sensors – history and current trends," *Sensors Actuators B Chem.*, vol. 370, p. 132363, Jul. 2022, doi: 10.1016/j.snb.2022.132363.

[14] S. Zhuiykov and N. Miura, "12 – Solid-state electrochemical gas sensors for emission control," in *Woodhead Publishing Series in Electronic and Optical Materials*, C. C. Sorrell, S. Sugihara, and J. B. T.-M. for E. C. D. Nowotny, Eds., Woodhead Publishing, 2005, pp. 303–335. https://doi.org/10.1533/9781845690915.2.303.

[15] U. Guth, P. Shuk, and C. McGuire, "Gas sensing technologies in combustion: A comprehensive review," *J. Electrochem. Sci. Eng.*, vol. 10, Feb. 2020, doi: 10.5599/jese.743.

[16] T. S. Dhahi *et al.*, "Advances in nano sensors for monitoring and optimal performance enhancement in photovoltaic cells," *iScience*, vol. 27, no. 4, p. 109347, 2024, https://doi.org/10.1016/j.isci.2024.109347.

[17] M. Hassan, R. Khan, and S. Andreescu, "Advances in electrochemical detection methods for measuring contaminants of emerging concerns," *Electrochem. Sci. Adv.*, vol. 2, Dec. 2021, doi: 10.1002/elsa.202100184.

[18] J. H. Pikul and J. W. Long, "Architected materials for advanced electrochemical systems," *MRS Bull.*, vol. 44, no. 10, pp. 789–795, 2019, doi: 10.1557/mrs.2019.229.

[19] J. Baranwal, B. Barse, G. Gatto, G. Broncová, and A. Kumar, "Electrochemical Sensor and their applications," *Chemosensors*, vol. 10, Sep. 2022, doi: 10.3390/chemosensors10090363.

[20] K. Nemčeková and J. Labuda, "Advanced materials-integrated electrochemical sensors as promising medical diagnostics tools: A review," *Mater. Sci. Eng. C*, vol. 120, p. 111751, 2021, https://doi.org/10.1016/j.msec.2020.111751.

[21] Y. Bae *et al.*, "Sensor impacts on building and HVAC controls: A critical review for building energy performance," *Adv. Appl. Energy*, 2021, [Online]. Available: https://api.semanticscholar.org/CorpusID:239679875

[22] S. Sivakumar, J. Logeshwaran, R. Kannadasan, M. Faheem, and D. Ravikumar, "A novel energy optimization framework to enhance the performance of sensor nodes in Industry 4.0," *Energy Sci. Eng.*, vol. 12, pp. 835–859, 2024, [Online]. Available: https://api.semanticscholar.org/CorpusID:266761130

[23] K. S. S. Raj, "Smart sensor technologies for performance optimization of power generating assets," *ASME 2019 Power Conf.*, 2019, [Online]. Available: https://api.semanticscholar.org/CorpusID:214025064

[24] R. L. Rausch, "Integration of smart sensors into utility operation and business practices," *2014 IEEE Rural Electr. Power Conf.*, pp. B5-1–B5-4, 2014, [Online]. Available: https://api.semanticscholar.org/CorpusID:223869

[25] "Online chemical monitoring," *Cond. Monit. Rotating Electr. Mach.*, 2020, [Online]. Available: https://api.semanticscholar.org/CorpusID:242444952

[26] J. Baranwal, B. Barse, G. Gatto, G. Broncová, and A. Kumar, "Electrochemical sensors and their applications: A review," *Chemosensors*, 2022, [Online]. Available: https://api.semanticscholar.org/CorpusID:252233758

[27] M. B. Tabacco, Q. Zhou, and B. N. Nelson, "Chemical sensors for environmental monitoring," in *Other Conferences*, 1992. [Online]. Available: https://api.semanticscholar.org/CorpusID:137131425

[28] A. Kroll, V. I. Smorchkov, and A. Y. Nazarenko, "Electrochemical sensors for hydrogen and hydrogen sulfide determination," *Sensors Actuators B Chem.*, vol. 21, pp. 97–100, 1994, [Online]. Available: https://api.semanticscholar.org/CorpusID:98669284

[29] S. Namuduri, B. N. Narayanan, V. S. P. Davuluru, L. Burton, and S. Bhansali, "Review—Deep learning methods for sensor based predictive maintenance and future perspectives for electrochemical sensors," *J. Electrochem. Soc.*, vol. 167, p. 37552, 2020, [Online]. Available: https://api.semanticscholar.org/CorpusID:214057166

[30] M. Albano *et al.*, "Sensors: The enablers for proactive maintenance in the real world," *2018 5th Int. Conf. Control. Decis. Inf. Technol.*, pp. 569–574, 2018, [Online]. Available: https://api.semanticscholar.org/CorpusID:49420322

[31] M. Taheri, I. A. Deen, M. Packirisamy, and M. J. Deen, "Metal oxide-based electrical/electrochemical sensors for health monitoring systems," *TrAC Trends Anal. Chem.*, 2024, [Online]. Available: https://api.semanticscholar.org/CorpusID:266876807

[32] R. Bogue, "Sensors for condition monitoring: A review of technologies and applications," *Sens. Rev.*, vol. 33, pp. 295–299, 2013, [Online]. Available: https://api.semanticscholar.org/CorpusID:111321915

[33] A. J. Bandodkar and J. Wang, "Non-invasive wearable electrochemical sensors: A review," *Trends Biotechnol.*, vol. 32 7, pp. 363–371, 2014, [Online]. Available: https://api.semanticscholar.org/CorpusID:205390728

[34] C. M. A. Brett, "Novel sensor devices and monitoring strategies for green and sustainable chemistry processes," *Pure Appl. Chem.*, vol. 79, pp. 1969–1980, 2007, [Online]. Available: https://api.semanticscholar.org/CorpusID:45387737

[35] M. M. Alam, V. Mitea, M. M. R. Howlader, P. R. Selvaganapathy, and M. J. Deen, "Analyzing electrochemical sensing fundamentals for health applications," *Adv. Sens. Res.*, vol. 3, 2023, [Online]. Available: https://api.semanticscholar.org/CorpusID:266299595

[36] P. R. Hauptmann, N. Hoppe, and A. Püttmer, "REVIEW ARTICLE: Application of ultrasonic sensors in the process industry," *Meas. Sci. Technol.*, 2002, [Online]. Available: https://api.semanticscholar.org/CorpusID:117965435

[37] P. Biechele, C. Busse, D. Solle, T. Scheper, and K. F. Reardon, "Sensor systems for bioprocess monitoring," *Eng. Life Sci.*, vol. 15, 2015, [Online]. Available: https://api.semanticscholar.org/CorpusID:110969033

[38] M. Doering *et al.*, "In situ performance monitoring of electrochemical oxygen and hydrogen peroxide sensors in an additively manufactured modular microreactor," *ACS Omega*, vol. 9, pp. 19700–19711, 2024, [Online]. Available: https://api.semanticscholar.org/CorpusID:269191986

[39] P. K. Sekhar, E. L. Brosha, R. Mukundan, and F. H. Garzon, "Chemical sensors for environmental monitoring and homeland security," *Electrochem. Soc. Interface*, vol. 19, pp. 35–40, 2010, [Online]. Available: https://api.semanticscholar.org/CorpusID:12253001

[40] P. Ananda, J. G. Manjunatha, H. Nagarajappa, E. D'Souza, C. Manjunath, and P. Nambudumada, "Voltammetric analysis of antihistamine drug cetirizine and paracetamol at poly (L-Leucine) layered carbon nanotube paste electrode," p. 101154, Jun. 2021, doi: 10.1016/j.surfin.2021.101154.

[41] G. Tigari and J. G. Manjunatha, "Poly(glutamine) film-coated carbon nanotube paste electrode for the determination of curcumin with vanillin: An electroanalytical approach," *Monatshefte Chem.*, vol. 151, no. 11, pp. 1681–1688, 2020, doi: 10.1007/s00706-020-02700-8.

[42] N. Hareesha and J. G. Manjunatha, "Electro-oxidation of formoterol fumarate on the surface of novel poly(thiazole yellow-G) layered multi-walled carbon nanotube paste electrode," *Sci. Rep.*, vol. 11, no. 1, p. 12797, 2021, doi: 10.1038/s41598-021-92099-x.

[43] P. Fabry and E. Siebert, "Electrochemical sensors," *CRC Handb. SOLID STATE Electrochem.*, 2019, [Online]. Available: https://api.semanticscholar.org/CorpusID:243681456

[44] H. Zhang *et al.*, "A review of sensor applications towards precise control of pyrolysis of solid waste and biomasses," *Renew. Sustain. Energy Rev.*, 2022, [Online]. Available: https://api.semanticscholar.org/CorpusID:252255218

[45] J. M. D. Pereira, J. Monge, and O. Postolache, "Measurement and applications: Electrochemical sensors and instruments: Main characteristics and applications," *IEEE Instrum. Meas. Mag.*, vol. 27, pp. 18–25, 2024, [Online]. Available: https://api.semanticscholar.org/CorpusID:267524970

[46] D. Giovanelli, N. S. Lawrence, and R. G. Compton, "Electrochemistry at high pressures: A review," *Electroanalysis*, vol. 16, pp. 789–810, 2004, [Online]. Available: https://api.semanticscholar.org/CorpusID:94720370

[47] R. Hanafi, R. D. Mayasari, Masmui, Agustanhakri B., J. Raharjo, and R. Nuryadi, "Electrochemical sensor for environmental monitoring system: A review," *8TH Natl. Phys. Semin. 2019*, 2019, [Online]. Available: https://api.semanticscholar.org/CorpusID:210254488

[48] K. P. Ramaiyan and R. M. (Mukund), "Electrochemical sensors for air quality monitoring," *Electrochem. Soc. Interface*, 2019, [Online]. Available: https://api.semanticscholar.org/CorpusID:210628941

[49] I. Cretescu, D. Lutic, and L. R. Manea, "Electrochemical sensors for monitoring of indoor and outdoor air pollution," 2017. [Online]. Available: https://api.semanticscholar.org/CorpusID:56032464

[50] G. Hanrahan, D. G. Patil, and J. Wang, "Electrochemical sensors for environmental monitoring: Design, development and applications," *J. Environ. Monit.*, vol. 6 8, pp. 657–664, 2004, [Online]. Available: https://api.semanticscholar.org/CorpusID:10594963

[51] Q. He *et al.*, "Research on the construction of portable electrochemical sensors for environmental compounds quality monitoring," *Mater. Today Adv.*, 2023, [Online]. Available: https://api.semanticscholar.org/CorpusID:255223386

[52] M. A. H. Khan, M. V Rao, and Q. Li, "Recent advances in electrochemical sensors for detecting toxic gases: NO_2, SO_2 and H_2S," *Sensors (Basel).*, vol. 19, 2019, [Online]. Available: https://api.semanticscholar.org/CorpusID:73496001

[53] M. A. Khan, "Recent trends in electrochemical detection of NH_3, H_2S and NO_x Gases," *Int. J. Electrochem. Sci.*, pp. 1711–1733, 2017, [Online]. Available: https://api.semanticscholar.org/CorpusID:102471074

[54] J. Fonollosa, A. Solórzano, and S. Marco, "Chemical sensor systems and associated algorithms for fire detection: A review," *Sensors (Basel).*, vol. 18, 2018, [Online]. Available: https://api.semanticscholar.org/CorpusID:3952064

[55] K. Kirkopulo and V. Tonkonogyi, "Analysis of the application of IoT and intelligent sensors for monitoring and control of technological processes," *Electr. Comput. Syst.*, 2024, [Online]. Available: https://api.semanticscholar.org/CorpusID:270033191

[56] A. Doke and A. Awate, "Smart sensors and their applications in IoT," *Int. J. Res. Appl. Sci. Eng. Technol.*, 2022, [Online]. Available: https://api.semanticscholar.org/CorpusID:247949004

[57] D. Sehrawat and N. S. Gill, "Smart sensors: Analysis of different types of IoT sensors," *2019 3rd Int. Conf. Trends Electron. Informatics*, pp. 523–528, 2019, [Online]. Available: https://api.semanticscholar.org/CorpusID:204230890

[58] R. Krishnamurthi, A. Kumar, D. Gopinathan, A. Nayyar, and B. Qureshi, "An overview of IoT sensor data processing, fusion, and analysis techniques," *Sensors (Basel).*, vol. 20, 2020, [Online]. Available: https://api.semanticscholar.org/CorpusID:226045236

[59] S. C. Mukhopadhyay, S. K. S. Tyagi, N. K. Suryadevara, V. Piuri, F. Scotti, and S. Zeadally, "Artificial intelligence-based sensors for next generation IoT applications: A review," *IEEE Sens. J.*, vol. 21, pp. 24920–24932, 2021, [Online]. Available: https://api.semanticscholar.org/CorpusID:234094773

[60] K. C. Honeychurch, "Nanosensors for chemical and biological applications: Sensing with nanotubes, nanowires and nanoparticles," 2014. [Online]. Available: https://api.semanticscholar.org/CorpusID:136371435

[61] A. B. Monnappa, J. G. Manjunatha, and A. S. Bhatt, "Design of a sensitive and selective voltammetric sensor based on a cationic surfactant-modified carbon paste electrode for the determination of alloxan," *ACS Omega*, vol. 5, no. 36, pp. 23481–23490, 2020, doi: 10.1021/acsomega.0c03517.

[62] R. Abdel-Karim, Y. Reda, and A. Abdel-Fattah, "Review—Nanostructured materials-based nanosensors," *J. Electrochem. Soc.*, 2020, [Online]. Available: https://api.semanticscholar.org/CorpusID:213541510

[63] Chenthattil Raril, J. G. Manjunatha, D. K. Ravishankar, S. Fattepur, G. Siddaraju, and L. Nanjundaswamy, "Validated electrochemical method for simultaneous resolution of tyrosine, uric acid, and ascorbic acid at polymer modified nano-composite paste electrode," *Surf. Eng. Appl. Electrochem.*, vol. 56, no. 4, pp. 415–426, 2020, doi: 10.3103/S1068375520040134.

[64] M. K. Patra, K. Manzoor, M. Manoth, S. Negi, S. R. Vadera, and N. H. Kumar, "Nanotechnology applications for chemical and biological sensors," *Def. Sci. J.*, vol. 58, pp. 636–649, 2008, [Online]. Available: https://api.semanticscholar.org/CorpusID:54497591

[65] W. Hu, S. Chen, J. Yang, Z. Li, and H. Wang, "Functionalized bacterial cellulose derivatives and nanocomposites," *Carbohydr. Polym.*, vol. 101, pp. 1043–1060, 2014, https://doi.org/10.1016/j.carbpol.2013.09.102.

[66] N. M. Noah and P. Ndangili, "Nanobiosensors and their applications in medical diagnosis and imaging," *Med. Imaging Methods*, 2021, [Online]. Available: https://api.semanticscholar.org/CorpusID:245048674

[67] Y. Saylan, S. Akgönüllü, and A. Denizli, "Nanosensors for medical diagnosis," *Nanotechnol. Hematol. Blood Transfusion, Artif. Blood*, 2022, [Online]. Available: https://api.semanticscholar.org/CorpusID:244182056

[68] S. A. Perdomo, J. M. Marmolejo-Tejada, and A. Jaramillo-Botero, "Review—Bio-nanosensors: Fundamentals and recent applications," *J. Electrochem. Soc.*, 2021, [Online]. Available: https://api.semanticscholar.org/CorpusID:244243938

[69] M. Swierczewska, G. Liu, S. Lee, and X. Chen, "High-sensitivity nanosensors for biomarker detection," *Chem. Soc. Rev.*, vol. 41, no. 7, pp. 2641–2655, 2012, [Online]. Available: https://api.semanticscholar.org/CorpusID:8521840

[70] "Highlights of the sensors nanotechnology signature initiative," 2022. [Online]. Available: https://api.semanticscholar.org/CorpusID:250242117

[71] A. B. Dahlin, "Size matters: Problems and advantages associated with highly miniaturized sensors," *Sensors (Basel).*, vol. 12, pp. 3018–3036, 2012, [Online]. Available: https://api.semanticscholar.org/CorpusID:1738335

[72] L. Soleymani and F. Li, "Mechanistic challenges and advantages of biosensor miniaturization into the nanoscale," *ACS sensors*, vol. 2, no. 4, pp. 458–467, 2017, [Online]. Available: https://api.semanticscholar.org/CorpusID:206763463

[73] G. W. Tregay, P. R. Calabrese, P. L. Kaplin, and M. J. Finney, "Optical fiber sensor for temperature measurement from 600 to 1900 C in gas turbine engines," in *Other Conferences*, 1991. [Online]. Available: https://api.semanticscholar.org/CorpusID:111044009

[74] S.-K. Liaw, "Introductory chapter: An overview the methodologies and applications of fiber optic sensing," *Fiber Opt. Sens. – Princ. Meas. Appl.*, 2019, [Online]. Available: https://api.semanticscholar.org/CorpusID:203129211

[75] S. Xu, X. Li, T. Wang, X. Wang, and H. Liu, "Fiber Bragg grating pressure sensors: A review," *Opt. Eng.*, vol. 62, p. 10902, 2023, [Online]. Available: https://api.semanticscholar.org/CorpusID:256297485

[76] F. Urban, J. Kadlec, R. Vlach, and R. Kuchta, "Design of a pressure sensor based on optical fiber Bragg grating lateral deformation," *Sensors (Basel).*, vol. 10, pp. 11212–11225, 2010, [Online]. Available: https://api.semanticscholar.org/CorpusID:10763209

[77] S. Ying, K. Shao, W. Jing, H. Feng, and G. Yuze, "Optical gas detection: Key technologies and applications review," *Opto-electronic Eng.*, vol. 47, p. 190280, 2020, [Online]. Available: https://api.semanticscholar.org/CorpusID:218805126

[78] J. Hodgkinson and R. P. Tatam, "Optical gas sensing: A review," *Meas. Sci. Technol.*, vol. 24, 2012, [Online]. Available: https://api.semanticscholar.org/CorpusID:122312714

[79] P. Skels, R. Janeliukstis, and V. Haritonovs, "Review on structural health interrogation using fiber bragg grating sensors," 2018. [Online]. Available: https://api.semanticscholar.org/CorpusID:54573278

[80] J. K. Sahota, N. Gupta, and D. Dhawan, "Fiber Bragg grating sensors for monitoring of physical parameters: A comprehensive review," *Opt. Eng.*, vol. 59, p. 60901, 2020, [Online]. Available: https://api.semanticscholar.org/CorpusID:220974669

[81] D. Inaudi, "Overview of fiber optic sensing technologies for structural health monitoring," 2017. [Online]. Available: https://api.semanticscholar.org/CorpusID:139269829

[82] X. Ye, Y. H. Su, and J. P. Han, "Structural health monitoring of civil infrastructure using optical fiber sensing technology: A comprehensive review," *Sci. World J.*, vol. 2014, 2014, [Online]. Available: https://api.semanticscholar.org/CorpusID:14825745

[83] W. Chen, Y. Zhu, Y. Fu, J. Feng, and S. Huang, "Fiber optic sensors for safety control of civil structures," in *Optics and Optoelectronic Inspection and Control*, 2000. [Online]. Available: https://api.semanticscholar.org/CorpusID:108579299

[84] T. Wu, G. Liu, S. Fu, and F. Xing, "Recent progress of fiber-optic sensors for the structural health monitoring of civil infrastructure," *Sensors (Basel).*, vol. 20, 2020, [Online]. Available: https://api.semanticscholar.org/CorpusID:221164221

[85] F. Delaine, B. Lebental, and H. Rivano, "In situ calibration algorithms for environmental sensor networks: A review," *IEEE Sens. J.*, vol. 19, pp. 5968–5978, 2019, [Online]. Available: https://api.semanticscholar.org/CorpusID:195832008

[86] Y. Kim, J. Gonzales, and Y. Zheng, "Sensitivity-enhancing strategies in optical biosensing," *Small*, vol. 17, no. 4 (2021): p. e2004988, 2020, [Online]. Available: https://api.semanticscholar.org/CorpusID:229692273

[87] U. Sennhauser, R. D. Broennimann, P. Mauron, and P. M. Nellen, "Reliability and durability of fiber grating sensors in structural monitoring applications," in *Other Conferences*, 1997. [Online]. Available: https://api.semanticscholar.org/CorpusID:137190504

[88] C. Jacq, T. Maeder, and P. Ryser, "Sensors and packages based on LTCC and thick-film technology for severe conditions," *Sadhana*, vol. 34, pp. 677–687, 2009, [Online]. Available: https://api.semanticscholar.org/CorpusID:110182694

[89] A. Pahl, K. Rathjen, and S. Dickmann, "Prevention of sensor disturbances caused by IEMI," *2023 Int. Symp. Electromagn. Compat. – EMC Eur.*, pp. 1–4, 2023, [Online]. Available: https://api.semanticscholar.org/CorpusID:263836634

[90] A. Khan, T. König, F. Liebgott, and T. Greiner, "External magnetic interference classification in magnetostrictive position sensors using neuro-symbolic AI with log-likelihood ratios," *2023 IEEE 21st Int. Conf. Ind. Informatics*, pp. 1–6, 2023, [Online]. Available: https://api.semanticscholar.org/CorpusID:261107290

[91] P. G. Mikhailov, Z. Ualiyev, and A. Kabdoldina, "Issues of development of capacitive pressure sensors operable under extreme operating conditions," *2020 Int. Russ. Autom. Conf.*, pp. 1112–1116, 2020, [Online]. Available: https://api.semanticscholar.org/CorpusID:222097625

11 Nutrient Detection by Carbon Composite Sensors

Khalid Ait Ben Brahim, Mohamed Bendany, Youssra El Hamdouni, Najoua Labjar, Hamid Nasrellah, Meryem Bensemlali, Halima Mortadi, and Souad El Hajjaji

11.1 INTRODUCTION

The detection and quantification of nutrients have become increasingly critical in different domains, such as agriculture, environmental monitoring, food science, and healthcare. As global challenges such as food security, water quality, and sustainable agriculture continue to grow, the need for efficient, accurate, and cost-effective nutrient sensing technologies has become paramount [1]. Recently, carbon composite sensors have gained attention as an effective method for detecting nutrients, offering a unique combination of sensitivity, versatility, and potential for widespread application [2]. Nutrients, defined as substances that provide nourishment essential for growth and life maintenance, play crucial roles in both environmental and biological systems.

The ability to accurately detect and measure these nutrients is fundamental to understanding and managing these systems effectively. Classic approaches to nutrient detection, such as spectrophotometry and chromatography, while precise, often require complex sample preparation, expensive equipment, and specialized expertise [3]. These limitations have driven the search for alternative sensing technologies that can provide rapid, on-site, and cost-effective nutrient analysis.

Carbon composite sensors, which integrate carbon-based materials with various matrices, have received significant recognition due to their superior electrical, mechanical, and chemical properties [4]. These sensors leverage the high surface area, excellent conductivity, and chemical stability of carbon materials, combined with the structural advantages of composite systems, to create highly sensitive and selective nutrient detection platforms [5].

The fundamental principles underlying carbon composite sensors are rooted in materials science and electrochemistry. The structure and composition of these sensors can be tailored to target specific nutrients, incorporating diverse carbon structures such as graphene and carbon nanotubes (CNTs), and activated carbon [6]. The working mechanisms of these sensors often involve electrochemical reactions, where the presence of target nutrients induces measurable changes in electrical properties [7].

Nutrient detection using carbon composite sensors encompasses a wide range of mechanisms, primarily focusing on electrochemical and optical methods. Electrochemical detection techniques, including amperometry, voltammetry, and impedance spectroscopy, have shown particular promise due to their high sensitivity and potential for miniaturization [8]. Optical detection methods, leveraging phenomena such as fluorescence quenching and surface-enhanced Raman spectroscopy, offer complementary approaches for nutrient sensing [9].

The applications of carbon composite sensors in nutrient detection span multiple domains. In agriculture, these sensors have been employed for soil nutrient monitoring, enabling precision farming practices and optimizing fertilizer use [10]. Water quality assessment has benefited from

DOI: 10.1201/9781003500322-11

carbon composite sensors capable of detecting nitrates, phosphates, and other nutrient pollutants in aquatic environments [11]. In the gastronomy and beverage industry, these sensors have found applications in quality control and safety monitoring [12]. Additionally, the biomedical field has seen the development of carbon composite sensors for detecting nutrients and metabolites in biological fluids, contributing to advances in personalized medicine [13].

11.1.1 Importance of Nutrient Detection

The detection and quantification of nutrients play a crucial role across various domains, underscoring its importance in addressing global challenges. In agriculture, precise nutrient management is essential for optimizing crop yields and quality, directly impacting food security and sustainable farming practices [14]. Environmental scientists rely on nutrient detection to monitor and mitigate the effects of excess nutrients on ecosystems, particularly in aquatic environments where nutrient runoff can lead to eutrophication and ecological imbalance [15]. In the realm of human health, nutrient detection is vital for assessing food quality, ensuring nutritional adequacy, and aiding in medical diagnostics [16]. Industries such as food production and wastewater treatment depend on accurate nutrient measurements for process optimization and regulatory compliance [17]. Furthermore, nutrient detection contributes significantly to scientific research, particularly in climate change studies where understanding nutrient cycles is crucial for developing accurate predictive models [18]. The economic implications of precise nutrient management are substantial, offering potential cost reductions in agriculture and industry while promoting environmental sustainability [17]. As technology advances, particularly in the field of sensor development, the prospects for more efficient, real-time nutrient monitoring continue to expand, promising to revolutionize our approach to nutrient management across multiple sectors [19].

11.1.2 Overview of Carbon Composite Sensors

Carbon composite sensors, incorporating materials like graphene and CNTs, offer exceptional sensitivity, selectivity, and stability for various sensing applications. These sensors leverage the uncommon properties of carbon nanomaterials, including high surface area, excellent conductivity, and chemical stability [20]. They can be tailored for specific uses through careful selection of carbon materials and matrix composition, enabling applications in environmental monitoring, food safety, and biomedical diagnostics [21]. The integration of polymers with carbon nanomaterials enhances dispersion, boosts features like redox behavior and biocompatibility, while providing added functionalities, including photoelectric or swelling capabilities [21]. Recent advancements in nanotechnology have expanded the capabilities of these sensors, allowing for the development of miniaturized, highly sensitive devices for real-time detection of various analytes, including environmental pollutants, pathogens, and biomolecules [21].

11.2 FUNDAMENTALS OF CARBON COMPOSITE SENSORS

11.2.1 Structure and Composition

Carbon composite sensors represent a cutting-edge approach in the field of chemical and biological sensing, combining the advantageous properties of carbon-based materials with the flexibility and robustness of composite structures. The primary composition of these sensors typically involves carbon-based nanomaterials like CNTs and graphene, or carbon black, which are dispersed within a polymer matrix or other binding agents. This combination allows for the creation of a conductive network that is responsive to various stimuli, including the presence of specific nutrients or chemical compounds [22].

Carbon nanomaterials, particularly CNTs and graphene, have emerged as exceptional components for chemical sensors due to their unique electrical, chemical, and mechanical properties [23, 24]. When combined with polymers, these nanocarbon/polymer composites offer tunable electrical, mechanical, and chemisensitive properties, making them ideal for chemiresistive sensing applications [25]. The incorporation of polymers not only improves dispersion of CNTs and graphene but also enhances their redox behavior, biocompatibility, and provides additional functionalities [23]. These nanocomposites have been utilized in various sensing applications, including gas sensors, strain sensors, and biosensors [26]. The performance of carbon nanomaterial-based sensors is evaluated based on factors such as sensitivity, selectivity, detection limit, and reproducibility [26]. Cutting-edge advancements in the synthesis and design of nanomaterials have led to the emergence of high-performance sensing systems with exceptional analytical performance, including better sensitivity, selectivity, and lower detection limits [27].

11.2.2 Working Principles

The working principles of carbon composite sensors are grounded in the unique electrical, mechanical, and chemical properties of carbon-based nanomaterials, such as CNTs, graphene, and carbon black, which are embedded within a composite matrix. These sensors primarily operate through changes in electrical conductivity that occur when the sensor material interacts with specific target analytes, including various nutrients. One of the most common mechanisms is the *chemiresistive principle*, where the sensor's electrical resistance changes in response to the adsorption of analyte molecules on the surface of the carbon nanomaterials. For example, when a nutrient like nitrate interacts with the sensor, it causes a charge transfer between the nutrient molecules and the carbon material, leading to a measurable change in resistance [27].

Another important mechanism is *electrochemical sensing*, where the sensor detects analytes through electrochemical reactions at the surface of the carbon composite. In this case, carbon nanomaterials, such as graphene, act as the electrode material, facilitating electron transfer reactions with the analyte. For instance, in the detection of phosphorus compounds, the electrochemical oxidation or reduction of these compounds at the sensor surface results in a change in current, which can be measured and correlated with the concentration of the target nutrient [27, 28]. *Capacitive sensing* is also employed in carbon composite sensors, particularly in non-invasive or label-free detection applications. In capacitive sensors, the interaction of the target analyte with the sensor surface alters the dielectric properties of the composite material, leading to a variation in capacitance that can be measured to evaluate the presence and concentration of specific nutrients [29]. Additionally, *optical sensing mechanisms* can be integrated with carbon composite sensors, in which the presence of an analyte triggers modifications in the optical properties of the sensor, such as absorption, fluorescence, or refractive index. Carbon nanomaterials can be functionalized with specific optical probes that interact with target nutrients, leading to detectable optical signals [30].

11.2.3 Types of Carbon Composite Sensors

Carbon composite sensors are versatile and may be sorted into various types based on the specific carbon nanomaterials used, the sensing mechanisms employed, and their application domains. The most prominent types include *chemiresistive sensors*, *electrochemical sensors*, *capacitive sensors*, and *optical sensors*, each leveraging the unique properties of carbon-based composites for detecting various analytes, including nutrients.

Chemiresistive sensors are among the most widely used types of carbon composite sensors. These sensors rely on changes in electrical resistance when carbon nanomaterials, such as CNTs or graphene, interact with target molecules. The electrical properties of these nanomaterials, particularly their high conductivity and large surface area, make them highly sensitive to changes in the chemical environment. For instance, CNT-based chemiresistive sensors have been successfully employed for detecting gases like ammonia and nitrogen dioxide, as well as for nutrient detection in agricultural and environmental applications [18]. The interaction between the target molecules and the carbon composite leads to a change in resistance, which can be precisely measured to determine the concentration of the analyte.

Electrochemical sensors are another critical type of carbon composite sensor, where the detection of analytes occurs through electrochemical reactions at the sensor's surface. These sensors utilize carbon nanomaterials such as graphene and CNTs as electrode materials, which provide a high surface area for electron transfer and boost the sensor's sensitivity. For example, graphene-based electrochemical sensors have been developed for the detection of heavy metals, glucose, medicament, and various nutrients [31], showing high sensitivity and selectivity due to the excellent electron transfer properties of graphene. These sensors are particularly useful in applications where low detection limits and rapid response times are required, such as in biomedical diagnostics and environmental monitoring [32].

Capacitive sensors leverage the dielectric properties of carbon composites to detect changes in capacitance caused by the interaction of target molecules with the sensor surface. These sensors are typically used in applications where non-invasive or label-free detection is preferred. The capacitive response is influenced by the specific carbon nanomaterials used, such as carbon black or CNTs, and the matrix in which they are embedded. For instance, capacitive sensors using carbon black-polymer composites have been employed for the analysis of volatile organic compounds and additional analytes, offering a simple and effective sensing mechanism that can be applied in various fields, including air quality monitoring and food safety [33].

Optical sensors represent a more specialized type of carbon composite sensor, when changes occur in the optical properties of the sensor, such as fluorescence, absorption, or refractive index, are used to detect target analytes. Carbon nanomaterials like graphene oxide and CNTs can be functionalized with specific molecules that change their optical behavior in the presence of particular analytes. For example, graphene oxide-based optical sensors have been developed for detecting heavy metals and organic pollutants, offering high sensitivity and the potential for real-time monitoring These sensors are particularly advantageous in applications requiring high specificity and the ability to detect multiple analytes simultaneously [34].

11.3 NUTRIENT DETECTION MECHANISMS

11.3.1 Electrochemical Detection

Electrochemical detection is one of the most effective and widely used mechanisms in nutrient detection, leveraging the unique properties of carbon composite sensors to achieve high sensitivity and selectivity. In this method, the detection of nutrients occurs through electrochemical reactions, such as oxidation or reduction, at the surface of the carbon-based electrode materials. These electrodes, typically made of carbon nanomaterials such as carbon nanotubes (CNTs), graphene, or carbon black, provide a high surface area and excellent conductivity, allowing efficient electron transfer between the analyte and the electrode surface [35]. The basic principle of electrochemical detection involves applying a potential across the sensor, which induces an electrochemical reaction when a target nutrient interacts with the sensor surface. This interaction results in a measurable current or voltage change that is directly related to the concentration of the analyte. For example, in the detection of nitrate, a common nutrient in soil and water, the nitrate ions undergo reduction at the electrode surface, leading to a current that can be measured and quantified [36]. The use of carbon nanomaterials in these electrodes enhances the sensitivity of the sensor, allowing for the detection of low concentrations of nutrients, which is crucial for applications in environmental monitoring and precision agriculture.

11.3.2 Optical Detection

Optical detection is a highly sensitive and versatile method employed in nutrient detection, leveraging the interaction of light with carbon composite sensors to produce measurable optical signals. This method is particularly advantageous due to its ability to provide rapid, real-time analysis and its suitability for non-invasive or remote sensing applications. Carbon nanomaterials, for instance, graphene oxide, CNTs, and carbon quantum dots, are frequently used in these sensors due to their unique optical properties, including strong absorption, fluorescence, and scattering characteristics [27].The basic principle of optical detection involves the interaction of light with the sensor material, which results in a change in one or more optical properties—such as absorbance, fluorescence, or refractive index—when a target nutrient is present. For example, in a fluorescence-based sensor, the nutrient molecules can quench or enhance the fluorescence of carbon nanomaterials, leading to a detectable change in the fluorescence intensity. Graphene oxide, due to its strong fluorescence quenching ability, is often used in such sensors to detect a variety of nutrients, including heavy metals and organic compounds, with high sensitivity [30].

Another widely used optical technique is *surface plasmon resonance (SPR)*, where carbon nanomaterials are integrated with metallic nanoparticles to augment the sensitivity of detection. In SPR-based sensors, the interaction of the nutrient with the sensor surface causes a change in the refractive index near the sensor, leading to a shift in the resonance angle or wavelength that can be precisely measured. This method is particularly useful for detecting trace amounts of nutrients in complex matrices, such as soil or water, where conventional techniques may fail to provide adequate sensitivity [37].

Raman spectroscopy is another powerful optical detection method that utilizes carbon composite materials to detect nutrients through the inelastic scattering of light. CNTs and graphene have been shown to enhance the Raman signals of certain molecules, making them suitable for detecting specific nutrients at very low concentrations. This amplification is recognized as surface-enhanced Raman scattering and it has been successfully applied to detect pesticides, heavy metals, and other contaminants in agricultural and environmental samples [38].

The integration of carbon nanomaterials in optical detection systems also enables the development of *multi-modal sensors*, which can simultaneously detect multiple nutrients by combining different optical techniques, such as fluorescence and absorbance, in a single device. This capability is particularly valuable in applications like precision agriculture, where comprehensive nutrient profiling is essential for optimizing crop management practices [37].

Overall, optical detection methods using carbon composite sensors offer a high degree of flexibility and can be tailored to meet the specific needs of various applications, from environmental monitoring to food safety and biomedical diagnostics. The ability to detect nutrients with high sensitivity and specificity, combined with the potential for miniaturization and integration with portable devices, makes these sensors a valuable tool in modern analytical chemistry.

11.4 APPLICATIONS IN NUTRIENT DETECTION

11.4.1 Soil Nutrient Monitoring

Soil nutrient monitoring is a critical application of carbon composite sensors, enabling precise management of agricultural practices. These sensors can detect key nutrients like nitrogen, phosphorus, and potassium directly in the soil, providing real-time data that helps optimize fertilizer use and improve crop yield. The high sensitivity of carbon nanomaterials, such as graphene and CNTs, allows for the detection of even trace amounts of nutrients, making them ideal for precision agriculture. By integrating these sensors into portable devices, farmers can perform on-site soil analysis, reducing the need for expensive and time-consuming lab tests [37].

11.4.2 Water Quality Assessment

In water quality assessment, carbon composite sensors play a vital role for detecting nutrients such as nitrates, phosphates, and heavy metals that can indicate pollution. These sensors are particularly useful in monitoring water bodies affected by agricultural runoff, where nutrient overloads can lead to harmful algal blooms. The rapid response and high specificity of carbon-based sensors enable continuous monitoring of water quality, ensuring that any changes in nutrient levels are quickly identified. This real-time monitoring capability is essential for maintaining the health of aquatic ecosystems and ensuring safe drinking water supplies [37].

11.4.3 Food and Beverage Analysis

Carbon composite sensors are increasingly used in the food and beverage industry to ensure product safety and quality. These sensors can detect nutrient levels, contaminants, and adulterants in various products, providing a quick and reliable means of quality control. For example, carbon-based electrochemical sensors have been used to measure glucose, amino acids, and vitamins in food products, ensuring that they meet nutritional standards. The versatility and sensitivity of these sensors make them suitable for both routine testing and the detection of specific nutrients in complex food matrices [39].

11.4.4 Biomedical Applications

In biomedical applications, carbon composite sensors are employed to monitor essential nutrients and metabolites in biological samples, such as blood, urine, and saliva. These sensors are crucial for managing health conditions that require regular monitoring of nutrient levels, such as diabetes, where glucose sensors are widely used. The biocompatibility and high sensitivity of carbon nanomaterials, like graphene, enhance their performance in detecting biomarkers at low concentrations, making them valuable tools in diagnostic devices and personalized medicine [40].

11.5 SPECIFIC NUTRIENTS AND THEIR DETECTION

11.5.1 Nitrogen Compounds

Nitrogen compounds, such as nitrates and ammonia, are critical nutrients in agriculture and environmental monitoring, and their accurate detection is essential for maintaining soil and water quality. Carbon composite sensors, particularly those based on CNTs and graphene, offer high sensitivity for detecting these compounds. For example, electrochemical sensors utilizing CNTs can effectively measure nitrate concentrations through reduction reactions at the electrode surface, providing rapid and accurate results. These sensors are valuable for managing fertilizer applications and preventing nutrient runoff that can lead to water pollution [41]. Additionally, the high surface area and conductivity of graphene-based sensors enhance the detection of ammonia, making them suitable for monitoring both soil and wastewater [41].

11.5.2 Phosphorus Compounds

Phosphorus compounds, such as phosphates, are vital for plant growth but can cause environmental issues when present in excess. Carbon composite sensors have been developed to detect phosphates with high precision, utilizing methods like colorimetric and electrochemical detection. For instance, carbon-based electrochemical sensors can measure phosphate levels through a reaction that generates a measurable current change [41]. Colorimetric sensors using carbon quantum dots offer a visual method for detecting phosphates, which is particularly useful for field applications where immediate results are needed. These sensors help in managing phosphorus levels in agricultural soils and monitoring wastewater treatment processes.

11.5.3 Potassium

Potassium is an essential nutrient for plant health, and its detection in soil and plant tissues is crucial for effective fertilization practices. Carbon composite sensors, including those based on graphene and carbon black, are used to detect potassium through various electrochemical and optical techniques. For example, electrochemical sensors with carbon black electrodes provide a high level of sensitivity for potassium ions, making them suitable for detailed soil nutrient analysis. Graphene-based sensors can also be employed to detect potassium by measuring changes in electrical conductivity when potassium ions interact with the sensor surface. These technologies enable precise monitoring of potassium levels, ensuring optimal crop growth and nutrient management [42].

11.5.4 Micronutrients

Micronutrients, including elements like iron, zinc, and copper, are required in trace amounts for plant and human health. Detecting these micronutrients requires highly sensitive sensors due to their low concentration levels. Carbon composite sensors, such as those using graphene oxide and CNTs, are particularly effective in this regard. For instance, electrochemical sensors incorporating graphene oxide can detect iron and zinc with high selectivity due to their enhanced electrochemical properties. CNT-based sensors are also employed to measure trace levels of copper and other micronutrients, offering high sensitivity and low detection limits. These sensors are essential for ensuring the adequate supply of micronutrients in agriculture and for monitoring their levels in food and environmental samples [43].

11.6 ADVANTAGES AND LIMITATIONS

11.6.1 Sensitivity and Selectivity

Carbon composite sensors offer exceptional sensitivity and selectivity for nutrient detection due to the unique properties of carbon nanomaterials. Materials such as graphene and CNTs provide a high surface area and excellent electrical conductivity, which enhance the sensitivity of the sensors. For instance, graphene-based electrochemical sensors can detect even trace levels of nutrients with high precision by exploiting the material's large surface area and high electron mobility [44]. Selectivity is achieved through functionalization, where specific chemical groups or receptors are attached to the sensor surface to preferentially interact with the target nutrients. This allows for accurate measurement of individual nutrients in complex samples, such as soil or water, minimizing interference from other substances [27].

11.6.2 Durability and Stability

The durability and stability of carbon composite sensors are crucial for their long-term performance and reliability in various environments. Carbon nanomaterials, such as CNTs and graphene, are known for their mechanical strength and chemical resistance, which contribute to the overall robustness of the sensors. For example, graphene-based sensors exhibit high stability even under harsh environmental conditions, making them suitable for long-term use in field applications [20]. Additionally, the chemical inertness of carbon composites helps maintain sensor performance over time, reducing degradation and ensuring consistent results. However, the stability can be influenced by factors such as environmental conditions and the presence of interfering substances, which must be managed through appropriate sensor design and calibration [44].

11.6.3 Cost-effectiveness

Cost-effectiveness is a significant consideration in the deployment of carbon composite sensors, particularly for large-scale applications. While the initial costs of fabricating sensors with advanced carbon nanomaterials may be high, the long-term benefits often outweigh these expenses. For instance, the use of scalable fabrication techniques, such as screen printing or inkjet printing, can reduce production costs while maintaining sensor performance. Moreover, the high sensitivity of these sensors often means that fewer sensors are needed for comprehensive monitoring, which can further reduce overall costs. Despite the relatively high cost of raw materials, advancements in manufacturing processes and material science are continuously improving the cost-effectiveness of carbon composite sensors [45].

11.6.4 Challenges and Limitations

Despite their advantages, carbon composite sensors face several challenges and limitations. One major challenge is the need for effective functionalization to ensure selectivity for specific nutrients. Inadequate functionalization can lead to cross-sensitivity and interference from other substances, affecting the accuracy of measurements [46]. Additionally, while carbon composites offer high sensitivity, they can sometimes exhibit variability in performance due to inconsistencies in material quality or sensor fabrication methods. Another limitation is the potential for sensor drift over time, which can impact long-term stability and accuracy. Addressing these challenges requires ongoing research and development to improve sensor design, enhance functionalization methods, and develop robust calibration protocols.

11.7 CONCLUSION AND FUTURE OUTLOOK

In summary, carbon composite sensors represent a cutting-edge technology with transformative potential in the field of nutrient detection. Their superior sensitivity, selectivity, and versatility make them invaluable tools for applications ranging from soil nutrient monitoring and water quality assessment to food safety and biomedical diagnostics. The integration of advanced nanomaterials, such as graphene and CNTs, has significantly enhanced sensor performance, providing unprecedented levels of accuracy and reliability.

Looking ahead, several exciting advancements are poised to further elevate the capabilities of carbon composite sensors. The incorporation of emerging nanomaterials and the development of multi-analyte detection systems promise to expand the range of detectable substances and improve the efficiency of monitoring processes. Additionally, the integration of IoT technology is set to revolutionize the field by enabling real-time data collection, remote monitoring, and automated decision-making, thereby increasing the accessibility and practicality of sensor applications.

However, challenges remain, including the need for continued improvement in sensor stability, cost-effectiveness, and functionalization methods. Addressing these challenges will be crucial for maximizing the impact of carbon composite sensors and ensuring their widespread adoption in both existing and new applications. Overall, the future of carbon composite sensors looks promising, with ongoing research and technological advancements likely to drive significant progress. As these sensors continue to evolve, they will play an increasingly important role in advancing our understanding of nutrient dynamics, safeguarding environmental quality, and enhancing human health. The ongoing exploration of novel materials, innovative detection methods, and smart technologies will undoubtedly shape the next generation of carbon composite sensors, paving the way for more precise and efficient nutrient detection in diverse fields.

REFERENCES

[1] D. K. Pandey and R. Mishra, "Towards sustainable agriculture: Harnessing AI for global food security," *Artif. Intell. Agric.*, vol. 12, pp. 72–84, 2024, doi: 10.1016/j.aiia.2024.04.003.

[2] M. Pan, Z. Yin, K. Liu, X. Du, H. Liu, and S. Wang, "Carbon-based nanomaterials in sensors for food safety," *Nanomaterials*, vol. 9, no. 9, pp. 1–23, 2019, doi: 10.3390/nano9091330.

[3] A. Bartošová, A. Michalíková, M. Sirotiak, and M. Soldán, "Comparison of two spectrophotometric techniques for nutrients analyses in water samples," *Res. Pap. Fac. Mater. Sci. Technol. Slovak Univ. Technol.*, vol. 20, no. 32, pp. 8–19, 2013, doi: 10.2478/v10186-012-0002-9.

[4] S. Verma, D. Thakur, C. M. Pandey, and D. Kumar, "Recent prospects of carbonaceous nanomaterials-based laccase biosensor for electrochemical detection of phenolic compounds," *Biosensors*, vol. 13, no. 3, pp. 1–26, 2023, doi: 10.3390/bios13030305.

[5] K. A. Ben Brahim *et al.*, "Electrochemical detection of sulfadiazine by sensors based on chemically modified carbon electrodes: A review," *Curr. Top. Med. Chem.*, vol. 23, no. 15, pp. 1464–1476, 2023, doi: 10.2174/1568026623666230210115740.

[6] V. Georgakilas *et al.*, "Functionalization of graphene: Covalent and non-covalent approach," *Chem. Rev.*, vol. 112, no. 11, pp. 6156–6214, 2012.

[7] M.-Y. Kim and K. H. Lee, "Electrochemical sensors for sustainable precision agriculture—A review," *Front. Chem.*, vol. 10, pp. 1–14, May 2022, doi: 10.3389/fchem.2022.848320.

[8] F. Arduini *et al.*, "Electrochemical biosensors based on nanomodified screen-printed electrodes: Recent applications in clinical analysis," *TrAC – Trends Anal. Chem.*, vol. 79, pp. 114–126, 2016, doi: 10.1016/j.trac.2016.01.032.

[9] Y. S. Li and J. S. Church, "Raman spectroscopy in the analysis of food and pharmaceutical nanomaterials," *J. Food Drug Anal.*, vol. 22, no. 1, pp. 29–48, 2014, doi: 10.1016/j.jfda.2014.01.003.

[10] T. H. S. Dhahi, U. D. A. Bin Hashim, N. M. Ahmed, and A. Mat Taib, "A review on the electrochemical sensors and biosensors composed of nanogaps as sensing material," *J. Optoelectron. Adv. Mater.*, vol. 12, no. 9, pp. 1857–1862, 2010.

[11] M. A. P. Mahmud *et al.*, "Recent progress in sensing nitrate, nitrite, phosphate, and ammonium in aquatic environment," *Chemosphere*, vol. 259, p. 127492, 2020, doi: 10.1016/j.chemosphere.2020.127492.

[12] P. D. Patel, "(Bio)sensors for measurement of analytes implicated in food safety: A review," *TrAC – Trends Anal. Chem.*, vol. 21, no. 2, pp. 96–115, 2002, doi: 10.1016/S0165-9936(01)00136-4.

[13] T. Lin, Y. Xu, A. Zhao, W. He, and F. Xiao, "Flexible electrochemical sensors integrated with nanomaterials for in situ determination of small molecules in biological samples: A review," *Anal. Chim. Acta*, vol. 1207, p. 339461, 2022, doi: 10.1016/j.aca.2022.339461.

[14] D. Godoy *et al.*, *The future of food and agriculture: trends and challenges*, vol. 4, no. 4, 2014. [Online]. Available: www.fao.org/publications%0A//www.fao.org/3/a-i6583e.pdf%0Ahttp://siteresources.worldbank.org/INTARD/825826-1111044795683/20424536/Ag_ed_Africa.pdf%0Awww.fao.org/cfs%0A//www.jstor.org/stable/4356839%0Ahttps://ediss.uni-goettingen.de/bitstream/han

[15] V. H. Smith, G. D. Tilman, and J. C. Nekola, "Eutrophication: Impacts of excess nutrient inputs on freshwater, marine, and terrestrial ecosystems," *Environ. Pollut.*, vol. 100, no. 1–3, pp. 179–196, 1999, doi: 10.1016/S0269-7491(99)00091-3.

[16] I. Elmadfa and A. L. Meyer, "Importance of food composition data to nutrition and public health," *Eur. J. Clin. Nutr.*, vol. 64, no. S3, pp. S4–S7, 2010, doi: 10.1038/ejcn.2010.202.

[17] A. Bah, S. K. Balasundram, and M. H. A. Husni, "Sensor technologies for precision soil nutrient management and monitoring," *Am. J. Agric. Biol. Sci.*, vol. 7, no. 1, pp. 43–49, 2012, doi: 10.3844/ajabssp.2012.43.49.

[18] N. Bin Chang, S. Imen, and B. Vannah, "Remote sensing for monitoring surface water quality status and ecosystem state in relation to the nutrient cycle: A 40-year perspective," *Crit. Rev. Environ. Sci. Technol.*, vol. 45, no. 2, pp. 101–166, 2015, doi: 10.1080/10643389.2013.829981.

[19] R. Prasad, A. Bhattacharyya, and Q. D. Nguyen, "Nanotechnology in sustainable agriculture: Recent developments, challenges, and perspectives," *Front. Microbiol.*, vol. 8, pp. 1–13, 2017, doi: 10.3389/fmicb.2017.01014.

[20] S. Sharma, S. Shekhar, S. Gautam, B. Sharma, A. Kumar, and P. Jain, *Carbon-based nanomaterials as novel nanosensors*. Elsevier Inc., 2020. doi: 10.1016/B978-0-12-820702-4.00014-3.

[21] P. Ramnani, N. M. Saucedo, and A. Mulchandani, "Carbon nanomaterial-based electrochemical biosensors for label-free sensing of environmental pollutants," *Chemosphere*, vol. 143, pp. 85–98, 2016, doi: 10.1016/j.chemosphere.2015.04.063.

[22] W. Siangproh, W. Dungchai, P. Rattanarat, and O. Chailapakul, "Nanoparticle-based electrochemical detection in conventional and miniaturized systems and their bioanalytical applications: A review," *Anal. Chim. Acta*, vol. 690, no. 1, pp. 10–25, 2011, doi: 10.1016/j.aca.2011.01.054.

[23] H. J. Salavagione, A. M. Díez-Pascual, E. Lázaro, S. Vera, and M. A. Gómez-Fatou, "Chemical sensors based on polymer composites with carbon nanotubes and graphene: The role of the polymer," *J. Mater. Chem. A*, vol. 2, no. 35, pp. 14289–14328, 2014, doi: 10.1039/c4ta02159b.

[24] N. Hareesha and J. G. Manjunatha, "Electro-oxidation of formoterol fumarate on the surface of novel poly(thiazole yellow-G) layered multi-walled carbon nanotube paste electrode," *Sci. Rep.*, vol. 11, no. 1, pp. 1–14, 2021, doi: 10.1038/s41598-021-92099-x.

[25] M. Ehsani, P. Rahimi, and Y. Joseph, "Structure–function relationships of nanocarbon/polymer composites for chemiresistive sensing: A review," *Sensors*, vol. 21, no. 9, 2021, doi: 10.3390/s21093291.

[26] A. M. Rabie, "New potential inhibitors of coronaviral main protease (CoV-Mpro): Strychnine bush, pineapple, and ginger could be natural enemies of COVID-19," *Int. J. New. Chem*, vol. 11, no. 1, pp. 48–57, 2022, [Online]. Available: www.ijnc.ir/article_247391_ccb98b0c7e358490d93275594b7bbbfd.pdf

[27] G. Speranza, "Carbon nanomaterials: Synthesis, functionalization and sensing applications," *Nanomaterials*, vol. 11, no. 4, 2021, doi: 10.3390/nano11040967.

[28] A. B. Monnappa, J. G. Manjunatha, and A. S. Bhatt, "Design of a sensitive and selective voltammetric sensor based on a cationic surfactant-modified carbon paste electrode for the determination of alloxan," *ACS Omega*, vol. 5, no. 36, pp. 23481–23490, 2020, doi: 10.1021/acsomega.0c03517.

[29] Y. Liu, X. Dong, and P. Chen, "Biological and chemical sensors based on graphene materials," *Chem. Soc. Rev.*, vol. 41, no. 6, pp. 2283–2307, 2012, doi: 10.1039/c1cs15270j.

[30] S. Kumar *et al.*, "Optically active nanomaterials and its biosensing applications—A review," *Biosensors*, vol. 13, no. 1, pp. 1–38, 2023, doi: 10.3390/bios13010085.

[31] P. A. Pushpanjali, J. G. Manjunatha, N. Hareesha, E. S. D. Souza, M. M. Charithra, and N. S. Prinith, "Voltammetric analysis of antihistamine drug cetirizine and paracetamol at poly(L-Leucine) layered carbon nanotube paste electrode," *Surfaces and Interfaces*, vol. 24, no. January, p. 101154, 2021, doi: 10.1016/j.surfin.2021.101154.

[32] C. B. Jacobs, M. J. Peairs, and B. J. Venton, "Review: Carbon nanotube based electrochemical sensors for biomolecules," *Anal. Chim. Acta*, vol. 662, no. 2, pp. 105–127, 2010, doi: 10.1016/j.aca.2010.01.009.

[33] M. A. Ryan, A. V. Shevade, H. Zhou, and M. L. Homer, "Polymer-carbon black composite sensors in an electronic nose for air-quality monitoring," *MRS Bull.*, vol. 29, no. 10, pp. 714–719, 2004, doi: 10.1557/mrs2004.208.

[34] B. N. Shivananju *et al.*, "The roadmap of graphene-based optical biochemical sensors," *Adv. Funct. Mater.*, vol. 27, no. 19, pp. 1–19, 2017, doi: 10.1002/adfm.201603918.

[35] S. Y. Won, C. H. Lee, H. S. Chang, S. O. Kim, S. H. Lee, and D. S. Kim, "Monitoring of 14 sulfonamide antibiotic residues in marine products using HPLC-PDA and LC-MS/MS," *Food Control*, vol. 22, no. 7, pp. 1101–1107, Jul. 2011, doi: 10.1016/j.foodcont.2011.01.005.

[36] P. Bansal *et al.*, "We are IntechOpen, the world's leading publisher of Open Access books built by scientists, for scientists TOP 1 %," *Intech*, vol. i, p. 15, 2016.

[37] D. Demarchi and A. Tagliaferro, "Carbon for sensing devices," *Carbon Sens. Devices*, pp. 1–264, 2015, doi: 10.1007/978-3-319-08648-4.

[38] D. Yılmaz, B. N. Günaydın, and M. Yüce, "Nanotechnology in food and water security: on-site detection of agricultural pollutants through surface-enhanced Raman spectroscopy," *Emergent Mater.*, vol. 5, no. 1, pp. 105–132, 2022, doi: 10.1007/s42247-022-00376-w.

[39] B. A. Pellerin *et al.*, "Emerging tools for continuous nutrient monitoring networks: Sensors advancing science and water resources protection," *J. Am. Water Resour. Assoc.*, vol. 52, no. 4, pp. 993–1008, 2016, doi: 10.1111/1752-1688.12386.

[40] M. A. Mustapa, S. Ambran, and A. Yuzir, "Application of carbon nanotubes and graphene to develop the heavy metal electrochemical sensor," *IOP Conf. Ser. Earth Environ. Sci.*, vol. 479, no. 1, 2020, doi: 10.1088/1755-1315/479/1/012036.
[41] D. Li, T. Wang, Z. Li, X. Xu, C. Wang, and Y. Duan, "Application of graphene-based materials for detection of nitrate and nitrite in water—a review," *Sensors (Switzerland)*, vol. 20, no. 1, pp. 1–35, 2020, doi: 10.3390/s20010054.
[42] E. J. Strand *et al.*, "Printed organic electrochemical transistors for detecting nutrients in whole plant sap," *Adv. Electron. Mater.*, vol. 8, no. 4, 2022, doi: 10.1002/aelm.202100853.
[43] D. Giust *et al.*, "Graphene oxide-upconversion nanoparticle based portable sensors for assessing nutritional deficiencies in crops," *ACS Nano*, vol. 12, no. 6, pp. 6273–6279, 2018, doi: 10.1021/acsnano.8b03261.
[44] K. Ait Ben Brahim *et al.*, "Electrochemical detection of sulfadiazine by sensors based on chemically modified carbon electrodes: A review," *Curr. Top. Med. Chem.*, vol. 23, pp. 1–13, 2023, doi: 10.2174/1568026623666230210115740.
[45] A. V. Bounegru and C. Apetrei, "Carbonaceous nanomaterials employed in the development of electrochemical sensors based on screen-printing technique—a review," *Catalysts*, vol. 10, no. 6, 2020, doi: 10.3390/catal10060680.
[46] I. Kim, J. S. Heo, and M. F. Hossain, "Challenges in design and fabrication of flexible/stretchable carbon-and textile-based wearable sensors for health monitoring: A critical review," *Sensors (Switzerland)*, vol. 20, no. 14, pp. 1–29, 2020, doi: 10.3390/s20143927.

12 Carbon Sensor Devices for Disease Detection

Vildan Sanko, Meltem Okan, Cem Erkmen, and Pattan Siddappa Ganesh

12.1 INTRODUCTION

Electrochemical sensors have become invaluable in modern diagnostics, enabling the precise and timely detection of various biomarkers associated with various diseases[1] Their intrinsic advantages, such as high sensitivity, specificity, rapid response, and the capability for miniaturization, make them ideal candidates for point-of-care testing (POCT).[2–4] Carbon-based materials stand out due to their exceptional electrical properties, chemical stability, and versatile surface chemistry among the materials explored for constructing these sensors.[5,6] For this purpose, this chapter focuses on designing and developing electrochemical carbon sensor devices for disease detection, highlighting the critical role these devices play in advancing diagnostic technologies.

Carbon materials, including graphene, carbon nanotubes (CNTs), graphite, fullerene, and carbon black (CB), offer unique physicochemical properties that can be fine-tuned to meet the specific demands of electrochemical sensing.[7,8] Graphene, for instance, with its two-dimensional structure and high surface area, provides an excellent platform for electron transfer, making it highly suitable for detecting electroactive species in biological samples.[9] Similarly, CNTs, with their remarkable mechanical strength and electrical conductivity, facilitate the development of sensors with enhanced sensitivity and stability. These properties enable carbon-based sensors to detect various biomarkers, ranging from small molecules like glucose and dopamine to larger macromolecules like proteins and nucleic acids.[10]

The application of carbon-based electrochemical sensors in disease detection is vast, encompassing the detection of infectious diseases, cancers, cardiovascular diseases (CVDs), and metabolic disorders, among others.[11–15] For instance, carbon-based sensors have been successfully employed in detecting cancer biomarkers like carcinoembryonic antigen and prostate-specific antigen, demonstrating their potential in early cancer diagnosis.[16,17] Additionally, carbon-based sensors are instrumental in monitoring glucose levels in diabetic patients, showcasing their utility in managing chronic conditions.[18] One of the significant advantages of using carbon-based materials in electrochemical sensors is their ability to be functionalized with various biomolecules, such as enzymes, antibodies, and aptamers.[19] This functionalization enhances the selectivity of the sensors toward specific biomarkers, allowing for precise disease detection even in complex biological matrices like blood, saliva or urine.[20] For example, immobilizing glucose oxidase on a multi-walled carbon nanotube (MWCNT)-based composite electrode surface has developed susceptible glucose sensors critical for diabetes management.[21] Similarly, integrating antibodies specific to cardiac troponin I on a graphene quantum dot (GQD)-based sensor platform has facilitated the early detection of myocardial infarction, thereby aiding in timely medical intervention.[22]

Recent advancements in nanotechnology and material science have further expanded the capabilities of carbon-based electrochemical sensors. The synthesis of novel carbon nanocomposites,

DOI: 10.1201/9781003500322-12

which combine carbon materials with metallic nanoparticles or polymers, has led to developing sensors with improved sensitivity, lower detection limits, and faster response times.[23,24] These innovations enhance the sensors' performance and open new avenues for their application in personalized medicine. Additionally, integrating these sensors into portable, user-friendly devices suitable for POCT requires further research and development.[25] Also, in addition to glassy carbon electrodes (GCEs), other types of carbon-based electrodes, such as screen-printed electrode (SPE), pencil graphite electrode (PGE), and carbon paste electrode (CPE), each offer unique advantages for electrochemical sensing. They have high electrical conductivity, allowing for efficient electron transfer and enhancing the sensor's sensitivity.[26] GCEs are also chemically inert and have a vast potential window, making them suitable for detecting various analytes without interference from other substances. Their smooth surface can be easily modified with nanomaterials, polymers, or biomolecules, improving selectivity and enabling the detection of specific biomarkers. Carbon-based electrodes are generally durable, reproducible, and relatively inexpensive, making them ideal for research, practical diagnostics, and environmental monitoring applications.[27] In conclusion, carbon-based electrochemical sensors represent a promising technology for disease detection, offering high sensitivity, specificity, and the potential for real-time monitoring of biomarkers. Their ability to be tailored for detecting a wide range of analytes, coupled with the advancements in material science and nanotechnology, positions them as critical tools in the future of diagnostics. This chapter provides an in-depth exploration of the design principles, functionalization strategies, and practical applications of these sensors, highlighting their potential to revolutionize disease detection and contribute significantly to the advancement of personalized medicine and public health.

12.2 OVERVIEW OF CARBON SENSOR DEVICES

Sensors designed with carbon-based materials continue to be the focus of attention due to their electronic, surface, and catalytic properties. These materials are used both as working electrodes and in electrode modification to improve sensor performance and enable the development of more sensitive sensors by increasing affinity and conductivity.[28,29] Figure 12.1 shows schematically the carbon-based materials used for both purposes, and the important points about each are discussed in the rest of this section. The other section focuses on studies of using these materials in electrochemical disease diagnosis. In modern sensor technologies, carbon-based materials and electrodes are observed to be used in different clinical, biomedical, and point-of-care diagnostic applications. However, this chapter will be helpful as we find both the potential of these materials in future studies and their role in perception worthy of further discussion.

12.2.1 Carbon-Based Electrodes

The electrodes used are an essential step in bench-type electrochemical disease detection. Some electrodes are designed for single use, but reusable electrodes are also available. In this section, the most commonly used carbon-based electrodes used in electrochemical disease diagnosis were selected and discussed, divided into four groups: GCE, CPE, PGE, and SPE. GCE, a Teflon-coated rod electrode, is one of the leading electrodes used in electrochemical sensing applications. GCE's popular features include low cost, electrochemical stability over a wide potential range, and the ability to allow surface modification. In addition, good conductivity is one of the essential features that enable it to be used in electrochemical sensors, and high hardness and chemical stability are other features that make it preferable.[30] However, this electrode is not disposable and is generally reusable after appropriate cleaning procedures. This cleaning procedure involves additional solvent usage, cost, and time constraints. Additionally, since it requires immersion in a high volume of electrolyte, other electrodes that can overcome these limitations need to be developed.[27]

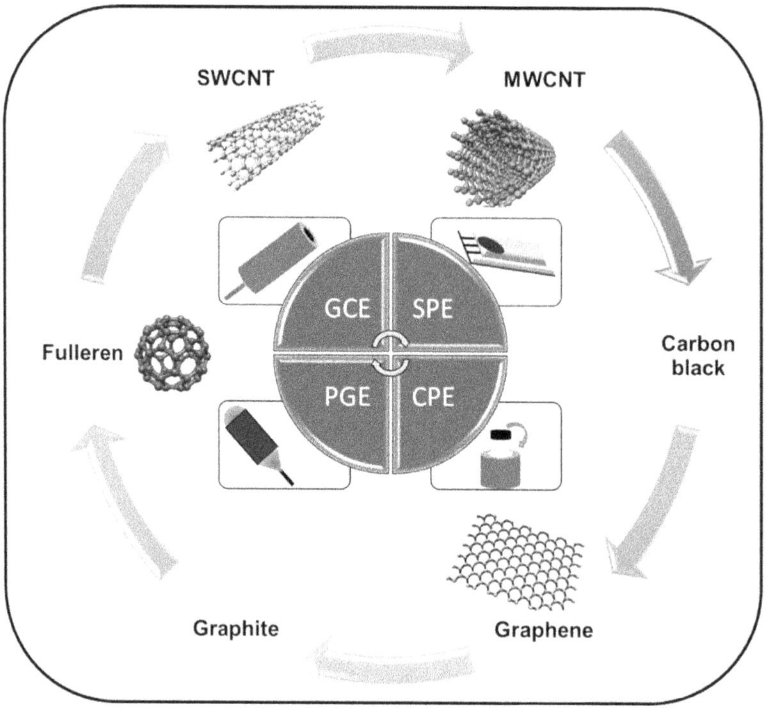

FIGURE 12.1 Schematic representation of carbon-based electrode and modification-based materials used in electrochemical disease diagnosis.

PGEs are commercially available and readily available pencil tips and are relatively inexpensive, making them a suitable alternative working electrode. In addition, PGE, a disposable electrode that does not require cleaning procedures, offers high sensitivity and reproducibility for electrochemical sensor studies. Thanks to these good properties, studies show that it can be applied in the electrochemical determination of many different analytes in recent years.[31,32] In addition, PGEs with high conductivity exhibit light, easily polarizable, heat, and chemical-resistant behavior.[33] However, it is necessary to mention the limitations, such as the different properties and fragility of different pencil groups, because these limitations make it difficult to use depending on the application.[34]

CPE, introduced by Ralph Norman Adams in 2008, consists of a mixture of carbon powder and a binder, and thanks to this famous structure, sensor platforms that enable the detection of various analytes have been created.[35] These CPEs, obtained due to the optimum combination of electrochemical and physicochemical properties of the carbon structure substrate, attract attention with their renewable, low resistance, and chemical inertness properties. If we talk about the limitations of this type of electrode, the most notable ones are low sensitivity, stability, heterogeneity, repeatability, limited life, and inefficient electron transfer. It has also been reported that the quality of graphite used and the carbon/adhesive ratio affect the kinetics of the electrode.[36,37] To overcome these limitations, the use of nanomaterials has been proposed and more efficient CPE platforms can be obtained with nanomaterial@CPE composite systems.[36]

The pace of advances in electrochemical sensors continues, and when the advantages and disadvantages of other carbon-based electrodes are considered, the use of SPEs that can provide customization in production has become widespread in recent years. One of the main features distinguishing it from other electrodes is that it can be used horizontally and, therefore, requires minimal reagent use in surface modification and measurement process steps.[26,27] The high repeatability and the fact that it can be designed from gold, platinum, or other metals other than carbon allow it to be used in a much more comprehensive range of applications.[38] It is also stated that SPEs, which provide ease of connection to portable devices and, therefore, point-of-care analysis, may experience problems such as ink dissolution when used with organic solvents.[26,39] When carbon-based electrodes are evaluated from a general point of view, it is possible to discuss the positive and limiting factors for each. The most suitable electrode type can be selected for the designed sensor and need, and there are a few literature examples that the electrochemical performance of each can be improved, especially using nanomaterials. For this purpose, the properties of carbon-based materials are discussed in the rest of this section, and their contributions to sensors in disease diagnosis are examined.

12.2.2 Carbon-Based Modified Materials

Single-walled carbon nanotubes (SWCNTs) are a class of carbon-based nanomaterials composed of a single layer of carbon atoms arranged in a cylindrical structure.[40] These materials exhibit exceptional electrical, mechanical, and thermal properties, making them highly suitable for electrochemical sensor applications, particularly in disease detection. The unique one-dimensional structure of SWCNTs allows for efficient electron transport, enhancing the performance of electrochemical sensors by improving electron transfer rates between the electrode surface and the analyte of interest. This property is critical for achieving high sensitivity and low detection limits in biosensors.[41] In addition to their excellent electrical conductivity, SWCNTs possess a high surface area, which increases the interaction sites available for analyte detection. This feature is particularly beneficial in sensor applications, as it allows for immobilizing various biomolecules, such as antibodies, enzymes, and aptamers, specific to the targeted disease biomarkers. Functionalizing SWCNTs with different chemical groups or biological recognition elements further enhances their versatility, making them adaptable to various sensing applications.[42]

MWCNTs are another prominent class of carbon nanomaterials, consisting of multiple concentric layers of graphene rolled into cylindrical structures. Due to their layered architecture, MWCNTs offer a combination of enhanced mechanical strength, electrical conductivity, and chemical stability, making them valuable materials in electrochemical sensing technologies for disease detection.[43] The distinct structure of MWCNTs allows for multiple electron pathways, resulting in improved electron transfer efficiency,[44] which is critical for increasing the sensitivity of electrochemical sensors.[45] MWCNTs, like their single-walled counterparts, possess a high surface area, providing numerous active sites for interaction with target analytes. This property is particularly beneficial in developing biosensors, where the immobilization of recognition elements such as enzymes, antibodies, or nucleic acids is essential for detecting specific disease biomarkers. MWCNT-modified electrodes have demonstrated significant improvements in sensor performance, with enhanced sensitivity and specificity due to the increased surface area and the ability to promote faster electron transfer rates between the electrode and the analyte.[46] One of the critical advantages of MWCNTs over single-walled nanotubes is their robust mechanical strength, which makes them more resilient in various sensor applications. This property ensures the durability and longevity of MWCNT-based sensors, allowing them to function effectively even under mechanical stress or prolonged use. Additionally, the multiple layers of MWCNTs can facilitate higher loading capacities for biorecognition molecules, further enhancing the sensor's ability to detect trace levels of disease-related biomarkers.[47,48]

Graphite, a crystalline form of carbon, is one of the most widely used carbon-based materials in electrochemical sensors due to its excellent electrical conductivity, chemical stability, and ease of modification.[49] Composed of stacked graphene layers, each layer in graphite consists of carbon atoms arranged in a hexagonal lattice, which allows for effective electron transfer, a key feature in electrochemical sensing applications.[50] Graphite has long been utilized in various types of electrodes, including those for disease detection, owing to its favorable electrochemical properties and low cost. Graphite's high surface area, derived from its layered structure, provides ample active sites for immobilizing biomolecules, such as enzymes, antibodies, and aptamers. This capability is critical in biosensor design, as it allows for detecting disease biomarkers in complex biological samples, such as blood or saliva, with improved accuracy. One of graphite's primary advantages is its cost-effectiveness and abundance, making it an attractive material for large-scale sensor production. Graphite-based electrodes are commonly used in SPEs, PGEs, and CPEs, each offering different benefits depending on the application. Graphite electrodes have been widely employed in sensors designed for the detection of cancer biomarkers,[51] glucose levels in diabetes management,[52] and various pathogens in infectious disease diagnostics.[53] Graphite's excellent conductivity is another critical factor in its widespread use in electrochemical sensors.[54]

Graphene, a single layer of carbon atoms arranged in a two-dimensional honeycomb lattice, has garnered significant attention in electrochemical sensors due to its exceptional electrical, mechanical, and chemical properties.[55] As a derivative of graphite, graphene maintains many of the advantageous features of carbon-based materials, such as conductivity and stability, while introducing unique attributes that make it particularly suitable for high-performance biosensors. In disease detection, graphene has proven to be a robust material for enhancing the sensitivity and selectivity of electrochemical sensors.[56] One of graphene's most notable properties is its high surface area, which provides many active sites for interaction with analytes.[57] This makes graphene ideal for biosensor applications, where the immobilization of biorecognition elements such as antibodies, enzymes, DNA, or aptamers is crucial for detecting specific disease biomarkers. The high surface area of graphene allows for a more significant loading of these biomolecules, enhancing the sensor's ability to detect even trace amounts of target analytes, which is essential for early disease diagnosis. Graphene also exhibits excellent electrical conductivity, facilitating rapid electron transfer between the electrode and the analyte.[58] This property is critical in electrochemical sensors, as it directly impacts the sensitivity and response time of the device. The superior electron mobility of graphene, coupled with its thin structure, ensures efficient electron transport, leading to highly sensitive sensors with low detection limits.[59] Graphene-modified electrodes have been shown to outperform many traditional materials in terms of sensitivity, making them ideal for detecting a wide range of diseases, including cancer, CVDs, and infectious diseases. Another critical advantage of graphene is its mechanical strength and flexibility. These properties allow graphene to be used in various sensor configurations, from flexible wearable devices to miniaturized point-of-care diagnostics.[60]

Fullerene's spherical shape also provides a large surface area relative to its size, allowing for the efficient immobilization of biomolecules. This makes fullerene an excellent candidate for use in biosensors, where the selective detection of disease biomarkers is crucial.[61] Modifying fullerenes with functional groups or nanomaterials further enhances their versatility, creating more selective and sensitive sensors tailored to specific disease detection applications. In addition to its electrochemical properties, fullerene is chemically stable and easily functionalized, making it adaptable for various sensor applications. Researchers have developed fullerene-based sensors for detecting cancer biomarkers, glucose in diabetic patients, and multiple pathogens responsible for infectious diseases. These sensors demonstrate enhanced performance compared to conventional sensors due to fullerene's ability to improve electron transfer and promote interactions between the target analyte and the electrode surface. Moreover, fullerenes exhibit unique photophysical properties, including their ability to undergo photochemical reactions when exposed to light.[62]

CB is a highly versatile form of amorphous carbon. It comprises fine particles with a high surface area and consists primarily of elemental carbon in the form of small, quasi-spherical particles.[63] Due to its unique properties, such as high conductivity, surface area, and chemical inertness, CB has become an essential material for modifying electrodes in biosensing devices.[64] One of the most important characteristics of CB is its excellent electrical conductivity,[65] which makes it an ideal material for enhancing the electron transfer processes in electrochemical sensors. This property allows CB to improve the sensitivity of sensors by facilitating the efficient movement of electrons between the analyte and the electrode.[66] CB's fine particle size and porous nature provide many active sites for immobilizing biomolecules. This large surface area increases the sensor's ability to capture target analytes, such as disease biomarkers, in biological samples, improving sensitivity and lower detection limits. CB is also chemically inert, ensuring its stability under various environmental conditions and in different biological samples. This chemical stability makes CB excellent for developing durable and long-lasting electrochemical sensors.[67] In addition, CB is relatively low cost and easy to produce on a large scale, making it an attractive option for developing disposable or cost-effective biosensors. CB can be used as a standalone material or combined with other nanomaterials to enhance performance in electrochemical sensors. For example, CB is often combined with graphene[68] or metal nanoparticles[69] to create hybrid materials with improved electrochemical properties. These combinations allow for better electron transfer, increased surface area, and enhanced selectivity for specific analytes.

12.3 RECENT APPLICATIONS OF CARBON SENSOR DEVICES FOR DISEASE DETECTION

Carbon-based sensor devices have emerged as powerful tools in disease detection, leveraging their unique properties such as high surface area, electrical conductivity, and biocompatibility. Recent advancements have focused on the development of graphene, CNTs, and other nanostructured carbon materials for detecting biomarkers related to cancers, CVDs, diabetes, and infectious diseases. These sensors offer rapid, sensitive, and non-invasive diagnostic solutions, making them increasingly valuable in POCT and personalized medicine applications. Table 12.1 presents recent applications, including different types of carbon electrodes, carbon-based modifications, the method used, and the relevant disease or biomarker, along with the linear range and limit of detection (LOD) for each.

Hatamluyi and Es'haghi[77] synthesized a novel quantum dot consisting of a ZnO core wrapped in a graphene shell (ZnO@GQD) and applied it to create an electrochemical sensor for the detection of 6-mercaptopurine (6-MP). 6-MP is a well-known chemotherapeutic drug used to treat various cancers and inflammatory bowel diseases, including ulcerative colitis and Crohn's disease. They fabricated the sensor utilizing a PGE coated with a molecularly imprinted polymer (MIP) reinforced with polypyrrole and ZnO@GQD core–shell nanoparticles, developed through a one-step electropolymerization process. They produced graphene oxide (GO) from graphite powder and employed cyclic voltammetry (CV) and differential pulse voltammetry (DPV) techniques to construct a selective electrochemical sensor for 6-MP. CV and electrochemical impedance spectroscopy (EIS) were also utilized to gather detailed information on the electrode surface modifications. During the first CV scan, in the presence of 6-MP, an oxidation peak appeared at 0.48 V, but this peak disappeared in subsequent CV cycles, indicating the oxidation of 6-MP during electropolymerization. As the number of cycles increased from 1 to 10, the peak current decreased, suggesting the formation of an insulating layer of MIP on the electrode surface. To test the practical application of the proposed MIP sensor, they created calibration curves using DPV measurements with 6-MP solutions in the concentration range of 0.01–700 mM. Two linear detection ranges were observed: 0.01–50.0 mM and 50.0–700.0 mM, with an LOD of 5.72 nM in the lower range. The fabricated MIP sensor demonstrated robustness and was suitable for 6-MP analysis, exhibiting acceptable levels of repeatability, reproducibility, and stability.

Li et al.[83] focused on both uric acid (UA) and ascorbic acid (AA) detection on the same platform. Elevated UA levels in human metabolism are linked to conditions such as hyperuricemia and gout, while AA plays a role in atherosclerosis. By utilizing different concentration ranges, their platform enabled the simultaneous detection of these two biomolecules. They prepared a three-dimensional (3D) nanostructured composite by self-assembling molybdenum disulfide (MoS_2) nanospheres and polyaniline (PANI) onto reduced graphene oxide (3D MoS_2-PANI/rGO) through a simple one-pot hydrothermal process. This 3D MoS_2-PANI/rGO nanocomposite demonstrated good functionality,

TABLE 12.1
Recent applications of carbon-based electrodes/modifications for several disease detection

E type	C-based modification	EC method	Detection type	Disease	Linear range	LOD	Ref
GCE	AgNP/poly(L-arginine)/GO composite	DPV	Electrocatalytic	Neuro-degenerative diseases	0.05–50 μmol/L	0.01 μmol/L	70
GCE	AuNP/rGO	DPV	Immunoassay	Lung cancer	0.1–2000 ng/mL	0.05 ng/mL	71
GCE	MWCNT	DPV	MIP	HIV-p24	10^{-4}–2 ng/cm^3	0.083 pg/cm^3	72
GCE	MWCNT-CTAB	LSV	Electrocatalytic	Inflammatory bowel disorders	50–300 μM	8.41 nM	73
GCE	rGO/MWCNT	DPV, EIS	Aptamer	Prostate cancer	0.005–100 ng/mL	1 pg/mL	74
GCE	CNT-whiskered nanofiber	CV	Immunoassay	CVD	1–25 ng/mL	~0.04 ng/mL	75
SPCE	Graphene/AuNP	LSV	Enzymatic	Hypercholesterolemia	0.01–5000 μg/mL	0.001 μg/mL	76
PGE	ZnO@GQD	DPV	MIP	Inflammatory bowel disorders	0.01–50 μM	5.72 nM	77
GCE	rGO	DPV	MIP	Neuro-degenerative diseases	10^{-7}–14×10^{-7} M	32.7 nM	78
GCE	Hemin functionalized graphene/PdNP	DPV	Aptamer	Prostate cancer	0.025–205 ng/mL	8 pg/mL	79
SPCE	AuNP array@magnetic graphene nanosheet	SWV	Aptamer	Cancer (general)	0–500 cells/mL	4 cells/mL	80
PGE	CB/AuNP	DPV	Oligonucleotide	Cancer (general)	2.9 fM–0.7 μM	1 fM	81
GCE	AuNP@fullerene (C_{60})/N-doped graphene nanosheet	DPV	Oligonucleotide	Tuberculosis	10 fM–10 nM	3 fM	82
GCE	MoS_2 nanosphere/PANI/rGO	DPV	Electrocatalytic	Neuro-degenerative diseases	5–500 μM	0.7 μM	83
GCE	MoS_2 nanosphere/PANI/rGO	DPV	Electrocatalytic	Atherosclerosis	50 μM–8 mM	22.20 μM	83
GCE	MoS_2 nanosphere/PANI/rGO	DPV	Electrocatalytic	Hyperuricemia	1–500 μM	0.36 μM	83
GCE	Graphitic C_3N_4/MWCNT	DPV	Electrocatalytic	Kidney disorders	0.2–4 μM, 420 μM	0.139 μM	84
GCE	Hexagonal BN nanosheet/MWCNT composite	DPV	Electrocatalytic	COPD, asthma	10^{-12}–10^{-8} M	10^{-13} M	85
GCE	rGO/AuNP	SWV	Immunoassay	Prostate cancer	0.0018–41 ng/mL	60 pg/mL	86
SPCE	AuNP/fullerene (C_{60})/chitosan-ionic liquid/MWCNT	EIS, DPV	Aptamer	Prostate cancer	1–200 pg/mL	0.5 pg/mL	87
GCE	rGO nanosheet/AuNP composite	CA	Immunoassay	Sepsis	0.05–100 ng/mL	0.1 pg/mL	88
GCE	PdNP/PANI/fullerene (C_{60})	CV	Immunoassay	Prostate cancer	1.6×10^{-4}–38 ng/m	1.95×10^{-5} ng/mL	89
CPE	Strongly oxidized MWCNT	A	Electrocatalytic	Inflammatory bowel disorders	1-100 μM	0.24 μM	90
GCE	Au nanocrystal/multiple graphene aerogel	DPV	Oligonucleotide	Cancer (general)	0.1–10^{6} fM	0.033 fM	91
GCE	MWCNT/PtNP	DPV	Electrocatalytic	Neuroblastoma	5×10^{-7}–8×10^{-5} M	1.73×10^{-7} M	92
GCE	MWCNT/PtNP	DPV	Electrocatalytic	Parkinson's disease	2×10^{-7}–8×10^{-5} M	8×10^{-8} M	92
SPCE	Fullerene-Pyrrole-Pyrrole-3-COOH	EIS	MIP	Neuro-degenerative diseases	25–250 ng/mL	8.77 ng/mL	93
GCE	COOH-MWCNT/PANI/AuNP	DPV	Immunoassay	Prostate cancer	1.66 ag/mL–1.3 ng/mL	0.5 pg/mL	94
GCE	Hybrid CNT	DPV	Electrocatalytic	Kidney disorders	2–150 μM	0.37 μM	95

GCE	Hybrid CNT	DPV	Electrocatalytic	Neuro-degenerative diseases	2–150 μM	0.61 μM	95
GCE	AuNP/$NiFe_2O_4$@GO nanosheet	DPV	Immunoassay	Alzheimer's disease	1 pg/mL–1 ng/mL	3.0 pg/mL	96
GCE	Dianix yellow/MWCNT	DPV	Electrocatalytic	Hyperuricemia, physiological disorders	0.100–7.500 μM	8.9 nM	97
CPE	CNT/FAacid/AgNP	CV, DPV, EIS, A	Electrocatalytic	Kidney disorders	1×10^{-8}–2×10^{-4} M	0.008 μM	98
GCE	N-doped porous C-SbNP	CV	Electrocatalytic	Prostate cancer	0.6–1 μM	0.74 μM	99
SPCE	Polymethylene blue/MWCNT	DPV	MIP	Myocardial infarction	0.10–8 pg/mL	0.040 pg/mL	100
SPCE	Graphene	EIS	Immunoassay	SARS-CoV-2	0.25 fg/mL–100 ng/mL	0.25 fg/mL	101
SPCE	AuNC@MWCNT-NH_2	DPV	Oligonucleotide	NSCLC	10^{-7}–10^{-14} M	42.8 fM	102
StPCE	GO	SWV	Immunoassay	CVDs	0.001–100 μg/mL	0.38 ng/mL	103
StPCE	GO	SWV	Immunoassay	CVDs	0.001–250 ng/mL	0.16 pg/mL	103
StPCE	GO	SWV	Immunoassay	Sepsis	0.0001–250 ng/mL	0.27 pg/mL	103
SPCE	GQD	CV	Aptamer	HIV	0.93 ng/mL–93 μg/mL	51.7 pg/mL	104
GCE	Prussian blue/MWCNT functionalized polypyrrole nanowire array	EIS	Oligonucleotide	NSCLC	0.1 pM–1 nM	33.4 fM	105
GCE	Covalent organic framework/MWCNT-NH_2/AuNP	DPV	Electrocatalytic	Neuro-degenerative diseases	0.7–108 μM	0.21 μM	106
GCE	Au/GO/COOH- MWCNT	DPV	Aptamer	Alzheimer's disease	0.1 pg/mL–1 ng/mL	0.088 pg/mL	107
SPCE	GO/Prussian blue nanocube	SWV	MIP	Alzheimer's disease	1.09–2.18 nmol/L	0.01 pmol/L	108
GCE	PtCoCu NP supported on N- and B-codoped rGO	A	Immunoassay	Alzheimer's disease	50.0 fg/mL–100 ng/mL	3.5 fg/mL	109
GCE	AuNPs/chitosan coated polyaniline functionalized activated C	DPV	Immunoassay	Sleep debt	1.95–1000 pg/mL	0.014 pg/mL	110

bioaffinity, and high electrochemical catalytic activity. GO was synthesized through the acid oxidation of graphite powder following the modified Hummers' method, while MoS_2 nanospheres were produced using a hydrothermal process. CV measurements assessed the electrochemical performance of electrodes modified with different materials. CV curves were recorded for bare GCE and those modified with MoS_2/rGO, PANI, and MoS_2-PANI/rGO. Compared to the bare GCE, the MoS_2/rGO and PANI-modified electrodes exhibited weak redox peaks, indicating that the films hindered electron transfer. However, for the MoS_2-PANI/rGO-modified GCE, the redox peaks were more pronounced, demonstrating that the MoS_2-PANI/rGO composite significantly enhances electron transfer at the electrolyte/electrode interface. DPV measurements revealed that the peak currents exhibited a linear relationship with the concentrations of AA and UA in the ranges of 50 μM–8.0 mM and 1.0 μM–500 μM, respectively, with low LOD of 22.20 μM for AA and 0.36 μM for UA. Furthermore, the 3D MoS_2-PANI/rGO-based electrochemical sensor showed high selectivity, good reproducibility, and stability for detecting small molecules.

Fu et al.[92] developed a novel electrochemical sensor combining multi-walled carbon nanotubes and platinum nanoparticles (MWCNTs-PtNPs) as peroxidase mimics for the simultaneous detection of homovanillic acid (HVA) and vanillylmandelic acid (VMA) in human urine. The simultaneous detection of HVA and VMA in urine is highly significant for the early diagnosis and identification of central nervous system disorders, including pheochromocytoma and neuroblastoma. They purchased MWCNTs and, to produce MWCNTs-PtNPs, dispersed MWCNTs in deionized water by ultrasound and then added H_2PtCl_6 and NaOH solutions. Excess NaBH4 solution was added dropwise in the final step, and the reaction was carried out for 12 hours. The modified electrode was characterized using DPV with 2×10^{-5} mol/L concentrations of HVA and VMA. A weak oxidation peak was observed on the bare GCE, and simultaneous detection of HVA and VMA was impossible. However, when the electrode was modified with PtNPs, MWCNTs, and MWCNTs-PtNPs, the oxidation peak currents for HVA and VMA increased progressively. This improvement is attributed to the large specific surface area of MWCNTs, combined with the excellent electrical conductivity and electrocatalytic performance of PtNPs. The combination of these materials displayed strong peroxidase-like activity, enhancing electron transfer at the electrode surface, allowing for the simultaneous detection of HVA and VMA with good resolution. The sensor demonstrated excellent resolution, with HVA and VMA being electrochemically separated by about 200 mV. Electrochemical studies showed that the MWCNTs-PtNPs-modified GCE provided superior separation performance for HVA and VMA with minimal overlap, indicating strong peroxidase-like catalytic activity. The sensor exhibited good linear detection ranges for HVA and VMA, from 2×10^{-7} to 8×10^{-5} mol/L and 5×10^{-7} to 8×10^{-5} mol/L, respectively, with LODs of 8×10^{-8} mol/L for HVA and 1.73×10^{-7} mol/L for VMA. The MWCNTs-PtNPs/GCE sensor was successfully applied to determine HVA and VMA in human urine samples simultaneously.

Fekry et al.[98] developed a sensitive electrochemical sensor by modifying a CPE with silver nanoparticles (AgNPs), MWCNTs, and folic acid (FA) to enhance the detection of creatinine. Creatinine is a byproduct of muscle metabolism constantly generated in the human body. Monitoring creatinine levels in blood and urine is clinically essential, as elevated blood creatinine levels indicate impaired kidney function. They prepared the CPE by grinding graphite powder with paraffin oil using a glass mortar. To create the sensor (SMFMCPE), CNTs, FA, and AgNPs were added to the mixture. They then ground the components for 10 minutes to ensure proper mixing. Afterward, they packed the modified carbon paste into a Teflon tube with a stainless-steel contact and pressed it onto filter paper to achieve a smooth surface. Techniques such as CV, DPV, chronoamperometry (CA), and EIS were used to study the sensor's electrocatalytic properties. The modified electrode showed excellent selectivity, stability, and a rapid response time of 1.5 seconds across a creatinine concentration range of 1×10^{-8}–2×10^{-4} M at pH 7.0, achieving an LOD of 0.008 mM. The sensor was also successfully applied to real human urine and serum sample analysis. Overall, the sensor demonstrated a low LOD, high sensitivity, and good reproducibility for creatinine detection.

Ehsan et al.[101] addressed the urgent need for a fast, accurate, and portable testing method for SARS-CoV-2, the virus responsible for the global pandemic that had caused over a million deaths and tens of millions of cases. They developed an electrochemical impedance biosensor (Figure 12.2A) for detecting the SARS-CoV-2 spike protein using an immunoglobulin G (IgG) anti-SARS-CoV-2 spike antibody. Their label-free sensor, based on SPEs, operates by measuring the impedance of a redox reaction and can directly detect antigen spikes in nasopharyngeal fluid and virus samples collected in the universal transport medium. To improve performance, they used high-conductivity graphene/carbon ink to minimize background impedance, thereby achieving a broader dynamic detection range. They fabricated single-strip sensing electrodes on cellulose fiber-based paper pads using a screen-printing and batch fabrication process that involved different masks and inks. The working electrode was printed using high-conductivity graphene/carbon hybrid ink, while the reference electrode was printed with silver/silver chloride ink, and the counter electrode was made with carbon ink. After printing each electrode, the strips were cured at a temperature of 100–140°C for 30 minutes. Conductive silver ink was used to print the connecting pads and wires, which were also temperature-cured. An ultraviolet (UV)-curable insulator ink was applied to cover the non-exposed areas, and after UV curing, the paper-based sensing strips were ready for the immobilization of biosensing elements. The antibody was immobilized onto the electrode surface using either a chemical or biological entity to assess their effects on sensor sensitivity. For direct COVID-19 antibody immobilization onto the electrode surface, they used 1-pyrenebutanoic acid succinimidyl ester (PBASE) as a probe linker. To prevent non-specific interactions, the free sites on the electrode surface were blocked using a 0.1% bovine serum albumin (BSA) solution in phosphate-buffered saline (PBS). Alternatively, for a more organized immobilization of the spike S1 IgG antibody, ProtA solution in PBS was applied to the graphene electrode surface for 1 hour. They analyzed the sensor fabrication process using CV and EIS. Starting with two electrodes, the peak currents and charge transfer resistance (Rct) were measured for both PBASE and ProtA-mediated binding. Both electrodes showed similar values for oxidation peak current (~50 μA) and Rct (~2 kΩ). The peak potentials for reduction and oxidation were close, indicating a reversible reaction with minimal redox conversion obstruction. After modifying the graphene sensor surfaces with PBASE, the redox peak current and Rct slightly shifted. In contrast, when modified with ProtA, a larger-sized biological entity with insulating properties, the redox peaks and Rct shifted significantly. These observations suggested that the added reagents acted as insulators, impeding electron transfer at the interface. However, this effect was partially reversed during antibody immobilization. The Rct continued to increase, confirming successful spike protein immobilization. The shift in Rct was more pronounced for the random antibody immobilization over PBASE compared to the more organized, ProtA-mediated process. Similar electrochemical behavior was observed when non-active sites were blocked with BSA, ensuring a lower impedance baseline for further quantitative tests. Biological immobilization was found to enhance antibody loading and improve sensitivity, leading to a wider linear detection range, though both methods resulted in a low limit of quantification of 0.25 fg/mL. The sensor's specificity was confirmed by testing against high concentrations of H1N1 flu antigens, which showed no significant response. The optimized sensors successfully and accurately distinguished between negative and positive COVID-19 samples.

Chen et al.[102] developed a novel and ultrasensitive electrochemical biosensor (Figure 12.2B) for detecting long non-coding RNAs (lncRNAs), which have been proposed as diagnostic biomarkers for non-small cell lung cancer (NSCLC) and disease progression monitoring. The biosensor was constructed using screen-printed carbon electrodes (SPCEs) decorated with gold nanocages and amidated multi-walled carbon nanotubes (AuNCs/MWCNT-NH_2). This design leveraged the large surface area, superior conductivity, and excellent materials biocompatibility. They purchased the MWCNTs and modified them by introducing active carboxyl groups at defect sites using an oxidizing acid, resulting in carboxyl-modified MWCNTs (MWCNTs-COOH). To prepare MWCNTs-NH_2, a condensation reaction was carried out between the –COOH and –NH_2 groups by adding

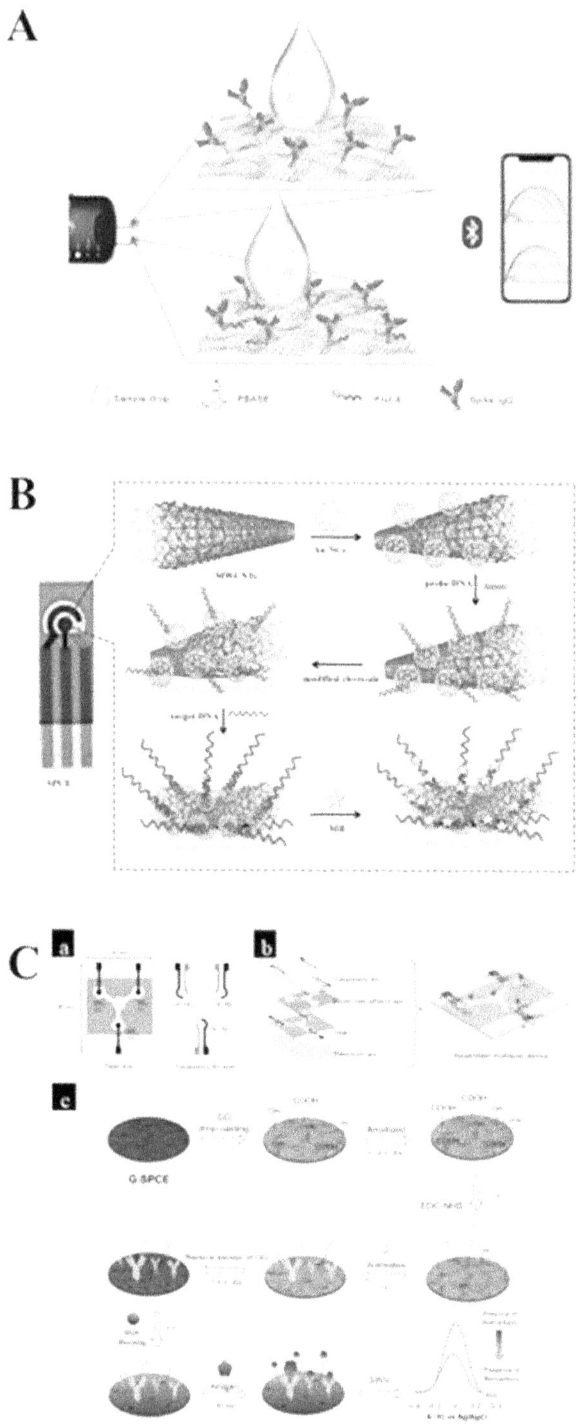

FIGURE 12.2 (A) A schematic illustration of the strategies adopted in this work for immunosensing of the coronavirus in the nasopharyngeal samples using paper substrates, and graphene electrode. (Reprinted with permission from Ehsan et al.[101]) (B) Schematic representation of the SPCE electrochemical DNA biosensor. (Reprinted with permission from Chen et al.[102]) (C) Design and fabrication of the multiplexed ePAD (a and b), and step-by-step preparation of the electrochemical immunosensor and detection of the antigen/biomarker on the sensor (c). (Reprinted with permission from Boonkaew et al.[103])

dicyclohexyl carbodiimide to a solution of MWCNTs-COOH in ethylenediamine. After washing with deionized water and drying the product in a vacuum at 35°C, MWCNTs-NH_2 were obtained. Next, the MWCNTs-NH_2 were dispersed in a solution containing synthesized AuNCs and stirred for 10 hours. The surface of the SPCE was cleaned with anhydrous ethanol and then coated with AuNCs/MWCNT-NH_2 suspension, which was air-dried at 25°C. To immobilize the 5′-thiolated RNA as the capture probe, 5 μL of lncRNA solution (1×10^{-8} M) was applied to the modified SPCE and incubated for 30 minutes at 30°C. CV was performed on various modified electrodes. Compared to the bare SPCE, AuNPs, AuNCs, and AuNP/MWCNTs-NH_2, the biosensor utilizing Au NCs/MWCNTs-NH_2 displayed a significantly higher electrochemical response. This enhanced performance was attributed to two main factors: (1) the novel hollow AuNCs offer a larger specific surface area and more active sites, resulting in excellent electrocatalytic activity, and (2) MWCNT-NH_2 serves as an effective support for AuNCs, creating a synergistic effect that boosts electrocatalytic performance. After assembling the capture probes onto the AuNCs/MWCNTs-NH_2 surface and hybridizing with the target lncRNA, the peak current decreased continuously as the negatively charged RNA biomolecules impeded electron transfer. This significant reduction in peak current confirmed the successful construction of the RNA biosensor. EIS was also employed to test the electrochemical properties of electrodes after each modification step. When AuNPs were attached to the bare SPCE, the electron transfer resistance (Ret) decreased. A further reduction in Ret was observed after modification with AuNCs, likely due to the hollow structure's ability to enhance electron transfer. The lowest Ret was recorded for AuNCs/MWCNTs-NH_2. However, when non-conducting RNA molecules were introduced and immobilized on the nanomaterial surface, the Ret increased significantly. The EIS results aligned with the CV data, indicating the successful fabrication of the SPCE-based biosensor. To evaluate the sensitivity of the proposed method for lncRNA detection, DPV was employed to measure the electrochemical response. The SPCE AuNCs/MWCNT-NH_2 lncRNA biosensor demonstrated a wide linear detection range (10^{-7}–10^{-14} M) and a LOD of 42.8 fM. Additionally, the biosensor exhibited excellent selectivity, stability, and overall performance. Compared to traditional real-time polymerase chain reaction (RT-PCR) methods, this biosensor provided similar stability and selectivity while being simpler to use, offering faster detection, and requiring less expensive materials. The proposed biosensor shows great potential as a powerful tool for detecting lncRNAs in clinical applications for efficient prognosis and cancer diagnosis.

Boonkaew et al.[103] addressed the global need for sensitive and affordable point-of-care devices for diagnosing and monitoring CVDs, the leading cause of death worldwide. They developed an electrochemical paper-based analytical device (ePAD) combined with a label-free immunoassay to simultaneously detect three key CVD biomarkers: C-reactive protein (CRP), troponin I (cTnI), and procalcitonin (PCT). The ePAD (Figure 12.2C) was fabricated by layering a wax-patterned paper and a transparency film using laser-cut double-sided adhesive, which defined the sample inlet, fluidic channels, and three separate detection zones for each biomarker. First, GO dispersed in water was drop-cast onto each working electrode through a small hole in the transparency film layer. The electrodes were then baked in an oven to evaporate the solvent, creating GO-modified electrodes (GO/G-SPCEs). To increase the number of carboxyl functional groups on the GO sheets, the GO/G-SPCEs were treated with NaOH, followed by two washes with water. This solution was applied through a central inlet in the transparency film. The target biomarker antibodies were immobilized on GO-modified carbon electrodes that were stencil-printed on the ePAD. For that, EDC/sulfo-NHS solution was added to each working electrode through a small hole in the transparency film to activate the carboxyl groups. Afterward, monoclonal antibodies were applied to the working electrodes and incubated for 1 hour at 2°C, followed by washing. The capture antibodies were covalently immobilized onto the GO-modified electrodes through a direct reaction between the amine groups (–NH_2) of the antibodies and the amine-reactive sulfo-NHS ester groups on the working electrodes, forming amide bonds (–CONH–) between them. The remaining unreacted carboxyl groups on the modified electrodes were reduced by applying –1.4 V for 30 seconds with PBS, followed by washing.

To reduce or eliminate non-specific binding on the electrodes, BSA was added to the working electrodes. Biomarker concentrations were measured using square wave voltammetry (SWV), which showed a significant decrease in current response in the presence of the cardiac biomarkers in a concentration-dependent manner. In contrast, no change in current response was observed in the absence of biomarkers. The device demonstrated excellent linearity ($R^2 > 0.99$), low LODs (0.38 ng/mL for CRP, 0.16 pg/mL for cTnI, and 0.27 pg/mL for PCT), and good reproducibility (RSD < 5%). Additionally, they applied the immunosensor successfully to detect the three biomarkers simultaneously in serum samples, yielding satisfactory results. The proposed immunosensor holds great promise for early detection of CVDs in point-of-care settings.

Hassine et al.[108] developed a sensing device (Figure 12.3A) for the selective and sensitive detection of the Tau-441 protein, a promising biomarker for the early diagnosis of Alzheimer's disease. To achieve this, they used SPCEs decorated with GO and Prussian blue nanocubes (GO/PBNCs). For the preparation of the biographene (bG), a suspension of graphite crystals was made containing protein. This mixture was ground using a kitchen blender. The concentration of bG was quantified by measuring the absorbance at 660 nmol/L after the non-exfoliated graphite was separated by centrifugation. With constant stirring, $FeCl_3$, poly(ethyleneimine) (PEI), and $K_3Fe(CN)_6$ were added to the bG dispersion. The mixture was heated and maintained under reflux, during which a color change from yellow to dark blue indicated the formation of the nanocomposite. A clean SPCE was then modified with the synthesized material. This modification involved applying 10 mg/mL of the rGO/PBNC nanocomposite onto the working electrodes and curing at 75°C for 30 minutes. A MIP was then constructed on the GO/PBNCs/SPCEs by electropolymerizing 3-aminophenol (3-AMP) in the presence of the target Tau-441 protein using CA. For this, the working electrodes of the SPCEs modified with the synthesized nanocomposites were exposed to CA-based polymerization at 1.2 V for 200 seconds. This was done after covering the three electrodes of the SPCEs with a solution containing AMP and Tau-441 in PBS. The electrodes were then washed with deionized water and incubated with oxalic acid. After this removal step, another wash with deionized water was performed, followed by electrochemical stabilization using CV in PBS. The target protein was then re-bound to the cavities formed in the film by incubating the electrodes for 30 minutes. The template was subsequently removed from the polymer matrix with oxalic acid. For comparison purposes, they also prepared a non-imprinted polymer (NIP) in the absence of the target protein. Each modification step resulted in changes to the surface's electroactive properties, which were monitored through CV, SWV, and EIS. They drop-cast the synthesized nanomaterial onto the working electrodes, which provided a conductive and electroactive surface due to the presence of GO and PBNCs. The electropolymerization of 3-AMP mixed with Tau-441 caused a decrease in current for the various polymer films. Both NIP and MIP resulted in non-conductive, electrically inactive films on the rGO/PBNCs-modified electrode surface, as indicated by the sharp current decrease in the CV data. The difference in current between the NIP and MIP after electropolymerization was due to the presence of the protein in the MIP layer, which interfered with the electropolymerization process. EIS was performed to track the electrical changes and confirm the presence of a barrier film on the nanocomposite surface. The Nyquist diagram they obtained showed two frequency ranges: a semicircle at high frequencies, indicating charge transfer processes, and linear behavior at low frequencies. The semicircle diameter, which represented the Rct value, indicated the electron transfer rate at the electrode-solution interface. They detected the concentration of the target protein during the rebinding step by monitoring the electroactivity of the Prussian blue redox signal using SWV for both MIP and NIP after drop-casting the target protein onto the working electrode and incubating it with increasing concentrations of Tau-441. The calibration curves showed a linear relationship between current and the logarithm of Tau-441 concentration, ranging from 1.09 to 2.18 nmol/L. Rebinding with the target protein resulted in a consistent current decrease for the MIP-based platform, which exhibited a strong linear response with a correlation coefficient greater than 0.98. The LOD for the MIP-based platform was below 0.01 pmol/L. In contrast, the NIP-based platform

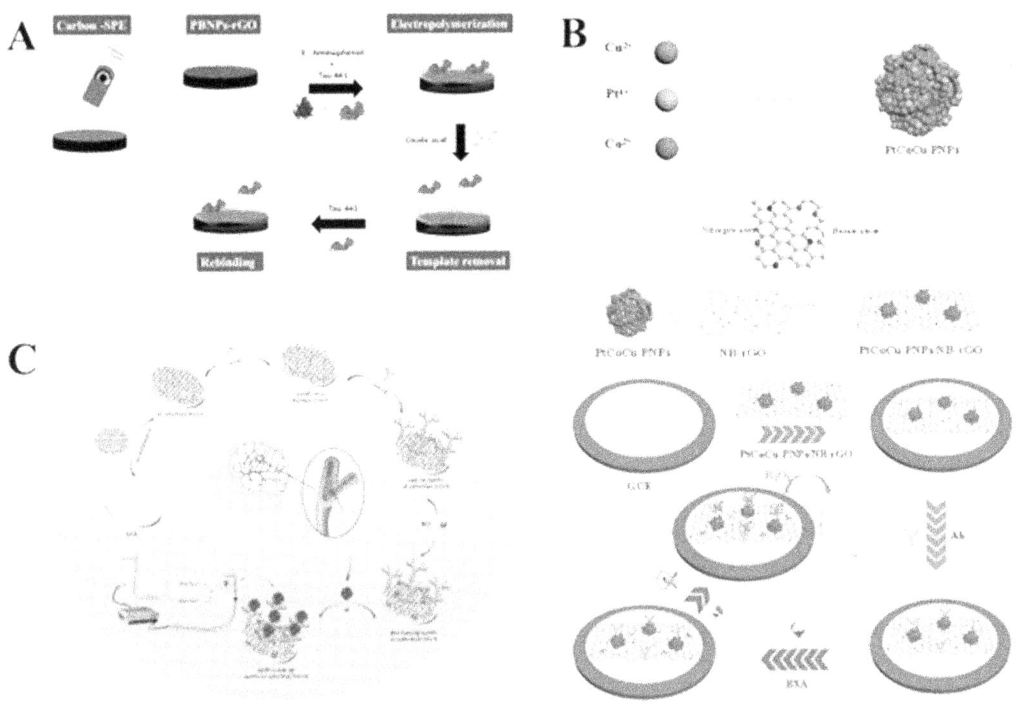

FIGURE 12.3 (A) Synthesis of the MIP-based sensor. (Reprinted with permission from Ben Hassine et al.[108]) (B) Schematic presentation of the immunosensor. (Reprinted with permission from Wang et al.[109]) (C) Schematic illustration of electrochemical immunosensor based on AuNPs/AC@PANI@CS for detection of amyloid-β (Aβ). (Reprinted with permission from Sun et al.[110])

showed a random behavior during incubation with the target. The MIP sensor demonstrated minimal interference from UA and BSA, highlighting its selectivity.

Wang et al.[109] developed a label-free electrochemical immunosensor (Figure 12.3B) for the sensitive detection of β-Amyloid1-42 oligomers (Aβ), the accumulation of which in the brain has been shown to initiate Alzheimer's disease. They utilized popcorn-shaped PtCoCu nanoparticles supported on nitrogen- and boron-codoped rGO (PtCoCu PNPs/NB-rGO). They drop-cast PtCoCu PNPs/NB-rGO onto the surface of the GCE using a pipetting gun. They evaluated the catalytic properties of the nanomaterials using amperometry. The initial GCE showed no current response. The current response of NB-rGO was nearly as high as that of noble metal platinum, while PtCoCu PNPs exhibited a higher current response than platinum at the same concentration, indicating that Cu and Co atoms enhanced catalytic efficiency at a lower cost. The combination of PtCoCu PNPs and NB-rGO resulted in a sharp increase in amperometric response, as the well-dispersed PtCoCu PNPs within the layered structure of NB-rGO improved conductivity and increased the number of catalytic active sites per unit area. CV curves of electrodes modified with PtCoCu PNPs and PtCoCu PNPs/NB-rGO were also presented, at scan rates ranging from 0.01 V/s to 0.1 V/s. As the scan rate increased, the redox peak potential shifted to more positive or negative values, with the current peak rising steadily due to increasing internal diffusion resistance. The linear relationship between peak current density and the square root of the scan rate indicated that a diffusion-controlled electrochemical process occurred at the electrode interface. The electroactive areas of PtCoCu PNPs and PtCoCu PNPs/NB-rGO were determined based on the CV data. Next, they applied antibody onto the GCE. To block non-specific active sites, BSA was added to the GCE. After drying at 4°C, 6.0 μL of Aβ in varying concentrations was incubated on the GCE for 60 minutes at 4°C, followed by rinsing off unbound antigens with PBS. EIS was used to confirm the successful construction of the immunosensor at each step. The initial GCE displayed almost no semicircle, indicating fast electron transfer. After modification with PtCoCu PNPs/NB-rGO, electron transfer rates were high due to low resistance at the electrode interface. However, as antibodies (Ab) and BSA were added, the semicircles expanded significantly, as these macromolecular proteins blocked electron transport channels and increased resistance. The maximum resistance was observed after Aβ was added to the BSA/Ab/PtCoCu PNPs/NB-rGO/GCE, confirming the precise recognition between the target antigen and the antibody, indicating successful construction of the immunosensor. The CV results corroborated the EIS data. The redox peak current of PtCoCu PNPs/NB-rGO/GCE was higher than that of the bare GCE and reached a maximum. However, the peak current decreased successively after Ab, BSA, and Aβ were added, further proving the successful assembly of the immunosensor. The PtCoCu nanoparticles exhibited excellent catalytic properties due to their popcorn-like structure, which increased the specific surface area and porosity, exposing more active sites and providing rapid ion/electron transport pathways. The NB-rGO, with its large surface area and pleated structure, facilitated the dispersion of PtCoCu PNPs via electrostatic adsorption and the formation of d-p dative bonds between the metal ions and pyridinic nitrogen in the NB-rGO. Overall, the doping of boron atoms further enhanced the catalytic capability of the rGO, amplifying the signal significantly. Additionally, both PtCoCu PNPs and NB-rGO effectively immobilized a large number of antibodies through M(Pt, Co, Cu)-N bonds and amide bonds, without requiring complex procedures such as carboxylation. This design enabled dual amplification of the electrocatalytic signal and efficient antibody immobilization. Under optimal conditions, the immunosensor demonstrated a wide linear detection range (50.0 fg/mL–100 ng/mL) and a low LOD of 3.5 fg/mL. The results indicated that immunosensor was highly promising for the sensitive detection of Alzheimer's disease biomarkers.

Sun et al.[110] developed an ultrasensitive electrochemical immunosensor (Figure 12.3C) for the detection of the sleep debt biomarker, amyloid-β (Aβ), with the aim of providing an early and objective assessment, which is linked to physiological and mental illness as well as unsafe work. The sensing platform was constructed using gold nanoparticles/chitosan-coated polyaniline-functionalized activated carbon (AuNPs/AC@PANI@CS) composites. The AC@PANI composites were synthesized through chemical oxidation polymerization. First, activated carbon (AC) was added to HCl solution and stirred. After cooling the solution below 5°C, aniline was added. Then, ammonium persulfate was prepared, and slowly added dropwise to the mixture, allowing the polymerization reaction to proceed for 24 hours. The resulting precipitate was filtered, washed, and dried to obtain the PANI@AC composite. Afterward, AC@PANI composite was mixed with chitosan solution and treated with ultrasound for 1 hour to obtain a uniform black dispersion, designated as AC@PANI@CS. The AC@PANI@CS dispersion was applied to the GCE to create the AC@PANI@CS-modified GCE (AC@PANI@CS/GCE). AuNPs were electrodeposited onto the AC@PANI@CS/GCE surface using CV in a potential range of –0.4–0.8 V at a scan rate of 60 mV/s in H_2SO_4 containing $HAuCl_4$. CV was used to monitor and compare the electrochemical behaviors of AC, PANI, AC@PANI, AC@PANI@CS, and AuNPs/AC@PANI@CS on the GCE. When AC@PANI was applied to modify the GCE, forming AC@PANI/GCE, the observed peak current was higher than that of bare GCE, AC-modified GCE (AC/GCE), and PANI-modified GCE (PANI/GCE). This is believed to be due to the combination of PANI and AC creating an effective conductive pathway, enhancing electrolyte penetration on the electrode surface and enabling rapid charge and ion transport. The electrochemical response of AC@PANI@CS/GCE further increased, due to the improved dispersion and increased specific surface area provided by the chitosan. The peak current of AuNPs/AC@PANI@CS/GCE was higher than that of AC@PANI@CS/GCE, attributed to the high specific surface area and affinity of AuNPs, which combined well with the high electrical conductivity of the base material. Additionally, EIS confirmed changes in the impedance of the modified electrode surface. Compared to GCE, AC/GCE, and PANI/GCE, AC@PANI/GCE exhibited lower Rct, demonstrating that AC@PANI is an excellent conductive material that facilitates electron transfer. When AC@PANI@CS/GCE was applied to the electrode, the Rct decreased further, due to its good conductivity. Upon modifying the electrode with AuNPs to form AuNPs/AC@PANI@CS/GCE, the Rct decreased even more, confirming the enhanced electron transfer. The modified electrode was next incubated in anti-Aβ solution for 4 hours at 4°C to immobilize the anti-Aβ. Any unbound anti-Aβ was removed by rinsing with PBST (PBS containing 0.05% Tween-20 at pH 7.0). Finally, the anti-Aβ/AuNPs/AC@PANI@CS/GCE was blocked with BSA solution containing 0.05% Tween-20 to block any remaining active sites and prevent non-specific adsorption. The DPV response of the immunosensor to Aβ showed a gradual decrease in DPV intensity as Aβ concentration increased. The current response exhibited a strong linear correlation with the logarithm of Aβ concentration, ranging from 1.95 pg/mL to 1000.00 pg/mL, with an LOD of 0.014 pg/mL. As the combination of PANI and AC is believed to have created an effective conductive pathway, it significantly enhanced the electrochemical response of the immunosensor.

12.4 CONCLUSION AND FUTURE EXPECTATIONS IN APPLICATIONS

In recent years, carbon sensor devices have demonstrated remarkable potential in the field of disease detection, offering enhanced sensitivity, selectivity, and biocompatibility. Their ability to detect biomarkers for a wide range of diseases, including cancer, CVDs, diabetes, and infectious diseases, has led to their growing popularity in both research and clinical settings. Carbon-based materials, such as graphene, CNTs, and carbon nanofibers, provide an excellent platform for developing efficient and cost-effective sensors that can be used in point-of-care diagnostics.

As these sensors continue to evolve, their integration with advanced technologies like microfluidics and wearable devices is expected to revolutionize early diagnosis and disease monitoring. The miniaturization of carbon-based sensors for portable, non-invasive devices offers exciting prospects for continuous monitoring and personalized healthcare. These devices could enable real-time tracking of biomarkers, allowing patients and healthcare providers to make informed decisions based on up-to-date data.

Looking ahead, future advancements will likely focus on improving the sensitivity, stability, and reproducibility of carbon sensors to meet the stringent requirements of clinical applications. Additionally, there is potential for the development of multifunctional carbon-based sensors capable of detecting multiple biomarkers simultaneously, increasing the diagnostic power of these devices.

However, one of the key challenges will be ensuring scalability and cost-effectiveness while maintaining high performance, particularly as these sensors transition from laboratory prototypes to commercial products. As research progresses, carbon sensor devices are expected to play an even more critical role in disease detection, opening up new possibilities for early diagnosis, personalized treatment, and improved patient outcomes. To sum up, with continued innovation, carbon-based sensors have the potential to significantly enhance the future landscape of medical diagnostics.

REFERENCES

(1) Campuzano, S.; Pedrero, M.; Yáñez-Sedeño, P.; Pingarrón, J. M. New challenges in point of care electrochemical detection of clinical biomarkers. *Sensors and Actuators B: Chemical* **2021**, *345*, 130349.

(2) Zhang, W.; Wang, R.; Luo, F.; Wang, P.; Lin, Z. Miniaturized electrochemical sensors and their point-of-care applications. *Chinese Chemical Letters* **2020**, *31* (3), 589–600.

(3) Şenocak, A.; Tümay, S. O.; Sarı, E.; Şanko, V.; Durmuş, M.; Demirbas, E. The simultaneously voltammetric determination of spinosad and chlorantraniliprole pesticides by carbazole-ferrocene functionalized carbon nanotube architecture. *Journal of The Electrochemical Society* **2021**, *168* (8), 087513.

(4) Erkmen, C.; Sanko, V.; Ozturk, B. O.; Quinchía, J.; Orozco, J.; Kuralay, F. Mesoporous silica-based electrochemical biosensors in the determination of cancer biomarkers: Current progress on analytical performance and future trends. *TrAC Trends in Analytical Chemistry* **2024**, *179*, 117876.
(5) Fu, S.; Zhu, Y.; Zhang, Y.; Zhang, M.; Zhang, Y.; Qiao, L.; Yin, N.; Song, K.; Liu, M.; Wang, D. Recent advances in carbon nanomaterials-based electrochemical sensors for phenolic compounds detection. *Microchemical Journal* **2021**, *171*, 106776.
(6) Cetinkaya, A.; Kaya, S. I.; Ozcelikay, G.; Budak, F.; Ozkan, S. A. Carbon nanomaterials-based novel hybrid platforms for electrochemical sensor applications in drug analysis. *Critical Reviews in Analytical Chemistry* **2024**, *54* (5), 1227–1242. DOI: 10.1080/10408347.2022.2109125.
(7) Gergeroglu, H.; Yildirim, S.; Ebeoglugil, M. F. Nano-carbons in biosensor applications: An overview of carbon nanotubes (CNTs) and fullerenes (C60). *SN Applied Sciences* **2020**, *2* (4), 603. DOI: 10.1007/s42452-020-2404-1.
(8) Gibi, C.; Liu, C.-H.; Barton, S. C.; Anandan, S.; Wu, J. J. Carbon materials for electrochemical sensing application – A mini review. *Journal of the Taiwan Institute of Chemical Engineers* **2024**, 154, 105071.
(9) Venkatesan, M.; Vusa, C. S. R.; Koyappayil, A.; Chavan, S. G.; Lee, M.-H. Biological recognition element anchored 2D graphene materials for the electrochemical detection of hazardous pollutants. *Electrochimica Acta* 2024, *494*, 144413.
(10) Meskher, H.; Ragdi, T.; Thakur, A. K.; Ha, S.; Khelfaoui, I.; Sathyamurthy, R.; Sharshir, S. W.; Pandey, A. K.; Saidur, R.; Singh, P.; et al. A review on CNTs-based electrochemical sensors and biosensors: Unique properties and potential applications. *Critical Reviews in Analytical Chemistry*, *54* (7), 2398–2421.
(11) Fahmy, H. M.; Abu Serea, E. S.; Salah-Eldin, R. E.; Al-Hafiry, S. A.; Ali, M. K.; Shalan, A. E.; Lanceros-Méndez, S. Recent progress in graphene- and related carbon-nanomaterial-based electrochemical biosensors for early disease detection. *ACS Biomaterials Science & Engineering* **2022**, *8* (3), 964–1000.
(12) Jiwanti, P. K.; Wardhana, B. Y.; Sutanto, L. G.; Chanif, M. F. A review on carbon-based electrodes for electrochemical sensor of quinolone antibiotics. *ChemistrySelect* **2022**, *7* (15), e202103997.
(13) Karimi, F.; Karimi-Maleh, H.; Rouhi, J.; Zare, N.; Karaman, C.; Baghayeri, M.; Fu, L.; Rostamnia, S.; Dragoi, E. N.; Ayati, A.; et al. Revolutionizing cancer monitoring with carbon-based electrochemical biosensors. *Environmental Research* **2023**, *239*, 117368.
(14) Chen, K.; Zhao, H.; Wang, Z.; Zhou, F.; Lan, M. Sandwich-type electrochemical aptasensor based on polydopamine-Au-metal ions as signal label and coralloid Au-conductive carbon architecture-modified electrode for the simultaneous detection of cardiac troponin I and myoglobin. *Sensors and Actuators B: Chemical* **2023**, *390*, 134044.
(15) Dervin, S.; Ganguly, P.; Dahiya, R. S. Disposable electrochemical sensor using graphene oxide–chitosan modified carbon-based electrodes for the detection of tyrosine. *IEEE Sensors Journal* **2021**, *21* (23), 26226–26233.
(16) Yan, M.; Fu, L.-l.; Feng, H.-c.; Namadchian, M. Application of Ag nanoparticles decorated on graphene nanosheets for electrochemical sensing of CEA as an important cancer biomarker. *Environmental Research* **2023**, *239*, 117363.
(17) Alnaimi, A.; Al-Hamry, A.; Makableh, Y.; Adiraju, A.; Kanoun, O. Gold nanoparticles-MWCNT based aptasensor for early diagnosis of prostate cancer. *Biosensors*, **2022**, *12*.
(18) Zheng, W.; Liu, Y.; Yang, P.; Chen, Y.; Tao, J.; Hu, J.; Zhao, P. Carbon nanohorns enhanced electrochemical properties of Cu-based metal organic framework for ultrasensitive serum glucose sensing. *Journal of Electroanalytical Chemistry* **2020**, *862*, 114018.
(19) Fritea, L.; Banica, F.; Costea, T. O.; Moldovan, L.; Dobjanschi, L.; Muresan, M.; Cavalu, S. Metal nanoparticles and carbon-based nanomaterials for improved performances of electrochemical (bio) sensors with biomedical applications. *Materials* **2021**, *14*.
(20) Elugoke, S. E.; Adekunle, A. S.; Fayemi, O. E.; Mamba, B. B.; Sherif, E.-S. M.; Ebenso, E. E. Carbon-based quantum dots for electrochemical detection of monoamine neurotransmitters—Review. *Biosensors* **2020**, *10*.
(21) Hadian, N. S.; Faridnouri, H.; Zare, E. N. Glucose biosensing based on glucose oxidase immobilization on carboxymethyl chitosan/polyaniline/multi-walled carbon nanotubes nanocomposite. *Diamond and Related Materials* **2024**, *148*, 111423.

(22) Hasabnis, G. K.; Altintas, Z. Cardiac troponin I-responsive nanocomposite materials for voltammetric monitoring of acute myocardial infarction. *ACS Omega* **2024**, *9* (28), 30737–30750. DOI: 10.1021/acsomega.4c03252.
(23) Akin, M.; Bayat, R.; Bekmezci, M.; Coguplugil, Z. K.; Sen, F.; Baghayeri, M.; Kaffash, A.; Tchranejad-Javazmi, F.; Sheikhshoaie, I. The use of polymer/carbon based material as an efficient and low-cost electrochemical sensor for rapid electrochemical detection of dopamine. *Carbon Letters* **2023**, *33* (7), 1993–2001.
(24) Saravanakumar, V.; Rajagopal, V.; Narayanan, K.; Nesakumar, N.; Kathiresan, M.; Suryanarayanan, V.; Anandan, S. Cu-MOF/Pd derived oxide nanoparticles based carbon composite – An innovative electrochemical sensing platform for Bisphenol A. *Journal of Alloys and Compounds* **2023**, *963*, 171216.
(25) Parihar, A.; Choudhary, N. K.; Sharma, P.; Khan, R. Carbon nanomaterials-based electrochemical aptasensor for point-of-care diagnostics of cancer biomarkers. *Materials Today Chemistry* **2023**, *30*, 101499.
(26) Sanko, V.; Erkmen, C.; Kuralay, F. The role of nanoparticles in non-invasive electrochemical immunosensor technology and recent developments. *Electroanalysis*, **2024**, *36*, e202400004.
(27) Ricci, F.; Adornetto, G.; Palleschi, G. A review of experimental aspects of electrochemical immunosensors. *Electrochimica Acta* **2012**, *84*, 74–83.
(28) Dhanjai; Sinha, A.; Lu, X.; Wu, L.; Tan, D.; Li, Y.; Chen, J.; Jain, R. Voltammetric sensing of biomolecules at carbon based electrode interfaces: A review. *TrAC Trends in Analytical Chemistry* **2018**, *98*, 174–189.
(29) Casanova, A.; Iniesta, J.; Gomis-Berenguer, A. Recent progress in the development of porous carbon-based electrodes for sensing applications. *Analyst* **2022**, *147* (5), 767–783.
(30) Abdel-Aziz, A. M.; Hassan, H. H.; Badr, I. H. A. Activated glassy carbon electrode as an electrochemical sensing platform for the determination of 4-nitrophenol and dopamine in real samples. *ACS Omega* **2022**, *7* (38), 34127–34135.
(31) David, I. G.; Popa, D.-E.; Buleandra, M. Pencil graphite electrodes: A versatile tool in electroanalysis. *Journal of Analytical Methods in Chemistry* **2017**, *2017* (1), 1905968.
(32) Sajeevan, B.; Gopika, M. G.; Murali, A. S.; Saraswathyamma, B. An electrochemical sensor based on a pencil graphite electrode modified with poly-riboflavin for 4-nitrophenol quantification. *Materials Chemistry and Physics* **2023**, *301*, 127568.
(33) Annu; Sharma, S.; Jain, R.; Raja, A. N. Pencil graphite electrode: An emerging sensing material. *Journal of The Electrochemical Society* **2020**, *167* (3), 037501.
(34) Ataide, V. N.; Arantes, I. V. S.; Mendes, L. F.; Rocha, D. S.; Baldo, T. A.; Coltro, W. K. T.; Paixão, T. R. L. C. a pencil drawing overview: From graphite to electrochemical sensors/biosensors applications. *Journal of the Electrochemical Society* **2022**, *169* (4), 047524.
(35) Švancara, I.; Vytřas, K.; Kalcher, K.; Walcarius, A.; Wang, J. Carbon paste electrodes in facts, numbers, and notes: A review on the occasion of the 50-years jubilee of carbon paste in electrochemistry and electroanalysis. *Electroanalysis: An International Journal Devoted to Fundamental and Practical Aspects of Electroanalysis* **2009**, *21* (1), 7–28.
(36) Tajik, S.; Beitollahi, H.; Nejad, F. G.; Safaei, M.; Zhang, K.; Van Le, Q.; Varma, R. S.; Jang, H. W.; Shokouhimehr, M. Developments and applications of nanomaterial-based carbon paste electrodes. *RSC advances* **2020**, *10* (36), 21561–21581.
(37) Milne, S. A.; Lasserre, P.; Corrigan, D. Fabrication of a graphite-paraffin carbon paste electrode and demonstration of its use in electrochemical detection strategies. *Analyst* **2024**.
(38) Mincu, N.-B.; Lazar, V.; Stan, D.; Mihailescu, C. M.; Iosub, R.; Mateescu, A. L. Screen-Printed Electrodes (SPE) for in vitro diagnostic purpose. *Diagnostics* **2020**, *10*.
(39) Pérez-Fernández, B.; Costa-García, A.; Muñiz, A. D. Electrochemical (bio)sensors for pesticides detection using screen-printed electrodes. *Biosensors* **2020**, *10*.
(40) Niyogi, S.; Hamon, M. A.; Hu, H.; Zhao, B.; Bhowmik, P.; Sen, R.; Itkis, M. E.; Haddon, R. C. Chemistry of single-walled carbon nanotubes. *Accounts of Chemical Research* **2002**, *35* (12), 1105–1113. DOI: 10.1021/ar010155r.
(41) Wang, J. Carbon-nanotube based electrochemical biosensors: A review. *Electroanalysis: An International Journal Devoted to Fundamental and Practical Aspects of Electroanalysis* **2005**, *17* (1), 7–14.

(42) Balasubramanian, K.; Burghard, M. Biosensors based on carbon nanotubes. *Analytical and Bioanalytical Chemistry* **2006**, *385* (3), 452–468.
(43) Kumar, S.; Sidhu, H. K.; Paul, A. K.; Bhardwaj, N.; Thakur, N. S.; Deep, A. Bioengineered multi-walled carbon nanotube (MWCNT) based biosensors and applications thereof. *Sensors & Diagnostics* **2023**, *2* (6), 1390–1413.
(44) Meredith, M. T.; Minson, M.; Hickey, D.; Artyushkova, K.; Glatzhofer, D. T.; Minteer, S. D. Anthracene-modified multi-walled carbon nanotubes as direct electron transfer scaffolds for enzymatic oxygen reduction. *ACS Catalysis* **2011**, *1* (12), 1683–1690.
(45) Wong, A.; Foguel, M. V.; Khan, S.; Oliveira, F. M. d.; Tarley, C. R. T.; Sotomayor, M. D. P. T. Development of an electrochemical sensor modified with MWCNT-COOH and MIP for detection of diuron. *Electrochimica Acta* **2015**, *182*, 122–130.
(46) Yang, Y.; Zhang, H.; Huang, C.; Jia, N. MWCNTs-PEI composites-based electrochemical sensor for sensitive detection of bisphenol A. *Sensors and Actuators B: Chemical* **2016**, *235*, 408–413.
(47) Pasinszki, T.; Krebsz, M.; Tung, T. T.; Losic, D. Carbon nanomaterial based biosensors for non-invasive detection of cancer and disease biomarkers for clinical diagnosis. *Sensors (Basel)* **2017**, *17* (8).
(48) Supraja, P.; Tripathy, S.; Krishna Vanjari, S. R.; Singh, S. G. Label-free, ultrasensitive and rapid detection of FDA-approved TBI specific UCHL1 biomarker in plasma using MWCNT-PPY nanocomposite as bio-electrical transducer: A step closer to point-of-care diagnosis of TBI. *Biosensors and Bioelectronics* **2022**, *216*, 114631.
(49) McCreery, R. L. Advanced carbon electrode materials for molecular electrochemistry. *Chemical Reviews* **2008**, *108* (7), 2646–2687.
(50) Ritzert, N. L.; Rodríguez-López, J.; Tan, C.; Abruña, H. D. Kinetics of interfacial electron transfer at single-layer graphene electrodes in aqueous and nonaqueous solutions. *Langmuir* **2013**, *29* (5), 1683–1694.
(51) Idris, A. O.; Mabuba, N.; Arotiba, O. A. An exfoliated graphite-based electrochemical immunosensor on a dendrimer/carbon nanodot platform for the detection of carcinoembryonic antigen cancer biomarker. *Biosensors* **2019**, *9*.
(52) Tiwari, C.; Jha, S. S.; Kumar, R.; Chhabra, M.; Malhotra, B. D.; Dixit, A. Exfoliated graphite carbon paper-based flexible nonenzymatic glucose sensor. *Materials Science and Engineering: B* **2022**, *285*, 115931.
(53) de Lima, L. F.; Ferreira, A. L.; Torres, M. D. T.; de Araujo, W. R.; de la Fuente-Nunez, C. Minute-scale detection of SARS-CoV-2 using a low-cost biosensor composed of pencil graphite electrodes. *Proceedings of the National Academy of Sciences* **2021**, *118* (30), e2106724118.
(54) Dokur, E.; Uruc, S.; Kurteli, R.; Gorduk, O.; Sahin, Y. Investigation of flexible electrochemical sensor properties of graphite conductive ink modified textile products and their applications in electrical circuits. *Materials Today Communications* **2023**, *37*, 106920.
(55) Geim, A. K.; Novoselov, K. S. The rise of graphene. *Nature Materials* **2007**, *6* (3), 183–191.
(56) Shao, Y.; Wang, J.; Wu, H.; Liu, J.; Aksay, I. A.; Lin, Y. Graphene based electrochemical sensors and biosensors: A review. *Electroanalysis: An International Journal Devoted to Fundamental and Practical Aspects of Electroanalysis* **2010**, *22* (10), 1027–1036.
(57) Navalon, S.; Dhakshinamoorthy, A.; Alvaro, M.; Antonietti, M.; García, H. Active sites on graphene-based materials as metal-free catalysts. *Chemical Society Reviews* **2017**, *46* (15), 4501–4529.
(58) Cao, M.; Xiong, D. B.; Yang, L.; Li, S.; Xie, Y.; Guo, Q.; Li, Z.; Adams, H.; Gu, J.; Fan, T. Ultrahigh electrical conductivity of graphene embedded in metals. *Advanced Functional Materials* **2019**, *29* (17), 1806792.
(59) Nag, A.; Mitra, A.; Mukhopadhyay, S. C. Graphene and its sensor-based applications: A review. *Sensors and Actuators A: Physical* **2018**, *270*, 177–194.
(60) Dhinakaran, V.; Vigneswari, K.; Lavanya, M.; Varsha Shree, M. Chapter Ten – Point-of-care applications with graphene in human life. In *Comprehensive Analytical Chemistry*, Hussain, C. M. Ed.; Vol. 91; Elsevier, **2020**; pp 235–262.
(61) Pilehvar, S.; De Wael, K. Recent advances in electrochemical biosensors based on fullerene-C60 nano-structured platforms. *Nanocarbons for Electroanalysis* **2017**, 173–196.
(62) Guldi, D. M.; Asmus, K.-D. Photophysical properties of mono- and multiply-functionalized fullerene derivatives. *The Journal of Physical Chemistry A* **1997**, *101* (8), 1472–1481. DOI: 10.1021/jp9633557.

(63) Reddy, S. S.; Shukla, B.; Chakraborty, S.; Srihari, V.; Bhalerao, G. M.; Shekar, N. V. C. Study on high-pressure behaviour of spherical carbon black nanoparticles with core–shell structure. *Carbon Letters* **2022**, *32* (5), 1337–1344.
(64) Silva, T. A.; Moraes, F. C.; Janegitz, B. C.; Fatibello-Filho, O. Electrochemical biosensors based on nanostructured carbon black: A review. *Journal of Nanomaterials* **2017**, *2017* (1), 4571614.
(65) Liu, C.-C.; Walters, A. B.; Vannice, M. A. Measurement of electrical properties of a carbon black. *Carbon* **1995**, *33* (12), 1699–1708.
(66) Bradley, R. H.; Sutherland, I.; Sheng, E. Carbon surface: Area, porosity, chemistry, and energy. *Journal of Colloid and Interface Science* **1996**, *179* (2), 561–569.
(67) Arduini, F.; Cinti, S.; Mazzaracchio, V.; Scognamiglio, V.; Amine, A.; Moscone, D. Carbon black as an outstanding and affordable nanomaterial for electrochemical (bio)sensor design. *Biosensors and Bioelectronics* **2020**, *156*, 112033.
(68) Baccarin, M.; Santos, F. A.; Vicentini, F. C.; Zucolotto, V.; Janegitz, B. C.; Fatibello-Filho, O. Electrochemical sensor based on reduced graphene oxide/carbon black/chitosan composite for the simultaneous determination of dopamine and paracetamol concentrations in urine samples. *Journal of Electroanalytical Chemistry* **2017**, *799*, 436–443.
(69) Arduini, F.; Zanardi, C.; Cinti, S.; Terzi, F.; Moscone, D.; Palleschi, G.; Seeber, R. Effective electrochemical sensor based on screen-printed electrodes modified with a carbon black-Au nanoparticles composite. *Sensors and Actuators B: Chemical* **2015**, *212*, 536–543.
(70) Tığ, G. A. Development of electrochemical sensor for detection of ascorbic acid, dopamine, uric acid and L-tryptophan based on Ag nanoparticles and poly(L-arginine)-graphene oxide composite. *Journal of Electroanalytical Chemistry* **2017**, *807*, 19–28.
(71) Wei, Z.; Zhang, J.; Zhang, A.; Wang, Y.; Cai, X. Electrochemical detecting lung cancer-associated antigen based on graphene-gold nanocomposite. *Molecules* **2017**, *22*.
(72) Ma, Y.; Shen, X.-L.; Zeng, Q.; Wang, H.-S.; Wang, L.-S. A multi-walled carbon nanotubes based molecularly imprinted polymers electrochemical sensor for the sensitive determination of HIV-p24. *Talanta* **2017**, *164*, 121–127.
(73) Gowda, J. I.; M, M.; Nandibewoor, S. T. CTAB functionalized multiwalled carbon nanotube composite modified electrode for the determination of 6-mercaptopurine. *Sensing and Bio-Sensing Research* **2017**, *12*, 1–7.
(74) Heydari-Bafrooei, E.; Shamszadeh, N. S. Electrochemical bioassay development for ultrasensitive aptasensing of prostate specific antigen. *Biosensors and Bioelectronics* **2017**, *91*, 284–292.
(75) Rezaei, B.; Shoushtari, A. M.; Rabiee, M.; Uzun, L.; Mak, W. C.; Turner, A. P. F. An electrochemical immunosensor for cardiac Troponin I using electrospun carboxylated multi-walled carbon nanotube-whiskered nanofibres. *Talanta* **2018**, *182*, 178–186.
(76) Huang, Y.; Tan, J.; Cui, L.; Zhou, Z.; Zhou, S.; Zhang, Z.; Zheng, R.; Xue, Y.; Zhang, M.; Li, S.; et al. Graphene and Au NPs co-mediated enzymatic silver deposition for the ultrasensitive electrochemical detection of cholesterol. *Biosensors and Bioelectronics* **2018**, *102*, 560–567.
(77) Hatamluyi, B.; Es'haghi, Z. Electrochemical biosensing platform based on molecularly imprinted polymer reinforced by ZnO–graphene capped quantum dots for 6-mercaptopurine detection. *Electrochimica Acta* **2018**, *283*, 1170–1177.
(78) Mohanan, V. M. A.; Kunnummal, A. K.; Biju, V. M. N. Selective electrochemical detection of dopamine based on molecularly imprinted poly (5-amino 8-hydroxy quinoline) immobilized reduced graphene oxide. *Journal of Materials Science* **2018**, *53* (15), 10627–10639.
(79) Zhang, G.; Liu, Z.; Fan, L.; Guo, Y. Electrochemical prostate specific antigen aptasensor based on hemin functionalized graphene-conjugated palladium nanocomposites. *Microchimica Acta* **2018**, *185*, 1–8.
(80) Dou, B.; Xu, L.; Jiang, B.; Yuan, R.; Xiang, Y. Aptamer-functionalized and gold nanoparticle array-decorated magnetic graphene nanosheets enable multiplexed and sensitive electrochemical detection of rare circulating tumor cells in whole blood. *Analytical Chemistry* **2019**, *91* (16), 10792–10799.
(81) Yammouri, G.; Mohammadi, H.; Amine, A. A highly sensitive electrochemical biosensor based on carbon black and gold nanoparticles modified pencil graphite electrode for microRNA-21 detection. *Chemistry Africa* **2019**, *2*, 291–300.
(82) Bai, L.; Chen, Y.; Liu, X.; Zhou, J.; Cao, J.; Hou, L.; Guo, S. Ultrasensitive electrochemical detection of Mycobacterium tuberculosis IS6110 fragment using gold nanoparticles decorated fullerene

nanoparticles/nitrogen-doped graphene nanosheet as signal tags. *Analytica Chimica Acta* **2019**, *1080*, 75–83.

(83) Li, S.; Ma, Y.; Liu, Y.; Xin, G.; Wang, M.; Zhang, Z.; Liu, Z. Electrochemical sensor based on a three dimensional nanostructured MoS 2 nanosphere-PANI/reduced graphene oxide composite for simultaneous detection of ascorbic acid, dopamine, and uric acid. *RSC advances* **2019**, *9* (6), 2997–3003.

(84) Lv, J.; Li, C.; Feng, S.; Chen, S.-M.; Ding, Y.; Chen, C.; Hao, Q.; Yang, T.-H.; Lei, W. A novel electrochemical sensor for uric acid detection based on PCN/MWCNT. *Ionics* **2019**, *25*, 4437–4445.

(85) Yola, M. L.; Atar, N. Simultaneous determination of β-agonists on hexagonal boron nitride nanosheets/multi-walled carbon nanotubes nanocomposite modified glassy carbon electrode. *Materials Science and Engineering: C* **2019**, *96*, 669–676.

(86) Assari, P.; Rafati, A. A.; Feizollahi, A.; Asadpour Joghani, R. An electrochemical immunosensor for the prostate specific antigen based on the use of reduced graphene oxide decorated with gold nanoparticles. *Microchimica Acta* **2019**, *186*, 1–9.

(87) Jalalvand, A. R. Fabrication of a novel and ultrasensitive label-free electrochemical aptasensor for detection of biomarker prostate specific antigen. *International Journal of Biological Macromolecules* **2019**, *126*, 1065–1073.

(88) Liu, P.; Li, C.; Zhang, R.; Tang, Q.; Wei, J.; Lu, Y.; Shen, P. An ultrasensitive electrochemical immunosensor for procalcitonin detection based on the gold nanoparticles-enhanced tyramide signal amplification strategy. *Biosensors and Bioelectronics* **2019**, *126*, 543–550.

(89) Suresh, L.; Bondili, J. S.; Brahman, P. K. Fabrication of immunosensor based on polyaniline, Fullerene-C60 and palladium nanoparticles nanocomposite: An electrochemical detection tool for prostate cancer. *Electroanalysis* **2020**, *32* (7), 1439–1448.

(90) Manasa, G.; Raj, C.; Satpati, A. K.; Mascarenhas, R. J. S (O) MWCNT/modified carbon paste–A non-enzymatic amperometric sensor for direct determination of 6-mercaptopurine in biological fluids. *Electroanalysis* **2020**, *32* (11), 2431–2441.

(91) Yuanfeng, P.; Ruiyi, L.; Xiulan, S.; Guangli, W.; Zaijun, L. Highly sensitive electrochemical detection of circulating tumor DNA in human blood based on urchin-like gold nanocrystal-multiple graphene aerogel and target DNA-induced recycling double amplification strategy. *Analytica Chimica Acta* **2020**, *1121*, 17–25.

(92) Fu, B.; Chen, H.; Yan, Z.; Zhang, Z.; Chen, J.; Liu, T.; Li, K. A simple ultrasensitive electrochemical sensor for simultaneous determination of homovanillic acid and vanillylmandelic acid in human urine based on MWCNTs-Pt nanoparticles as peroxidase mimics. *Journal of Electroanalytical Chemistry* **2020**, *866*, 114165.

(93) Ertuğrul Uygun, H. D.; Demir, M. N. A novel fullerene-pyrrole-pyrrole-3-carboxylic acid nanocomposite modified molecularly imprinted impedimetric sensor for dopamine determination in urine. *Electroanalysis* **2020**, *32* (9), 1971–1976.

(94) Assari, P.; Rafati, A. A.; Feizollahi, A.; Joghani, R. A. Fabrication of a sensitive label free electrochemical immunosensor for detection of prostate specific antigen using functionalized multi-walled carbon nanotubes/polyaniline/AuNPs. *Materials Science and Engineering: C* **2020**, *115*, 111066.

(95) Guan, J.-F.; Zou, J.; Liu, Y.-P.; Jiang, X.-Y.; Yu, J.-G. Hybrid carbon nanotubes modified glassy carbon electrode for selective, sensitive and simultaneous detection of dopamine and uric acid. *Ecotoxicology and Environmental Safety* **2020**, *201*, 110872.

(96) Devi, R.; Gogoi, S.; Dutta, H. S.; Bordoloi, M.; Sanghi, S. K.; Khan, R. Au/$NiFe_2O_4$ nanoparticle-decorated graphene oxide nanosheets for electrochemical immunosensing of amyloid beta peptide. *Nanoscale Advances* **2020**, *2* (1), 239–248.

(97) Hatefi-Mehrjardi, A.; Karimi, M. A.; Soleymanzadeh, M.; Barani, A. Highly sensitive detection of dopamine, ascorbic and uric acid with a nanostructure of dianix yellow/multi-walled carbon nanotubes modified electrode. *Measurement* **2020, 163**, 107893.

(98) Fekry, A. M.; Abdel-Gawad, S. A.; Tammam, R. H.; Zayed, M. A. An electrochemical sensor for creatinine based on carbon nanotubes/folic acid /silver nanoparticles modified electrode. *Measurement* **2020**, *163*, 107958.

(99) Jamil, M.; Fatima, B.; Hussain, D.; Chohan, T. A.; Majeed, S.; Imran, M.; Khan, A. A.; Manzoor, S.; Nawaz, R.; Ashiq, M. N.; et al. Quantitative determination of creatinine from serum of prostate cancer patients by N-doped porous carbon antimony (Sb/NPC) nanoparticles. *Bioelectrochemistry* **2021**, *140*, 107815.

(100) Phonklam, K.; Wannapob, R.; Sriwimol, W.; Thavarungkul, P.; Phairatana, T. A novel molecularly imprinted polymer PMB/MWCNTs sensor for highly-sensitive cardiac troponin T detection. *Sensors and Actuators B: Chemical* **2020**, *308*, 127630.
(101) Ehsan, M. A.; Khan, S. A.; Rehman, A. Screen-printed graphene/carbon electrodes on paper substrates as impedance sensors for detection of coronavirus in nasopharyngeal fluid samples. *Diagnostics* **2021**, *11*.
(102) Chen, M.; Wu, D.; Tu, S.; Yang, C.; Chen, D.; Xu, Y. A novel biosensor for the ultrasensitive detection of the lncRNA biomarker MALAT1 in non-small cell lung cancer. *Scientific Reports* **2021**, *11* (1), 3666.
(103) Boonkaew, S.; Jang, I.; Noviana, E.; Siangproh, W.; Chailapakul, O.; Henry, C. S. Electrochemical paper-based analytical device for multiplexed, point-of-care detection of cardiovascular disease biomarkers. *Sensors and Actuators B: Chemical* **2021**, *330*, 129336.
(104) Gogola, J. L.; Martins, G.; Gevaerd, A.; Blanes, L.; Cardoso, J.; Marchini, F. K.; Banks, C. E.; Bergamini, M. F.; Marcolino-Junior, L. H. Label-free aptasensor for p24-HIV protein detection based on graphene quantum dots as an electrochemical signal amplifier. *Analytica Chimica Acta* **2021**, *1166*, 338548.
(105) Yang, L.; Wang, J.; Lü, H.; Hui, N. Electrochemical sensor based on Prussian blue/multi-walled carbon nanotubes functionalized polypyrrole nanowire arrays for hydrogen peroxide and microRNA detection. *Microchimica Acta* **2021**, *188*, 1–12.
(106) Guan, Q.; Guo, H.; Xue, R.; Wang, M.; Zhao, X.; Fan, T.; Yang, W.; Xu, M.; Yang, W. Electrochemical sensor based on covalent organic frameworks-MWCNT-NH2/AuNPs for simultaneous detection of dopamine and uric acid. *Journal of Electroanalytical Chemistry* **2021**, *880*, 114932.
(107) Negahdary, M.; Buoro, R. M.; Bacil, R. P.; Santos, B. G.; Angnes, L. Design of an electrochemical aptasensor in the presence of an array of gold nanostructure and a GO-MWCNTs nanocomposite: Application in diagnosis of Alzheimer's disease. *Microchimica Acta* **2023**, *190* (10), 409.
(108) Ben Hassine, A.; Raouafi, N.; Moreira, F. T. C. Novel biomimetic Prussian blue nanocubes-based biosensor for Tau-441 protein detection. *Journal of Pharmaceutical and Biomedical Analysis* **2023**, *226*, 115251.
(109) Wang, Y.; Wang, P.; Zhao, Z.; Ye, S.; Wang, W.; Liu, Q.; Li, Y.; Zhang, D.; Li, Y. A label-free electrochemical immunosensor based on PtCoCu PNPs/NB-rGO as a dual signal amplification platform for sensitive detection of β-Amyloid1-42. *Bioelectrochemistry* **2023**, *152*, 108405.
(110) Sun, B.; Kan, L.; Gao, C.; Shi, H.; Yang, L.; Zhao, T.; Ma, Q.; Shi, X.; Sang, C. Investigating immunosensor for determination of SD-biomarker-Aβ based on AuNPs/AC@PANI@CS modified electrodes with amplifying the signal. *Analytical Biochemistry* **2023**, *661*, 114996.

13 Analysis of Clinical Samples by Carbon Sensor

Nutan Rani, Sapna Yadav, Kalawati Saini, and Dipak Maity

13.1 INTRODUCTION

Improving people's health and lives is unquestionably one of the top priorities for healthcare organizations, politicians, and leaders throughout the world. To address the primary reason of illness and mortality caused by infections and chronic illnesses, it is critically required to use various types of analytical methods for clinical samples [1]. A dramatic shift in digital technology, from traditional to smart healthcare, has the potential to completely transform healthcare systems throughout the world. Smart healthcare links a wide range of participants in the health system, including patients, healthcare professionals, organizations, and regulators [2]. Sensors play a key role in the healthcare sector. It is crucial to recognize various biomolecules quickly, sensitively, and accurately, as early disease detection enhances the probability of successful treatment. Biosensors are devices that assess both organic and inorganic chemicals present in living organisms. A biosensor consists of three parts: a detector that detects the target analyte signal, a converter that turns the detected signal into a usable output, and a signal processor that processes and analyzes the output signal [3]. Biosensors have found extensive use in clinical analysis due to their specificity, sensitivity, and rapid response. Some common applications include blood glucose monitoring, pregnancy tests, infectious disease detection, cancer biomarker detection, etc. [1–5].

In recent decades, an extensive variety of methods and approaches have been developed, resulting in significant advances in the search for fast, sensitive, accurate, reliable, and efficient clinical sample analysis methods [4]. An ideal sensor should have a wide detection range, high sensitivity, selectivity, resolution, repeatability, reproducibility, reaction speed, and enough flow [5]. Nanomaterials have superior physiochemical qualities than their bulk counterparts, such as conductivity, strength, and reactivity, because of the high surface-to-volume ratio and various other physical and electrical features [6]. Carbonaceous nanomaterials, including carbon nanotubes (CNTs), carbon dots, graphene oxides (GOs), etc., are at the top of advanced materials research [7, 8]. The surface characteristics and particle composition of carbon nanostructures influence their behavior and characterization. Furthermore, the chemical and reactive capabilities of these carbon nanostructures can be improved by strengthening their surface features. These nanoparticles can be employed alone or in combination with other materials such as electrodes, membranes, and structural media [7].

The remarkable physical and chemical properties of CNTs have prompted substantial basic investigation. The potential economic uses of CNTs have brought back interest of researches in them. CNTs have a variety of biological applications, including biosensors, nanoinjectors, drug delivery, phototherapy, imaging, and artificial muscles [8]. CNTs have various benefits, including a high surface-to-volume ratio, outstanding electrical characteristics, and the existence of edge-plane-like defects, which make them an appealing material for the development of electrochemical (bio)

DOI: 10.1201/9781003500322-13

sensors. These properties of CNTs are responsible for enzymatic recognition, catalysis, etc. The direct charge transfer enabled the design of several biosensing systems for the very sensitive and selective measurement of biomarkers [9].

Carbon dots are organic nanosized particles. Carbon dots are semiconductor in nature. They are made of spherical, nanometric-sized carbon atoms arranged in a crystalline structure. Due to their functionalized surface with oxygen groups caused by synthesis/precursor, carbon dots exhibit significant water dispersion. To get desired qualities, carbon dots can be functionalized, passivated, or doped with heteroatoms. Carbon dots may require some further steps, such as surface passivation for enhanced optical properties, functionalization for improved chemical reactivity, or heteroatom doping for improved electrical and optical properties [10]. Because of these above-mentioned surface properties carbon dots and its composites can be used as potential immunosensor, colorimetric biosensor, fluorescent sensors, and electrochemical sensors [11–13].

Graphene and its derivatives, particularly GO, reduced graphene oxide, and graphene nanoribbons, have gained global attention for biosensor applications. Graphene-based biosensors are using several mechanisms for clinical sample analysis such as optical, electrochemical, and electrical analyses [14]. Graphene has special qualities that make it useful for electrochemical biosensors, including exceptional thermal conductivity, rapid electron transport, superior mechanical flexibility, and good biocompatibility [15]. GO has various advantages. For example, it can be synthesized on large scale by simple and low-cost synthesis methods; it has a large surface area, fluorescence-quenching ability, and high energy transfer efficiency. Moreover, GO is water dispersible, biocompatible, and readily chemically modifiable. Because of these advantages GO has been seen as having endless potential for biosensor applications [16].

Recent advances in robotics, wearable devices, the Internet, and sensor and electronics size reduction have significantly increased the range of applications for smart sensors. The evolution of these technologies has had a dramatic impact on various disciplines, including medicine and healthcare, where they are considerably advancing a wide range of applications. Wearable technologies let patients and healthcare providers deal with a variety of health conditions at a minimal cost. These devices acquire patient data by integration with various sensors and wearable accessories, including smartwatches, cellphones, necklaces, and wristbands. The information is uploaded to servers and viewed via smartphone. In reality, smart sensors provide unique answers to a variety of important healthcare challenges, such as early pathology detection, less intrusive therapy, and disease prevention [17, 18]. Toxicity is a serious concern in the field of wearable devices in healthcare sector. Carbon black, CNTs, graphene, and their composites are widely used as smart wearable sensors because of their non-toxic nature [19].

In this chapter, we give a brief review on analysis of clinical samples by carbon sensor. The chapter aims to provide a detailed introduction to researchers who are new to this topic, as well as a quick viewpoint description of the obstacles of these biosensors toward practical use.

13.2 PRINCIPLES OF CARBON SENSORS IN CLINICAL ANALYSIS

As we have already mentioned, biosensors are analytical devices that translate biological responses into electrical signals [3]. They are comprised of three major components: bioreceptor, transducer and signal processor. A bioreceptor is a biological element that has a particular interaction with an analyte. Enzymes, antibodies, nucleic acids, cells, etc., can all function as bioreceptors, recognizing and binding to the target analyte [20, 21]. Transducer is a component that transforms the interaction between the analyte and the bioreceptor into a quantifiable signal, which is often electrical, optical, thermal, etc. The kind of signal depends on the transducer utilized [22, 23]. The output signal from the transducer is frequently weak and requires further processing. The signal processor amplifies, filters, and transforms the signal into an understandable format [21, 23]. The major components of biosensors are presented in Figure 13.1.

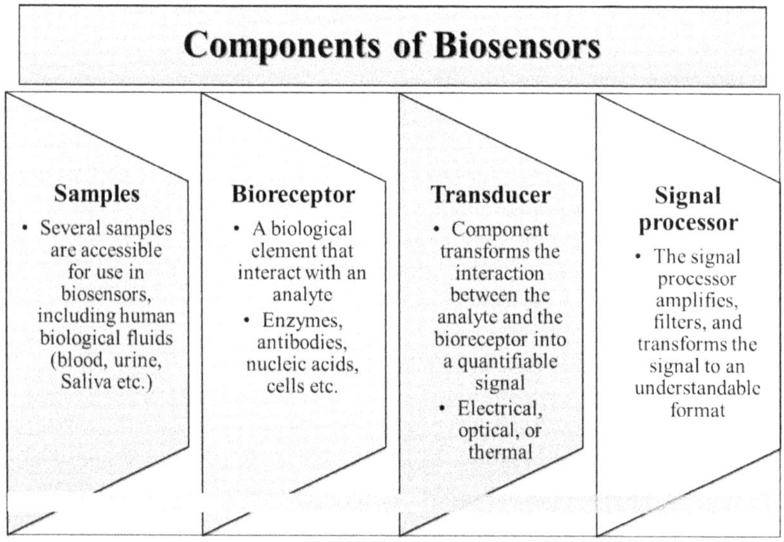

FIGURE 13.1 Major components of biosensors.

Biosensors can be categorized based on the type of transducer used:

- **Electrochemical Biosensors**: Measure electrical signals generated by the chemical reaction between the analyte and the bioreceptor. These are widely used forblood glucose monitoring [24, 25].
- **Optical Biosensors**: Detect changes in light absorption, fluorescence, or phosphorescence in response to the analyte's interaction with the bioreceptor [3, 26].
- **Thermal Biosensors**: Measure changes in heat or temperature during the analyte–bioreceptor interaction [5, 6, 27, 28].
- **Piezoelectric Biosensors**: Detect changes in mass or mechanical properties, often used for detecting small molecules [5, 29]. The symmetric representation of classification of biosensors based on the type of transducer is given in Figure 13.2.

The working principle of electrochemical sensors is based on the measurement of an electrical signal that is generated by the interaction between the sensor's electrode surface and a target analyte in a solution. This interaction typically involves an electrochemical reaction, such as oxidation or reduction, which produces a measurable current, voltage, or change in impedance [5, 24, 25]. Electrochemical sensors involve four major elements: working electrode (WE), reference electrode (RE), counter electrode (CE), and electrolyte solution [30]. WE is where the electrochemical reaction occurs. In the electrochemical biosensors, the WE serves as the transduction element. This is often made of a conductive material like carbon, gold, or platinum, and it interacts directly with the analyte. The RE provides a constant reference potential against which the potential of WE is monitored. The two most commonly used reference electrodes are Ag/AgCl and saturated calomel electrodes. The CE completes the electrical circuit and allows current to pass through the solution. It is often made of an inert material, such as platinum. Electrolyte solution contains the target analyte and acts as a medium for ions to move between electrodes [31–33].

The working principle of electrochemical biosensor involves analyte interaction followed by electron transfer. When the target analyte comes into contact with the WE, it undergoes an electrochemical reaction, typically oxidation (loss of electrons) or reduction (gain of electrons). This

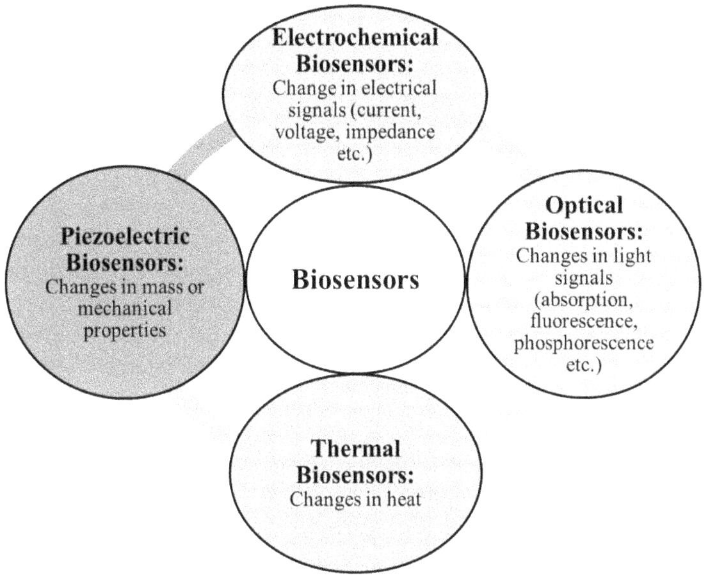

FIGURE 13.2 The symmetric representation of classification of biosensors based on the type of transducer.

reaction causes the transfer of electrons between the analyte and the electrode surface, generating an electrical signal. Electrochemical biosensors can be classified into three types according to their working principle [5, 32, 34, 35].

Potentiometric Sensors: Measure the potential (voltage) difference between the working and reference electrodes, which is related to the concentration of the analyte [32, 36, 37].

Amperometric Sensors: Measure the current generated by the redox reaction of the analyte at the WE, which is proportional to the concentration of the analyte [32, 36, 37].

Conductometric/Impedimetric Sensors: Measure the change in electrical conductivity or impedance of the solution due to the interaction with the analyte. In conductometric/impedimetric sensors, the change in the solution's conductivity or impedance is measured as a function of analyte concentration [32, 36, 37].

Therefore, the core of an electrochemical sensor's working principle is the electrochemical reaction between the target analyte and the sensor's electrode. This reaction generates a measurable electrical signal, which is then correlated with the concentration of the analyte. The design and material choice for the electrodes, along with the type of electrochemical measurement used (potentiometric, amperometric, or conductometric), determine the sensor's performance characteristics, including its sensitivity, selectivity, and response time.

Optical biosensors are analytical devices that use light to detect biological interactions, making them highly valuable in clinical analysis. Optical transducers are utilized by optical biosensors to make the analysis. These biosensors rely on the principles of optics, such as fluorescence, absorbance, or refractive index changes, to detect and quantify biomolecules like proteins, nucleic acids, or small molecules in clinical samples. The transducer converts the biorecognition event (binding of the analyte to the biorecognition element) into an optical signal. The type of the optical signal depends on the sensor design (e.g., fluorescence and absorbance) [38–41].

Optical biosensors can operate on different principles depending on how they convert the biological interaction into an optical signal.

Fluorescence-Based Biosensors: Fluorescence biosensors rely on the emission of light by a fluorophore (a fluorescent molecule) upon excitation by light of a specific wavelength. The intensity of the emitted light is proportional to the concentration of the target analyte [5, 41–43].

Surface Plasmon Resonance (SPR) Biosensors: SPR biosensors detect changes in the refractive index at the surface of a metal (typically gold) when the target analyte binds to the biorecognition element immobilized on the sensor surface. This binding event changes the angle at which light reflects off the sensor surface. Carbon-based materials such as graphene have been used for the coating of this gold and silver metal for enhanced SPR biosensors performance [40, 44–46].

Absorbance-Based Biosensors: Absorbance biosensors measure changes in the absorbance of light as it passes through a sample containing the target analyte [47, 48].

Optical Waveguide Biosensors: These biosensors use optical waveguides, which confine light within a specific path, to detect changes in the light signal as it interacts with the biorecognition element on the waveguide surface [49, 50].

Optical biosensors are integral to modern clinical analysis, offering precise, rapid, and non-invasive methods for detecting a wide range of biomarkers and pathogens.

Thermal biosensors, also known as calorimetric or thermometric biosensors, are devices that detect biological interactions by measuring temperature changes. These biosensors are based on the principle that biochemical reactions often involve heat exchange, either absorption or release of heat, which can be detected and correlated with the concentration of the target analyte. The transducer converts the heat generated or absorbed during the biochemical reaction into a measurable temperature change. The temperature sensor, often a thermistor, thermocouple, or infrared detector, measures the temperature change resulting from the biochemical reaction [51, 52]. Thermal biosensors operate based on the heat generated or absorbed during a biochemical reaction. The change in heat during a biological process is proportional to its molar enthalpy and the number of product molecules [51, 52]. Overall, thermal biosensors offer a versatile and powerful tool for clinical analysis, with potential applications in a wide range of diagnostic tests.

Piezoelectric biosensors leverage the piezoelectric effect to detect biological interactions and are commonly used in clinical analysis for the detection of various biomolecules, such as proteins, nucleic acids, and pathogens. The piezoelectric effect is the generation of an electrical charge in certain materials (such as quartz crystals) when they are mechanically deformed. Conversely, when an electrical field is applied to these materials, they can deform or vibrate. This is a reversible process [53–55].

The core component of a piezoelectric biosensor is a piezoelectric crystal, typically made of quartz. This crystal is connected to electrodes to measure changes in electrical signals. The surface of the piezoelectric crystal is functionalized with a layer of biomolecules (e.g., antibodies, enzymes, or nucleic acids) that specifically bind to the target analyte [55]. When the target molecule from a sample (such as blood or urine) interacts with the functionalized surface, it binds to the biomolecules on the crystal. This binding event causes a change in the mass on the crystal's surface [53, 55]. Due to the piezoelectric effect, this mass change results in a shift in the crystal's natural oscillation frequency. The frequency shift is proportional to the amount of target molecules bound to the sensor surface. The change in frequency is measured by the electrodes and is correlated to the concentration of the target analyte in the sample. The frequency shift is converted into an electrical signal, which is then processed and analyzed to provide quantitative information about the presence and concentration of the analyte. [53, 56, 57].

Emerging two-dimensional nanomaterials are rapidly expanding in biological applications. The unique features of graphene, including high exibility and stretchability, make it a promising material for piezoelectric biosensors [54].

Piezoelectric biosensors are powerful tools in clinical diagnostics, offering rapid, sensitive, and specific detection of a wide range of biomolecules.

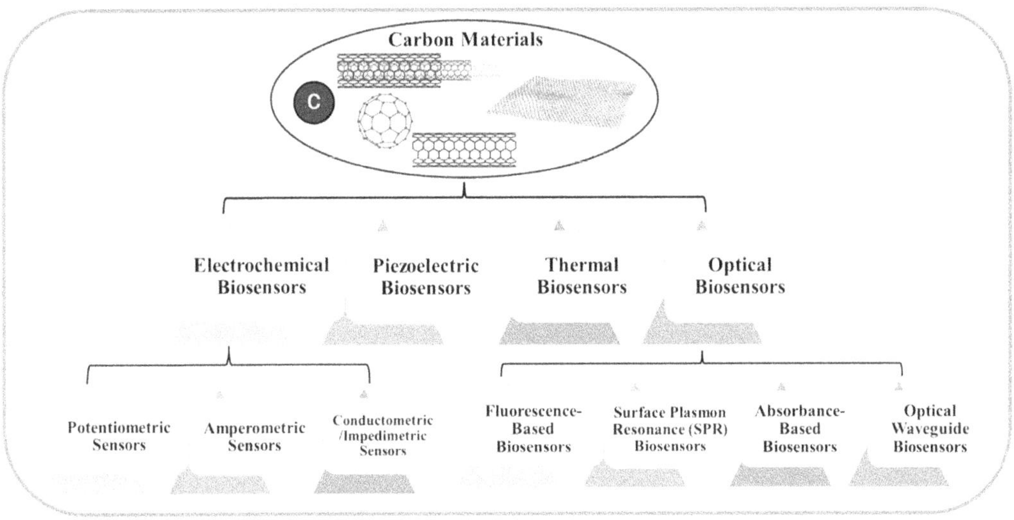

FIGURE 13.3 The symmetric representation of principles of carbon sensors in clinical analysis (Part of this figure is created with BioRender.com.).

Carbon-based materials, such as graphene, CNTs, and carbon nanofibers, play a crucial role in the development of biosensors in the healthcare sector due to their unique physical, chemical, and electronic properties. Carbon materials have excellent electrical conductivity, which allows them to act as effective transducers in biosensors. They can convert the biological interaction (like binding of a biomolecule) into an electrical signal. The large surface area of carbon nanomaterials allows for a high density of biomolecule immobilization, which increases the sensitivity of the biosensor. Carbon materials can be chemically functionalized with various groups to enhance their interaction with specific biomolecules, such as enzymes, antibodies, or nucleic acids. This selective binding enables the detection of specific analytes. Functionalized carbon materials are biocompatible, making them suitable for in vivo applications and long-term monitoring in the human body [58–61].

In these sensors, carbon materials are used as electrodes. The binding of a target molecule (e.g., glucose) to the sensor surface leads to a change in current, potential, or impedance, which is then measured and correlated with the concentration of the analyte [59, 62]. Carbon materials can also be used in optical biosensors. For example, graphene's fluorescence quenching ability can be used to detect fluorescently labelled biomolecules. The binding of these molecules to the carbon material alters the optical signal [26, 63]. In piezoresistive or piezoelectric biosensors, the interaction of biomolecules with carbon-based materials can induce mechanical stress, leading to a measurable change in resistance or voltage [54, 64–67].

Overall, carbon-based materials are key components in the next generation of biosensors, offering promising solutions for personalized healthcare and early disease detection. The symmetric representation of principles of carbon sensors in clinical analysis is given in Figure 13.3.

13.3 APPLICATIONS OF CARBON SENSORS IN CLINICAL SAMPLE ANALYSIS

13.3.1 Nano-Carbon Based Sensors for Clinical Sample Analysis

13.3.1.1 Carbon Dots Sensing in Biomedical Analysis

In the past few decades, carbon quantum dots (also known as CDs) have drawn a lot of interest in the field of nanotechnology [68]. Zero-dimensional carbon nanomaterials are called carbon dots (CDs). These are made up of carbon atoms having a size less than 10 nm. CDs are quasi-spherical, luminous nanoparticles consisting of carbon shells that are primarily turbostratic or graphitic, and amorphous to nanocrystalline core [69–71]. Due to its remarkable characteristics, which include water solubility, high photostability, resistance to photobleaching, minimal toxicity, low cost, environment-friendliness, excellent biocompatibility, simple synthesis, and facile bioconjugation, there is a growing demand for this carbon material [71–73].

Saliva contains a wide variety of clinically significant biomarkers and chemicals, which make it an intriguing alternative biological medium for clinical analysis. As compared to collecting blood samples, collecting saliva samples is a simple, non-invasive, and stress-free process. It enables the collection of multiple samples in less time by operators with only minimum experience. As a result, more research on saliva diagnostics has been published, with investigations on analytes such as tyrosine, glucose, lactate, nitrite, and hormones among others.

Rossini et al. reported a paper microfluidic device (μPADs) using CDs for the detection of glucose and lactate in saliva sample. It was an easy, fast, low-cost, sensitive, and safe method. The authors obtained a limit of detection (LOD) value of 8.14×10^{-7} mol L^{-1} and 2.60×10^{-6} mol L^{-1} for lactate and glucose, respectively. The method of detection depends on the fluorescence quenching of CDs by the hydrogen peroxidase enzyme present in the saliva [74]. Similarly, Wang et al. reported the application of multicolor luminescent CDs for glucose estimation in dilute blood samples [75].

The concentration of glucose in the fingertip blood was estimated by Li et al. using ionic liquid-functionalized carbon nanodots (IL-CDs) modified with GOx. These IL-CDs were prepared by electrochemically exfoliating graphite rods in the presence of an amino-terminated ionic liquid ([apmim][BF_4]). The IL-CDs served as a good matrix for the immobilization and activity of the GOx enzyme. This could be related to the strong electrostatic interaction that exists between the negatively charged GOx and the positively charged [apmim] moiety. Testing revealed that the glucose content in fingertip blood was 6.4 mM, which is comparable to the 6.7 mM measured in commercial blood glucose monitors [76]. Similarly, Zheng et al. reported the determination of glucose in human blood serum samples using a carbon dot-based molecularly imprinted electrochemical sensor (MIECS) [77]. The developed novel MIECS demonstrated high stability, reliability, and remarkable selectivity. The distinctive characteristics of CDs, which enhanced electron transfer and increased electroactive surface area, maybe the reason for the remarkable sensitivity. The presented MIECS also had the benefits of being inexpensive, highly reproducible, and perfectly selective for glucose even in the presence of prevalent interfering substances such as dopamine (DA), ascorbic acid (AA), etc.

Tuteja et al. developed a highly sensitive immunosensor based on graphene quantum dots for the detection of cardiac biomarker myoglobin (cMyo) [78]. An LOD of 0.01 ng/mL has been provided by an EIS-based immunosensor, which is over 400 times better than the results of the enzyme linked immunosorbent assay (ELISA) testing. A curcumin-quantum dot-based dual DNA sensor was developed by Mars et al. for the sensing of APOe4 DNA. This DNA(APOe4) is responsible for the Alzheimers disease [79]. Clinical fluids, such as human blood plasma, were used to test the applicability of the designed system. A high sensitivity response to APO e4 DNA was found by amperometric analysis, having a linear range of 20–400 pg/mL. Furthermore, the fluorescence data revealed a similar sensitivity of about 47.58 (a.u.) mL/pg.

13.3.1.2 Carbon Nanotube Sensing in Biomedical Analysis

Carbon nanotubes are rolled cylindrical sheets of graphene. Each carbon atom of CNTs is sp^2 hybridized. In 1991, carbon nanotubes were discovered by Iijima [80]. These exhibit unique properties such as high electrical and thermal conductivity, high aspect ratio, lightweight, and greater mechanical strength. All these properties made CNTs an important distinctive nanomaterial with a wide range of potential applications, particularly in the medical sector. The geometry and characteristics of CNTs can vary depending on how the graphene sheet folds up. Thus, the alignment of the hexagons along the tube's axis governs the electrical properties of the CNTs [81].

The structure of CNTs imparts to them an array of interesting and potential advantages for a range of applications. This material exhibits excellent elasticity and can withstand strain and bending without losing its structural integrity. Moreover, CNTs exhibit enhanced thermal and electrical conductivity and tensile strength compared to steel. Furthermore, CNTs offer several advantages such as low weight, high specific surface area, chemical stability, strong electrochemical performance, and so on. These properties make CNTs a promising research subject for biomolecular detection studies in medicine.

CNTs are an ideal material for the fabrication of sensors because of their high surface area, potential adsorption capacity, and rapid electron transfer velocity. These characteristics make it possible to modify CNTs with different molecules, such as enzymes, and various metal complexes. The analyte-to-electrode electron transfer can be further enhanced by these alterations, leading to a significant improvement in the ability to respond. In addition, sensitive and label-free biosensing is possible with the use of CNTs as biosensors since electrical conductivity is highly responsive to biomarker absorption. It is very easy to rapidly identify biomolecules even at very low concentrations owing to the high surface-to-volume ratio of CNTs.

Numerous sectors have conducted research on CNTs, primarily in the areas of environment, agriculture, food, energy, and medicine. The healthcare sector has been greatly impacted by the most recent and fast development of CNTs, as these materials are highly sensitive to biomolecules. The high hydrophobicity of CNTs makes them insoluble in water and various other solvents. When CNTs are functionalized, their cytotoxicity is significantly decreased while their biocompatibility is simultaneously enhanced. Due to functionalization, the more water soluble CNTs became non-toxic and safe and can be easily used in various biomedical applications.

An optical biosensor based on CNTs was reported by Barone et al. for the detection of glucose [82]. Herein, the enzyme glucose oxidase, which degrades glucose molecules, was coated on the CNTs. Subsequently, the surfaces of the nanotubes were sprayed with ferricyanide molecules. When stimulated by infrared light, the potential of CNTs to glow was quenched by ferricyanide, which takes electrons out of them. The CNTs produced higher infrared fluorescence when the glucose level was high. Cha et al. reported the detection of insulin with the help of an optical nanosensor made up of single-walled CNTs (SWCNTs) coated with DNA aptamer [83]. The insulin binding aptamer was used as a molecular recognition element. Numerous studies have been conducted and focus has been drawn to the CNT electrochemical biosensors for nucleic acid detection [84–87].

SWCNT-based imaging investigations have been developing in the last few years. The use of single-walled CNTs in molecular imaging was reviewed by Hong et al. and summarizes the most recent uses of SWCNTs as contrast agents in several imaging modalities [88]. Gd^{3+}-functionalized SWCNTs resulted in good tissue penetration and high resolution in the magnetic resonance imaging (MRI).

CNTs have been considered a better approach for identifying the expression of provocative tumor biomarkers at an early stage of cancer owing to their unique mechanical, thermal, and electrical properties. Several significant tumor markers have been used extensively in clinical trials to diagnose different cancers. These markers include carcinoma antigen 125 (CA125), alpha-fetoprotein (AFP), prostate-specific antigen (PSA), carcinoembryonic antigen (CEA) human chorionic gonadotropin (hCG), and carbohydrate antigen 19-9 (CA19-9). Therefore, the fundamental role of biomarkers is prognostic assessment, therapeutic response prediction, and monitoring after remedial surgery [89]. PSA is one of the many biomarkers that has been extensively employed in the identification of cancer. It has been revealed that PSA in blood samples can be detected more accurately and selectively through DNA strand functionalized MWCNTs and SWCNTs [90]. A brief overview and report have been provided on the electrochemical biosensors utilized to identify a biomarker for prostate cancer [91].

Several studies have been reported wherein CNTs were utilized for the detection of cancer biomarkers [92–95]. Therefore, future clinical studies of cancer biomarkers may employ CNT-based detection measurement as a substitute technique. Numerous studies have been conducted and focus has been drawn to the CNT electrochemical biosensors for nucleic acid detection [84–87]. Majidi et al. carried out additional research employing CNTs to modify electrochemical sensors [96]. They designed a sensor that detected tryptophan (Trp) in biological materials, such as saliva, blood serum, and urine. Because it is a small, lightweight, portable, and effective device, this sensor has a lot of potential promise. Tryptophan is an essential amino acid. Therefore, sensitive, and efficient analytical techniques have become vital for measuring tryptophan levels in clinical and medical research, especially in medicine and neurology. The authors reported the detection limit of the designed sensor as 4.9×10^{-12} mol/L with excellent selectivity and sensitivity.

Hočevar et al. reported an electrochemical sensor for the determination of DA at physiological pH in the presence of ascorbate. This carbon fiber microelectrode was fabricated using multi-walled CNTs. Under the 2–20 μM DA range under investigation, the microsensor exhibits extremely linear behavior, with a detection limit of 70 nM [97]. Similarly, Swamy et al. reported *in vivo* simultaneous detection of serotonin (5-HT) and DA using single-walled CNT-modified carbon fiber microelectrode [98]. Xiang et al. presented a novel approach for real-time, *in vivo* ascorbate monitoring with high selectivity in the rat brain utilizing vertically aligned carbon nanotube-sheathed carbon fibers (VACNT-CFs) as pristine microelectrodes [99].

Compared to chemical self-assembly of CNTs, growing aligned CNTs is simpler, more reliable, and appropriate for batch production. As a result, several approaches for aligning CNTs have shown to be effective when applied in actual clinical analysis measurements, indicating that this strategy is reliable enough for everyday application. A helical carbon nanotube (HCNT) has good electrical performance and a large specific surface area due to its 3D helical structure. The simultaneous electrochemical measurement of DA, AA, and uric acid (UA) has been done using HCNTs [100]. Similarly, the detection of DA, UA, and AA was reported by Zhang et al. in fetal bovine serum samples using HCNTs [101].

CNT yarns consist of several CNT strands coiled jointly. CNT yarn microelectrodes (CNTYMEs) were employed by Schmidt et al. to identify fast neurotransmitter (DA) variations in brain slices with fast-scan cyclic voltammetry [102]. The outcomes of the study show that enhanced selectivity, spatial temporal resolution, and sensitivity are the consequence of the unique structural and electrical characteristics of the fabricated CNTs. Therefore, these CNTYMEs are appropriate for *in vivo* imaging. Similarly, Jacob et al. reported the use of CNTYMEs for the detection of DA [103].

Carbon nanofibres (CNFs) have also been extensively used for the monitoring of neurochemicals [104]. In another study, Rand et al. reported vertically aligned CNFs for the measurement of 5-HT and DA simultaneously in the presence of AA [105]. Also, in terms of sensitivity and selectivity, the CNF electrode functions better than the usual glassy carbon electrode (GCE). The limit of detection was reported to be 250 nM for 5-HT and 50 nM for DA, respectively, using a CNF electrode. Three distinctive peaks for the CNF electrode were observed at 0.13, 0.45, and 0.70 V for the mixture of DA, 5-HT, and AA while performing differential pulse voltammetry analysis. On the other hand, a GCE cannot differentiate the three analytes (DA, 5-HT, and AA). The numerous active sites along the sidewall and the structure of carbon nanofibers were responsible for this efficiency in detecting different neurotransmitters.

DA and 5-HT are two important neurotransmitters found in the brain. Any disturbance in the level of these two neurotransmitters leads to the occurrence of a variety of human diseases. Epilepsy, Parkinson's disease, addiction, and many other illnesses have all been caused by alterations in the DA levels. The disturbances in the 5-HT levels are associated with several biological processes, including appetite, sleep disorder, addiction, and depression.

13.3.1.3 Graphene-Based Sensor in Biomedical Analysis

Enormous research works on clinical analysis using sensors based on graphene and graphene-related materials, notably reduced GO and GO, have been published in the literature [106]. Because of its enormous surface area and unique electrical characteristics, graphene is highly sensitive to variations in the surroundings in which it is placed. Due to its ability to recognize molecular changes at ease, this material is very useful in the fabrication of improved and extremely accurate sensors [107]. Furthermore, graphene may physically adsorb or chemically interact with other materials, including biomolecules like DNA and enzymes, polymers, and various nanoparticles, to produce composites.

A biosensor for prostate cancer detection based on electrodes modified with graphene-poly (3-aminobenzoic acid) was reported by Pothipor et al. [108]. At 0.13 pg mL^{-1} of detection limit value for prostate specific antigen, which is higher than the clinically significant threshold in human serum, the designed biosensor demonstrated an extremely acceptable performance. The authors have indicated that the sensor exhibits potential in the diagnosis of prostate cancer and for other medical purposes. DNA detection is very essential for gene expression studies as well as clinical diagnosis. Luo et al. reported GO-based chemiluminescence biosensor for the detection of DNA [109]. Zou et al. demonstrated the simultaneous detection of DA, UA, and AA using GO sheets decorated with hemin [110].

Graphene and its compounds are widely used as quenchers and carriers for fluorescence recognition targets or probes. This is due to their high surface-to-volume ratio, electronic properties, high mechanical strength, and very distance-dependent fluorescence quenching capabilities based on fluorescence resonance energy transfer. The sensitive and selective detection of DA using near-IR graphene-based biosensors was reported by Chen et al. in biological systems [111]. DA self-assembles on the surface of GO through a variety of noncovalent interactions, providing the basis for this detection method. The authors reported a detection limit of 94 nM for DA using this biosensor. Furthermore, the graphene quantum dots-based fluorescence biosensor was reported by Quan et al. for the highly sensitive and selective detection of ssDNA, with a detection limit of 75 pM DNA [112]. A graphene-modified electrochemical biosensor was reported by Castillo et al. for the detection of human cervical cancer cells. This biosensor was modified with peptide nanotube-folic acid [113].

In biological fluids like blood and urine, mostly these molecules (amino acid, UA, and DA) coexist together. These have almost similar structures due to which their oxidation occurs at similar potentials and hence their identification is difficult in the presence of each other. Furthermore, DA concentrations in body fluids range from 0.01 to 1 μM, which is significantly lower than AA and UA concentrations, making it challenging to achieve a highly selective assessment. Keshavan and John reported a biosensor based on GO with 2,4-diamino-1,3,5-triazine (aminotriazine-AT) for the simultaneous determination of paracetamol (PA) and UA in the human urine samples [114]. When administered at standard therapeutic dosages, PA undergoes full metabolism and is excreted in urine. Conversely, severe liver damage, pancreatic inflammation, and renal failure can result from consuming large amounts of PA. It was possible to selectively detect the PA even in the presence of a very high concentration of UA. The limit of detection for PA was reported to be 6.8×10^{-10} mol/L.

For sensitive acetaminophen detection in pharmaceutical formulations and human body fluids, Adhikari et al. have developed a high-performance electrochemical biosensor utilizing reduced graphene oxide (rGO). Electrochemical bio-sensing devices based on rGO are highly effective in detection of acetaminophen due to their oxygen-related defects, excellent electrical conductivity, and large surface area. The sensor based on rGO exhibited a linear range of 5.0 nM to 800.0 μM, with a low detection limit of 2.13 nM, indicating high sensitivity and selectivity. The authors reported that the biosensor based on ERG showed a remarkable recovery rate, ranging from 96.08% to 103.2% [115].

A potential candidate for the targeted identification and capture capacity for cancer cells is functional graphene. Jing et al. designed a reliable and effective electrochemical cytosensor for the label-free detection of cancer cells (HCT-116) [116]. The highly stable and biocompatible label-free electrochemical cytosensor was developed by self-assembling HG (hyaluronate functionalized graphene) with ethylene diamine and then with sodium hyaluronate. The detection limit reported was 100 cells·mL^{-1} (S/N = 3).

13.3.2 Carbon Paste Sensors for Clinical Sample Analysis

From the clinical point of view, the identification of various neurochemicals such as DA, 5-HT, epinephrine, norepinephrine, adenosine, adenosine triphosphate, and acetylcholine and amino acids such as acetylcholine and tyrosine from biological samples is substantially important. These chemicals play a critical role in maintaining nerve signal balance and promoting healthy brain function. Alzheimer's disease, Parkinson's disease, dementia, addiction, depression, and schizophrenia are among the physical, psychiatric, and neurological diseases related to abnormal neurochemical levels [117]. Identifying these neurochemicals from real-life samples thus demands extensive investigation of these molecules, and sophisticated analytical instruments may potentially be useful in this regard. Much interest has been generated, particularly in early diagnosis and clinical studies, by the development of an accurate and reliable method for the quantitative determination of analytes in biological samples. Carbon paste electrodes (CPEs) have been applied to determine several neurotransmitters electrochemically. Atta et al. reported the applicability of a sensor having carbon paste electrodes for the determination of L-DOPA, a catecholamine in the human urine sample [118].

Considering the low cost of the electrode material and improved electron transfer kinetics compared to other carbon-based electrodes, carbon paste-based electrochemical sensors are attracting a lot of interest. CPEs offer a number of benefits and drawbacks. Both their fabrication and electrode surface cleaning are simple processes. In real sample analysis, it appears that confounding effects have been eliminated by performing modifications of the CPE [119].

Chandra et al. have reported the detection of DA at neutral pH using poly (calmagite) film-coated CPE [120]. This polymer film-coated CPE performed exceptionally well when investigating DA simultaneously in the presence of high concentration of UA and AA. The limit of detection for DA was reported to be 1×10^{-8} M. The proposed CPE was used to find DA concentration in injection samples. The use of CPE fabricated with carbon nano dot-(3-aminopropyl) triethoxysilane (CDs-APTES) was reported by Bodur et al. for the detection of acetylcholine (ACh), which is an important neurotransmitter. The deficiency of this neurotransmitter in the brain is linked to Alzheimer's disease. A synthetic blood solution was prepared with the following interferes: UA (3.0×10^{-4} M), methanol (5.0×10^{-5} M), ethanol (0.0109 M), AA (1.0×10^{-4} M), PA (1.0×10^{-4} M), cysteine (1.0×10^{-4} M), glucose (5.0×10^{-3} M), and cystine (6.5×10^{-5} M). The concentration of ACh in the prepared blood solution was kept at 5.0×10^{-6} M. The concentration of ACh in the synthetic blood was found to be $(5.45 \pm 0.22) \times 10^{-6}$ M using calibration graph. The authors measured the amount of ACh less than 0.49 μM with the prepared ACh biosensor [121]. Furthermore, the simultaneous detection of DA, adrenaline, AA, and UA in real dilute urine and serum samples was performed by Salimi et al. using ceramic carbon electrode [122]. The electrode was prepared using carbon powder and methyltrimethoxysilane using sol gel method. To avoid the matrix impact of the real samples prior to measurement, dilution of the urine and serum samples was done 83 and 10 times, respectively, using phosphate buffer solution.

Athiroh et al. developed a zeolite-modified CPE for the determination of creatine in blood samples potentiometry. Creatine is a nitrogen containing compound made up from amino acids [123]. Blood creatine levels typically range from 10^{-5} mg/L. Overconsumption of creatine is linked to hypertension since it can increase blood vessel density; therefore, it is important to manage creatine levels in the body as soon as possible. The modified CPE showed high selectivity for creatine compared to urea, glucose, and UA, and their response time was shorter than 4 minutes. The electrodes demonstrated an extended life span and a detection limit that was approximately 1,000 times lower than the typical levels of creatine in blood serum. Recovery rates of 76.8% and 81.8% were obtained by applying the modified electrode using the spiking approach to two serum samples. A 73% accuracy rate was observed in tests that were compared to the normal clinical laboratory procedure.

13.3.3 Fullerene-Based Sensors for Clinical Sample Analysis

Fullerene has a fused structure made up of pentagon and hexagon carbon rings. The occurrence of several atomic Cn clusters ($n > 20$) of carbon atoms on a sphere-shaped surface is a unique trait of fullerenes. With its unique structure made up of sp^2 carbons, fullerene provides a highly symmetrical lattice in a variety of sizes (C60, C70, and so forth). Fullerene is highly compatible with different types of biomolecules. Owing to its hydrophobic nature, fullerene is easily soluble in organic (particularly aromatic) solvents including benzene, toluene, and chloroform; however, its solubility in polar solvents is extremely low, which is significant for biological and medicinal applications. Due to the lack of aromaticity in C60, weak delocalization of electrons occurs. Hence, it behaves like an electron-deficient moiety and readily combines or interacts with electron-rich species. The C60 has an ionization potential of 7.8 eV and an excellent electron affinity of 2.7 eV. These values indicate the high electrochemical behavior of fullerenes, which makes them excellent materials for biosensing applications [124, 125]. Additionally, they can readily exchange electrons with other species that encircle the biomolecules. Fullerene has a spherical structure due to which some angle strain is present in the molecule. The angle strain causes fullerenes to undergo addition reactions across the double bond very easily. The discovery of fullerene (C60) has transformed the field of biosensor production [126]. Gavalas and Chaniotakis were the first to use fullerene for glucose estimation in the designing of various biosensors [127]. In their work, they utilized various amounts of fullerene to produce an amperometric biosensor for the identification of glucose oxidase (GOx) enzyme. The two most prevalent fullerene types are C60 and C70.

In another study, Wei and Shih reported a piezoelectric-fullerene C60-cryptand 22 biosensor for the detection of urea in aqueous samples [128]. A novel potentiometric urea biosensor has been developed by Saeedfar et al. by effectively modifying fullerene. Prior to accomplishing the immobilization method, they functionalized the fullerene using heat, sonication, and ultraviolet light. This modified fullerene was then used as a matrix for the immobilization of the urease enzyme [129]. The authors reported the linear range for the biosensor was 2.31×10^{-3} M to 8.28×10^{-5} M. Also, the sensitivity of the designed biosensor was 59.67 ± 0.91 mV/decade, which was very near to the theoretical value. The stability of the designed urea sensor was up to 140 days.

Much research has been conducted on the application of C60-based immunosensors in the diagnosis of a wide range of diseases arising from alterations in the metabolism process of our body.

In a study carried out by Pan and Shih, antibody-immobilized fullerene was used to identify immunoglobulin G (IgG) and hemoglobin (Hb) [130]. An anti-IgG and anti-Hb immobilization onto fullerene C60 has been investigated using a C60-coated piezoelectric crystal detecting technique. The detection limits of immunosensors for IgG and Hb in aqueous solutions were less than 10^{-4} mg/mL. The various organic and inorganic species like UA, cysteine, urea, AA, tyrosine, and metal ions present in the blood showed nearly no interference with the IgG piezoelectric biosensor.

Goyal et al. reported a fast and accurate electroanalytical method for determining the amount of dexamethasone in human blood plasma samples and pharmaceutical formulations [131]. Dexamethasone has been identified as a doping substance since some athletes regularly misuse it to increase performance in sports. Dexamethasone is an adrenocortical steroid (synthetic) that is water-insoluble. The authors used the fullerene-C_{60}-modified edge plane pyrolytic graphite electrode for the detection of dexamethasone. The detection limit reported was 5.5×10^{-8} M using square wave voltammetry.

13.3.4 Carbon Composite Sensors for Clinical Sample Analysis

Carbon is a low-cost, adaptable, safe, and widely available material for electrochemical sensors. Electron transfer kinetics, stability, low ohmic resistance, biocompatibility, and increased interfacial adsorption qualities distinguish these materials as ideal for sensing applications. Carbon is commonly used as a modifier because of its wide potential window, low background current, chemical inertness, low cost, and suitability for a variety of sensing applications. The potential of carbon to interact with other nanomaterials has a synergistic impact, resulting in a range of composites and hybrid materials that improve the sensitivity and selectivity of the electrochemical sensor [66, 67].

Carbon composite sensors are an emerging technology used in the analysis of clinical samples due to their unique properties and advantages. Carbon composite sensors are made from a combination of carbon materials (such as GO, graphene, and CNTs) with other components like polymers, metal, metal oxide, etc. These sensors leverage the electrical conductivity, chemical stability, and biocompatibility of carbon materials, making them ideal for detecting various analytes in clinical samples. Carbon composites provide a large surface area and excellent electrical properties, allowing for the detection of low concentrations of biomolecules like glucose, cholesterol, and other markers. The carbon materials used are generally biocompatible, reducing the risk of interference from the body's immune response. Carbon composite sensors are resistant to corrosion and degradation, which is essential for reliable and long-term clinical monitoring. These sensors can be engineered with specific functional groups to target particular molecules, improving selectivity and reducing false positives in clinical tests. Carbon composite sensors were developed for a variety of uses in clinical sample analysis, such as blood glucose monitoring, cancer biomarker detection, detection of pathogens, environmental monitoring, etc. [25, 132–136].

Thadathil et al. reported polyindole-derived nitrogen-doped graphene quantum dots-based electrochemical sensor for DA detection, with a limit of detection of 0.15 nM and a linear range of 0.001–1000 μM [137]. Jayaraman et al. fabricated a ruthenium-anchored carbon sphere-based amperometric detection of melatonin. The calculated values of lower detection limit and sensitivity were found to be 0.27 μM and 0.85 $\mu A\ \mu M^{-1}\ cm^{-2}$, respectively [138]. Istrate et al. reported gold nanoparticles/reduced graphene oxide/polyallylamine hydrochloride nanocomposite-based electrochemical biosensor for ethanol detection, with a limit of detection of 10 μM and a high sensitivity of ~44 $\mu A\ mM^{-1}\ cm^{-2}$, respectively [139]. Fall et al. developed highly efficient non-enzymatic electrochemical glucose sensor based on CNTs functionalized by molybdenum disulfide and decorated with nickel nanoparticles. They reported high sensitivity (1212 $\mu A \cdot mM^{-1} cm^{-2}$), excellent limit of detection (0.197 μM), and a wide linear range (0.05–0.65 mM) [140]. Manjunatha et al. reported phthalic acid and TritonX-100 (TX-100) modified CPE for DA detection [141]. Manjunatha et al. modified CPE with poly (nicotinic acid) for selective DA detection in the presence of AA [142]. Pushpanjali et al. modified a CNT with poly(L-arginine) for voltammetric curcumin detection [143]. Raril et al. fabricated a poly threonine-modified graphite-CNT paste electrode for tyrosine detection [144]. Tigari et al. successfully fabricated octoxynol-9 and poly (titan yellow) modified CNT paste electrodes for electrochemical detection of vanillin [145]. For Alizarin Carmine detection, Pushpanjali and Manjunatha have modified CNT paste electrodes with poly(glycine) [146].

Carbon composite sensors hold significant promise in clinical sample analysis due to their sensitivity, stability, and versatility. Ongoing research is focused on overcoming challenges related to manufacturing and integration, with the potential to revolutionize diagnostics and patient monitoring [25, 132, 134].

13.4 CHALLENGES AND FUTURE PERSPECTIVES

Since biological fluids (blood, serum, urine, sweat, etc.) contain a wide variety of organic and inorganic compounds, biosensors often cannot distinguish between responses of substances with comparable electro-active functional groups in their structures. This is particularly prevalent in the case of biological fluids, which have a very complicated nature. As per reports, modified electrodes based on carbon materials can significantly enhance the chemical response of chemicals, such as epinephrine, DA, norepinephrine, and other compounds. Hence, it is impossible to accurately quantify them simultaneously in a mixed solution, like human blood or urine. As seen, carbon nanostructure-based materials have considerably increased analytical sensitivity. Yet, they are still capable of responding or reacting to a large variety of compounds present simultaneously in the biological matrices/fluids [147].

- The identification and measurement of these compounds in biological samples demand the designing of a very basic point-of-care technology.
- Accurate, fast, and affordable sensor systems are required for medical diagnostics. In addition to allowing highly sensitive and selective detection, carbon-based sensors offer clinical analysis with near-real-time monitoring capacity.
- Many biosensors should ideally have the same extent of consistency as many traditional devices.
- In the realm of clinical analysis, producing sensors with enhanced storage capacity and operational reliability is crucial from a commercial standpoint.
- A requirement for a highly accurate sensing device has become vital since therapeutic fluids/clinical fluids include multiple interfering proteins/substances.
- The use of carbon paste-based electrodes demands sufficient user convenience, which is the main drawback of CPEs and restricts their application in clinical analysis.
- By integrating a metal-organic framework and nanoparticle-modified electrodes, high sensitivity, selectivity, and very low detection limit may be achieved.

13.5 CONCLUSION

This chapter intends to expedite present advancements in the clinical analysis of real samples by exploring the benefits of carbon-based materials. In the medical sector, diagnosing the disease before it becomes severe and timely detection can save the number of lives. In earlier times, research examined the distinctions between healthy organisms and diseased ones to identify pathological variables or disease biomarkers. Nowadays, in medical detection laboratory testing is frequently utilized. Biosensors now have a new avenue for disease evaluation owing to the rapid developments of nanotechnology and the production of various semiconductor materials. Higher electrochemical qualities and great biocompatibility are displayed by carbon nanomaterial-based sensors.

Clinical diagnosis and therapy greatly benefit from the availability of biosensors for early disease identification and medical monitoring. Biosensors offer low-cost, highly accurate, quick, sensitive, and reproducible findings, requiring minimal technical expertise. Currently, a wide range of sensing methods are employed to identify biomarkers. Over the past few years, graphene and CNTs have become vital components of sensor platforms because of their greater optical, mechanical, and electrical properties. Furthermore, their application potential is likely to be improved by conceptual functionalization. Similarly, future research on carbon quantum dots will be expected as they represent a great promising potential for use in various applications.

Research on carbon-based biosensors for the detection of biomolecules, including biomarkers for cancer and other diseases, is still in its early stages in the laboratory. Nevertheless, it has already shown promise for the future of health monitoring and clinical diagnosis. Further advancements in this field are expected to lead to the development of biomarker sensors for common clinical applications. Based on current advancements, carbon nanomaterial-based biosensors have a promising future for applications in biomedical analysis. In summary, carbon-based materials can be effectively used in clinical analysis.

REFERENCES

1. Uddin, M.; Syed-Abdu, S. Data analytics and applications of the wearable sensors in healthcare: An overview. *Sensors* **2020**, *20*, 1379.
2. Mbunge, E.; Muchemwa, B.; Jiyane, S.; Batani, J. Sensors and healthcare 5.0: Transformative shift in virtual care through emerging digital health technologies. *Glob. J. Health Sci.* **2021**, *5*, 169–177.
3. Ji, C.; Zhou, Y.; Leblanc, R. M.; Peng, Z. Recent developments of carbon dots in biosensing: A review. *ACS Sens.* **2020**, *5*, 2724–2741.
4. Fakayode, S. O.; Lisse, C.; Medawala, W.; Brady, P. N.; Bwambok, D. K.; Anum, D.; Alonge, T.; Taylor, M. E.; Baker, G. A.; Mehari, T. F.; Rodriguez, J. D.; Elzey, B.; Siraj, N.; Macchi, S.; Le, T.; Forson, M.; Bashiru, M.; Narcisse, V. E. F.; Grant, C. Fluorescent chemical sensors: Applications in analytical, environmental, forensic, pharmaceutical, biological, and biomedical sample measurement, and clinical diagnosis. *Appl. Spectrosc. Rev.* **2024**, *59(1)*, 1–89.
5. Naresh, V.; Lee, N. A review on biosensors and recent development of nanostructured materials-enabled biosensors. *Sensors* **2021**, *21*, 1109.
6. Celik, N.; Balachandran, W.; Manivannan, N. Graphene-based biosensors: Methods, analysis and future perspectives. *IET Circuits Devices Syst.* **2015**, *9(6)*, 434–445.
7. Ansari, S. Combination of molecularly imprinted polymers and carbon nanomaterials as a versatile biosensing tool in sample analysis: Recent applications and challenges. *Trends Anal. Chem.* **2017**, *93*, 134–151.
8. Kruss, S.; Hilmer, A. J.; Zhang, J.; Reuel, N. F.; Mu, B.; Strano, M. S.; Carbon nanotubes as optical biomedical sensors. *Adv. Drug Deliv. Rev.* **2013**, *65*, 1933–1950.
9. Rivas, G. A.; Rodríguez, M. C.; Rubianes, M. D.; Gutierrez, F. A.; Eguílaz, M.; Dalmasso, P. R.; Primo, E. N.; Tettamanti, C.; Ramírez, M. L.; Montemerlo, A.; Gallay, P.; Parradoc, C. Carbon nanotubes-based electrochemical (bio)sensors for biomarkers. *Appl. Mater. Today* **2017**, *9*, 566–588.
10. Barrientos, K.; Arango, J. P.; Moncada, M. S.; Placido, J.; Pati˜no, J.; Macías, S. L.; Maldonado, C.; Torijano, S.; Bustamante, S.; Londo˜no, M. E.; Jaramillo, M. Carbon dot-based biosensors for the detection of communicable and non-communicable diseases. *Talanta* **2023**, *251,* 123791.
11. Jin, J.; Li, L.; Zhang, L.; Luan, Z.; Xin, S.; Song, K. Progress in the application of carbon dots-based nanozymes. *Front. Chem.* **2021**, *9*, 748044.
12. Campuzano, S.; Yáñez-Sedeño, P.; Pingarrón, J. M. Carbon dots and graphene quantum dots in electrochemical biosensing. *Nanomaterials* **2019**, *9*, 634.
13. Shellaiah, M.; Sun, K. W. Review on carbon dot-based fluorescent detection of biothiols. *Biosensors* **2023**, *13*, 335.
14. Szunerits, S.; Boukherroub, R. Graphene-based biosensors. *Interface Focus* **2018**, *8*, 20160132.
15. Kuila, T.; Bose, S.; Khanra, P.; Mishra, A. K.; Kim, N. H.; Lee, J. H. Recent advances in graphene-based biosensors. *Biosens. Bioelectron.* **2011**, *26*, 4637–4648.
16. Lee, J.; Kim, J.; Kim, S.; Min, D.-H. Biosensors based on graphene oxide and its biomedical application. *Adv. Drug Deliv. Rev.* **2016**, *105*, 275–287.
17. Formica, D.; Schena, E. Smart sensors for healthcare and medical applications. *Sensors* **2021**, *21*, 543.
18. Al-kahtani, M. S.; Khan, F.; Taekeun, W. Application of Internet of Things and sensors in healthcare. *Sensors* **2022**, *22*, 5738.
19. Wang, F.; Liu, S.; Shu, L.; Tao, X-M. Low-dimensional carbon-based sensors and sensing network for wearable health and environmental monitoring. *Carbon* **2017**, *121*, 353–367.
20. Shavanova, K.; Bakakina, Y.; Burkova, I.; Shtepliuk, I.; Viter, R.; Ubelis, A.; Beni, V.; Starodub, N.; Yakimova, R.; Khranovskyy, V.; Application of 2D non-graphene materials and 2D oxide nanostructures for biosensing technology. *Sensors* **2016**, *16*, 223.

21. Katey, B.; Voiculescu, I.; Penkova, A. N.; Untaroiu, A. A review of biosensors and their applications. *ASME Open J. Eng.* **2023**, *2*, 020201-1-020201-16.
22. Polat, E. O.; Cetin, M. M.; Tabak, A. F.; Güven, E. B.; Uysal, B. Ö.; Arsan, T.; Kabbani, A.; Hamed, H.; Gül, S. B. Transducer technologies for biosensors and their wearable applications. *Biosensors* **2022**, *12*, 385.
23. Sethi, R. S.; Transducer aspects of biosensors. *Biosens. Bioelectron.* **1994**, *9*, 243–264.
24. Wang, Z.; Yu, J.; Gui, R.; Jin, H.; Xia, Y. Carbon nanomaterials-based electrochemical aptasensors. *Biosens. Bioelectron.* **2016**, *79*, 136–149.
25. Kour, R.; Arya, S.; Young, S.-J.; Gupta, V.; Bandhoria, P.; Khosla, A. Review-recent advances in carbon nanomaterials as electrochemical biosensors. *J. Electrochem. Soc.* **2020**, *167*, 037555.
26. Lei, Z.-L.; Guo, B. 2D material-based optical biosensor: Status and prospect. *Adv. Sci.* **2022**, *9*, 2102924.
27. Vasuki, S.; Varsha, V.; Mithra; Dharshni, R. A.; Abinaya, S.; Dharshini, R. D.; Sivarajasekar, N. Thermal biosensors and their applications. *Am. Int. J. Res. Sci., Technol., Eng. Math.*, Special issue of National Conference on Innovations in Bio, Chemical and Food Technology (IBCFT-2019), **2019**, 262–264.
28. Polshettiwar, S. A.; Deshmukh, C. D.; Baheti, A. M.; Wani, M. S.; Bompilwar, E.; Jambhekar, D.; Choudhari, S.; Tagalpallewar, A. Recent trends on biosensors in healthcare and pharmaceuticals: An overview. *Int. J. Pharm. Investigation* **2021**, *11(2)*, 131–136.
29. Skládal, P. Piezoelectric biosensors. *Trends Anal. Chem.* **2016**, *79*, 127–133.
30. Gupta, S.; Murthy, C. N.; Prabha, C. R. Recent advances in carbon nanotube based electrochemical biosensors. *Int. J. Biol. Macromol.* **2018**, *108*, 687–703.
31. Damiati, S.; Schuster, B. Electrochemical biosensors based on S-layer proteins. *Sensors* **2020**, *20*, 1721.
32. Grieshaber, D.; MacKenzie, R.; Vörös, J.; Reimhult, E. Electrochemical biosensors-sensor principles and architectures. *Sensors* **2008**, *8*, 1400–1458.
33. Shinwari, M. W.; Zhitomirsky, D.; Deen, I. A.; Selvaganapathy, P. R.; Deen, M. J.; Landheer, D. Microfabricated reference electrodes and their biosensing applications. *Sensors* **2010**, *10*, 1679–1715.
34. Porto, L. S.; Silva, D. N.; de Oliveira, A. E. F.; Pereira, A. C.; Borges, K. B. Carbon nanomaterials: Synthesis and applications to development of electrochemical sensors in determination of drugs and compounds of clinical interest. *Rev. Anal. Chem.* **2019**, *38*, 20190017.
35. Shanbhag, M. M.; Manasa, G.; Mascarenhas, R. J.; Mondal, K.; Shetti, N. P. Fundamentals of bio-electrochemical sensing. *Chem. Eng. J. Adv.* **2023**, *16*, 100516.
36. Erduran, V.; Bekmezci, M.; Bayat, R.; Sena, F. Functionalized carbon material-based electrochemical sensors for day-to-day applications. Editor(s): Hussain, C. M.; Manjunatha, J. G. In *Woodhead Publishing Series in Electronic and Optical Materials, Functionalized Nanomaterial-Based Electrochemical Sensors*, Woodhead Publishing, Elsevier **2022**, 97–111. https://doi.org/10.1016/B978-0-12-823788-5.00017-X
37. Muniandy, S.; Teh, S. J.; Thong, K. L.; Thiha, A.; Dinshaw, I. J.; Lai, C. W.; Ibrahim, F.; Leo, B. F. Carbon nanomaterial-based electrochemical biosensors for foodborne bacterial detection. *Crit. Rev. Anal. Chem.* **2019**, *49(6)*, 510–533.
38. Singh, A. K.; Mittal, S.; Das, M.; Saharia, A.; Tiwari, M. Optical biosensors: A decade in review. *Alex. Eng. J.* **2023**, *67*, 673–691.
39. Mostufa, S.; Rezaei, B.; Ciannella, S.; Yari, P.; Gómez-Pastora, J.; He, R.; Wu, K. Advancements and perspectives in optical biosensors. *ACS Omega* **2024**, *9*, 24181–24202.
40. Damborsky´, P.; Sˇ vitel, J.; Katrlı´k, J. Optical biosensors. *Essays Biochem.* **2016**, *60*, 91–100.
41. Borisov, S. M.; Wolfbeis, O. S. Optical biosensors. *Chem. Rev.* **2008**, *108*, 423–461.
42. Govindaraju, R.; Govindaraju, S.; Yun, K.; Kim, J. Fluorescent-based neurotransmitter sensors: Present and future perspectives. *Biosensors* **2023**, *13*, 1008.
43. Strianese, M.; Staiano, M.; Ruggiero, G.; Labella, T.; Pellecchia, C.; D'Auria, S. Fluorescence-based biosensors. Editor(s): Bujalowski, W. M. *Spectroscopic Methods of Analysis: Methods and Protocols, Methods in Molecular Biology*, Springer **2012**, *875*, 193–216. DOI 10.1007/978-1-61779-806-1_9.
44. Eddin, F. B. K.; Fen, Y. W. The principle of nanomaterials based surface plasmon resonance biosensors and its potential for dopamine detection. *Molecules* **2020**, *25*, 2769.

45. Szunerits, S.; Maalouli, N.; Wijaya, E.; Vilcot, J.-P.; Boukherroub, R. Recent advances in the development of graphene-based surface plasmon resonance (SPR) interfaces. *Anal. Bioanal. Chem.* **2013**, *405*,1435–1443.
46. Kushwaha, A. S.; Kumar, A.; Kumar, R.; Srivastava, S. K. A study of surface plasmon resonance (SPR) based biosensor with improved sensitivity. *Photonics Nanostructures: Fundam. Appl.* **2018**, *31*, 99–106.
47. Choi, M. M. F. Progress in enzyme-based biosensors using optical transducers. *Microchim. Acta* **2004**, *148*, 107–132.
48. Acha, N. D.; Socorro-Leránoz, A. B.; Elosúa, C.; Matías, I. R. Trends in the design of intensity-based optical fiber biosensors (2010–2020). *Biosensors* **2021**, *11*, 197.
49. Passaro, V. M. N.; Dell'Olio, F.; Casamassima, B.; Leonardis, F. D. Guided-wave optical biosensors. *Sensors* **2007**, *7*, 508–536.
50. Mukundan, H.; Anderson, A. S.; Grace, W. K.; Grace, K. M.; Hartman, N.; Martinez, J. S.; Swanson, B. I. Waveguide-based biosensors for pathogen detection. *Sensors* **2009**, *9*, 5783–5809.
51. Ramanathan, K.; Danielsson, B. Principles and applications of thermal biosensors. *Biosens. Bioelectron.* **2001**, *16*, 417–423.
52. Xie, B.; Ramanathan, K.; Danielsson, B. Mini/micro thermal biosensors and other related devices for biochemical / clinical analysis and monitoring. *Trends Anal. Chem.* **2000**, *19(5)*, 340–349.
53. Pohanka, M. Overview of piezoelectric biosensors, immunosensors and DNA sensors and their applications. *Materials* **2018**, *11*, 448.
54. Huang, S.; Gao, Y.; Hu, Y.; Shen, F.; Jin, Z.; Cho, Y. Recent development of piezoelectric biosensors for physiological signal detection and machine learning assisted cardiovascular disease diagnosis. *RSC Adv.* **2023**, *13*, 29174.
55. Skládal, P. Piezoelectric biosensors: Shedding light on principles and applications. *Microchim. Acta* **2024**, *191*, 184.
56. Pohanka, M. The piezoelectric biosensors: Principles and applications, a review. *Int. J. Electrochem. Sci.* **2017**, *12*, 496–506.
57. Tîlmaciu, C.-M.; Morris, M. C. Carbon nanotube biosensors. *Front. Chem.* **2015**, *3*, 59.
58. Joshi, P.; Mishra, R.; Narayan, R. J. Biosensing applications of carbon-based materials. *Curr. Opin. Biomed. Eng.* **2021**, *18*, 100274.
59. Eivazzadeh-Keihan, R.; Noruzi, E. B.; Chidar, E.; Jafari, M.; Davoodi, F.; Kashtiaray, A.; Gorab, M. G.; Hashemi, S. M.; Javanshir, S.; Cohan, R. A.; Maleki, A.; Mahdavi, M.; Applications of carbon-based conductive nanomaterials in biosensors. *J. Chem. Eng.* **2022**, *442*, 136183.
60. Castro, K. P. R.; Colombo, R. N. P.; Iost, R. M.; da Silva, B. G. R.; Crespilho, F. N. Low-dimensionality carbon-based biosensors: The new era of emerging technologies in bioanalytical chemistry. *Anal. Bioanal. Chem.* **2023**, *415*, 3879–3895.
61. Safari, M.; Moghaddam, A.; Moghaddam, A. S.; Absalan, M.; Kruppke, B.; Ruckd¨aschel, H.; Khonakdar, H. A. Carbon-based biosensors from graphene family to carbon dots: A viewpoint in cancer detection. *Talanta* **2023**, *258*, 124399.
62. Karimi, F.; Karimi-Maleh, H.; Rouhi, J.; Zare, N.; Karaman, C.; Baghayeri, M.; Fu, L.; Rostamnia, S.; Dragoi, E. N.; Ayati, A.; Krivoshapkin, P. Revolutionizing cancer monitoring with carbon-based electrochemical biosensors. *Environ. Res.* **2023**, *239*, 117368.
63. Liu, P.; Jiao, Y.; Chai, X.; Ma, Y.; Liu, S.; Fang, X.; Fan, F.; Xue, L.; Han, J.; Liu, Q. High-performance electric and optical biosensors based on single-walled carbon nanotubes. *J. Lumin.* **2022**, *250*, 119084.
64. Nag, A.; Mukhopadhyay, S. C. Fabrication and implementation of carbon nanotubes for piezoresistive-sensing applications: A review. *J. Sci.: Adv. Mater. Devices* **2022**, *7*, 100416.
65. Yi, Y.; Samara, A.; Wang, B. A new approach for an ultra-thin piezoresistive sensor based on solidified carbon ink film. *J. Mater. Sci.* **2021**, *56*, 607–614.
66. Gibi, C.; Liu, C.-H.; Barton, S. C.; Anandan, S.; Wu, J. J. Carbon materials for electrochemical sensing application - a mini review. *J. Taiwan Inst. Chem. Eng.* **2024**, *154*, 105071.
67. Khan, A.; DeVoe, E.; Andreescu, S. Carbon-based electrochemical biosensors as diagnostic platforms for connected decentralized healthcare. *Sens. Diagn.* **2023**, *2*, 529.
68. Chen, J.; Wei, J. S.; Zhang, P.; Niu, X. Q.; Zhao, W.; Zhu, Z. Y.; Ding, H.; Xiong, H. M. Red-emissive carbon dots for fingerprints detection by spray method: Coffee ring effect and unquenched fluorescence in drying process. *ACS Appl. Mater. Interfaces* **2017**, *9(22)*, 18429–18433.

69. Baker, S. N.; Baker, G. A. Luminescent carbon nanodots: Emergent nanolights. *Angew. Chem., Int. Ed.* **2010**, *49(38)*, 6726–6744.
70. Li, H.; Kang, Z.; Liu, Y.; Lee, S. T. Carbon nanodots: Synthesis, properties and applications. *J. Mater. Chem.* **2012**, *22(46)*, 24230–24253.
71. Lim, S. Y.; Shen, W.; Gao, Z. Carbon quantum dots and their applications. *Chem. Soc. Rev.* **2015**, *44(1)*, 362–381.
72. Choi, Y.; Choi, Y.; Kwon, O. H.; Kim, B. S. Carbon dots: Bottom-up syntheses, properties, and light-harvesting applications. *Chem. Asian J.* **2018**, *13(6)*, 586–598.
73. Hu, S.; Trinchi, A.; Atkin, P.; Cole, I. Tunable photoluminescence across the entire visible spectrum from carbon dots excited by white light. *Angew. Chem., Int. Ed.* **2015**, *54(10)*, 2970–2974.
74. Rossini, E. L.; Milani, M. I.; Lima, L. S.; Pezza, H. R. Paper microfluidic device using carbon dots to detect glucose and lactate in saliva samples. *Spectrochim. Acta A Mol. Biomol. Spectrosc.* **2021**, *248*, 119285.
75. Wang, X.; Qu, K.; Xu, B.; Ren, J.; Qu, X. Multicolor luminescent carbon nanoparticles: Synthesis, supramolecular assembly with porphyrin, intrinsic peroxidase-like catalytic activity and applications. *Nano Res.* **2011**, *4*, 908–920.
76. Li, H.; Chen, L.; Wu, H.; He, H.; Jin, Y. Ionic liquid-functionalized fluorescent carbon nanodots and their applications in electrocatalysis, biosensing, and cell imaging. *Langmuir* **2014**, *30*, 15016–15021.
77. Zheng, W.; Wu, H.; Jiang, Y.; Xu, J.; Li, X.; Zhang, W.; Qiu, F. A molecularly-imprinted-electrochemical-sensor modified with nano-carbon-dots with high sensitivity and selectivity for rapid determination of glucose. *Anal. Biochem.* **2018**, *555*, 42–49.
78. Tuteja, S. K.; Chen, R.; Kukkar, M.; Song, C. K.; Mutreja, R.; Singh, S.; Paul, A. K.; Lee, H.; Kim, K. -H.; Deep, A.; Suri, C. R. A label-free electrochemical immunosensor for the detection of cardiac marker using graphene quantum dots (GQDs). *Biosens. Bioelectron.* **2016**, *86*, 548–556.
79. Mars, A.; Hamami, M.; Bechnak, L.; Patra, D.; Raouafi, N. Curcumin-graphene quantum dots for dual mode sensing platform: Electrochemical and fluorescence detection of APOe4, responsible for Alzheimer's disease. *Anal. Chim. Acta* **2018**, *1036*, 141–146.
80. Iijima, S.; Helical microtubules of graphitic carbon. *Nature* **1991**, *354(6348)*, 56–58.
81. Terrones, M.; Science and technology of the twenty-first century: Synthesis, properties, and applications of carbon nanotubes. *Annu. Rev. Mater. Res.* **2003**, *33(1)*, 419–501.
82. Barone, P. W.; Baik, S.; Heller, D. A.; Strano, M. S. Near-infrared optical sensors based on single-walled carbon nanotubes. *Nat. Mater.* **2005**, *4(1)*, 86–92.
83. Cha, T. G.; Baker, B. A.; Sauffer, M. D.; Salgado, J.; Jaroch, D.; Rickus, J. L.; Porterfield, D. M.; Choi, J. H. Optical nanosensor architecture for cell-signaling molecules using DNA aptamer-coated carbon nanotubes. *ACS nano*, **2011**, *5(5)*, 4236–4244.
84. Heller, D. A.; Jeng, E. S.; Yeung, T. K.; Martinez, B. M.; Moll, A. E.; Gastala, J. B.; Strano, M. S. Optical detection of DNA conformational polymorphism on single-walled carbon nanotubes. *Science* **2006**, *311(5760)*, 508–511.
85. Jeng, E. S.; Nelson, J. D.; Prather, K. L.; Strano, M. S. Detection of a single nucleotide polymorphism using single-walled carbon-nanotube near-infrared fluorescence. *Small* **2010**, *6(1)*, 40–43.
86. Zhang, X.; Jiao, K.; Liu, S.; Hu, Y.; Readily reusable electrochemical DNA hybridization biosensor based on the interaction of DNA with single-walled carbon nanotubes. *Anal. Chem.* **2009**, *81(15)*, 6006–6012.
87. Jeng, E. S.; Moll, A. E.; Roy, A. C.; Gastala, J. B.; Strano, M. S. Detection of DNA hybridization using the near-infrared band-gap fluorescence of single-walled carbon nanotubes. *Nano Lett.* **2006**, *6(3)*, 371–375.
88. Hong, H.; Gao, T.; Cai, W. Molecular imaging with single-walled carbon nanotubes. *Nano today* **2009**, *4(3)*, 252–261.
89. Duffy, M. J. Role of tumor markers in patients with solid cancers: A critical review. *Eur. J. Intern. Med.* **2007**, *18(3)*, 175–184.
90. Shobha, B. N.; Muniraj, N. J. R. Design, modeling and performance analysis of carbon nanotube with DNA strands as biosensor for prostate cancer. *Microsyst. Technol.* **2015**, *21*, 791–800.
91. Singh, S.; Gill, A. A.; Nlooto, M.; Karpoormath, R. Prostate cancer biomarkers detection using nanoparticles based electrochemical biosensors. *Biosens. Bioelectron.* **2019**, *137*, 213–221.

92. Yu, X.; Munge, B.; Patel, V.; Jensen, G.; Bhirde, A.; Gong, J. D.; Kim, S. N.; Gillespie, J.; Gutkind, J. S.; Papadimitrakopoulos, F.; Rusling, J. F. Carbon nanotube amplification strategies for highly sensitive immunodetection of cancer biomarkers. *J. Am. Chem. Soc.* **2006**, *128(34)*, 11199–11205.
93. Panini, N. V.; Messina, G. A.; Salinas, E.; Fernández, H.; Raba, J. Integrated microfluidic systems with an immunosensor modified with carbon nanotubes for detection of prostate specific antigen (PSA) in human serum samples. *Biosens. Bioelectron.* **2008**, *23(7)*, 1145–1151.
94. Rusling, J. F.; Sotzing, G.; Papadimitrakopoulosa, F. Designing nanomaterial-enhanced electrochemical immunosensors for cancer biomarker proteins. *Bioelectrochemistry* **2009**, *76(1-2)*, 189–194.
95. Ho, J. A. A.; Lin, Y. C.; Wang, L. S.; Hwang, K. C.; Chou, P. T. Carbon nanoparticle-enhanced immunoelectrochemical detection for protein tumor marker with cadmium sulfide biotracers. *Anal. Chem.* **2009**, *81(4)*, 1340–1346.
96. Majidi, M. R.; Omidi, Y.; Karami, P.; Johari-Ahar, M. Reusable potentiometric screen-printed sensor and label-free aptasensor with pseudo-reference electrode for determination of tryptophan in the presence of tyrosine. *Talanta*, **2016**, *150*, 425–433.
97. Hočevar, S. B.; Wang, J.; Deo, R. P.; Musameh, M.; Ogorevc, B. Carbon nanotube modified microelectrode for enhanced voltammetric detection of dopamine in the presence of ascorbate. *Electroanalysis* **2005**, *17(5-6)*, 417–422.
98. Swamy, B. K.; Venton, B. J. Carbon nanotube-modified microelectrodes for simultaneous detection of dopamine and serotonin in vivo. *Analyst* **2007**, *132(9)*, 876–884.
99. Xiang, L.; Yu, P.; Hao, J.; Zhang, M.; Zhu, L.; Dai, L.; Mao, L. Vertically aligned carbon nanotube-sheathed carbon fibers as pristine microelectrodes for selective monitoring of ascorbate in vivo. *Anal. Chem.* **2014**, *86(8)*, 3909–3914.
100. Cui, R.; Wang, X.; Zhang, G.; Wang, C. Simultaneous determination of dopamine, ascorbic acid, and uric acid using helical carbon nanotubes modified electrode. *Sens. Actuators B Chem.* **2012**, *161(1)*, 1139–1143.
101. Zhang, B.; Huang, D.; Xu, X.; Alemu, G.; Zhang, Y.; Zhan, F.; Shen, Y.; Wang, M. Simultaneous electrochemical determination of ascorbic acid, dopamine and uric acid with helical carbon nanotubes. *Electrochim. Acta* **2013**, *91*, 261–266.
102. Schmidt, A. C.; Wang, X.; Zhu, Y.; Sombers, L. A. Carbon nanotube yarn electrodes for enhanced detection of neurotransmitter dynamics in live brain tissue. *ACS Nano*, **2013**, *7(9)*, 7864–7873.
103. Jacobs, C. B.; Ivanov, I. N.; Nguyen, M. D.; Zestos, A. G.; Venton, B. J. High temporal resolution measurements of dopamine with carbon nanotube yarn microelectrodes. *Anal. Chem.* **2014**, *86(12)*, 5721–5727.
104. Zhang, D. A.; Rand, E.; Marsh, M.; Andrews, R. J.; Lee, K. H.; Meyyappan, M.; Koehne, J. E. Carbon nanofiber electrode for neurochemical monitoring. *Mol. Neurobiol.* **2013**, *48*, 380–385.
105. Rand, E.; Periyakaruppan, A.; Tanaka, Z.; Zhang, D. A.; Marsh, M. P.; Andrews, R. J.; Lee, K. H.; Chen, B.; Meyyappan, M.; Koehne, J. E. A carbon nanofiber based biosensor for simultaneous detection of dopamine and serotonin in the presence of ascorbicacid. *Biosens. Bioelectron.* **2013**, *42*, 434–438.
106. Justino, C. I.; Gomes, A. R.; Freitas, A. C.; Duarte, A. C.; Rocha-Santos, T. A. Graphene based sensors and biosensors. *Trends Analyt. Chem.* **2017**, *91*, 53–66.
107. Chen, X. M.; Wu, G. H.; Jiang, Y. Q.; Wang, Y. R.; Chen, X. Graphene and graphene-based nanomaterials: The promising materials for bright future of electroanalytical chemistry. *Analyst*, **2011**, *136(22)*, 4631–4640.108. Pothipor, C.; Wiriyakun, N.; Putnin, T.; Ngamaroonchote, A.; Jakmunee, J.; Ounnunkad, K.; Laocharoensuk, R.; Aroonyadet, N. Highly sensitive biosensor based on graphene–poly (3-aminobenzoic acid) modified electrodes and porous-hollowed-silver-gold nanoparticle labelling for prostate cancer detection. *Sens. Actuators B: Chem.* **2019**, *296*, 126657.
109. Luo, M.; Chen, X.; Zhou, G.; Xiang, X.; Chen, L.; Ji, X.; He, Z. Chemiluminescence biosensors for DNA detection using graphene oxide and a horseradish peroxidase-mimicking DNAzyme. *ChemComm.* **2012**, *48(8)*, 1126–1128.
110. Zou, H. L.; Li, B. L.; Luo, H. Q.; Li, N. B. A novel electrochemical biosensor based on hemin functionalized graphene oxide sheets for simultaneous determination of ascorbic acid, dopamine and uric acid. *Sens. Actuators B: Chem.* **2015**, *207*, 535–541.

111. Chen, J. L.; Yan, X. P.; Meng, K.; Wang, S. F. Graphene oxide based photoinduced charge transfer label-free near-infrared fluorescent biosensor for dopamine. *Anal. Chem.* **2011**, *83*, 8787–8793.
112. Qian, Z. S.; Shan, X. Y.; Chai, L. J.; Ma, J. J.; Chen, J. R.; Feng, H. A universal fluorescence sensing strategy based on biocompatible graphene quantum dots and graphene oxide for the detection of DNA. *Nanoscale* **2014**, *6*, 5671–5674.
113. Castillo, J. J.; Svendsen, W. E.; Rozlosnik, N.; Escobar, P.; Martinez, F.; Castillo-Leon, J. Detection of cancer cells using a peptide nanotube-folic acid modified graphene electrode. *Analyst* **2013**, *138*, 1026–1031.
114. Kesavan, S.; John, S. A. Stable determination of paracetamol in the presence of uric acid in human urine sample using melamine grafted graphene modified electrode. *J. Electroanal. Chem.* **2016**, *760*, 6–14.
115. Adhikari, B. R.; Govindhan, M.; Chen, A. Sensitive detection of acetaminophen with graphene-based electrochemical sensor. *Electrochim. Acta* **2015**, *162*, 198–204.
116. Jing, A.; Zhang, C.; Liang, G.; Feng, W.; Tian, Z.; Jing, C. Hyaluronate-functionalized graphene for label-free electrochemical cytosensing. *Micromachines* **2018**, *9(12)*, 669.
117. Madhurantakam, S.; Karnam, J. B.; Brabazon, D.; Takai, M.; Ahad, I. U.; Balaguru Rayappan, J. B.; Krishnan, U. M. "Nano": An emerging avenue in electrochemical detection of neurotransmitters. *ACS Chem. Neurosci.* **2020**, *11(24)*, 4024–4047.
118. Atta, N. F.; Galal, A.; Azab, S. M. Electrochemical determination of neurotransmitters using gold nanoparticles on nafion/carbon paste modified electrode. *J. Electrochem. Soc.* **2012**, *159(10)*, H765.
119. Wang, J.; Walcarius, A. Zeolite-modified carbon paste electrode for selective monitoring of dopamine. *J. Electroanal. Chem.* **1996**, *407(1-2)*, 183–187.
120. Chandra, U.; Swamy, B. K.; Gilbert, O.; Sherigara, B. S. Voltammetric resolution of dopamine in the presence of ascorbic acid and uric acid at poly (calmagite) film coated carbon paste electrode. *Electrochim. Acta* **2010**, *55(24)*, 7166–7174.
121. Bodur, O. C.; Dinç, S.; Özmen, M.; Arslan, F. A sensitive amperometric detection of neurotransmitter acetylcholine using carbon dot-modified carbon paste electrode. *Biotechnol. Appl. Biochem.* **2021**, *68*, 20–29.
122. Salimi, A.; MamKhezri, H.; Hallaj, R. Simultaneous determination of ascorbic acid, uric acid and neurotransmitters with a carbon ceramic electrode prepared by sol–gel technique. *Talanta* **2006**, *70(4)*, 823–832.
123. Athiroh, A.; Fadillah, T. Damayanti, D. F.; Widati, A. A.; Abdulloh, A.; Khasanah, M. Carbon paste electrode modified imprinted zeolite as a selective sensor for creatine analysis by potentiometry. *IOP Conf. Ser.: Earth Environ. Sci.* **2019**, *217(1)*, 012003.
124. Pilehvar, S.; De Wael, K. Recent advances in electrochemical biosensors based on fullerene-C60 nano-structured platforms. *Biosensors* **2015**, *5(4)*, 712–735.
125. Maduraiveeran, G.; Sasidharan, M.; Ganesan, V. Electrochemical sensor and biosensor platforms based on advanced nanomaterials for biological and biomedical applications. *Biosens. Bioelectron.* **2018**, *103*, 113–129.
126. Afreen, S.; Muthoosamy, K.; Manickam, S.; Hashim, U. Functionalized fullerene (C60) as a potential nanomediator in the fabrication of highly sensitive biosensors. *Biosens. Bioelectron.* **2015**, *63*, 354–364.
127. Gavalas, V. G.; Chaniotakis, N. A. Fullerene-mediated amperometric biosensors. *Anal. Chim. Acta* **2000**, *409(1-2)*, 131–135.
128. Wei, L. F.; Shih, J. S. Fullerene-cryptand coated piezoelectric crystal urea sensor based on urease. *Anal. Chim. Acta* **2001**, *437(1)*, 77–85.
129. Saeedfar, K.; Heng, L. Y.; Ling, T. L.; Rezayi, M. Potentiometric urea biosensor based on an immobilised fullerene-urease bio-conjugate. *Sensors* **2013**, *13(12)*, 16851–16866.
130. Pan, N. Y.; Shih, J. S. Piezoelectric crystal immunosensors based on immobilized fullerene C60-antibodies. *Sens. Actuators B: Chem.* **2004**, *98(2-3)*, 180–187.
131. Goyal, R. N., Gupta, V. K.; Chatterjee, S.. Fullerene-C60-modified edge plane pyrolytic graphite electrode for the determination of dexamethasone in pharmaceutical formulations and human biological fluids. *Biosens. Bioelectron.* **2009**, *24(6)*, 1649–1654.

132. Gupta, S.; Tai, N.-H. Carbon nanomaterials and their composites for electrochemical glucose biosensors: A review on fabrication and sensing properties. *J. Taiwan Inst. Chem. Eng.* **2024**, *154*, 104957.
133. Goldoni, A.; Alijani, V.; Sangaletti, L.; D'Arsi, L. Advanced promising routes of carbon/metal oxides hybrids in sensors: A review. *Electrochim. Acta* **2018**, *266*, 139–150.
134. Fritea, L.; Banica, F.; Costea, T. O.; Moldovan, L.; Dobjanschi, L.; Muresan, M.; Cavalu, S. Metal nanoparticles and carbon-based nanomaterials for improved performances of electrochemical (bio) sensors with biomedical applications. *Materials* **2021**, *14*, 6319.
135. Manjunatha, J. G.; Raril, C.; Hareesha, N.; Charithra, M. M.; Pushpanjali, P. A.; Tigari, G.; Ravishankar, D. K.; Mallappaji, S. C.; Gowda, J. Electrochemical fabrication of poly (niacin) modified graphite paste electrode and its application for the detection of riboflavin. *Open Chem. Eng. J.* **2020**, *14*, 90–98.
136. Manjunatha, J. G. Highly sensitive polymer Based sensor for determination of the drug mitoxantrone. *J. Surf. Sci. Technol.* **2018**, *34(1-2)*, 74–80.
137. Thadathil, A.; Thacharakkal, D.; Ismail, Y. A.; Periyat, P. Polyindole-derived nitrogen-doped graphene quantum dots-based electrochemical sensor for dopamine detection. *Biosensors* **2022**, *12*, 1063.
138. Jayaraman, S.; Rajarathinam, T.; Jang, H.-G.; Thirumalai, D.; Lee, J.; Paik, H.-J.; Chang, S.-C. Ruthenium-anchored carbon sphere-customized sensor for the selective amperometric detection of melatonin. *Biosensors* **2023**, *13*, 936.
139. Istrate, O.-M.; Bala, C.; Rotariu, L. A new highly sensitive electrochemical biosensor for ethanol detection based on gold nanoparticles/reduced graphene oxide/polyallylamine hydrochloride nanocomposite. *Biosensors* **2023**, *13*, 954.
140. Fall, B.; Sall, D. D.; H´emadi, M.; Diaw, A. K. D.; Fall, M.; Randriamahazaka, H.; Thomas, S. Highly efficient non-enzymatic electrochemical glucose sensor based on carbon nanotubes functionalized by molybdenum disulfide and decorated with nickel nanoparticles (GCE/CNT/MoS_2/NiNPs). *Sens. Actuators Rep.* **2023**, *5*, 100136.
141. Manjunatha, J. G.; Swamy, B. E. K.; Mamatha, G. P.; Shankar, S. S.; Gilbert, O.; Chandrashekar, B. N.; Sherigara, B. S. Electrochemical response of dopamine at phthalic acid and Triton X-100 modified carbon paste electrode: A cyclic voltammetry study. *Int. J. Electrochem. Sci.* **2009**, *4*, 1469–1478.
142. Manjunatha, J. G.; Swamy, B. E. K.; Shreenivas, M. T.; Mamatha, G. P. Selective determination of dopamine in the presence of ascorbic acid using a poly (nicotinic acid) modified carbon paste electrode. *Anal. Bioanal. Electrochem.* **2012**, *4(3)*, 225–237.
143. Pushpanjali, P. A.; Manjunatha, J. G.; Amrutha, B. M.; Hareesha, N. Development of carbon nanotube-based polymer-modified electrochemical sensor for the voltammetric study of Curcumin. *Mater. Res. Innov.* **2020**, *25(7)*, 412–420.
144. Raril, C.; Manjunatha, J. G.; Ravishankar, D. K.; Fattepur, S.; Siddaraju, G.; Nanjundaswamy, L. Validated electrochemical method for simultaneous resolution of tyrosine, uric acid, and ascorbic acid at polymer modified nano-composite paste electrode. *Surf. Eng. Appl. Electrochem.* **2020**, *56(4)*, 415–426.
145. Tigari, G.; Manjunatha, J. G.; D'Souza, E. S.; Sreeharsha, N. Surfactant and polymer composite modified electrode for the sensitive determination of vanillin in food sample. *ChemistrySelect* **2021**, *6(11)*, 2700–2708.
146. Pushpanjali, P. A.; Manjunatha, J. G. Development of polymer modified electrochemical sensor for the determination of alizarin carmine in the presence of tartrazine. *Electroanalysis* **2020**, *32(11)*, 2474–2480.
147. Sanghavi, B. J.; Wolfbeis, O. S.; Hirsch, T.; Swami, N. S. Nanomaterial-based electrochemical sensing of neurological drugs and neurotransmitters. *Microchim. Acta* **2015**, *182*, 1–41.

14 Electro Analysis of Vitamins Using Carbon Sensor

Sümeyye Güney-Kalkan, Selenay Sadak, Ipek Kucuk, Meltem Ceylan-Ünlüsoy, and Bengi Uslu

14.1 INTRODUCTION

Vitamins are organic molecules that can be found in nature from animal or plant sources. Unlike fats, carbohydrates, and proteins, they are categorized based on their chemical composition and functions (Maqbool et al., 2018). Trace amounts of vitamins are needed in the body for growth, development, maintenance of health, and biochemical and physiological events. Vitamins were first discovered by biochemist Casimir Funk (Suychinov et al., 2019). There are vitamins that can be synthesized in very small amounts in the body; apart from these, the majority must be obtained from external sources through nutrition. The amount of vitamins needed by the body varies depending on the person's age, gender, physiological condition, personal characteristics, and other factors (Omonnazarova et al., 2022).

Vitamins can be added to foods externally (fortified foods) because their deficiencies cause serious health problems and have protective effects against diseases. The compatibility of vitamin content in these foods should be monitored (Blake, 2007). It is important to verify the amount of vitamins declared on the food label, to minimize waste in the production process, and to analyze the appropriate and legal amounts for human health in a sensitive, reliable, and most accurate way (Blake, 2007). Most of the methods used to analyze vitamins in foods date from the 1990s. Advances in nutritional science, analytical technology, and developments in legal regulations have increased the need for new and complementary standards regarding vitamins (Blake, 2007). In addition, the existence of various forms of vitamins (vitamin A, provitamin A, vitamin D_2, vitamin D_3, tocopherol, tacotrienes, vitamin K_1, and vitamin K_2), different actisve forms in the body, and the presence of metabolites make the analysis of vitamins more complicated (Fanali et al., 2017). The presence of vitamins in very low concentrations in the body (such as picomolar and nanomolar) and together with other substances requires the method to be used for analyses from biological fluids to be more sensitive and selective (Bates, 1997).

The method chosen for vitamin analysis in foods or biological fluids should be fast, economical, and reliable for the aforementioned reasons. Spectrophotometric methods, microbiological analyses, liquid chromatography, capillary electrophoresis, analytical and biosensors are among the analytical methods used for water- or fat-soluble vitamins (Blake, 2007; Fanali et al., 2017; Fatima et al., 2019). One of the vitamin analysis methods frequently used in the industry is high-performance liquid chromatography. This method is relatively time-consuming and costly and requires technical operator skill. On the other hand, the use of simpler and lower-cost carbon electrodes has become a developing field in both the food and pharmaceutical industries (Westmacott et al., 2018). The "carbon paste" sensor, which Adams declared as a new type of electrode in 1958, has undergone many changes until today. These sensors have gained an important place in the analysis of inorganic

DOI: 10.1201/9781003500322-14

ions or molecules, organic substances, biologically important compounds and pharmaceuticals (Švancara et al., 2001). The use of carbon materials as biosensors or electrochemical sensors has become widespread due to their good electrical conductivity, chemical inertness, and wide potential range (Corb et al., 2007). Carbon-based nanomaterials provide high selectivity and specificity for the recognition of molecules in the analysis of environmental and pharmaceutical wastes, biological samples, and food samples (Pandey et al., 2020; Oliveira et al., 2019; Jagannathan & Badhulika, 2024). Carbon sensors used in vitamin analysis offer reproducibility and repeatability, which are important for an analysis. In this context, carbon-based sensors constitute a good alternative in addition to the methods used for vitamin analyses (Antherjanam et al., 2021; Westmacott et al., 2018; Zeybek & Üğe, 2021).

14.2 VITAMINS

Vitamins are organic molecules that must be taken to maintain metabolic events in the body and to maintain health. They are not energy sources or basic building blocks. They can be synthesized insufficiently in the body or taken with food (Kayaalp, 1996). After their discovery at the beginning of the 18th century, this group of compounds were called vitamins (from Latin *vita* meaning "life") because they were necessary for the continuity of life, growth, and development. They are separated into two groups according to their solubility: fat-soluble vitamins and water-soluble vitamins (Akgün et al., 2004).

14.2.1 Fat-Soluble Vitamins

Vitamins A, D, E, and K are fat-soluble vitamins. They have different chemical structures. They exist in different chemical forms. Vitamin A is divided into two groups: retinoids and vitamin A precursor carotenoids. Vitamin D is divided into two groups: ergosterol (D_2) and cholecalciferol (D_3). Vitamin E is divided into two main groups: tocotrienols and tocopherols. Vitamin K is divided into two groups: phylloquinone and menaquinone (Stevens, 2021).

14.2.1.1 Vitamin A

The first vitamin to be found was vitamin A. Structurally, it contains a β-ionone ring and an isoprene ring (Figure 14.1). Vitamin A is obtained from carotenoids found as pigments in plants. These structures are precursors of vitamin A. Once they enter the organism, they are converted into vitamin A (Sikorski & Kołakowska, 2003). Its active metabolite is retinoic acid (Ross, 2010).

Vitamin A has an important role in vision, reproduction, cell proliferation and differentiation, epithelial cell growth and differentiation, and antioxidant metabolism. Its deficiency causes night blindness and dry eyes. In addition, its deficiency causes immunosuppression and mortality increases in cases of infection (Hashemi, Pourmohammadi et al., 2018).

Carotenoids are found mainly in orange-red fruits and vegetables (carrots, tomatoes, red peppers, etc.). Apart from this, animal vitamin A is also found in milk, eggs, and fish. Although vitamin A toxicity is rare, it may occur during vitamin A intake or treatment with retinoids. Toxicity may occur orally or topically (Carazo et al., 2021).

14.2.1.2 Vitamin D

Considering its effects on the endocrine system, vitamin D is considered a hormone in the latest classifications (Figure 14.2). 7-dehydrocholesterol, the precursor of vitamin D_3, is synthesized from cholesterol by the biosynthetic pathway (Holick, 2018). There are two main forms of vitamin D: D_3 (cholecalciferol) and D_2 (ergocalciferol). Vitamin D_3 is formed as a result of the interaction of 7-dehydrocholesterol with sunlight or ultraviolet light. Vitamin D_2 is formed from the interaction of

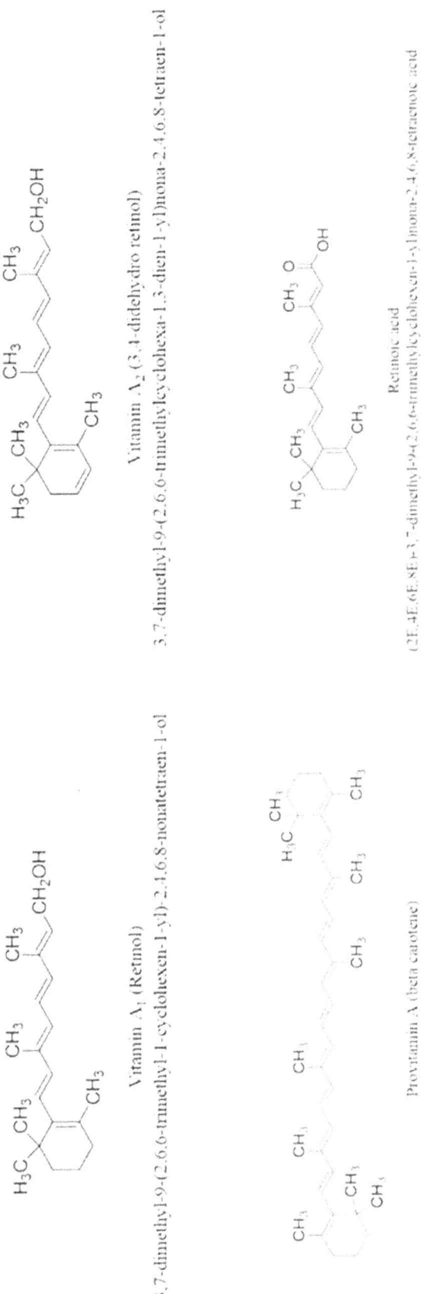

FIGURE 14.1 Chemical structure of vitamin A.

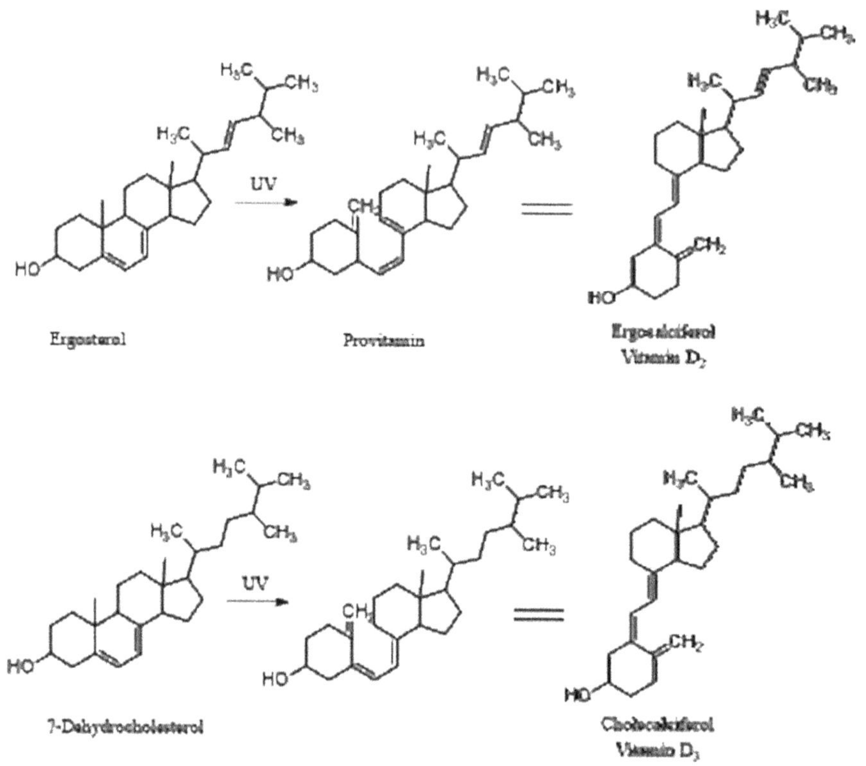

FIGURE 14.2 Chemical structure of vitamin D.

ergosterol with sunlight (Lips, 2006). After undergoing hydroxylation, they transform into their active forms, 25-hydroxycholecalciferol and 1,25-dihydroxyvitamin D_3 (Nair & Maseeh, 2012). Vitamin D has a key role in maintaining calcium balance, promoting cell growth, and mediating cell differentiation (Kato, 2000). In its deficiency, disorders in the skeletal-muscular system (e.g., rickets, bone softening), insufficiency in antimicrobial activity, and increased mortality caused by cardiovascular diseases have been observed. Additionally, epidemiological studies have shown that there is a correlation between vitamin D deficiency and metabolic diseases (Adams & Hewison, 2010).

Vitamin D_2 is found in plants and mushrooms, while the D3 form is found in fatty fish, egg yolk, meat, liver, and kidney (Wilson et al., 2017). In hypervitamin D or vitamin D toxicity, hyperphosphatemia, hypercalcemia, and associated calcification in the kidneys or other soft tissues are observed (Maji, 2012).

14.2.1.3 Vitamin E

Vitamin E (α-tocopherol) was discovered approximately 100 years ago (Figure 14.3). There are eight different tocopherols synthesized by plants related to vitamin E in the daily diet (Niki & Traber, 2012). Vitamin E has an antioxidant effect by scavenging peroxyl radicals in the body and preventing the formation of free radicals mediated by lipid peroxidation. Vitamin E clears lipid peroxyl radicals before they attack lipid molecules (fatty acids, phospholipids, acylglycerols, cholesterol, and cholesteryl esters) (Niki, 2021).

Vitamin E deficiency causes fat malabsorption and disorders in lipoprotein metabolism (Torquato et al., 2020). In addition, spinocerebral ataxia, myopathy, peripheral neuropathy, and vitamin E-related anemia may develop in humans due to vitamin E deficiency (Brigelius-Flohé & Traber, 1999).

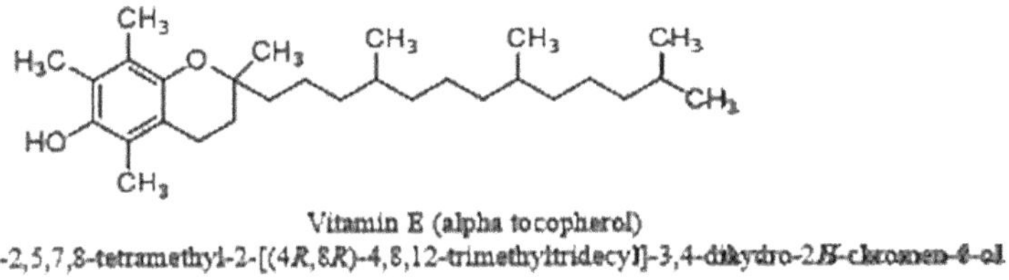

FIGURE 14.3 Chemical structure of vitamin E.

Vitamin E is found in coconut, cottonseed, olive, palm, peanut, canola, soybean and sunflower oils, milk, meat, and butter (Herting & Drury, 1963; Shahidi et al., 2021). Hypervitamin E mainly causes an increase in the risk of bleeding. Apart from this, hepatobiliary dysfunction and malabsorption of other fat-soluble vitamins develop (Owen & Dewald, 2023).

14.2.1.4 Vitamin K

Vitamin K, known as antihemorrhagic or coagulant factor, consists of two main forms: phylloquinone (K_1) and menaquinone (K_2) (Shearer, 1995). There are 14 known vitamin K-related compounds. Vitamin K is involved in the regulation of hemostasis in metabolism, calcium metabolism, control of cellular growth, apoptosis, and protection of vascular health (Gröber et al., 2015). Another well-known function of vitamin K is that it takes part in the γ-carboxylation of many proteins as a vital coenzyme (Mladenka et al., 2021) (Figure 14.4).

Vitamin K deficiency may cause spontaneous cutaneous purpura, nose bleeding, gastrointestinal, genitourinary, gum, and other bleeding, deterioration in bone mineralization, arterial calcification, and an increase in cardiovascular diseases (Vermeer & Schurgers, 2000).

Vitamin K_1 is found mainly in green leafy plants, olive oil, and soybean oil, while vitamin K_2 is found in small amounts in chicken, butter, egg yolk, cheese, and fermented soybeans (Dinicolantonio et al., 2015). Although vitamin K toxicity has not been precisely defined, it has been suggested that it may be similar to menadione, which has a similar structure. These are hemolytic anemia, jaundice, kernicterus, and hyperbilirubinemia (Sikorski, n.d.).

14.2.2 Water-Soluble Vitamins

This group of vitamins is basically called water-soluble vitamins because they are soluble in water. These vitamins have excellent dispersion in the intestines due to the presence of polar groups in their properties, and their absorption occurs also on the surface of the intestines (Hashemi, Es, et al., 2018). Their physico-chemical properties are different from fat-soluble vitamins, and they are not retained within the body. Their elimination is higher, so their deficiencies are more frequently encountered. The situation is the opposite for fat-soluble vitamins. Fat-soluble vitamins can cause toxicity because they can be stored in the body. These vital compounds show different pharmacokinetic properties due to the fact that their physicochemical properties are quite different from each other in two groups. In this regard, quantity determinations and monitoring are important (Nicolov et al., 2021).

14.2.2.1 Vitamin B_1

The first discovered vitamin B_1 (thiamine) is one of the B group vitamins. It can be found in the body with an added phosphate group (Figure 14.5). Its active form is thiamine pyrophosphate (Collie et al., 2017). It is present as a cofactor in energy metabolism. It is involved in the transmission of nerve impulses and the repair of the myelin sheath. It is an important part of the cardiovascular, nervous, and immune systems. Its deficiency causes beriberi (Wiley & Gupta, 2019).

Vitamin K_1 (Phylloquinone)
2-methyl-3-[(E,7*R*,11*R*)-3,7,11,15-tetramethylhexadec-2-enyl]naphthalene-1,4-dione

Vitamin K_2 (Menaquinone-4)
2-methyl-3-[(2E,6E,10E)-3,7,11,15-tetramethylhexadeca-2,6,10,14-tetraenyl]naphthalene-1,4-dione

FIGURE 14.4 Chemical structure of vitamin K.

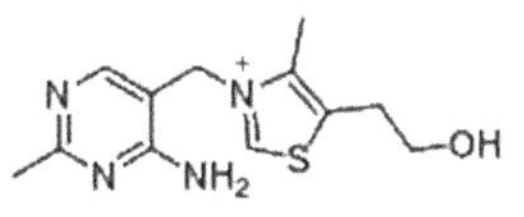

Vitamin B_1 (Thiamine)
3-[(4-amino-2-methylpyrimidin-5-yl)methyl]-5-(2-hydroxyethyl)-4-methyl-1,3-thiazol-3-ium

FIGURE 14.5 Chemical structure of vitamin B_1.

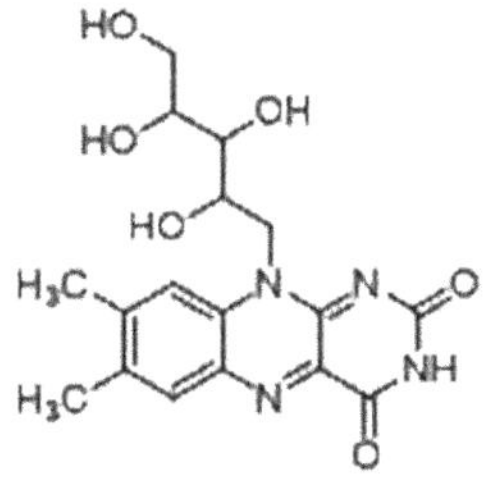

Vitamin B_2 (Riboflavin)
7,8-dimethyl-10-[(2S,3S,4R)-2,3,4,5-tetrahydroxypentyl]benzo[g]pteridine-2,4-dione

FIGURE 14.6 Chemical structure of vitamin B_2.

It is abundant in bread and cereals. Apart from this, it is found in peas, beans, hazelnuts, brown rice, beef, and pork (Fattal-valevski, 2011). The human body eliminates excess thiamine through urine. Therefore, there is no definitive information regarding thiamine toxicity in the literature (Martel et al., 2020).

14.2.2.2 Vitamin B_2

Vitamin B2, often known as riboflavin, is a natural vitamin synthesized by plants and the majority of microorganisms (Figure 14.6). It plays a crucial role in the development and reproduction of both humans and animals. As the precursor of flavin adenine dinucleotide and flavin mononucleotide, it takes part in many redox reactions in the body (Averianova et al., 2020). It is necessary for DNA repair in the body, energy production, fatty acid and amino acid synthesis, folic acid activation, and glutathione production, which scavenges free radicals. Deficiency of vitamin B2 causes lesions on the lips and scaly rashes on the scrotum or vulva. Its deficiency is also associated with night blindness, photosensitivity, cataracts, migraines, anemia, and depression (Buehler, 2011).

Vitamin B_2 is found in yeast, grains, meat, fatty fish, and green leafy vegetables (Burgess et al., 2009). Since the absorption of vitamin B_2 taken through the diet is low and the storage of flavins in the body is limited, it is unlikely to show toxicity according to general opinion (Rucker et al., 2001).

14.2.2.3 Vitamin B_3

Vitamin B_3 (niacin) consists of two forms: nicotinic acid and nicotine amide (Figure 14.7). Cellular niacin can be taken into the body in two different ways. It may be biosynthesized in the body from tryptophan acquired from amino acid-rich diets or derived directly from foods containing niacin (Denu, 2005). Niacin is converted into nicotinamide adenine dinucleotide by biosynthesis. Therefore, it has an important role as a cofactor for energy metabolism. Apart from this, it has been reported to be effective in preventing atherosclerosis, diabetes, and hypercholesterolemia. It is traditionally known to prevent pellagra (Sauve, 2008).

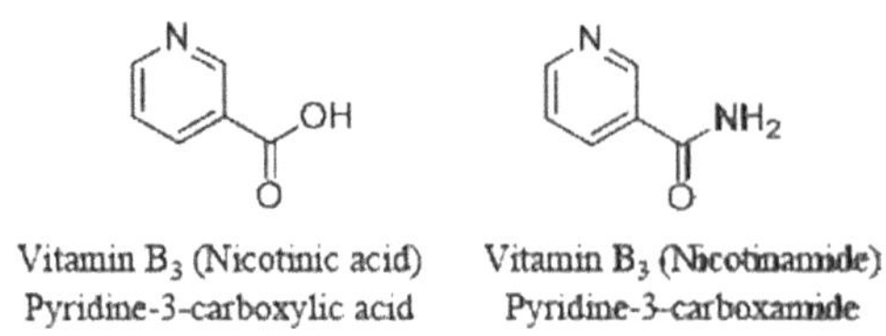

FIGURE 14.7 Chemical structure of vitamin B_3.

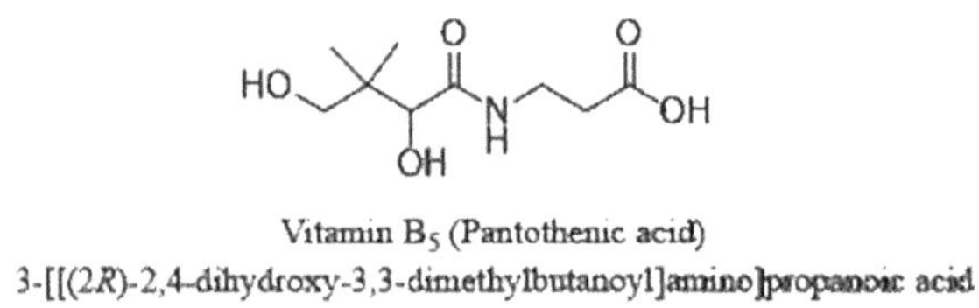

FIGURE 14.8 Chemical structure of vitamin B_5.

Liver, fish products, wheat bran, fermented broth, distillate of cereals, and yeast are rich in vitamin B_3. It has been reported that it causes abnormalities in liver functions when high doses are used for the treatment of hypercholesterol (Rader et al., 1992).

14.2.2.4 Vitamin B_5

Vitamin B_5 (pantothenic acid) serves as a precursor in the synthesis of coenzyme A (Figure 14.8). This enzyme has a key role in energy-producing pathways in cells. It is necessary for energy metabolism and especially the activation of fatty acids. Only plants and some organisms can synthesize vitamin B_5 (Ottenhof et al., 2004). Since it can be found in many sources, its deficiency is rare. Vitamin B_5 deficiency causes dermatitis, hypoglycemia, convulsions, and encephalopathy due to liver failure (Depeint et al., 2006).

Vitamin B5 is found in liver, kidney, peanut butter, almonds, wheat bran, milk, eggs, and vegetables (K. Zhao et al., 2022). Although vitamin B_5 does not generally cause hypervitaminosis with daily nutrition, some side effects may occur when taken externally. Excessive intake may cause joint and muscle pain, hypersensitivity reactions, and an increase in alanine transaminase (Sanvictores & Chauhan, 2020).

14.2.2.5 Vitamin B_6

Vitamin B_6 includes pyridoxine, pyridoxal, pyridoxamine, and their phosphate derivatives (Figure 14.9). Its active form, pyridoxal phosphate, acts as a cofactor in 160 different reactions in the body. It is not produced in humans or other organisms. It is produced by yeast and bacteria in different ways (Stach et al., 2021). It is involved in aldol cleavage, α-decarboxylation, racemization, β- and γ- elimination, amino acid biosynthesis and degradation, and sugar and fatty acid metabolism in the body (Hellmann & Mooney, 2010). It is also recommended to relieve nausea in pregnant women (Wibowo et al., 2012). Its deficiency causes oral stomatitis, glossitis, and cheilosis. It can cause irritability, confusion, and depression. Adults with severe deficiency of vitamin B_6 may experience seborrheic dermatitis, microcytic anemia, and seizures (Brown et al., 2017).

Vitamin B_6 is found in wheat bran, brown rice, beans, walnuts, red meat, and chicken breast (Rucker et al., 2001). Although vitamin B_6 is a water-soluble vitamin, toxicity may occur when taken externally for the treatment of some diseases. It causes neuropathy when used for a long time (Betti et al., 2022).

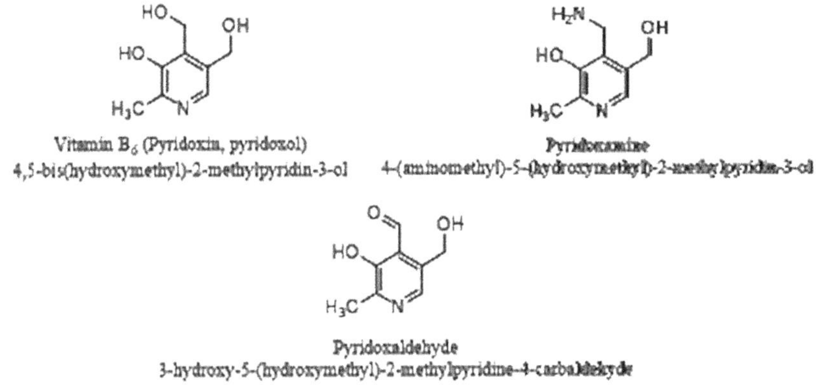

FIGURE 14.9 Chemical structure of vitamin B_6.

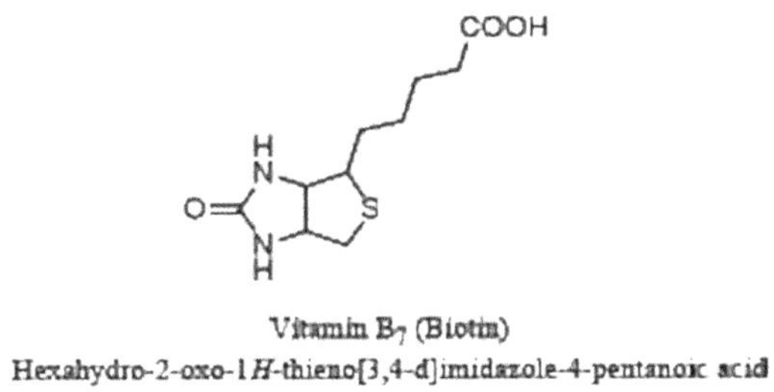

FIGURE 14.10 Chemical structure of vitamin B_7.

14.2.2.6 Vitamin B_7

Vitamin B_7 (biotin, vitamin H, and coenzyme R) is a heterocyclic structure that carries sulfur in its structure (Figure 14.10). It is necessary for the carboxylase enzyme. It takes part in fat and amino acid metabolism, fatty acid production, cell growth, and gluconeogenesis. It helps regulate blood sugar (Yadav, 2015). Biotin is important in maintaining hair, nail, and skin health (Beigmohammadi et al., 2021). Its deficiency causes neurological and dermatological problems. Deficiency symptoms include alopecia, spasticity in the hind legs, dermatitis, skin ulcerations, a brown leak on the skin and around the eyes, cracking and bleeding on the soles of the feet, diarrhea, and inflammation of the mucosa of the mouth (Dove, 2011; Saleem & Simpson, 2023).

Vitamin B_7 is relatively abundant in liver, kidney, egg yolk, milk, hazelnuts, grains, and fibrous green leafy vegetables (Coerdt et al., 2021; Yadav, 2015). There is no evidence regarding its toxicity (Carling & Turner, 2019).

14.2.2.7 Vitamin B_9

Vitamin B_9 (folic acid, folate) serves as a coenzyme for the synthesis of purine and pyrimidine nucleotides in the body (Figure 14.11). Folate is a general expression and represents a group of molecules containing tetrahydrofolate (5,6,7,8-tetrahydropteroylpolyglutamate) in its structure. Chemically, they contain pteridine ring, p-aminobenzoic acid, and glutamate residue in their structure. Folates are involved in single carbon transfer reactions, also known as C1 metabolism (Rébeillé et al., 2007). In its deficiency, many metabolic pathways are affected. The most problematic situation is that nucleotide synthesis is affected. Due to vitamin B_9 deficiency, megaloblastic anemia, cardiovascular diseases, neurocognitive disorders, cancer, and neuronal tube defects in the embryo may develop (Corfield et al., 2023; Rucker, 2001).

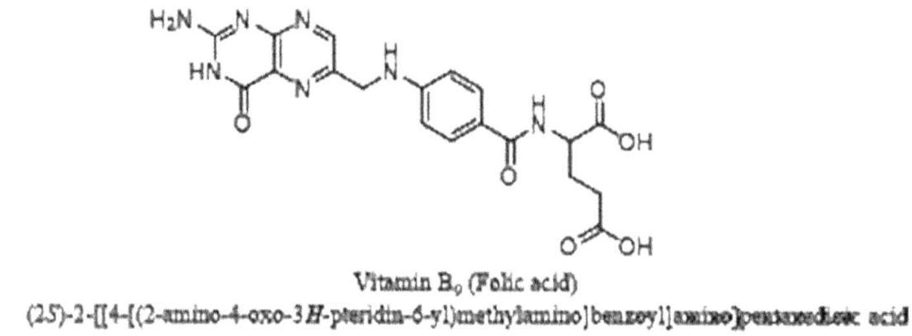

FIGURE 14.11 Chemical structure of vitamin B_9.

Vitamin B_9 is abundant in mainly dark green leafy vegetables, spinach, white beans, asparagus, Brussels sprouts, soybeans, kidney beans, black-eyed peas, broccoli, avocado, peas, bananas, melons, tomatoes, and liver (Singh, 2021). There is no definitive data on whether folic acid taken from foods can cause toxicity. However, synthetic derivatives taken as supplements can cause toxicity (Titcomb & Tanumihardjo, 2019). High doses of folic acid rarely show acute toxicity. It may occur as a gastrointestinal and central nervous system disorder. Chronic toxicity may occur in patients with pernicious anemia who use folic acid medications (Pradeep Devnath et al., 2017).

14.2.2.8 Vitamin B_{12}

The molecular formula of vitamin B_{12} (cyanocobalamin) is $C_{63}H_{88}N_{14}O_{14}PCo$. It can only be biosynthesized by bacteria through 30 different enzymes. The net load of vitamin B_{12} containing cobalt is one. It has a cyano group attached to the cobalt molecule in its center. This group can be replaced by various other smaller groups to form hydroxycobalamin, nitrocobalamin, thiocyanocobalamin, and chlorocobalamin (Halver, 2003; Smith et al., 2018). Vitamin B_{12} serves as a cofactor for methionine synthase and methylmalonic CoA mutase. Homocysteine accumulation without vitamin B_{12} causes toxicity in the body. Vitamin B_{12} is important for DNA synthesis and neurological functions. In its deficiency, hematological and neuropsychiatric anomalies may develop. Its deficiency may also cause skin hyperpigmentation, glossitis, taste disturbance, oral ulcers, stomatitis, and cheilitis (Oh & Brown, 2003).

Vitamin B_{12} is found in foods of animal origin such as red meat, liver, seafood, milk, and eggs. Cyanocobalamin and hydrocobalamin are the main forms of vitamin B_{12} in clinical use. Some of the conditions arising from high dose use of these are transient chromaturia, erythema, acne-like rash, increase in blood pressure, nausea, and headache (Morales-Gutierrez et al., 2019; Pawlak et al., 2012).

14.2.2.9 Vitamin C

Vitamin C consists of two isomers in the redox system. It is ascorbic acid (vitamin C) in its reduced state and dehydroascorbic acid in its oxidized state (Figure 14.12). The functioning of most vitamins in the body depends on vitamin C's role as an electron donor (Schlueter & Johnston, 2011). Vitamin C serves as a cofactor in the synthesis of catecholamines, such as norepinephrine, amino acids, and several peptide hormones (Carr & Maggini, 2017). Vitamin C is important in repairing skin, cartilage, bones, and teeth. Its deficiency causes scurvy. Due to its role in collagen metabolism, its deficiency also affects the wound healing process. Vitamin C deficiency is also reported in patients with depression, confusion, cognitive disorders, or who are at risk of psychiatric diseases (Maxfield, 2019; Plevin & Galletly, 2020)

Vitamin C is abundant in raw fruits and vegetables. Citrus fruits, potatoes, tomatoes, kiwi, strawberries, blackberries, blueberries, watermelon, red and green peppers, broccoli, Brussels sprouts, cauliflower, spinach, cabbage, and turnips are among the foods containing vitamin C (Anastassakis,

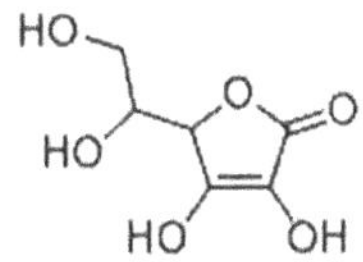

Vitamin C (Ascorbic acid)
(2*R*)-2-[(1*S*)-1,2-dihydroxyethyl]-3,4-dihydroxy-2*H*-furan-5-one

FIGURE 14.12 Chemical structure of vitamin C.

2022). Vitamin C is partially metabolized to oxalate in the human body. Therefore, depending on the dose, vitamin C increases urinary oxalate levels. It is thought that it may cause kidney stone formation, especially in patients prone to stone formation in the urinary system. Some studies have reported that this risk is significant. Apart from this, it has been stated that vitamin C toxicity, vitamin B_{12} deficiency, and hyperuricosuria may occur (Dosedel et al., 2021; Harris, 2013).

14.3 ELECTROCHEMICAL ANALYSIS OF VITAMINS

Electrochemical sensors are commonly used to investigate the redox properties of water-soluble vitamins. These vitamins consist of various molecules that facilitate electron transfer in aqueous solutions (Yang et al., 2022). There has been significant interest in the use of electrochemical sensors for detecting water-soluble vitamins. These devices have a simple design, are affordable, highly sensitive, can be made smaller, and consistently deliver accurate results when quickly and efficiently detecting various samples (Kiamiloglou & Girousi, 2023). Water-soluble vitamins are excreted quickly in urine, which prevents them from being stored in the body for long periods of time. Since the widespread presence of deficiencies in water-soluble vitamins, they may have adverse impacts on human health and cause various disorders. Excessive intake of these vitamins frequently results in the development of additional problems. It is essential to detect the amounts of vitamins in the body and determine their dietary sources (Huang et al., 2021).

Electrochemical sensors can operate by detecting the electrochemical activity of a wide range of antioxidant vitamins. Due to the dependence of the electrochemical redox process on the current in the working electrode, electrochemical sensors can accurately measure the concentration of the analyte (in this case, vitamins) in the electrolyte solution. This is achieved by detecting the current in the working electrode. The connection between the current reaction and the concentration of the analyte is typically directly proportional. Presently, amperometry and voltammetry are the preferred electrochemical methods. There are two voltammetry processes that are widely recognized for their simplicity: cyclic voltammetry (CV) and linear scanning voltammetry. Improved sensitivity can be achieved by utilizing two methodologies: square wave voltammetry (SWV) and differential pulse voltammetry (DPV). Both of these methodologies require the adjustment of specific parameters (Martinsen & Heiskanen, 2023).

14.3.1 The Role of Carbon Sensors in Vitamin Analysis

Carbon-based electrochemical sensors are at the forefront of the analysis of fat- and water-soluble vitamins. In particular, they are preferred for analyzing the redox properties of water-soluble vitamins. This is because water-soluble vitamins consist of diverse molecules that are capable of electron transfer in an aqueous electrolyte. Electrochemical sensors for detecting water-soluble vitamins have gained significant popularity in recent decades. Carbon-based electrodes offer several notable benefits, including simplified catalyst involvement, the ability to regenerate the surface, and an expanded range of electrical potential with reduced ohmic resistance. The procedures performed on a working electrode vary based on the specific vitamin under investigation. The electron transfer in vitamins B_2, B_3, B_7, C, and D is straightforward, involving both oxidation and reduction. On the other hand, vitamins B_1, B_5, B_6, B_9, and E undergo simultaneous oxidation and reduction, or two consecutive reductions in the case of vitamins K and B_{12}. Vitamin A, however, exhibits intricate electrochemical behavior (Salvigni et al., 2023).

Carbon is a very adaptable element that exists in several forms, including graphite, graphene, diamond (cubic), Q carbon (amorphous), and carbon nanotubes (CNTs). It is a widely supported component used in the manufacture of electrodes (W. Zhang et al., 2016). The reason for this is that it is a durable material with a broad spectrum of potential for oxidation, it can be readily shaped, it is cost-effective, and the surface of the electrode can be quickly restored. Variable forms of carbon are used as electrode materials (McCreery, 2008). Voltammetry using carbon electrodes enables a broad range of operations in both the oxidation and reduction regions, approximately ranging from –1.8 V to +1.8 V in conditions of water. Carbon may be used in electrochemical applications in several forms including carbon fiber, one-crystalline graphite, powdered graphite, glassy or ceramic carbon, carbon black, carbon paste, and carbon film (Vicentini et al., 2024). These carbon materials are extensively used in the creation of sensors because they are readily available in nature, inexpensive, have low background current, can be easily modified on the surface, and are simple to fabricate (McCreery et al., 1994).

14.3.1.1 Glassy Carbon Electrodes (GCEs)

Given its unique combination of physical and chemical characteristics, glassy carbon (GC) has quickly gained popularity as an electrode material. Its chemical inertness is quite great, and its oxidation rate is very low (Engstrom, 1982). Glassy or ceramic carbon is a substance that lacks a definite shape and is impermeable to water or air. It is often used as a carbon electrode in many analytical applications. GC has distinct physical features compared to other carbon forms. Due to its smaller surface pores, this carbon type is more often utilized compared to others (Dekanski et al., 2001). The GC structure was discovered to be composed of randomly distributed and intermixed aromatic ribbon molecules. Like other solid electrodes, several pre-treatments have been devised to assure activation and achieve consistent outcomes with GCE. The methods may be categorized as polishing, chemical and electrochemical treatments, radiofrequency, low-pressure temperature application, vacuum-temperature application, laser beam excitation, and deposition of metal oxide layers on the electrode surface. At the moment, there is no standard activation method that is accepted internationally. It is possible to link the variety of the activation process to the particular solution that was used as well as the chemical that was being investigated. The activation process in electron transfer serves to eliminate surface contaminants, generate functional groups on the surface, and increase the surface area (Wightman et al., 1984).

Electrode surfaces are thoroughly prepared by polishing with alumina for permanent cleanliness and precise definition. Several other methods have been used to generate a GC surface that is both consistent and reactive. These methods include laser therapy, ultrasonic irradiation of GC, and the stimulation of GC using a carbon arc.

Over the past few years, the use of GCE in the fields of electrochemistry and surface modification has gained great importance and has inspired many researchers with various goals. There are certain rationales for altering an electrode. These components facilitate self-assembly during stripping analysis, prepare the electrode environment by immobilizing molecules on the surface, enable catalytic effects, and alter the surface characteristics of the electrodes (Dekanski et al., 2001). Surface modification of GCE may effectively inhibit undesired reactions and facilitate the desired reactions. The unaltered surface of solid electrodes is inherently unstable and exhibits the drawback of undergoing changes over time. The intended reactions on the surface of the GC electrode may be hindered over time owing to oxidation and contamination of the surface, or they may occur via various processes. In order to avoid this, the surface of the solid electrode is altered. Chemically modified electrodes only alter the surface while leaving the underlying substance of the electrode unaffected. Through alteration, the surface of the electrodes is tailored to the intended response and maintains its integrity over time. There are many different alterations that may be made to carbon electrodes. These processes include the physical adsorption of catalysts onto surfaces or chemical changes that bind certain chemical groups to surfaces. Chemical alteration allows for the acquisition of surfaces that are more stable and long-lasting (Downard, 2000).

14.3.1.2 Carbon Paste Electrodes (CPEs)

Graphite has a uniform layered structure that consists of a honeycomb hexagonal pattern of carbon atoms. This enables various forms of carbon to be positioned between the layers in diverse ways. Powdered graphite is a frequently used form of carbon in electrodes. Graphite is a flexible and permeable substance that has a notable ability to absorb. Graphite powder with excellent purity and micrometer-sized particles is the crucial constituent of CPEs (Švancara et al., 2001). CPEs are electrodes created by combining graphite powder with a binder (such as paraffin or mineral oil) and then inserting the resulting paste into different electrode structures. The electrode also contains a binder liquid that is inert, electroinactive, and has a high viscosity. The desired properties for these liquids are high viscosity, hydrophilicity, non-volatility, resistance to organic solvents, and non-interference with agglomeration (J. Wang et al., 2001). Additionally, they should possess electroinactivity and inertness. Nujol and uvasol, often referred to as mineral oil or paraffin, are extensively used in CPEs due to their inherent characteristics. Alternative binder liquids, such as organophosphate liquids, electrolytic liquids, or silicone oil, may be used with nujol.

A high signal-to-noise ratio is achieved by the CPEs that are produced due to their low ohmic resistance and high conductivity. The water-insolubility of the binding fluid has endowed the CPEs with lipophilic characteristics. One drawback of these electrodes is their heterogeneous nature, which makes them vulnerable to damage from organic solvents (McCreery et al., 1994).

CPEs have a very wide potential range. Their construction and renovation are not time-consuming, and the ground currents are very low. If there is absorbed oxygen left on the carbon paste, a significant amount of residual current is generated due to its reduction. It is useful to eliminate this with a pre-electrolysis. CPEs are a type of electrode frequently used in electrochemical studies due to their properties such as low ground current, wide anodic potential range, chemical stability, easy modification with nanomaterials, and low cost preparation procedure (Malode et al., 2022).

CPEs may be modified in several ways. The alteration of electrodes can be achieved through various techniques. These include adsorption or covalent bonding of the changes to the electrode surface, applying a thin film of the modifier polymer onto the electrode, and incorporating the undissolved or minimally soluble modifier into a conductive matrix like carbon paste (Švancara et al., 2001). Various methods may be used to incorporate modification material into the mixture. The modification material can be dissolved directly in the binder liquid, added to the mixture consisting of graphite and binder liquid, the modification material, graphite, and binder liquid can be mixed in a suitable solvent, the solvent can be removed, and finally, after the carbon paste electrode is prepared, the modification material can be coated onto the electrode surface (Vytřas et al., 2009). There are many materials used for the modification process. These can be chemical compounds and analytical reagents, ion exchangers, clay minerals, humic substances, silica and silica-containing substances, biological materials, and nanoparticles. Recently, there has been a notable rise in the utilization of CPEs that have been modified with nanoparticles. This is mainly owing to their numerous advantages, including their preparation simplicity, low cost, and simple construction and regeneration of their surfaces. Additionally, these electrodes have the ability to contain multiple biosensor materials simultaneously, offer a wide potential range, and exhibit very low ground currents (ADAMS, 1963).

14.3.1.3 Screen-Printed Carbon Electrodes (SPCEs)

By incorporating the advantages of carbon materials into the disposable, low-cost, miniaturization properties of SPEs, high-precision, biocompatible, and specific SPCEs have been developed. SPCEs are widely used in industrial applications and especially at the point of maintenance. The emergence of screen printing technology has opened up an abundant number of opportunities for utilizing electrochemical methods in environmental analysis, eliminating the requirement for a centralized laboratory. Especially in modern times, it is used for mass production of electronic circuits and to meet the increasing demand for miniaturization in the global world. Innovative printing technologies are facilitating the development of simple and portable flexible devices, such as implantable biomedical devices and wearable biosensors. SPE is commonly utilized because of its affordability, user-friendly nature, and consistent results. Additionally, there is no requirement to clean its surfaces (Fletcher, 2015). SPEs are emerging as innovative printing technologies in the field of sensors, offering disposable, low-cost, flexible, and portable electrochemical sensors depending on the substrate used. This technology is mainly produced by printing conductive ink on mesh screens to create the reference, counter, and working electrodes on a substrate (polyethylene terephthalate, polyimide, plastic, or ceramic). Typically, conductive inks are prepared by combining a conductive material, like metal or carbonaceous materials, with a binder, such as resins or polymers, along with additives (Niu et al., 2012). Next, the prepared conductive ink is printed on the flexible substrate, and the electrodes are coated with insulating material to determine the electrode measurement area and contact points. Compared to conventional electrodes, SPEs have many benefits, such as the capacity to be employed in tiny quantities and the capacity to produce repeatable, precise, and sensitive sensors. Modifications to SPE surfaces can be tailored to suit the requirements of a specific analyte. Using metallic nanoparticles to modify SPEs leads to a boost in surface area and electrical conductivity. The modifications lead to higher surface activity, resulting in faster reaction speeds and improved detection sensitivity. Studies have shown that it is possible to directly electrodeposit nanoparticles onto SPCE strips (Wring & Hart, 1992).

14.4 CARBON SENSOR APPLICATIONS FOR VITAMIN ANALYSIS

Table 14.1 presents the findings from electrochemical vitamin studies carried out over the last decade. In the literature, it has been noticed that carbon sensors are mostly used for the analysis of water-soluble vitamins. The developed methods have mostly been applied in pharmaceutical dosage forms, fruit juice, and milk samples.

14.4.1 Applications of Carbon Sensors in Vitamin Analysis

CNTs are commonly employed in the modification of carbon sensors. CNTs are nanometer-sized materials with distinctive forms. Nanotubes feature exceptional mechanical and electrical capabilities due to their semi-one-dimensional structure and arrangement of carbon atoms (Riu et al., 2006). Initially renowned for their remarkable physical qualities, CNTs subsequently garnered significant interest due to their noteworthy chemical capabilities. Multiwalled carbon nanotubes (MWCNTs) are a type of CNTs that are currently receiving significant attention in this particular setting. Crystalline and hollow tubes, which were discovered in 1991, were subsequently designated as MWCNT (Iijima, 1991). MWCNTs are nanomaterials characterized by their distinctive structure, exceptional endurance, and large surface area relative to their volume (Ma et al., 2017). Due to these factors, they are strongly favored in the modification of carbon sensors. Nickel oxide (NiO) nanoparticles also attract the attention of researchers due to their highly developed electrochemical and magnetic properties (Carbone et al., 2017). These nanoparticles exhibit rapid reaction kinetics and possess exceptional chemical resistance (L. Zhao et al., 2011). Ionic liquids are notable for their exceptional thermal stability, extensive electrochemical applicability, and broad liquid phase range

TABLE 14.1
Analysis of various vitamins with electrochemical carbon sensors

Vitamin	Transducer	Method	Linear Range	LOD	Application	References
D_3	NCNT/GCE	CV	0–9 nM	16 pM	Pharmaceutical tablets	(Bora et al., 2022)
C	NiO/MWCNTs/$[B_{mim}]BF_4$/CPE	SWV	0.1–1000 μM	0.06 μM	Fresh juice	(Khaleghi et al., 2016)
B_1	SPCE	CV	15–110 μg/ml	0.1 μg/ml	Food and pharmaceutical analysis	(Westmacott et al., 2018)
B_2			0.1–20 μg/mL	3.5 μg/ml		
B_6			2–80 μg/mL	0.4 μg/ml		
C	AuNPs/RGO/SPCE	Amp	2–20 μM	0.088 mg/L	Commercial milk	(Bettazzi et al., 2021)
C	2FE/ZC/IL/CPEs	SWV	$4.0\times 10^{-8} – 1.0\times10^{-3}$ M	2.0×10^{-8} M	Vitamin C	(Motaghi et al., 2016)
B_6			1.0×10^{-6}- 1.2×10^{-3} M	8.0×10^{-7} M	injection, effervescent C tablets, multivitamin syrup and urine samples	
B_2	Gd_2O_3/GCE	DPV	0.1–550 μM	10 nM	Vitamin B_2 tablet and milk powder sample	(Vijayaprasath et al., 2020)
D_3	MIP/SPCE	SWV	1.0×10^{-11}-2.0×10^{-9} M	1.0×10^{-12}M	Human plasma samples	(Kia et al., 2019)
B_2	44-DABP/GCE	DPV	13 nM–660 μM	8.52 nM	Multivitamin tablets and orange juice	(Tabanlıgil Calam, 2021)
B_6			15.7–2210 μM	4.70 μM		
C	MoS_2/f-MWCNTs/ZnO/GCE	CV	10–100 μM	1.65 μM	Serum samples	(Murugan et al., 2019)
D_3	MB-TAD@Apt@MoS_2@ErGO@GCE	DPV	0.1–150 ng/mL	0.02 ng/mL	Saliva	(Park et al., 2022)
B_1	ds-DNA/PMWCNTPE	DPV	1.0–80 μg/mL	0.44 μg/mL	Human blood	(Brahman et al., 2013)
D_3	CuNPs-NiNPs@reduced-fullerene-C_{60}/GCE	SWV	1.25–475 μM	0.0025 μM	Vitamin D_3 tablet	(Anusha et al., 2020)
B_6	NiO-CNTs/MOHFPE/CPE	DPV	4.0 nM–500 μM	0.9 nM	Energy drinks, drinking water, and orange juice	(Sadeghi et al., 2020)
C	β-CD-GQDs-GCE	SWV	0.01–170 μM	0.49 μM	-	(Shadjou et al., 2016)
D	Ab-25$(OH)D_3$/BSA/ Ab-25$(OH)D_3$/ EDC-NHS/GCN–b-CD@Au/GCE	EIS	0.1–500 mg/mL	0.01 ng/mL	Serum samples	(Anusha et al., 2022)
B_2	VMSF/p-GCE	DPV	20 nM–7 μM and 7 μM–20 μM	11 nM	Milk and tablet leachate	(Wu et al., 2024)
B_2	PEDOT/ZrO_2NPs/GCE	DPV	0.05–300 μM	0.0120 μM	Honey samples	(Nie et al., 2014)
B_6			0.5–1000 μM	0.20 μM		
C			1–1500 μM	0.450 μM		

(*continued*)

TABLE 14.1 (Continued)
Analysis of various vitamins with electrochemical carbon sensors

Vitamin	Transducer	Method	Linear Range	LOD	Application	References
B_2	P(ASP)-GQDs-GCE	DPV	1–1000 µM	0.22 µM	Commercial multivitamin	(Shadjou et al., 2017)
B_9			10–100 µM	0.1 µM		
C			0.22–1000 µM	0.1 µM		
C	rGO/Fe_3O_4/GCE	CV	10 µM–100 mM	18.5 µM	Commercial vitamin C tablets	(Niranjana et al., 2023)
B_2	PdCu@NSC/SPCE	DPV	0.02–9 µM	7.6 pM	Raw milk samples	(Sangili et al., 2019)
K_1	poly(ARS)/MWCNTs/GCE	SWV	0.5–80 µM	0.06 µM	Pharmaceutical products	(Filik & Avan, 2017)
D_3			8–160 µM	5 µM		
D_3	GCE	DPV	5.0×10^{-6}–5.0×10^{-5} M	1.18×10^{-7}M	Tablets	(Cincotto et al., 2014)
B_2	MWCNTs-Mn^{III}salen/GCE	DPV	1.0–400 µM	0.73 µM	Pharmaceutical formulations	(Sonkar et al., 2017)
B_6			1.0–300 µM	0.42 µM		
D_3	MIP/SPCE	CV	1–100 nM	0.10 ng/mL	Human plasma	(SINGH et al., 2020)
C	MoS_2/GCE	Amp	90–590 nM	0.041 µM	Artificial urine samples	(Sampathkumar et al., 2021)
C	AuNP/GO/GCE	Amp	0.1–0.6 mM	100 nM	Real serum sample and vitamin C tablet	(J. Song et al., 2014)
B_{12}	poly(PBHQ)/MWCNTs/GCE	SWV	0.1–10 µM	0.01 µM	Milk samples	(Filik et al., 2016)
C	PMO/GCE	DPV	10–40 µM	0.575 µM	Fresh citrus juices	(Abd-Elsabour et al., 2022)
B_2			4–85 µM	0.922 µM		
C	CdO/NPs/ILs/CPE	SWV	0.07–480 µM	0.03 µM	Fresh juices	(Arabali et al., 2016)
A	Pt:Co/IL/CPE	SWV	0.1–100 µM	0.04 µM	Vitamin A tablets and food specimen	(Lv et al., 2017)

Note: Abbreviations: $[B_{mim}]BF_4$: 1-butyl-3-methylimidazolium tetrafluoroborate; 2FE: 2-(ferrocenylethynyl)fluoren-9-one; 44-DABP: 4,4′-diamino benzophenone; Ab-25(OH)D_3: Serum 25-hydroxyvitamin antibody; Amp: Amperometry; Apt: Aptamer; Au: Gold; AuNPs: Gold nanoparticles; CdO: Cadmium oxide; CNTs: Carbon nanotubes; Co: Cobalt; CPE: Carbon paste electrode; CuNPs: Copper nanoparticles; CV: Cyclic voltammetry; ds-DNA: double stranded DNA; EDC: 1-Ethyl-3-(3-dimethyl aminopropyl) carbodiimide; ErGO: Electrochemically reduced graphene oxide; Fe_3O_4: Iron (II, III) oxide; GCN: Graphitic carbon nitride; GQDs: Graphene quantum dots; IL: Ionic liquid; MB-TAD: Methylene blue-labeled 1,2,4-triazole-3,5-dione; MIP: Molecularly imprinted polymer; Mn^{III}salen: Manganese salen; MnO: Manganese oxide; MOHFPE: 1-methyl-3-octylimidazolium hexafluorophosphate; MoS_2: Molybdenum disulfide; MWCNT: Multi-walled carbon nanotube; NCNT: Nitrogen-doped carbon nanotube; NHS: N-hydroxysuccinimide; NiNPs: Nickel nanoparticles; NiO: Nickel (II) oxide; NSC: Nutshell carbon; P(ASP): Poly aspartic acid; PdCu: Palladium copper; PEDOT: Poly(3,4-ethylene-dioxythiophene); PMO: Poly(methyl orange); PMWCNTPE: Pretreated multiwalled carbon nanotube paste electrode; poly(ARS): Poly (Alizarin red S); poly(PBHQ): Poly(2,2′-(1,4-phenylenedivinylene)bis-8-hydroxyquinaldine); Pt: Platinum; RGO: Reduced-graphene oxide; SPCE: Screen-printed carbon electrode; SWV: Square wave voltammetry; VMSF: Vertically ordered mesoporous silica film; ZC: Zinc oxide/Copper oxide; ZnO: Zinc oxide; ZrO_2NPs: Zirconia nanoparticles; β-CD: β-Cyclodextrin.

(Y. Zhang et al., 2021). $[B_{mim}]BF_4$ is one of the most popular ionic liquids. Surface modification with MWCNT, NiO, and $[B_{mim}]BF_4$ was performed in a 2016 study by Khaleghi et al. using a CPE electrode (Khaleghi et al., 2016). The objective of the study was to create a sensor for detecting vitamin C, and the SWV approach was used for this purpose. The sensor demonstrated a linear concentration range of 0.1–1000 μM for the determination of vitamin C. Additionally, successful results were recorded at a detection limit of 0.06 μM. This method has facilitated the accurate detection of vitamin C in both food and pharmaceutical dosage forms. Also, it allows for analysis with exceptional selectivity in situations when vitamins B_9 and C are present together.

GCEs are highly favored electrodes because of their user-friendly nature, ability to be reused, and consistent outcomes. Vitamin C was also determined in a separate research by means of GCE (Murugan et al., 2019). Murugan et al. conducted the study where they made modifications to the GCE by incorporating MoS_2, MWCNT, and ZnO. MoS_2 is a representative example of a transition metal dichalcogenide and possesses a layered structure that closely resembles graphite (H. Song et al., 2018). The material's preference is mostly due to its characteristic traits, including its high surface area, adjustable structure, and electrical properties (Dai et al., 2021). There are several types of nanostructured MoS_2, including nanowires, nanoribbons, and nanosheets (T. Wang et al., 2015). Based on its properties and characteristics, MoS_2 appears to be an excellent choice for incorporation into sensor applications (Cheng et al., 2017). Several studies have demonstrated that the catalytic effect of this nanomaterial is due to its metallic edges (Vilian et al., 2019). Due to its catalytic property, it is appropriate to use it in sensor studies. The study also used ZnO, a nanomaterial that is one of the most common metal oxides, for modification purposes. However, their use in biosensor studies is limited by their semiconductor properties. For this reason, it is more advantageous to use them in combination with other nanomaterials (Salih et al., 2016). Due to its excellent electrical, chemical, and optical properties, ZnO is a highly desirable nanomaterial for electrochemical sensor studies. One of the reasons it is favored in the manufacture of electrochemical sensors is its wide bandgap, which gives it stable semiconductor properties. In addition, its high biocompatibility and high affinity to oxygen make it a desirable choice (Marie et al., 2015). In the study, Murugan et al. conducted surface characterization using field-emission scanning electron microscopy and energy-dispersive X-ray spectroscopy. Murugan et al. employed a CV method to detect vitamin C. The concentration range for determining vitamin C was between 10 and 100 μM, with a detection limit of 1.65 μM. The findings demonstrate that the developed sensor exhibited considerable surface area. This modification resulted in an acceleration of electron transfer and the emergence of a multitude of electroactive sites. Furthermore, the method was utilized for the determination of vitamin C levels in human serum samples. In real sample studies, the recovery rate was found to be within the range of 98%–101.3%.

Another study in which vitamin C was determined using GCE was carried out by Niranjana et al. (2023). The GCE surface was altered using reduced graphene oxide (rGO) and Fe_3O_4 as shown in Figure 14.13. The exceptional surface area, electrical conductivity, and mechanical conductivity of graphene have collectively attracted the attention of researchers. The employment of two-dimensional nanoplates of graphene in the examination of electrochemical sensors confers advantages by facilitating electron transfer (You et al., 2023). GO is a widely utilized material in sensor studies due to its compatibility with biological systems and its advantageous electronic properties (Verma et al., 2017). Furthermore, the integration of metal nanoparticles into the GO surface results in the enhancement of the electrochemical properties of the composite (Gutés et al., 2012). Fe_3O_4 is utilized in a number of applications, comprising sensors and drug carrier systems, due to its distinctive magnetic and electrochemical properties (Han et al., 2016). This nanomaterial is notable for its convenience and low-toxicity properties (Xie et al., 2016). However, when utilized independently in the development of sensors, they have the potential to impair electron transfer as a result of their susceptibility to oxidation and tendency to aggregate. Consequently, their application in the shape of composites alongside materials such as graphene and conductive polymers is favored

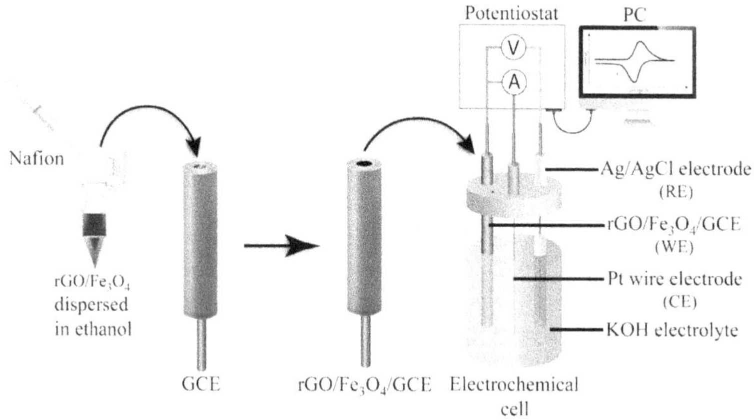

FIGURE 14.13 Schematic diagram of an electrochemical sensing experimental setup. (Reprinted with permission from Niranjana et al., 2023.)

(Han et al., 2016). Niranjana et al. preferred to use rGO and Fe_3O_4 together in their study. The results of the surface characterization conducted using electrochemical impedance spectroscopy (EIS) method indicated that rGO/Fe_3O_4/GCE exhibited a lower resistance in comparison to Fe_3O_4/GCE. This lower resistance assisted the transfer of charges between the surface and $[Fe(CN)_6]^{3-/4-}$. Evidence suggests that a synergistic effect exists between rGO and Fe_3O_4. The determination of vitamin C was conducted using the CV method within a linear concentration range of 10 μM–100 mM. The detection limit was found to be 18.5 μM. The developed sensor demonstrates an exceptional sensitivity of 92 μA mM^{-1} cm^{-2} toward vitamin C. Furthermore, the method facilitated the successful determination of vitamin C in commercial vitamin C tablets. The findings of the study demonstrate that the rGO/Fe_3O_4 composite represents an extremely accurate and unique approach for the detection of vitamin C.

SPCEs are the electrodes that are preferred to be used in studies at least as widely as GCEs. An instance study was conducted by Sangili et al. (2019), which involved the modification of the SPCE surface using PdCu and Nutshell carbon (NSC). In recent years, bimetals have been preferred by researchers compared to monometals. The reason behind this is the superior catalytic properties exhibited by bimetals in comparison to their monometallic counterparts (Li et al., 2018). Bimetallic systems incorporating palladium have the potential to demonstrate enhanced electrocatalytic effects due to the modifiable surface characteristics and morphologies of the constituent materials. The bimetal PdCu exemplifies this property. PdCu is notable for its exceptional sensitivity, stability, and capacity to function at low detection thresholds (Veerakumar et al., 2023). In their investigation, Sangili et al. (2019) utilized a carbon material derived from pistachio nutshell with PdCu. They referred to the resulting composite as PdCu@NSC. This composite exhibits a porous construction, which provides several active sites for the binding of reactants. The sensor developed in the study was employed for the purpose of detecting vitamin B_2. The DPV method was used to determine the vitamin, which was detected with a detection limit of 7.6 pM within a concentration range of 0.02–9 μM. The sensor, which has a sensitivity of 163.4 μA μM^{-1} cm^{-2}, enabled the successful determination of the vitamin in raw milk samples. Upon evaluation of the results, it was observed that the method demonstrated consistent outcomes, high selectivity, and stability.

14.5 CONCLUSION

Vitamins are essential molecules for metabolism. Many diseases could occur due to their deficiency. Most vitamins cannot be synthesized in the body. Consequently, vitamins can be taken into the body through food and supplements (Godswill et al., 2020). This situation makes important the analysis of vitamins. The fact that most vitamins are electrochemically active makes it possible to analyze them by electrochemical methods. This method, which is basically based on oxidation and reduction reactions, allows the analysis of water- and fat-soluble vitamins (Huang et al., 2021). Carbon-based electrodes that can show selectivity for biomolecules offer important advantages, such as reproducibility, repeatability, and stability, for the analyses (Bora et al., 2022). Especially when compared to other methods (e.g., high-performance liquid chromatography, immuno-affinity columns, fluorescence, ultraviolet spectroscopy, and capillary electrophoresis), the use of carbon-based electrodes has important advantages, such as being cheap, fast, easy, and diversifiable (Sharma & Jayaprakas, 2022).

ACKNOWLEDGMENTS

Ipek Kucuk thanks TUBITAK for the financial support under the BIDEB/2211-A doctoral scholarship program. Selenay Sadak thanks TUBITAK for the financial support under the BIDEB/2210-A master scholarship program.

REFERENCES

Abd-Elsabour, M., Abd-Elsabur, K. M., Assaf, F. H., & Hasan, I. M. A. (2022). An Electrochemical sensor based on poly(methyl orange) modified glassy carbon electrode for simultaneous determination of vitamins B2 and C in aqueous solution. *Analytical and Bioanalytical Chemistry Research*, *9*(3), 259–268. https://doi.org/10.22036/abcr.2022.306152.1682

Adams, R. N. (1963). Carbon paste electrodes. *Review of Polarography*, *11*(1–2), 71–78. https://doi.org/10.5189/revpolarography.11.71

Adams, J. S., & Hewison, M. (2010). Update in vitamin D. *The Journal of Clinical Endocrinology & Metabolism*, *95*(2), 471–478. https://doi.org/10.1210/jc.2009-1773

Akgün, H., Balkan, A., & Bilgin, A. (2004). *Farmasötik kimya.*

Anastassakis, K. (2022). *Androgenetic Alopecia from A to Z: Vol. 2. Drugs, herbs, nutrition and supplements.* Springer International Publishing. https://books.google.com.tr/books?id=nWkszwEACAAJ

Antherjanam, S., Saraswathyamma, B., Krishnan, R. G., & Gopakumar, G. M. (2021). Electrochemical sensors as a versatile tool for the quantitative analysis of vitamin B12. *Chemical Papers*, *75*(7), 2981–2995. https://doi.org/10.1007/s11696-021-01574-2

Anusha, T., Bhavani, K. S., Kumar, J. V. S., & Brahman, P. K. (2020). Designing and fabrication of electrochemical nanosensor employing fullerene-C60 and bimetallic nanoparticles composite film for the detection of vitamin D3 in blood samples. *Diamond and Related Materials*, *104*, 107761. https://doi.org/10.1016/j.diamond.2020.107761

Anusha, T., Bhavani, K. S., Shanmukha Kumar, J. V, Brahman, P. K., & Hassan, R. Y. A. (2022). Fabrication of electrochemical immunosensor based on GCN-β-CD/Au nanocomposite for the monitoring of vitamin D deficiency. *Bioelectrochemistry*, *143*, 107935. https://doi.org/10.1016/j.bioelechem.2021.107935

Arabali, V., Ebrahimi, M., Abbasghorbani, M., Gupta, V. K., Farsi, M., Ganjali, M. R., & Karimi, F. (2016). Electrochemical determination of vitamin C in the presence of NADH using a CdO nanoparticle/ionic liquid modified carbon paste electrode as a sensor. *Journal of Molecular Liquids*, *213*, 312–316. https://doi.org/10.1016/j.molliq.2015.10.001

Averianova, L., Balabanova, L., Son, O., Podvolotskaya, A., & Tekutyeva, L. (2020). Production of vitamin B2 (Riboflavin) by microorganisms: An overview. *Frontiers in Bioengineering and Biotechnology*, *8*. https://doi.org/10.3389/fbioe.2020.570828

Bates, C. J. (1997). Vitamin analysis. *Annals of Clinical Biochemistry*, *34*(6), 599–626. https://doi.org/10.1177/000456329703400604

Beigmohammadi, M. T., Bitarafan, S., Hoseindokht, A., Abdollahi, A., Amoozadeh, L., & Soltani, D. (2021). The effect of supplementation with vitamins A, B, C, D, and E on disease severity and inflammatory responses in patients with COVID-19: A randomized clinical trial. *Trials*, *22*(1), 802. https://doi.org/10.1186/s13063-021-05795-4

Bettazzi, F., Ingrosso, C., Sfragano, P. S., Pifferi, V., Falciola, L., Curri, M. L., & Palchetti, I. (2021). Gold nanoparticles modified graphene platforms for highly sensitive electrochemical detection of vitamin C in infant food and formulae. *Food Chemistry*, *344*, 128692. https://doi.org/10.1016/j.foodchem.2020.128692

Betti, N., Al-Amiery, A., & al-azzawi, W. (2022). Experimental and quantum chemical investigations on the anticorrosion efficiency of a nicotinehydrazide derivative for mild steel in HCl. *Molecules (Basel, Switzerland)*, *27*. https://doi.org/10.3390/molecules27196254

Blake, C. (2007). Analytical procedures for water-soluble vitamins in foods and dietary supplements: A review. *Analytical and Bioanalytical Chemistry*, *389*, 63–76. https://doi.org/10.1007/s00216-007-1309-9

Bora, H., Mandal, D., & Chandra, A. (2022). High-performance, nitrogen-doped, carbon-nanotube-based electrochemical sensor for vitamin D3 detection. *ACS Applied Bio Materials*, *5*(4), 1721–1730. https://doi.org/10.1021/acsabm.2c00094

Brahman, P. K., Dar, R. A., & Pitre, K. S. (2013). DNA-functionalized electrochemical biosensor for detection of vitamin B1 using electrochemically treated multiwalled carbon nanotube paste electrode by voltammetric methods. *Sensors and Actuators B: Chemical*, *177*, 807–812. https://doi.org/10.1016/j.snb.2012.11.073

Brigelius-Flohé, R., & Traber, M. G. (1999). Vitamin E: Function and metabolism. *The FASEB Journal*, *13*(10), 1145–1155. https://doi.org/10.1096/fasebj.13.10.1145

Brown, M. J., Ameer, M. A., Daley, S. F., & Beier, K. (2017). *Vitamin B6 deficiency*.

Buehler, B. (2011). Vitamin B2: Riboflavin. *Journal of Evidence-Based Complementary & Alternative Medicine*, *16*, 88–90. https://doi.org/10.1177/1533210110392943

Burgess, C. M., Smid, E. J., & van Sinderen, D. (2009). Bacterial vitamin B2, B11 and B12 overproduction: An overview. *International Journal of Food Microbiology*, *133*(1), 1–7. https://doi.org/10.1016/j.ijfoodmicro.2009.04.012

Carazo, A., Macáková, K., Matou\vsová, K., Kr\vcmová, L. K., Protti, M., & Mladěnka, P. (2021). Vitamin A update: Forms, sources, kinetics, detection, function, deficiency, therapeutic use and toxicity. *Nutrients*, *13*. https://api.semanticscholar.org/CorpusID:235228474

Carbone, M., Maria Bauer, E., Micheli, L., & Missori, M. (2017). NiO morphology dependent optical and electrochemical properties. *Colloids and Surfaces A: Physicochemical and Engineering Aspects*, *532*, 178–182. https://doi.org/10.1016/j.colsurfa.2017.05.046

Carling, R. S., & Turner, C. (2019). Chapter 10 – Methods for assessment of biotin (Vitamin B7). In V. S. Harrington (Ed.), *Laboratory assessment of vitamin status* (pp. 193–217). Academic Press. https://doi.org/10.1016/B978-0-12-813050-6.00010-3

Carr, A., & Maggini, S. (2017). Vitamin C and immune function. *Nutrients*, *9*, 1211. https://doi.org/10.3390/nu9111211

Chand, T., & Savitri, B. (2016). Vitamin B 3, niacin. In E.J. Vandamme & J.L. Revuelta (Eds.), *Industrial biotechnology of vitamins, biopigments, and antioxidants* (pp. 41–66). Wiley-VCH. https://doi.org/10.1002/9783527681754.ch3

Cheng, M., Zhang, X., Wang, M., Huang, H., & Ma, J. (2017). A facile electrochemical sensor based on well-dispersed graphene-molybdenum disulfide modified electrode for highly sensitive detection of dopamine. *Journal of Electroanalytical Chemistry*, *786*, 1–7. https://doi.org/10.1016/j.jelechem.2017.01.012

Cincotto, F. H., Canevari, T. C., & Machado, S. A. S. (2014). Highly sensitive electrochemical sensor for determination of vitamin D in mixtures of water-ethanol. *Electroanalysis*, *26*(12), 2783–2788. https://doi.org/10.1002/elan.201400451

Coerdt, K. M., Goggins, C. A., & Khachemoune, A. (2021). Vitamins A, B, C, and D: A short review for the dermatologist. *Alternative Therapies in Health & Medicine*, *27*(4), 41–49. https://research.ebsco.com/linkprocessor/plink?id=f1a642b6-021e-3051-a20b-4c91a356414a

Collie, J., Greaves, R., Jones, O., Lam, Q., Eastwood, G., & Bellomo, R. (2017). Vitamin B1 in critically ill patients: Needs and challenges. *Clinical Chemistry and Laboratory Medicine (CCLM)*, *55*. https://doi.org/10.1515/cclm-2017-0054

Corb, I., Manea, F., Radovan, C., Pop, A., Burtica, G., Malchev, P., Picken, S., & Schoonman, J. (2007). Carbon-based composite electrodes: Preparation, characterization and application in electroanalysis. *Sensors*, *7*(11), 2626–2635. https://doi.org/10.3390/s7112626

Corfield, R., Lalou, G., Di Lella, S., Martínez, K. D., Schebor, C., Allievi, M. C., & Pérez, O. E. (2023). Experimental and modeling approaches applied to the whey proteins and vitamin B9 complexes study. *Food Hydrocolloids*, *142*, 108834. https://doi.org/10.1016/j.foodhyd.2023.108834

Dai, J., Lv, Y., Zhang, J., Zhang, D., Xie, H., Guo, C., Zhu, A., Xu, Y., Fan, M., Yuan, C., & Dai, L. (2021). Effect of morphology and phase engineering of MoS2 on electrochemical properties of carbon nanotube/polyaniline@MoS2 composites. *Journal of Colloid and Interface Science*, *590*, 591–600. https://doi.org/10.1016/j.jcis.2021.01.051

Dekanski, A., Stevanović, J., Stevanović, R., Nikolić, B. Ž., & Jovanović, V. M. (2001). Glassy carbon electrodes: I. Characterization and electrochemical activation. *Carbon*, *39*(8), 1195–1205. https://doi.org/10.1016/S0008-6223(00)00228-1

Denu, J. M. (2005). Vitamin B3 and sirtuin function. *Trends in Biochemical Sciences*, *30*(9), 479–483. https://doi.org/10.1016/j.tibs.2005.07.004

Depeint, F., Bruce, W. R., Shangari, N., Mehta, R., & O'Brien, P. J. (2006). Mitochondrial function and toxicity: Role of the B vitamin family on mitochondrial energy metabolism. *Chemico-Biological Interactions*, *163*(1), 94–112. https://doi.org/10.1016/j.cbi.2006.04.014

Dinicolantonio, J., Bhutani, J., & O'Keefe, J. (2015). The health benefits of Vitamin K. *Open Heart*, *2*, e000300. https://doi.org/10.1136/openhrt-2015-000300

Dosedel, M., Jirkovský, E., Macakova, K., Krcmova, L., Javorská, L., Pourová, J., Mercolini, L., Remião, F., Nováková, L., & Mladenka, P. (2021). Vitamin C—Sources, physiological role, kinetics, deficiency, use, toxicity, and determination. *Nutrients*, *13*, 615. https://doi.org/10.3390/nu13020615

Dove, C. R. (2011). Vitamins: Water soluble: Biotin, choline, niacin, and ascorbic acid. In *Encyclopedia of animal science* (2nd ed., pp. 1073–1075). CRC Press.

Downard, A. J. (2000). Electrochemically assisted covalent modification of carbon electrodes. *Electroanalysis*, *12*(14), 1085–1096. https://doi.org/10.1002/1521-4109(200010)12:14<1085::AID-ELAN1085>3.0.CO;2-A

Engstrom, R. C. (1982). Electrochemical pretreatment of glassy carbon electrodes. *Analytical Chemistry*, *54*(13), 2310–2314. https://doi.org/10.1021/ac00250a038

Fanali, C., D'Orazio, G., Fanali, S., & Gentili, A. (2017). Advanced analytical techniques for fat-soluble vitamin analysis. *TrAC Trends in Analytical Chemistry*, *87*, 82–97. https://doi.org/10.1016/j.trac.2016.12.001

Fatima, Z., Jin, X., Zou, Y., Kaw, H. Y., Quinto, M., & Li, D. (2019). Recent trends in analytical methods for water-soluble vitamins. *Journal of Chromatography A*, *1606*, 360245. https://doi.org/10.1016/j.chroma.2019.05.025

Fattal-valevski, A. (2011). Thiamine (vitamin B1). *Complementary Health Practice Review*, *16*, 12–20. https://doi.org/10.1177/1533210110392941

Filik, H., & Avan, A. A. (2017). Simultaneous electrochemical determination of vitamin K1 and vitamin D3 by using poly (Alizarin red S)/multi-walled carbon nanotubes modified glassy electrode. *Current Analytical Chemistry*, *13*(5), 350–360. https://doi.org/10.2174/1573411013666170105143113

Filik, H., Avan, A. A., & Aydar, S. (2016). Electrochemical determination of vitamin B-12 in food samples by poly(2,2′-(1,4-phenylenedivinylene) bis-8 hydroxyquinaldine)/multi-walled carbon nanotube-modified glassy carbon electrode. *Food Analytical Methods*, *9*(8), 2251–2260. https://doi.org/10.1007/s12161-016-0420-y

Fletcher, S. (2015). Screen-printed carbon electrodes. In *Electrochemistry of carbon electrodes* (pp. 425–444). Wiley. https://doi.org/10.1002/9783527697489.ch12

Godswill, A. G., Somtochukwu, I. V., Ikechukwu, A. O., & Kate, E. C. (2020). Health benefits of micronutrients (vitamins and minerals) and their associated deficiency diseases: A systematic review. *International Journal of Food Sciences*, *3*(1), 1–32. https://doi.org/10.47604/ijf.1024

Gröber, U., Reichrath, J., Holick, M., & Kisters, K. (2015). Vitamin K: An old vitamin in a new perspective. *Dermato-Endocrinology*, *6*, e968490. https://doi.org/10.4161/19381972.2014.968490

Gutés, A., Hsia, B., Sussman, A., Mickelson, W., Zettl, A., Carraro, C., & Maboudian, R. (2012). Graphene decoration with metal nanoparticles: Towards easy integration for sensing applications. *Nanoscale*, *4*(2), 438–440. https://doi.org/10.1039/C1NR11537E

Halver, J. E. (2003). 2 – The vitamins. In J. E. Halver & R. W. Hardy (Eds.), *Fish nutrition* (3rd ed., pp. 61–141). Academic Press. https://doi.org/10.1016/B978-012319652-1/50003-3

Han, Q., Shen, X., Zhu, W., Zhu, C., Zhou, X., & Jiang, H. (2016). Fe_3O_4 is employed in a number of applications, including sensors and drug carrier systems, due to its distinctive magnetic and electrochemical properties. *Biosensors and Bioelectronics*, *79*, 180–186. https://doi.org/10.1016/j.bios.2015.12.017

Harris, J. R. (2013). *Subcellular biochemistry: Ascorbic acid: Biochemistry and biomedical cell biology*. Springer US. https://books.google.com.tr/books?id=okHaBwAAQBAJ

Hashemi, S. M. B., Es, I., Gholamhosseinpour, A., Loizzo, M., Giardinieri, A., Pacetti, D., Pourmohammadi, K., & Ferreira, D. (2018). *Water-soluble vitamins*. https://doi.org/10.1016/B978-0-12-814174-8.00008-1

Hashemi, S. M. B., Pourmohammadi, K., Gholamhosseinpour, A., Es, I., & Ferreira, D. (2018). *Fat-soluble vitamins*. https://doi.org/10.1016/B978-0-12-814174-8.00009-3

Hellmann, H., & Mooney, S. (2010). Vitamin B6: A molecule for human health? *Molecules (Basel, Switzerland)*, *15*, 442–459. https://doi.org/10.3390/molecules15010442

Herting, D. C., & Drury, E.-J. E. (1963). Vitamin E content of vegetable oils and fats. *The Journal of Nutrition*, *81*(4), 335–342. https://doi.org/10.1093/jn/81.4.335

Holick, M. F. (2018). Chapter 4 – Photobiology of vitamin D. In D. Feldman (Eds.), *Vitamin D* (4th ed., pp. 45–55). Academic Press. https://doi.org/10.1016/B978-0-12-809965-0.00004-5

Huang, L., Tian, S., Zhao, W., Liu, K., & Guo, J. (2021). Electrochemical vitamin sensors: A critical review. *Talanta*, 222, 121645. https://doi.org/10.1016/j.talanta.2020.121645

Iijima, S. (1991). Helical microtubules of graphitic carbon. *Nature*, *354*(6348), 56–58. https://doi.org/10.1038/354056a0

Jagannathan, M., & Badhulika, S. (2024). Graphitic carbon nitride and its nanocomposites-based sensors for detection of pharmaceutical effluents in food, ecological and biological samples: A mini-review. *Sensors and Actuators Reports*, *7*, 100183. https://doi.org/10.1016/j.snr.2023.100183

Kato, S. (2000). The function of vitamin D receptor in vitamin D action. *Journal of Biochemistry*, *127*, 717–722. https://doi.org/10.1093/oxfordjournals.jbchem.a022662

Kayaalp, O. (1996). *Tıbbi farmakoloji*. Hacettepe-Taş.

Khaleghi, F., Arab, Z., Gupta, V. K., Ganjali, M. R., Norouzi, P., Atar, N., & Yola, M. L. (2016). Fabrication of novel electrochemical sensor for determination of vitamin C in the presence of vitamin B9 in food and pharmaceutical samples. *Journal of Molecular Liquids*, *221*, 666–672. https://doi.org/10.1016/j.molliq.2016.06.061

Kia, S., Bahar, S., & Bohlooli, S. (2019). A novel electrochemical sensor based on plastic antibodies for vitamin D3 detection in real samples. *IEEE Sensors Journal*, *19*(13), 4752–4757. https://doi.org/10.1109/JSEN.2019.2903090

Kiamiloglou, D., & Girousi, S. (2023). Different aspects of the voltammetric detection of vitamins: A review. *Biosensors*, *13*(6). https://doi.org/10.3390/bios13060651

Li, Z.-H., Zhao, X.-L., Jiang, X.-C., Wu, Y.-H., Chen, C., Zhu, Z.-G., Marty, J.-L., & Chen, Q.-S. (2018). An enhanced nonenzymatic electrochemical glucose sensor based on copper-palladium nanoparticles modified glassy carbon electrodes. *Electroanalysis*, *30*(8), 1811–1819. https://doi.org/10.1002/elan.201800017

Lips, P. (2006). Vitamin D physiology. *Progress in Biophysics and Molecular Biology*, *92*(1), 4–8. https://doi.org/10.1016/j.pbiomolbio.2006.02.016

Lv, X., Zhao, J., Dong, X., Tian, H., Qi, S., Jiang, Y., & Ping, Y. (2017). Electrochemical determination of vitamin A based on a novel nanoalloy modified carbon paste electrode. *International Journal of Electrochemical Science*, *12*(9), 8457–8466. https://doi.org/10.20964/2017.09.23

Ma, Y., Shen, X.-L., Zeng, Q., Wang, H.-S., & Wang, L.-S. (2017). A multi-walled carbon nanotubes based molecularly imprinted polymers electrochemical sensor for the sensitive determination of HIV-p24. *Talanta*, *164*, 121–127. https://doi.org/10.1016/j.talanta.2016.11.043

Maji, D. (2012). Vitamin D toxicity. *Indian Journal of Endocrinology and Metabolism*, *16*, 295–296. https://doi.org/10.4103/2230-8210.93773

Malode, S. J., Sharma, P., Hasan, M. R., Shetti, N. P., & Mascarenhas, R. J. (2022). 4 – Carbon and carbon paste electrodes. In G. Maruccio & J. Narang (Eds.), *Electrochemical sensors* (Woodhead publishing series in electronic and optical materials, pp. 79–114). Woodhead Publishing. https://doi.org/10.1016/B978-0-12-823148-7.00004-0

Maqbool, M. A., Aslam, M., Akbar, W., & Iqbal, Z. (2018). Biological importance of vitamins for human health: A review. *Journal of Agriculture and Basic Science*, *2*.

Marie, M., Mandal, S., & Manasreh, O. (2015). An electrochemical glucose sensor based on zinc oxide nanorods. *Sensors*, *15*(8), 18714–18723. https://doi.org/10.3390/s150818714

Martel, J., Kerndt, C., & Franklin, D. (2020). *Vitamin B1 (Thiamine)* (p. PMID: 29493982).

Martinsen, Ø. G., & Heiskanen, A. (2023). Chapter 7 – Electrodes. In Ø. G. Martinsen & A. Heiskanen (Eds.), *Bioimpedance and bioelectricity basics* (4th ed., pp. 175–248). Academic Press. https://doi.org/10.1016/B978-0-12-819107-1.00005-4

Maxfield, L. (2019). *Vitamin C deficiency (scurvy)*. ACS Publishing.

McCreery, R. L. (2008). Advanced carbon electrode materials for molecular electrochemistry. *Chemical Reviews*, *108*(7), 2646–2687.

McCreery, R. L., Cline, K. K., McDermott, C. A., & McDermott, M. T. (1994). Control of reactivity at carbon electrode surfaces. *Colloids and Surfaces A: Physicochemical and Engineering Aspects*, *93*, 211–219. https://doi.org/10.1016/0927-7757(94)02899-0

Mladenka, P., Macakova, K., Krcmova, L., Javorská, L., Mrštná, K., Carazo, A., Protti, M., Remião, F., & Nováková, L. (2021). Vitamin K – sources, physiological role, kinetics, deficiency, detection, therapeutic use, and toxicity. *Nutrition Reviews*, *80*. https://doi.org/10.1093/nutrit/nuab061

Morales-Gutierrez, J., Díaz-Cortés, S., Montoya Giraldo, M., & Zuluaga, A. (2019). Toxicity induced by multiple high doses of vitamin B 12 during pernicious anemia treatment: A case report. *Clinical Toxicology*, *58*, 1–3. https://doi.org/10.1080/15563650.2019.1606432

Motaghi, M. M., Beitollahi, H., Tajik, S., & Hosseinzadeh, R. (2016). Nanostructure electrochemical sensor for voltammetric determination of vitamin C in the presence of vitamin B6: Application to real sample analysis. *International Journal of Electrochemical Science*, *11*(9), 7849–7860. https://doi.org/10.20964/2016.09.60

Murugan, N., Kumar, T. H. V., Devi, N. R., & Sundramoorthy, A. K. (2019). A flower-structured MoS2-decorated f-MWCNTs/ZnO hybrid nanocomposite-modified sensor for the selective electrochemical detection of vitamin C. *New Journal of Chemistry*, *43*(38), 15105–15114. https://doi.org/10.1039/C9NJ02993A

Nair, R., & Maseeh, A. (2012). Vitamin D: The "sunshine" vitamin. *Journal of Pharmacology & Pharmacotherapeutics*, *3*, 118–126. https://doi.org/10.4103/0976-500X.95506

Nicolov, M., Mioara, C., Buda, V., Danciu, C., Duse, A., Watz, C., & Borcan, F. (2021). Hydrosoluble and liposoluble vitamins: New perspectives through ADMET analysis. *Medicina*, *57*, 1204. https://doi.org/10.3390/medicina57111204

Nie, T., Zhang, K., Xu, J., Lu, L., & Bai, L. (2014). A facile one-pot strategy for the electrochemical synthesis of poly(3,4-ethylenedioxythiophene)/Zirconia nanocomposite as an effective sensing platform for vitamins B2, B6 and C. *Journal of Electroanalytical Chemistry*, *717–718*, 1–9. https://doi.org/10.1016/j.jelechem.2014.01.006

Niki, E. (2021). Lipid oxidation that is, and is not, inhibited by vitamin E: Consideration about physiological functions of vitamin E. *Free Radical Biology and Medicine*, *176*, 1–15. https://doi.org/10.1016/j.freeradbiomed.2021.09.001

Niki, E., & Traber, M. (2012). A history of vitamin E. *Annals of Nutrition & Metabolism*, *61*, 207–212. https://doi.org/10.1159/000343106

Niranjana, J. S., Koyakutty, H., Thomas, A. A., & Bushiri, M. J. (2023). Novel synthesis of multi-layered rGO/Fe_3O_4 nanocomposite in a single step and its efficient electrochemical sensing of vitamin C. *Materials Science and Engineering: B*, *290*, 116283. https://doi.org/10.1016/j.mseb.2023.116283

Niu, X., Chen, C., Zhao, H., Tang, J., Li, Y., & Lan, M. (2012). Porous screen-printed carbon electrode. *Electrochemistry Communications*, *22*, 170–173. https://doi.org/10.1016/j.elecom.2012.06.020

Oh, R., & Brown, D. (2003). Vitamin B12 deficiency. *American Family Physician*, *67*, 979–986.

Oliveira, T., Ribeiro, F., Sousa, C., Salazar Banda, G., Lima-Neto, P., Correia, A., & Morais, S. (2019). Current overview and perspectives on carbon-based (bio)sensors for carbamate pesticides electroanalysis. *TrAC Trends in Analytical Chemistry*, *124*, 115779. https://doi.org/10.1016/j.trac.2019.115779

Omonnazarova, M., Muysinov, A., Mardonov, M., Tojiddinov, D., & Ubaydullayev, D. (2022). Vitamins. Concept of vitamin. Importance of vitamins in human life today. *Theoretical Aspects in the Formation of Pedagogical Sciences*, *1*(6), 270–274.

Ottenhof, H., Ashurst, J., Whitney, H., Saldanha, S., Schmitzberger, F., Gweon, H., Blundell, T., Abell, C., & Smith, A. (2004). Organisation of the pantothenate (vitamin B 5) biosynthesis pathway in higher plants. *The Plant Journal : For Cell and Molecular Biology*, *37*, 61–72. https://doi.org/10.1046/j.1365-313X.2003.01940.x

Owen, K. N., & Dewald, O. (2023). *Vitamin E toxicity*. StatPearls Publishing, Treasure Island (FL). http://europepmc.org/abstract/MED/33232043

Pandey, H., Khare, P., Singh, S., & Singh, S. P. (2020). Carbon nanomaterials integrated molecularly imprinted polymers for biological sample analysis: A critical review. *Materials Chemistry and Physics*, *239*, 121966. https://doi.org/10.1016/j.matchemphys.2019.121966

Park, J., Kim, M., Kim, W., Jo, S., Kim, W., Kim, C., Park, H., Lee, W., & Park, J. (2022). Ultrasensitive detection of 25-hydroxy vitamin D3 in real saliva using sandwich-type electrochemical aptasensor. *Sensors and Actuators B: Chemical*, *355*, 131239. https://doi.org/10.1016/j.snb.2021.131239

Pawlak, R., Parrott, J., Raj, S., Cullum-Dugan, D., & Lucus, D. (2012). Understanding vitamin B12. *American Journal of Lifestyle Medicine*, *7*, 60–65. https://doi.org/10.1177/1559827612450688

Plevin, D., & Galletly, C. (2020). The neuropsychiatric effects of vitamin C deficiency: A systematic review. *BMC Psychiatry*, *20*. https://doi.org/10.1186/s12888-020-02730-w

Pradeep Devnath, G., Kumaran, S., R., R., Shaha, K., & Nagaraj, A. (2017). Fatal folic acid toxicity in humans. *Journal of Forensic Sciences*, *62*. https://doi.org/10.1111/1556-4029.13489

Rader, J. I., Calvert, R. J., & Hathcock, J. N. (1992). Hepatic toxicity of unmodified and time-release preparations of niacin. *The American Journal of Medicine*, *92*(1), 77–81. https://doi.org/10.1016/0002-9343(92)90018-7

Rébeillé, F., Ravanel, S., Marquet, A., Mendel, R., Webb, M., Smith, A., & Warren, M. (2007). Roles of vitamins B5, B8, B9, B12 and molybdenum cofactor at cellular and organismal levels. *Natural Product Reports*, *24*, 949–962. https://doi.org/10.1039/b703104c

Riu, J., Maroto, A., & Rius, F. X. (2006). Nanosensors in environmental analysis. *Talanta*, *69*(2 SPEC. ISS.), 288–301. https://doi.org/10.1016/j.talanta.2005.09.045

Ross, A. (2010). Vitamin A. In *Bioactive compounds and cancer. Nutrition and health* (pp. 335–356). Humana Press.

Rucker, R. B. (2001). *Handbook of vitamins*. Marcel Dekker. https://books.google.com.tr/books?id=YYGpzwEACAAJ

Rucker, R. B., Suttie, J. W., & McCormick, D. B. (2001). *Handbook of vitamins* (3rd ed.). Taylor & Francis. https://books.google.com.tr/books?id=6WcrnwEACAAJ

Sadeghi, H., Shahidi, S.-A., Raeisi, S. N., Ghorbani-HasanSaraei, A., & Karimi, F. (2020). Electrochemical determination of vitamin B6 in water and juice samples using an electrochemical sensor amplified with NiO/CNTs and Ionic liquid. *International Journal of Electrochemical Science*, *15*(10), 10488–10498. https://doi.org/10.20964/2020.10.51

Saleem, H., & Simpson, B. (2023). *Biotinidase deficiency*. StatPearls Publishing, Treasure Island (FL). http://europepmc.org/abstract/MED/32809442

Salih, E., Mekawy, M., Hassan, R. Y. A., & El-Sherbiny, I. M. (2016). Synthesis, characterization and electrochemical-sensor applications of zinc oxide/graphene oxide nanocomposite. *Journal of Nanostructure in Chemistry*, *6*(2), 137–144. https://doi.org/10.1007/s40097-016-0188-z

Salvigni, L., Mariani, F., Gualandi, I., Decataldo, F., Tessarolo, M., Tonelli, D., Fraboni, B., & Scavetta, E. (2023). Selective detection of liposoluble vitamins using an organic electrochemical transistor. *Sensors and Actuators B: Chemical*, *393*, 134313. https://doi.org/10.1016/j.snb.2023.134313

Sampathkumar, P., Sudalaimani, S., Giribabu, K., & Suresh, C. (2021). Nanomolar detection of vitamin C in artificial urine using a glassy carbon electrode modified with molybdenum disulfide. *Journal of The Electrochemical Society*, *168*(8), 87507. https://doi.org/10.1149/1945-7111/ac1973

Sangili, A., Veerakumar, P., Chen, S.-M., Rajkumar, C., & Lin, K.-C. (2019). Voltammetric determination of vitamin B2 by using a highly porous carbon electrode modified with palladium-copper nanoparticles. *Microchimica Acta*, *186*(5), 299. https://doi.org/10.1007/s00604-019-3396-7

Sanvictores, T., & Chauhan, S. (2020). *Vitamin B5 (pantothenic acid)*.

Sauve, A. (2008). NAD+ and vitamin B3: From metabolism to therapies. *The Journal of Pharmacology and Experimental Therapeutics*, *324*, 883–893.

Schlueter, A., & Johnston, C. (2011). Vitamin C: Overview and update. *Complementary Health Practice Review*, *16*, 49–57. https://doi.org/10.1177/1533210110392951

Shadjou, N., Hasanzadeh, M., & Omari, A. (2017). Electrochemical quantification of some water soluble vitamins in commercial multi-vitamin using poly-amino acid caped by graphene quantum dots nanocomposite as dual signal amplification elements. *Analytical Biochemistry*, *539*, 70–80. https://doi.org/10.1016/j.ab.2017.10.011

Shadjou, N., Hasanzadeh, M., Talebi, F., & Marjani, A. P. (2016). Integration of β-cyclodextrin into graphene quantum dot nano-structure and its application towards detection of Vitamin C at physiological pH: A new electrochemical approach. *Materials Science and Engineering: C*, *67*, 666–674. https://doi.org/10.1016/j.msec.2016.05.078

Shahidi, F., da Costa Pinaffi Langley, A. C., Fuentes, J., Speisky, H., & de Camargo, A. (2021). Vitamin E as an essential micronutrient for human health: Common, novel, and unexplored dietary sources. *Free Radical Biology and Medicine*, *176*. https://doi.org/10.1016/j.freeradbiomed.2021.09.025

Sharma, V., & Jayaprakas, G. K. (2022). Fabrications of electrochemical sensors based on carbon paste electrode for vitamin detection in real samples. *Journal of Electrochemical Science and Engineering*, *12*(3), 421-430. http://dx.doi.org/10.5599/jese.1313 421

Shearer, M. J. (1995). Vitamin K. *The Lancet*, *345*(8944), 229–234. https://doi.org/10.1016/S0140-6736(95)90227-9

Sikorski, Z. E. (n.d.). *Kolakowska (2003). Chemical and functional properties of food lipids*. United States of America: CRC PRESS.

Sikorski, Z. E., & Kołakowska, A. T. A.-T. T.-. (2003). *Chemical and functional properties of food lipids* (NV-1 o). CRC Press. https://doi.org/LK - https://worldcat.org/title/326582975

Singh, J. (2021). Vitamin B9 in dark green vegetables: Deficiency disorders, bio-availability, and fortification issues. In J. G. LeBlanc (Ed.), *B-complex vitamins – sources, intakes and novel applications*. IntechOpen.

Singh, S., Batra, N., Ansari, M. A., & Urooj, S. (2020). Electrochemical sensing for examining vitamin D3 based on MIP using NOVA 1.7 and Autolab PGSTAT 302N. *Journal of Clinical & Diagnostic Research*, *14*(7), 17–20. https://doi.org/10.7860/JCDR/2020/44347.13847

Smith, A. D., Warren, M. J., & Refsum, H. (2018). Chapter Six – Vitamin B12. In M. N. A. Eskin (Ed.), *New research and developments of water-soluble vitamins* (Vol. 83, pp. 215–279). Academic Press. https://doi.org/10.1016/bs.afnr.2017.11.005

Song, H., Tang, A., Xu, G., Liu, L., Pan, Y., & Yin, M. (2018). Hydrothermal synthesis and electrochemical properties of MoS2/C nanocomposite. *International Journal of Electrochemical Science*, *13*(7), 6708–6716. https://doi.org/10.20964/2018.07.54

Song, J., Xu, L., Xing, R., Li, Q., Zhou, C., Liu, D., & Song, H. (2014). Synthesis of Au/graphene oxide composites for selective and sensitive electrochemical detection of ascorbic acid. *Scientific Reports*, *4*(1), 7515. https://doi.org/10.1038/srep07515

Sonkar, P. K., Ganesan, V., Sen Gupta, S. K., Yadav, D. K., Gupta, R., & Yadav, M. (2017). Highly dispersed multiwalled carbon nanotubes coupled manganese salen nanostructure for simultaneous electrochemical sensing of vitamin B2 and B6. *Journal of Electroanalytical Chemistry*, *807*, 235–243. https://doi.org/10.1016/j.jelechem.2017.11.050

Stach, K., Stach, W., & Augoff, K. (2021). Vitamin B6 in health and disease. *Nutrients*, *13*, 3229. https://doi.org/10.3390/nu13093229

Stevens, S. (2021). Fat-soluble vitamins. *Nursing Clinics of North America*, *56*, 33–45. https://doi.org/10.1016/j.cnur.2020.10.003

Suychinov, A., Rebezov, M., Maksimiuk, N., Khayrullin, M., Kulikov, D., Konovalov, S., Konovalova, O., Penkova, I., & Moldabayeva, Z. (2019). Vitamins and their role in human body. *International Journal of Pharmaceutical Research (09752366)*, *11*(3), 1246–1248. https://doi.org/10.31838/ijpr/2019.11.03.018

Švancara, I., Vytřas, K., Barek, J., & Zima, J. (2001). Carbon paste electrodes in modern electroanalysis. *Critical Reviews in Analytical Chemistry*, *31*(4), 311–345. https://doi.org/10.1080/20014091076785

Tabanlıgil Calam, T. (2021). A novel, efficient and sensitive method for the simultaneous determination of riboflavin (vitamin B2) and pyridoxine hydrochloride (vitamin B6) in food and pharmacological samples using an electrochemical sensor based on 4,4′-diamino benzophenone. *Microchemical Journal*, *169*, 106557. https://doi.org/10.1016/j.microc.2021.106557

Titcomb, T., & Tanumihardjo, S. (2019). Global concerns with B vitamin statuses: Biofortification, fortification, hidden hunger, interactions, and toxicity. *Comprehensive Reviews in Food Science and Food Safety*, *18*. https://doi.org/10.1111/1541-4337.12491

Torquato, P., Marinelli, R., Bartolini, D., & Galli, F. (2020). Chapter 23 – Vitamin E: Nutritional aspects. In V. B. Patel (Ed.) *Molecular nutrition* (pp. 447–485). Academic Press. https://doi.org/10.1016/B978-0-12-811907-5.00019-1

Veerakumar, P., Koventhan, C., & Chen, S. M. (2023). Copper-palladium alloy nanoparticles immobilized over porous carbon for voltammetric determination of dimetridazole. *Journal of Alloys and Compounds*, *931*, 167474. https://doi.org/10.1016/j.jallcom.2022.167474

Verma, S., Singh, A., Shukla, A., Kaswan, J., Arora, K., Ramirez-Vick, J., Singh, P., & Singh, S. P. (2017). Anti-IL8/AuNPs-rGO/ITO as an immunosensing platform for noninvasive electrochemical detection of oral cancer. *ACS Applied Materials & Interfaces*, *9*(33), 27462–27474. https://doi.org/10.1021/acsami.7b06839

Vermeer, C., & Schurgers, L. J. (2000). A comprehensive review of vitamin K and vitamin K antagonists. *Hematology/Oncology Clinics of North America*, *14*(2), 339–353. https://doi.org/10.1016/S0889-8588(05)70137-4

Vicentini, F. C., Silva, T. A., & Fatibello-Filho, O. (2024). Carbon black electrodes applied in electroanalysis. *Current Opinion in Electrochemistry*, *43*, 101415. https://doi.org/10.1016/j.coelec.2023.101415

Vijayaprasath, G., Habibulla, I., Dharuman, V., Balasubramanian, S., & Ganesan, R. (2020). Fabrication of Gd_2O_3 nanosheet-modified glassy carbon electrode for nonenzymatic highly selective electrochemical detection of vitamin B2. *ACS Omega*, *5*(29), 17892–17899. https://doi.org/10.1021/acsomega.9b04284

Vilian, A. T. E., Dinesh, B., Kang, S.-M., Krishnan, U. M., Huh, Y. S., & Han, Y.-K. (2019). Recent advances in molybdenum disulfide-based electrode materials for electroanalytical applications. *Microchimica Acta*, *186*(3), 203. https://doi.org/10.1007/s00604-019-3287-y

Vytřas, K., Švancara, I., & Metelka, R. (2009). Carbon paste electrodes in electroanalytical chemistry. *Journal of the Serbian Chemical Society*, *74*(10), 1021–1033. https://doi.org/10.2298/JSC0910021V

Wang, J., Anik Kirgöz, Ü., Mo, J.-W., Lu, J., Nasser Kawde, A., & Muck, A. (2001). Glassy carbon paste electrodes. *Electrochemistry Communications*, *3*(4), 203–208. https://doi.org/10.1016/S1388-2481(01)00142-4

Wang, T., Du, K., Liu, W., Zhang, J., & Li, M. (2015). *Electrochemical sensors based on molybdenum disulfide nanomaterials* (pp. 2091–2097). https://doi.org/10.1002/elan.201500117

Westmacott, K. L., Crew, A., Doran, O., & Hart, J. P. (2018). A novel electroanalytical approach to the measurement of B vitamins in food supplements based on screen-printed carbon sensors. *Talanta*, *181*, 13–18. https://doi.org/10.1016/j.talanta.2017.12.074

Wibowo, N., Purwosunu, Y., Sekizawa, A., Farina, A., Tambunan, V., & Bardosono, S. (2012). Vitamin B6 supplementation in pregnant women with nausea and vomiting. *International Journal of Gynecology & Obstetrics*, *116*(3), 206–210. https://doi.org/10.1016/j.ijgo.2011.09.030

Wightman, R. M., Deakin, M. R., Kovach, P. M., Kuhr, W. G., & Stutts, K. J. (1984). Methods to improve electrochemical reversibility at carbon electrodes. *Journal of The Electrochemical Society*, *131*(7), 1578. https://doi.org/10.1149/1.2115913

Wiley, K. D., & Gupta, M. (2019). *Vitamin B1 thiamine deficiency*.

Wilson, L. R., Tripkovic, L., Hart, K. H., & Lanham-New, S. A. (2017). Vitamin D deficiency as a public health issue: Using vitamin D2 or vitamin D3 in future fortification strategies. *Proceedings of the Nutrition Society*, *76*(3), 392–399. https://doi.org/DOI: 10.1017/S0029665117000349

Wring, S. A., & Hart, J. P. (1992). Chemically modified, screen-printed carbon electrodes. *Analyst*, *117*(8), 1281–1286. https://doi.org/10.1039/AN9921701281

Wu, Y., Shi, Z., Liu, J., Luo, T., Xi, F., & Zeng, Q. (2024). Simple fabrication of electrochemical sensor based on integration of dual signal amplification by the supporting electrode and modified nanochannel array for direct and sensitive detection of vitamin B(2). *Frontiers in Nutrition*, *11*, 1352938. https://doi.org/10.3389/fnut.2024.1352938

Xie, Z., Xu, J., Xie, F., & Xiong, S. (2016). Electrochemical detection of As(III) by a rGO/Fe_3O_4-modified screen-printed carbon electrode. *Analytical Sciences*, *32*(10), 1053–1058. https://doi.org/10.2116/analsci.32.1053

Yadav, R. A. (2015). Experimental and quantum chemical investigations on vibrational and molecular structural characteristics of vitamin B 7: Biotin molecule. *Asian Journal of Physics*, *24*(5), 635–656.

Yang, Y., Hu, N., Deng, J., & Yang, J. (2022). Electrochemical sensing for vitamins. *Chemosensors*, *10*(11). https://doi.org/10.3390/chemosensors10110494

You, J., Li, J., Wang, Z., Baghayeri, M., & Zhang, H. (2023). Application of Co_3O_4 nanocrystal/rGO for simultaneous electrochemical detection of cadmium and lead in environmental waters. *Chemosphere*, *335*, 139133. https://doi.org/10.1016/j.chemosphere.2023.139133

Zeybek, B., & Üğe, A. (2021). Development of carbon-based sensors for electrochemical quantification of vitamins B2 and B6 at nanomolar levels. *Chemical Papers*, *75*(4), 1323–1339. https://doi.org/10.1007/s11696-020-01387-9

Zhang, Y., Zhang, X., Lang, X., & Zhang, Q. (2021). Study on the themodynamic properties and electrochemical performance of mixture electrolyte for supercapacitor composed of ionic liquid [BMIM][BF4] and SBPBF4/PC. *Ionics*, *27*(9), 4003–4011. https://doi.org/10.1007/s11581-021-04145-3

Zhang, W., Zhu, S., Luque, R., Han, S., Hu, L., & Xu, G. (2016). Recent development of carbon electrode materials and their bioanalytical and environmental applications. *Chemical Society Reviews*, *45*(3), 715–752. https://doi.org/10.1039/c5cs00297d

Zhao, L., Su, G., Liu, W., Cao, L., Wang, J., Dong, Z., & Song, M. (2011). Optical and electrochemical properties of Cu-doped NiO films prepared by electrochemical deposition. *Applied Surface Science*, *257*(9), 3974–3979. https://doi.org/10.1016/j.apsusc.2010.11.160

Zhao, K., Tang, H., Zhang, B., Zou, S., Liu, Z., & Zheng, Y. (2022). Microbial production of vitamin B5: Current status and prospects. *Critical Reviews in Biotechnology*, *43*, 1–21. https://doi.org/10.1080/07388551.2022.2104690

15 Carbon Sensors for Herbicides, Insecticides, Fungicides and Other Eco-toxic Chemicals

Murat Çelik and Bengi Uslu

15.1 INTRODUCTION

Pesticides are chemicals used in agriculture, industry and the military. The vast majority of pesticides are used in agriculture, and this sector is also the global sector with the highest use (Bhambri et al., 2023). The most common pests controlled by pesticides in agriculture include insects, fungi, bacteria, rodents, weeds and other pests (Clesceri & Clesceri, 2021). The utilisation of pesticides in agriculture is increasing primarily due to numerous benefits, such as enhancing agricultural yields and preventing crop loss, which in turn increases food safety and food production (D. Wang et al., 2022). Pesticide use has led to a reduction in crop loss by about 30–35%. By 2050, the global population is projected to reach 9.2 billion, which will require a 70% increase in food production (Popp et al., 2013). It is undeniable that this significant increase in food demand will lead to a parallel increase in pesticide use in the future. In light of these conditions, it is becoming increasingly evident that effective control, determination, identification and characterisation of pesticides are of greater importance (Tudi et al., 2021).

The most commonly used analytical methods for the identification of pesticides include gas chromatography-mass spectrometry (Afify et al., 2022), high-performance liquid chromatography (Barut et al., 2023; Kowalska et al., 2020), liquid chromatography-mass spectrometry (Gao et al., 2022), surface-enhanced Raman spectroscopy (Moldovan et al., 2021) and fluorescence spectroscopy (Li et al., 2024). While these techniques offer sensitive and accurate detection of pesticides, they also have a number of significant disadvantages. These include high costs of the instruments required, the need for trained and qualified personnel, and lengthy and complex pre-sample preparation processes. In order to overcome these limitations, alternative methods have been developed and applied in pesticide analyses (Songa & Okonkwo, 2016; H. Zhang et al., 2024).

Electrochemical sensors are among the newest and most promising methods used in pesticide analysis. These sensors operate by measuring the electrochemical reactions of pesticides, thereby allowing for rapid, precise and cost-effective pesticide detection. Electrochemical sensors are typically advantageous in a number of ways, including their ability to achieve low detection limits, high selectivity, simplicity of use and suitability for field applications (Celik et al., 2023, 2024; Erkmen et al., 2022; Hossain & Hasnat, 2023; Huang et al., 2024).

The working principle of electrochemical sensors is based on redox reactions occurring at the electrode surface. Pesticide molecules adsorb onto the electrode surface and undergo redox reactions, generating measurable electrical signals. These signals provide direct information about

 DOI: 10.1201/9781003500322-15

the concentration of pesticides. Carbon-based nanomaterials, including carbon nanotubes (CNTs), graphene (GR), graphene oxide (GO), reduced graphene oxide (rGO), carbon nanowires (CNW), graphene nanoribbons (GNR), carbon quantum dots and graphene quantum dots (GQD), play a critical role in the development of electrochemical sensors due to their high reactivity, selectivity and excellent electrocatalytic properties. These materials offer a large specific surface area, facilitating the detection of biomolecules with high sensitivity and selectivity. Carbon-based nanocomposites are widely used as sensing electrode materials in biomedical applications for detecting various biomarkers, exhibiting high sensitivity, low detection limits and good selectivity (Annamalai et al., 2023; Chuasontia, 2023; Silah et al., 2022; Wahyudi et al., 2023).

In specific relation to the detection of pesticides in soil, the adaptation of flexible screen-printed electrodes through composite materials comprising carbon allotropes, polymers and metallic nanoparticles has facilitated the accurate and selective identification of these contaminants. These methods demonstrate good stability, repeatability and high selectivity against interfering species. These developments highlight the substantial impact of carbon-based materials in optimising the performance of electrochemical sensors for a range of vital bioanalytical and environmental monitoring applications (Abedeen et al., 2024).

The incorporation of carbon-based nanomaterials into electrochemical sensors has been demonstrated to enhance sensor performance, enabling more rapid, precise and reliable detection. The exceptional electrical conductivity and expansive surface area of GR result in the formation of a greater number of active sites on the electrode surface, thereby facilitating the adsorption of analytes with greater ease. Similarly, CNTs, which possess distinctive mechanical, electrical and thermal characteristics, enhance the overall performance of electrochemical sensors (Pengsomjit et al., 2024).

In conclusion, carbon-based electrochemical sensors present a revolutionary approach in pesticide analysis, offering rapid, sensitive and cost-effective detection, and holding great potential for both scientific research and industrial applications. The development of this technology will significantly contribute to the monitoring and management of pesticide contamination (Huang et al., 2024).

The present chapter is devoted to an examination of the utilisation of carbon-based electrochemical sensors for the purposes of pesticide detection, monitoring and control. These sensors have been developed with the objective of addressing the limitations of conventional methods, including high cost, complex device configurations and the necessity for specialised personnel. They offer the benefits of low cost, portability and sensitivity, while overcoming the shortcomings of traditional approaches. This study examines the potential of carbon nanomaterials in the construction of sensors for the detection of pesticides, as well as their suitability for environmental monitoring and agricultural applications. The superior physical and chemical properties of CNT, GR, GO and other carbon-based nanomaterials enhance the performance of these sensors, enabling fast, sensitive and reliable detection of pesticides. This review examines recent works in the detection of pesticides in real samples using electrochemical sensors, focusing on sensor architecture, electrochemical method, linearity range, sensitivity, selectivity and application areas.

15.2 PESTICIDES

Pesticides are chemical compounds used in agriculture, horticulture and other sectors to control or eradicate harmful organisms such as bacteria, viruses and pests. While pesticides have many benefits, they also pose certain risks, including environmental harm and human health concerns. Pesticides are used to protect crops and other agricultural produce from disease, and to prevent the growth of weeds. They are also used to combat mosquitoes, which transmit a number of diseases, including malaria, Japanese encephalitis and dengue fever. In addition to the aforementioned beneficial effects, pesticides also pose a number of environmental hazards, including toxicity. Pesticides,

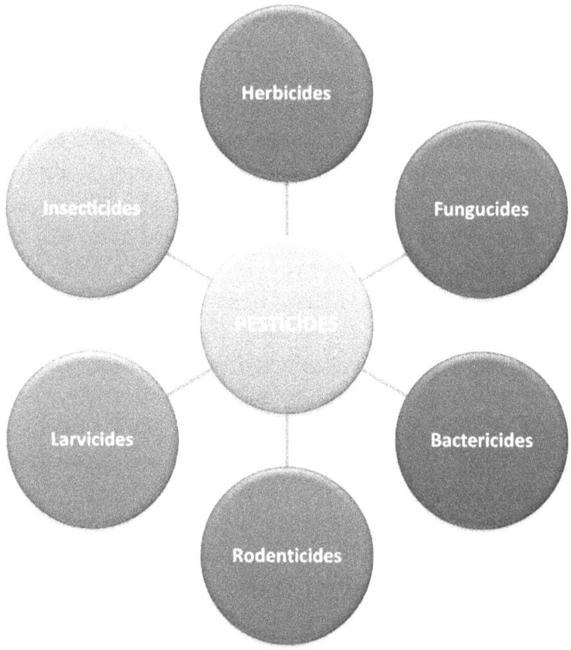

FIGURE 15.1 Classification of pesticides on the basis of use.

in particular those that are used to protect plants, act on enzymatic systems that are similar to those of humans. This can result in toxic poisoning in humans, with children being particularly vulnerable (Kadiru et al., 2022; Khan et al., 2023; Sood, 2024).

The term "pesticide" is a broad one, encompassing a diverse range of chemical agents. These include insecticides, fungicides, herbicides, rodenticides, garden chemicals, wood preservatives and household disinfectants. These agents are used for the purpose of eradicating pests. Pesticides display a range of distinctive structural, physical and chemical properties. Synthetic pesticides are classified in a number of ways. In general, three main methods of classification are employed: (1) by mode of action, (2) by the type of pest targeted and (3) by the chemical composition of the pesticides (Kaur et al., 2019).

Pesticides can be classified into various categories based on their specific applications, including herbicides, insecticides, fungicides, bactericides, and more. Figure 15.1 depicts these categories.

A further common method for classifying substances is based on their chemical structure. This is demonstrated in Figure 15.2. In this classification system, pesticides are categorised according to their chemical structure and active ingredients. This classification is of particular utility to researchers engaged in the study of pesticides and the environment. It provides insights into the efficacy, physical and chemical properties of the pesticides under consideration, as well as the precautions to be taken during application and the recommended application rates (Malhotra et al., 2023).

The first synthetic pesticides were introduced in 1940, marking the advent of a novel approach to agricultural pest control. There has been a consistent and ongoing increase in the utilisation of pesticides on a yearly basis. At present, the global usage of these chemicals has reached 3 million tons. It can be reasonably estimated that approximately 25% of global pesticide consumption can be attributed to developing countries. The most prevalent pesticides used globally are herbicides and insecticides (Bhatt et al., 2022).

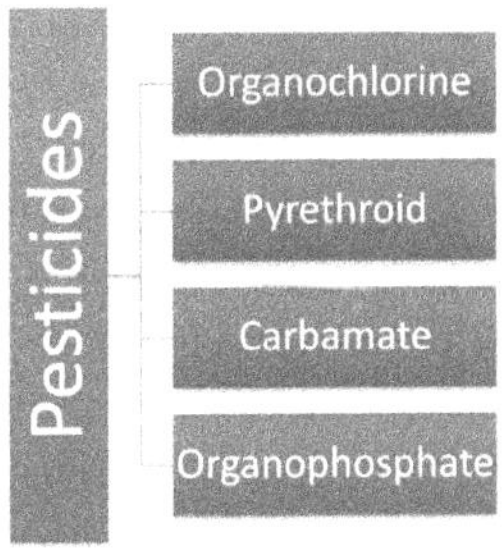

FIGURE 15.2 Classification of pesticides on the basis of chemical structure.

TABLE 15.1
Toxicity classification

Group	Level of Toxicity	LD_{50} (mg/kg by Weight of a Laboratory Mouse)	
		Oral	Dermal
I	extremely toxic	<5	<50
II	strongly toxic	5–50	50–200
III	moderately toxic	50–2000	200–2000
IV	low toxicity	<5000	-

Source: WHO (2019).

15.2.1 Effects of Pesticides on Human Health

The utilisation of pesticides on a global scale is occurring at an increasingly rapid rate, which is a cause for concern. Consequently, pesticide concentrations have reached extremely high levels in foods and ecosystems. This situation has brought about a number of negative effects on human health. A significant number of pesticide-induced poisonings are observed every year (Khan et al., 2023).

Pesticides act on target organisms in different ways through highly complex biochemical processes. Poisoning occurs in two ways: chronic and acute toxicity (Sood, 2024).

- Chronic toxicity is a toxic effect that occurs over a long period of time with repeated exposure below a lethal dose.
- Acute toxicity is the toxicity that occurs in a short time after a single dose and can be defined by its symptoms. Symptoms may vary depending on the dose and the properties of the substance entering the body. The measure of acute toxicity is LD_{50}, defined as the dose that causes 50% mortality in the population (K. Sadasivan Pilla, 2023).

As shown in Table 15.1, pesticides are classified into groups such as very toxic, toxic, and non-toxic according to their LD_{50} value. The lower the LD_{50}, the more toxic the compound.

The LD_{50} value is an important criterion for assessing the safety of a pesticide and its effects on human health (K. Sadasivan Pilla, 2023).

Pesticides can affect a large segment of the human population due to factors such as household use, proximity of agricultural fields and consumption of contaminated food and water. In this regard, they differ from other industrial chemicals (Lynch et al., 2006).

Approximately eight hundred thousand people worldwide die by suicide every year. Of these, approximately 30% of suicides are due to pesticide poisoning, with the majority occurring in rural agricultural areas in low- and middle-income countries. In low- and middle-income countries, self-poisoning with pesticides represents a significant public health concern. Those engaged in agricultural activities are becoming increasingly prone to suicide by self-poisoning with pesticides (Alex et al., 2007).

One of the most significant challenges posed by pesticides is their persistence. Unlike many other chemical substances, pesticides do not disappear over time after use. In particular, organophosphorus pesticides can persist in the soil for an extended period due to their hydrophobic structure and durability. In particular, resistant pesticides can result in the partial or complete destruction of microorganisms in the soil (Parra-Arroyo et al., 2022)

The most prevalent diseases associated with pesticide exposure are cancer, diabetes, genetic disorders and neurological diseases such as Alzheimer's and Parkinson's disease. Furthermore, they can also induce acute and transient conditions such as chronic fatigue, dizziness and nausea (Srivastava, 2019).

The impact of pesticides is contingent upon the extent of exposure. Moreover, pesticides exert effects on the human brain and nervous system. Given that the central nervous system of children does not reach full maturity until the age of 12, they are more susceptible to the effects of pesticides (Bakirhan et al., 2018).

15.2.2 Effects of Pesticides on the Environment

Pesticides are used extensively in agriculture with the dual objective of enhancing yield and protecting approximately one-third of global agricultural products. Nevertheless, the widespread use of pesticides has resulted in considerable negative impacts on the environment and ecosystems. The release of pesticides into the environment and their effects on various organisms have been extensively researched and documented (Ali et al., 2021). The indiscriminate and misguided use of organochlorine pesticides between the 1960s and 1970s resulted in a significant decline in wildlife populations worldwide (Köhler, 2013). In particular, toxic organochlorine pesticides such as dichloro-diphenyl-trichloroethane (DDT), dieldrin and similar toxic organochlorine pesticides had a severe impact on birds and other wild species at that time. Despite the eventual prohibition of these pesticides in agricultural applications, the continued decline in bird, wild bee and aquatic organism populations since that time indicates that the full impact of these pesticides on the environment has not yet been fully realised (Kennedy et al., 2013).

The utilisation of pesticides has a detrimental impact on a multitude of living organisms within the environment. Pesticides can affect not only pests but also beneficial insect species. This can result in the disruption of natural pest control mechanisms. Furthermore, pesticides can have detrimental effects on beneficial microorganisms and earthworms in the soil. Such effects have the potential to negatively impact soil fertility and plant growth. Some pesticides have the capacity to persist in the environment for extended periods of time, which can result in bioaccumulation. For example, DDT, a pesticide that was banned 30 years ago, can still be detected in soil and water samples (Mahmood et al., 2016; Rani et al., 2021). Such pesticides can have deleterious effects on a multitude of organisms by entering the food chain.

Pesticides can accumulate in the soil, where they can weaken plant root and immune systems. Furthermore, pesticides can disrupt the structure of the soil and the cycling of nutrients, which can result in a reduction in essential plant nutrients such as nitrogen and phosphorus. Pesticides can be transported to surface water and groundwater by rain and irrigation water. Such contamination can have adverse effects on aquatic ecosystems and cause pollution of water resources. Additionally, some pesticides are capable of being transported through the atmosphere over long distances. This can result in the contamination of the atmosphere and the subsequent adverse effects on ecosystems in remote areas. The extensive utilisation of pesticides may result in a decline in biodiversity. It is possible that particularly sensitive species may be more vulnerable to pesticides, with a consequent decline in their populations (Tallapragada, 2022).

Consequently, pesticides exert a multitude of far-reaching and intricate effects on the environment. While enhancing agricultural yield, it is crucial to recognise the detrimental effects of pesticides on the natural environment and biological systems. It is therefore of great importance that the utilisation of pesticides should be meticulously supervised and that their identification and determination should be conducted with utmost care.

15.3 CARBON SENSORS FOR PESTICIDE DETERMINATION AND THEIR APPLICATIONS

The rising prevalence of pesticide use on an annual basis signifies a considerable threat to environmental sustainability and human health. This situation thus requires the monitoring, detection, and determination of pesticides in nature. The disadvantages of classical instrumental methods used for pesticide determination, such as high device cost, large device configurations, and the need for qualified personnel, limit the use of these devices. The use of electrochemical sensors represents a more effective alternative to overcome the aforementioned disadvantages. Carbon nanomaterials are widely employed in the construction of sensors due to their distinctive physical and chemical properties. Carbon-based nanomaterials, including CNT, GR, GO, rGO, CNW, GNR, carbon and GQD, have been successfully employed in the determination of pesticides, including herbicides, insecticides and fungicides. These sensors play a pivotal role in the detection and monitoring of pesticides due to their low cost, portability, and sensitivity (Bolat et al., 2023; Hu et al., 2023; Pratap et al., 2024; Rajapaksha et al., 2021; Wahyudi et al., 2023).

15.3.1 Carbon–Graphene Nanotube-Based Sensors

CNTs are highly suitable materials for various applications, including pesticide detection, due to their excellent structural, optical and electrical properties. As a highly conductive material, CNTs significantly improve signal transmission in electronic devices and sensors. Furthermore, CNTs are added to the electrode surface to increase the surface area, which allows electrochemical reactions to take place more efficiently and improves detection sensitivity. Their excellent structural properties offer mechanical stability, due to their strong and durable structure, ensuring that the sensors are long-lasting and reliable. Because of all these properties, CNT-based sensors have become highly effective tools for the detection and analysis of pesticides. Increasing the roughness of the electrode surface allows the formation of more reactive sites, which increases the detection capacity and accuracy (Hareesha & Manjunatha, 2021; Hu et al., 2023; Meskher et al., 2023; Tigari & Manjunatha, 2020).

Zhenzhen Yang (2024) and colleagues have developed an innovative electrochemical sensor for the detection of carbosulfan, which is used as an insecticide. The sensor was modified with the addition of silver nanoparticles (AgNPs) and carboxylated multi-walled carbon nanotubes (c-MWCNTs) on a glassy carbon electrode (GCE). This modification was undertaken with the objective of increasing the electrochemical active surface area of the sensor and enhancing the analytical signal. The electrode surface was characterised structurally and morphologically using scanning electron microscopy (SEM) and X-ray diffraction (XRD) methods. The SEM analysis confirmed the presence of AgNPs and c-MWCNTs in a homogeneous distribution on the electrode surface and revealed the surface roughness. XRD analysis demonstrated the crystalline structure of the AgNPs and confirmed the successful completion of the modification process. The electrochemical characterisation was conducted using cyclic voltammetry (CV), which revealed that the modification process markedly enhanced the electrochemical activity. As part of the process of optimising the experimental parameters, the ambient pH, the quantity of AgNPs and c-MWCNTs were optimised. Optimisation studies enabled the determination of the optimum pH and material concentrations, thus optimising the sensor performance. Subsequently, the sensor's performance was assessed using differential pulse voltammetry (DPV) to determine the optimal conditions for analysis. The results obtained showed that the sensor can operate with high accuracy over a wide linear range from 1 to 726 μM and has a low detection limit of 0.009 μM. In tests on real samples such as tomato, the sensor achieved a high recovery rate of 97.00–99.33%, proving that the sensor is suitable for practical applications. Moreover, the sensor demonstrated resilience to potential interfering substances, maintaining its performance. This study demonstrates the successful design and fabrication of an electrochemical sensor with high selectivity, stability and sensitivity for the determination of carbosuflan pesticide. The developed sensor is promising for use in agricultural fields and environmental monitoring applications (Zhenzhen Yang, 2024).

The study by Han (2024) and colleagues details the development of an electrochemical sensor based on a composite of single-walled carbon nanotubes (SWCNTs) and cerium oxide (CeO_2) nanoparticles. This sensor is designed for the ultrasensitive detection of the pesticide methyl parathion (MP). The unique one-dimensional structure and excellent conductivity of SWCNTs enhance charge diffusion, while CeO_2 nanoparticles, known for their biocompatibility and high catalytic capacity, aid in the adsorption of MP molecules due to their affinity for phosphate groups. This combination effectively compensates for the inherent electrical conductivity limitations of CeO_2 nanoparticles. The GCE was modified with a suspension of SWCNT and CeO_2 nanoparticles, creating the SWCNT@CeO_2/GCE sensor. The preparation involved dispersing CeO_2 nanoparticles in a dimethylformamide solution, followed by the addition of SWCNTs and ultrasonic-assisted dispersion. The modified electrode was then dried to complete the fabrication. The fabrication of the sensor is summarised in Figure 15.3.

Characterisation techniques such as SEM, energy-dispersive X-ray spectroscopy, X-ray photoelectron spectroscopy, and XRD confirmed the successful creation of the SWCNT@CeO_2 nanocomposite. The electrochemical properties were evaluated using CV and DPV. A wide linear

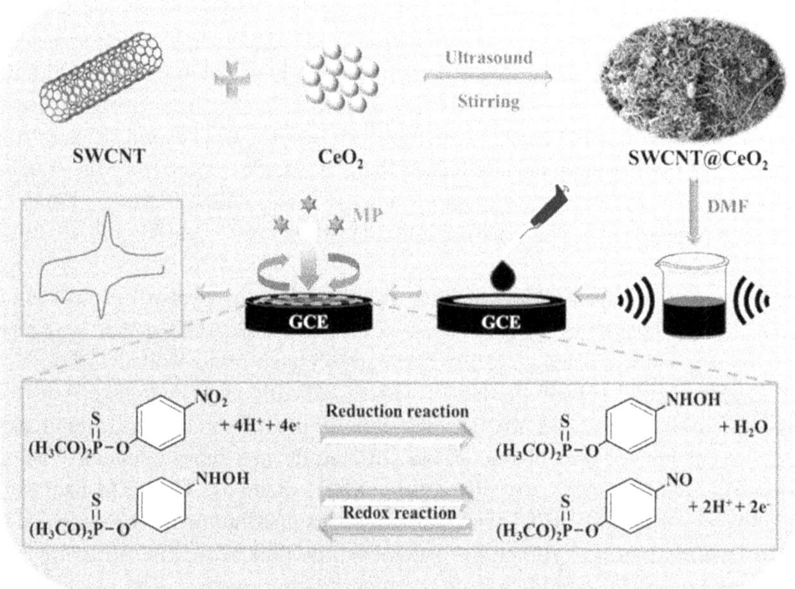

FIGURE 15.3 Schematic representation of the design of the SWCNT@CeO2/GCE sensor.(Adapted with permission from Han et al., 2024), Copyright 2024 Elsevier.

range between 0.01 and 10 μM was detected. The sensor exhibited a high sensitivity for MP detection with a limit of detection (LOD) of 9.12 nM. It also demonstrated excellent repeatability, reproducibility and anti-interference capabilities. For practical applications, the sensor was tested with real food samples (peach and cabbage juice), achieving satisfactory recovery rates of 87.68–107.31% and low relative standard deviations (RSDs) of 1.09–2.65%, indicating its potential for effective pesticide residue analysis in food safety monitoring. In summary, the SWCNT@CeO_2/GCE sensor represents a promising tool for the selective, stable and sensitive detection of MP, offering significant advantages for environmental and food safety applications (Han et al., 2024). Table 15.2 summarises the carbon–graphene nanotube-based electrochemical sensing platforms developed for the determination of pesticides.

15.3.2 Graphene-Based Sensors

The role of GR-based electrochemical sensors in the determination of pesticides is becoming increasingly significant due to their high sensitivity, portability and rapid analytical capabilities. These sensors are designed using advanced materials, including GR, GO, rGO and graphdiyne composites, with the objective of accurately assessing the effects of pesticides. The incorporation of these composite materials enhances the electrocatalytic activity and surface area of electrochemical sensors, thereby markedly improving the sensitivity and accuracy of the detection process (Gopal et al., 2023; Pratap et al., 2024). For instance, electrodes modified with nitrogen-doped graphene (N-Gr) and chitosan (CS) demonstrated superior performance in the detection of MP residues in Chinese herbal medicine. The sensor was capable of accurately detecting the presence of MP residues due to its low detection limit and high sensitivity. Such modifications enhance the utility of GR-based sensors in pesticide detection, facilitating crucial applications in agriculture and food safety (Sun et al., 2024). The development of GR-based electrochemical sensors not only enables sensitive and accurate detection of pesticides, but also offers broad application potential in areas such as in situ analysis, environmental monitoring and food safety. These sensors offer effective solutions in critical areas such as monitoring and control of environmental hazards (Ramirez et al., 2024). In conclusion, electrochemical sensors developed with the use of GR and its derivatives occupy a significant position in the scientific and industrial fields, serving as innovative and powerful tools for the detection of pesticides. Further development and dissemination of these sensors will contribute to the creation of a safer and more sustainable environment and food system in the future.

Wang et al. (2023) developed a portable, three-electrode electrochemical sensor for the detection of carbendazim (CBZ) residues in water. CBZ is a widely used fungicide that presents considerable environmental and health risks due to its persistence and stability. The sensor employs laser-induced graphene (LIG), which is formed by exposing the polyimide film to a laser, resulting in the formation of a porous graphene structure with enhanced electrocatalytic activity and a substantial surface area. To improve the sensor's sensitivity, platinum nanoparticles (PtNPs) were electrodeposited onto the LIG surface, as demonstrated in Figure 15.4.

The performance of the optimised LIG/Pt sensor was evaluated by square wave voltammetry (SWV), which demonstrated a good linear relationship with CBZ concentrations in the range of 1–40 μM (Figure 15.5).

The sensor demonstrated a low detection limit of 0.67 μM. Moreover, the sensor demonstrated recovery rates spanning from 97.7% to 104.7% for the detection of CBZ in wastewater samples. This high recovery rate serves to reinforce the reliability and accuracy of the sensor, thereby providing a rapid, sensitive and efficient method for the real-time analysis of CBZ. This approach has significant potential for environmental monitoring and food safety due to its simplicity, portability and on-site analysis capability, which are advantageous in a variety of contexts (L. Wang et al., 2023) (Table 15.3).

TABLE 15.2
Selected carbon–graphene nanotube-based electrochemical sensor studies for the determination of pesticides in real samples

Pesticides	Type of Pesticides	Technique	Design of sensor	Linear range	LOD	Recovery %	Applications	Referance
chlorantraniliprole	insecticide	DPV	E-FT@CNT/GCE	0.01–7.00 μM	8.1 nM	97.21–104.64	soil and foods	(Tümay et al., 2021)
carbendazim	fungicide	DPV	ZSHPC/MWCNT/SPE	0.4–20.6 μM	0.06 μM	98.77–109.32	tea and rice	(X. Wang et al., 2023)
diuron	herbicide	DPV	MWCNTs-CS@NGQDs/GCE	0.08-12 μg/mL	0.04μg/mL	90–104	river water and soil	(Zhu et al., 2023)
parathion	insecticide	SWSV	MWCNTPE	3.74–25.8 μM	0.73 μM	103.7–105.9	apple juice, tap water and grape juice	(Murathan et al., 2024)
methyl parathion	insecticide	DPV	β-CD@SCN/GCE	0.01–20 μM	7.23 nM	96.54–98.18	grape juice	(Guo et al., 2023)
carbendazim	fungicide	DPV	MIP/MWCNT	0–20 ppM	1 ppm	94–101	river water	(Júnior et al., 2024)
4-cyanophenol	herbicide	CA	SPCE/DMPPy/MWCNT	0.001–400 μM	0.8 nM	99.35–99,99	land soil, irrigation water, land water	(Rajkumar & Kim, 2023)
methyl parathion	insecticide	DPV	SCN@UIO-66/GCE	0.01–10 μM	8.05 nM	98.28–101.95	orange juice and lettuce samples	(Han et al., 2024)
dichlorvos	insecticide	DPV	Au@MWCNTs/GCE	1–120 μM	5 nM	90.75–97.50	lettuce	(Chen et al., 2021)
paraoxon	insecticide	DPV	MWCNTs/SiC/AgNP@SPE	0.001–0.1 μg/mL	0.0038 μg/mL	-	-	(Chuasontia, 2023)
chlorpyrifos	insecticide	DPASV	Co,N-HPNC@CNT/GCE	0.1 pmol/L–10.0 μmol/L	0.03 pmol/L	96.0–102.3	strawberry, apple, grapes	(J. Liu et al., 2024)

Note: Abbreviations: E-FT, ferrocene-thiophene; CNT, carbon nanotube; ZSHPC, nano-ZnS-decorated hierarchically porous carbon; MWCNT, multi-walled carbon nanotubes; NGQDs, nitrogen-doped graphene quantum dots; CS, chitosan; SWSV, square wave stripping voltammetry; MWCNTPE, multi-walled carbon nanotube paste electrode; β-CD, β-cyclodextrin; SCN, single-wall carbon nanotubes; MIP, molecularly imprinted polymer; UIO-66, Zr-based metal–organic framework nanoparticles; CA, Chronoamperometry; DMPPy, dual-microporous polypyrrole nanoparticle; Co,N-HPNC, Cobalt,N-doped hollow porous carbon nanocage; Differential pulse anodic stripping voltammetry.

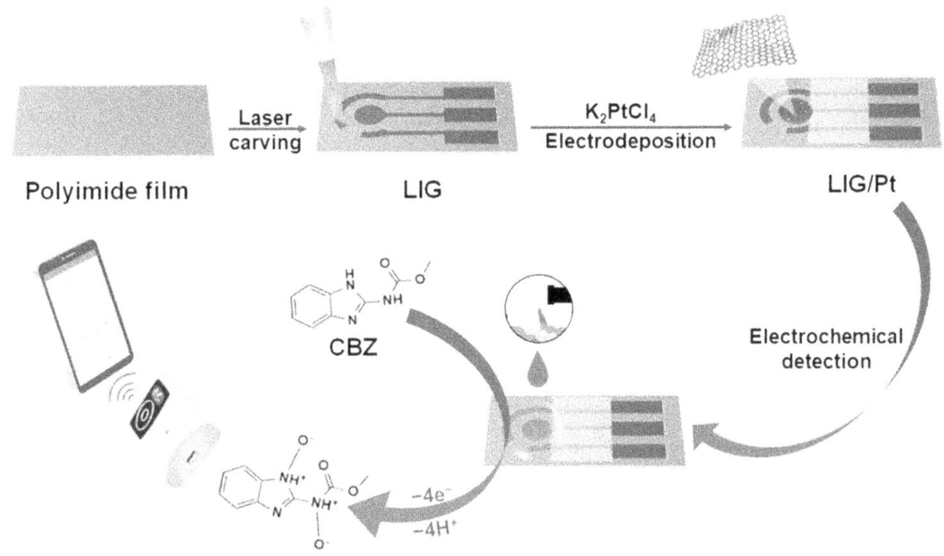

FIGURE 15.4 The following schematic representation depicts the design of the LIG/Pt sensor. (Adapted with permission from L. Wang et al., 2023.)

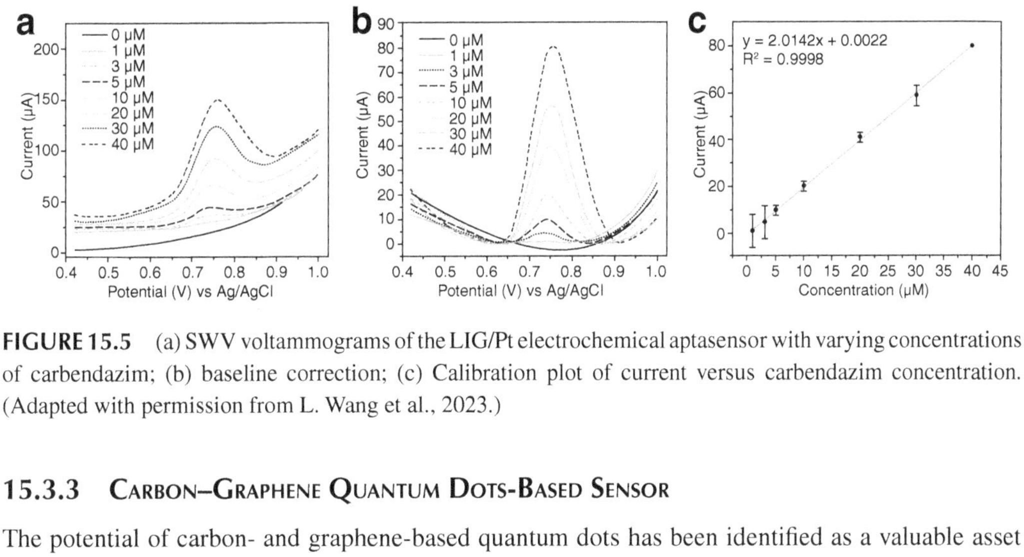

FIGURE 15.5 (a) SWV voltammograms of the LIG/Pt electrochemical aptasensor with varying concentrations of carbendazim; (b) baseline correction; (c) Calibration plot of current versus carbendazim concentration. (Adapted with permission from L. Wang et al., 2023.)

15.3.3 Carbon–Graphene Quantum Dots-Based Sensor

The potential of carbon- and graphene-based quantum dots has been identified as a valuable asset in the advancement of electrochemical sensors for the determination of pesticides. This is due to their demonstrated biocompatibility, photostability and low toxicity. These quantum dots are typically formed by arranging carbon atoms at the nanometre scale. These nanometre-scale materials are composed of carbon structures that range in size from 1 to 10 nm and exhibit distinctive optical and electronic properties, which can be attributed to the quantum confinement effect. The small size, high surface area and abundance of reactive groups on the surface of quantum dots render them ideal for use in sensor applications. The electronic properties of these dots facilitate the rapid and precise transmission of electrical signals, thereby conferring upon them a pivotal role in the field of sensor technology (Facure et al., 2021; Tanwar et al., 2023; Uslu, 2021).

TABLE 15.3
Graphene-based electrochemical sensor studies for the determination of pesticides in real samples

Pesticides	Type of Pesticides	Technique	Design of sensor	Linear range	LOD	Recovery %	Applications	Referance
carbofuran	insecticide	DPV	AuNPs/rGO/GCE	1.0×10^{-9}–1.0×10^{-5} mol/L	3.3×10^{-10}	95.20–105.45	tangerine, potato, cowpea, cornmeal	(Tian et al., 2022)
pirimicarb	insecticide	SWV	AuNPs@GO/IL@SPCE	50–1500 µM	4.49 µM	97.38–103.36	groundwater and surface water	(Weheabby et al., 2024)
thiram	fungicide	DPV	GO/WS_2@SPCE	0.083–0.33 µM	0.02 µM	-	-	(Abid et al., 2024)
chlorpyrifos	insecticide	LSV	BSA/AChE/GA/CIS/rGO-9/SPCE	0.5–470 ng/mL	0.023 ng/mL	96–106	lettuce	(Itsoponpan et al., 2021)
fenitrothion	insecticide	SWV	PGO/SPCE	0.02–250 µM	0.061 µM	97.8–99.9	orange, grapes, tap water	(Balasubramanian et al., 2024)
methyl parathion	insecticide	DPV	AChE/ZnO-rGO/GCE	0.01–1000 ng/mL	0.00463 ng/mL	90.94–108.4	apple-cucumber	(Y. Liu et al., 2024)
paraquat	herbicide	SWV	SPCE/GO/AuNPs/P3ABA	0.001–100 µM	0.45 nM	96.5–116.1	tapped water, natural water	(Pimalai et al., 2025)
glyphosate	herbicide	DPV	CGE/CG/CuONP	5.00–25.00 $\times 10^{-12}$ M	0.30 $\times 10^{-12}$ M	96.2–100.0	water samples	(Gonçalves et al., 2023)
Maleic hydrazide	herbicide	DPV	ZnS:SnS@rGO/GCE	0.1 – 2359 µM	0.02 µM	94.2–99.6	potato, water	(Disouza et al., 2023)
4-nitrophenol	herbicide	DPV	Ni@CuO/rGO/PtE	0.09–105 µM	0.0054 µM	97.0 –101.0	river and tap water	(Buledi et al., 2024)

Note: Abbreviations: rGO, reduced graphene oxide; IL, Ionic liquid; AChE, acetylcholinesterase; WS_2 tungsten disulphide BSA/AChE/GA/CIS/rGO-9/SPCE, bovine serum albumin/acetylcholinesterase/ glutaraldehyde/$CuInS_2$/reduced graphene oxide/screen-printed carbon electrode; LSV, linear sweep voltammetry; PGO, reduced porous graphene oxide; P3ABA, poly(3-aminobenzoic acid); CG, Crumpled graphene; PtE, platinum electrode.

In contrast, GR-based quantum dots offer enhanced electrical conductivity and mechanical strength, attributes derived from the two-dimensional structure of GR. Graphene is a one-atom thick structure comprising carbon atoms arranged in a honeycomb configuration. It is well known for its exceptional electrical conductivity, flexibility and chemical resistance. In addition to these characteristics, GR-based quantum dots also display optical properties associated with the quantum confinement effect. This combination renders GR-based quantum dots an attractive proposition for a plethora of applications, including the detection of pesticides and a vast array of biomedical and electronic applications (Abdelsalam & Zhang, 2022; Khosropour et al., 2023).

Electrochemical sensors are typically low-cost, portable devices that facilitate high-precision measurements. The sensors were developed using carbon and GR-based quantum dots, which enable them to reach the low detection limits required for the detection of pesticides. To illustrate, these sensors can reliably detect low concentrations of pesticide residues in water and food samples. Electrochemical sensors operate by measuring the redox reactions occurring between the analyte and the electrode. The integration of carbon and GR-based quantum dots on the surface of the electrodes enhances the sensitivity and selectivity of the sensors. This enables the detection of extremely low concentrations of pesticides (Rasheed, 2023).

The incorporation of carbon quantum dots with nanomaterials and molecularly imprinted polymers serves to augment the functionality of the sensor, offering a dependable and efficacious approach for the determination of pesticides in diverse matrices. The incorporation of nanomaterials into the sensor design enhances its sensitivity, as these materials possess a large surface area and high reactivity. Molecularly imprinted polymers, on the other hand, increase the specificity of the sensor by selectively recognising target molecules. Nanomaterials provide a greater number of active sites due to their elevated surface area, which optimises the interaction between the analyte and the sensor surface. Molecularly imprinted polymers comprise cavities that are shaped and sized to align with the characteristics of the target molecule. This enables the selective recognition of the target molecule. These characteristics serve to further emphasise the significance of such sensors in the context of environmental monitoring and food safety applications. Consequently, electrochemical sensors supported by carbon and GR-based quantum dots represent a high-performance, reliable and effective solution for pesticide detection. These sensors play a pivotal role in the management of environmental and health risks, thereby contributing to the development of a safer and more sustainable future (Facure et al., 2021; Z. Wang et al., 2022).

In one of the many studies based on GQDs for pesticide determination, Tanwar et al. (2023) developed an electrochemical sensor based on a modified GCE utilising GQDs. The structure and surface characterisation of the electrode were investigated using SEM and EDS, which revealed that the GQDs were homogeneously distributed on the GCE surface, resulting in a highly modified surface. Electrochemical characterisation was conducted using CV and electrochemical impedance spectroscopy (EIS), which revealed that the GQDs/GCE electrode exhibited high conductivity and low resistivity. The analytical performance of the sensor was evaluated by DPV, which demonstrated the ability to detect malathion in a linear range from 0.05 μM to 70 μM. The LOD of the sensor was determined to be 0.017 μM. Recovery and selectivity studies evaluated the sensor's capacity to detect malathion in diverse matrices. The recovery rate ranged from 95.3% to 104.2%, indicating that the sensor operates with high accuracy and reliability. The results demonstrate that the GQDs/GCE-based electrochemical sensor exhibits high performance in malathion detection, thereby establishing itself as an effective tool for monitoring pesticide residues in agricultural settings (Tanwar et al., 2023).

Adeyemi Oladipo et al. (2024) developed a carbon quantum dot-based electrochemical sensor for glyphosate, a chemical compound used as an agricultural herbicide. A summary of the work is provided in Figure 15.6. The electrode design entails the modification of Cu^0-doped graphitic carbon quantum dots (Cu^0-OGQ) on the surface of a GCE. The modification of the electrode surface with

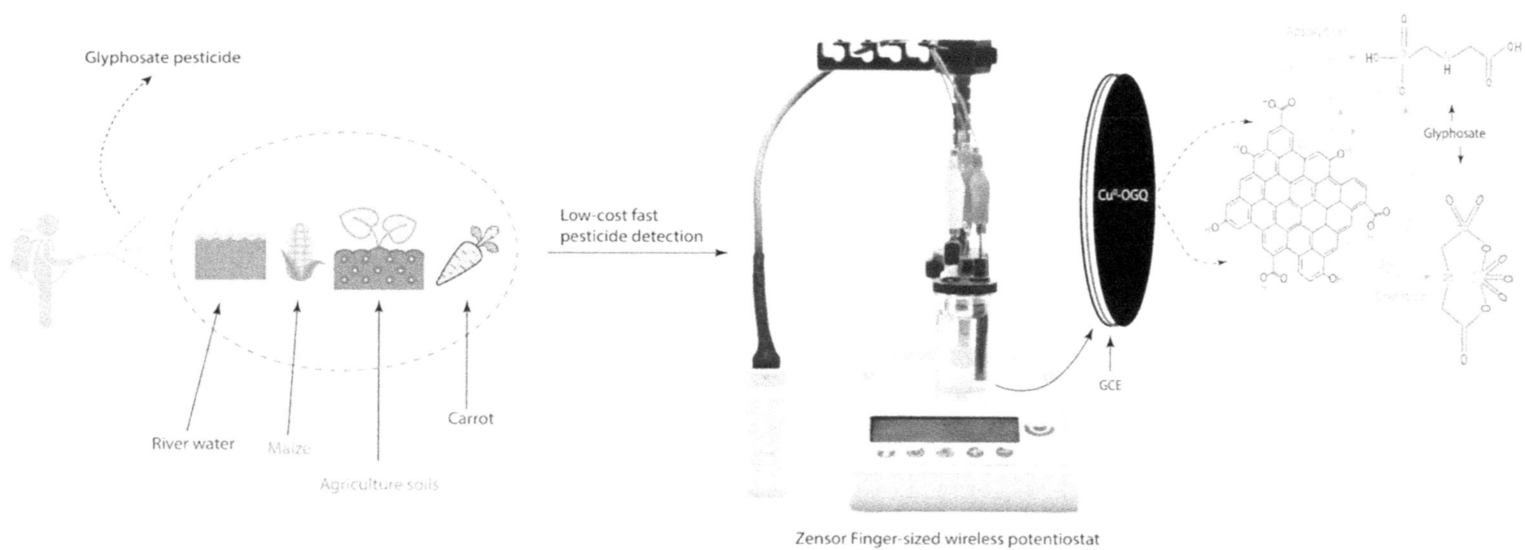

FIGURE 15.6 Schematic representation of the preparation of Cu^0-OGQ/GCE sensor. (Adapted with permission from Adeyemi Oladipo, 2024. Copyright 2024, Elsevier.)

Cu^0-OGQ led to an enhancement in the electrochemical active area, thereby improving the sensor performance.

Surface characterization was performed using SEM, transmission electron microscopy and Brunauer-Emmett-Teller method (BET). The average particle size of Cu^0-OGQ was determined to be 6.56 ± 0.38 nm, and the specific surface area was found to be 757.6 m²/g. As determined by BET analysis, Cu^0-OGQ exhibits micropore and mesopore structures. Electrochemical characterisation was conducted using CV and EIS. The Cu^0-OGQ/GCE electrode exhibited pronounced redox reactions in a $[Fe(CN)_6]^{3-/4-}$ solution, thereby enhancing its electrochemical performance. The incorporation of Cu^0 resulted in an improvement in sensor performance, largely due to an enhancement in the electrochemical active site and electron transfer. The sensor performance was evaluated in terms of sensitivity, selectivity and stability. The sensitivity of the Cu^0-OGQ electrochemical sensor was found to be 179.7 μA ng/mL, exhibiting a broad linear range and low detection limit for the quantification of glyphosate. The recovery rates ranged from 90% to 102%, while the RSD values were between 0.2% and 2.5%. The sensor demonstrated high stability over a period of 30 days. The Cu^0-OGQ/GCE sensor demonstrated the ability to selectively detect glyphosate without interference from other potential contaminants. The sensor performance was not affected by major metabolites of glyphosate, such as AMPA. However, an antagonistic effect was observed with some pesticides (Adeyemi Oladipo, 2024).

15.3.4 Diamond-Based Sensor

Nanodiamonds possess outstanding qualities that surpass those of other carbon-based nanomaterials, such as CNTs, GR and graphite, commonly used in electrochemical sensors. They offer several benefits, including extremely high resistivity, a broad potential window, low background current, excellent chemical and electrochemical stability, biocompatibility, low toxicity and superior corrosion resistance. These attributes enhance the sensor's ability to provide precise and durable measurements, unaffected by parasitic currents (Nunn et al., 2017; Y. Zhang et al., 2018).

The combination of nanodiamonds with metal nanoparticles or polymers results in a further enhancement of sensor performance, particularly in critical parameters such as selectivity and sensitivity. Consequently, nanodiamond-based sensors are extensively employed in a multitude of fields, including biomedical, environmental, food safety and industrial applications. For example, they are employed in biomedical sensors for the detection of glucose, biological molecules and drugs, in environmental sensors for the detection of gases, heavy metals and pollutants, in food safety sensors for the detection of toxins and harmful substances in food, and in industrial sensors for the monitoring and control of chemical processes (Freitas et al., 2022; Lee et al., 2014; Shahrokhian & Khafaji, 2010; Su et al., 2020).

The analysis of pesticides represents a significant application area for nanodiamond-based electrochemical sensors. The detection of pesticides is of paramount importance for the reduction of the detrimental environmental and health impacts associated with these chemicals, which are extensively utilised in agricultural practices. Nanodiamond sensors are capable of detecting the electrochemical oxidation and reduction reactions of pesticides with high sensitivity. These sensors can rapidly, accurately and reliably ascertain the presence and concentration of pesticides, thereby making substantial contributions to both environmental monitoring and food safety (Elugoke et al., 2021; Şener et al., 2024).

The incorporation of nanodiamonds into electrochemical sensors markedly enhances their performance, which is attributed to the superior properties of these nanoparticles. The sensitivity, selectivity and longevity of these sensors are all improved by the addition of nanodiamonds. The utilisation of nanodiamond-based sensors in critical applications, such as pesticide analysis, facilitates the implementation of safer and more effective monitoring systems. The combination of nanodiamonds with other materials and the development of new manufacturing techniques will facilitate the utilisation of these sensors in a broader range of applications (Silva et al., 2023).

Zambianco et al. (2019) developed a novel electrochemical sensor based on nanodiamonds and manioc starch for the detection of a herbicide diquat (DQ) in environmental samples. The sensor was prepared by modifying the surface of a GCE with a dispersion of manioc starch and nanodiamonds. The surface characterisation of the sensor was conducted using Fourier transform infrared (FTIR) and XRD techniques, which facilitated the elucidation of the chemical functional groups and the crystal structures of the materials. The electrochemical characterisation entailed an investigation of the electrochemical response of DQ utilising CV and SWV. SWV optimisation was conducted with regard to the composition and pH of the supporting electrolyte, the parameters of the SWV (frequency, pulse amplitude and step potential) and the pre-concentration times. Optimal conditions were identified as a 0.10 mol L^{-1} acetate buffer solution (pH 5.0), a frequency of 100 Hz, a pulse amplitude of 90 mV, a step potential of 8 mV and a pre-concentration time of 60 seconds. Analytical performance was evaluated by conducting SWV measurements at varying concentrations of DQ. The linear range was determined to be between 5.0×10^{-7} mol L^{-1} and 4.6×10^{-5} mol L^{-1}, with an LOD of 1.1×10^{-7} mol L^{-1}. The accuracy and recovery of the sensor were evaluated in natural water samples, with recovery percentages ranging from 96.0% to 103.0%. To evaluate the potential impact of interfering substances, interference studies were conducted by introducing different concentrations of these substances to the DQ solution. These studies demonstrated that the sensor is selective and that the presence of other compounds has a negligible effect on the detection of DQ (Zambianco et al., 2019).

15.4 CONCLUSIONS AND FUTURE PERSPECTIVES

As outlined in this chapter, the emphasis is on the advancement and utilisation of carbon-based electrochemical sensors, specifically engineered for the detection of pesticides. The distinctive attributes of diverse carbon materials, including nanodiamonds, graphite, CNTs and GR, have been assessed, establishing their exceptional suitability for sensor applications. These properties include a high surface area, which enables greater interaction between the sensor and the analyte; excellent electrical conductivity, which facilitates efficient electron transfer during the detection process; and superior chemical stability, which ensures the durability and reliability of the sensor under different environmental conditions. Carbon-based sensors are employed for the detection of a range of pesticides, including organophosphates, carbamates and neonicotinoids, which are extensively utilised in agricultural contexts. The capacity of sensors to identify these pesticides at low concentrations is of significant consequence for the assurance of food safety and the protection of the environment. The study demonstrated that these sensors can reliably detect trace amounts of pesticides in a variety of samples, including water, soil and agricultural products. This research highlights the potential of carbon-based electrochemical sensors in environmental monitoring and food safety applications. It would be beneficial for future research to focus on integrating these sensors into portable devices for on-site testing, thereby enhancing their practicality and accessibility. It is recommended that further investigation should be conducted into additional nanomaterials and polymers with a view to increasing sensitivity and selectivity. An expansion of the scope of application of these sensors to encompass the detection of a more diverse range of pesticides and pollutants will enhance their utility in comprehensive environmental monitoring. Moreover, the integration of wireless data transfer technologies can facilitate real-time monitoring and analysis, thereby enabling the implementation of proactive measures in the fields of environmental protection and public health. Continued innovation in carbon-based sensor technology promises significant advances in both sensitivity and application diversity.

REFERENCES

Abdelsalam, H., & Zhang, Q. F. (2022). Properties and applications of quantum dots derived from two-dimensional materials. *Advances in Physics: X, 7*(1). https://doi.org/10.1080/23746149.2022.2048966

Abedeen, M. Z., Sharma, M., Singh Kushwaha, H., & Gupta, R. (2024). Sensitive enzyme-free electrochemical sensors for the detection of pesticide residues in food and water. *TrAC Trends in Analytical Chemistry, 176*, 117729. https://doi.org/10.1016/J.TRAC.2024.117729

Abid, K., Iannazzo, D., Celesti, C., Khaskhoussi, A., Foti, A., Maalej, R., Gucciardi, P. G., & Neri, G. (2024). A novel 2D-GO@WS2 electrochemical platform for the determination of thiram fungicide. *Journal of Environmental Sciences, 136*, 226–236. https://doi.org/10.1016/J.JES.2022.11.018

Adeyemi Oladipo, A. (2024). Cu0-doped graphitic carbon quantum dots for rapid electrochemical sensing of glyphosate herbicide in environmental samples. *Microchemical Journal, 200*, 110294. https://doi.org/10.1016/J.MICROC.2024.110294

Afify, A. S., Abdallah, M., Ismail, S. A., Ataalla, M., Abourehab, M. A. S., Al-Rashood, S. T., & Ali, M. A. (2022). Development of GC–MS/MS method for environmental monitoring of 49 pesticide residues in food commodities in Al-Rass, Al-Qassim region, Saudi Arabia. *Arabian Journal of Chemistry, 15*(11), 104199. https://doi.org/10.1016/J.ARABJC.2022.104199

Alex, R., Prasad, J., Kuruvilla, A., & Jacob, K. S. (2007). Self-poisoning with pesticides in India [12]. *British Journal of Psychiatry, 190*, 274–275. https://doi.org/10.1192/bjp.190.3.274a

Ali, S., Ullah, M. I., Sajjad, A., Shakeel, Q., & Hussain, A. (2021). *Environmental and Health Effects of Pesticide Residues. In* Inamuddin, Mohd Imran Ahamed, & Eric Lichtfouse (Eds.), *Sustainable Agriculture Reviews 48: Pesticide Occurrence, Analysis and Remediation Vol. 2 Analysis* (pp. 311–336). Springer. https://doi.org/10.1007/978-3-030-54719-6_8

Annamalai, P., Ramalingam, M., Eswaran, M., Chokkiah, B., Saravanan, P., Chang, S.-C., Valdes, H., & Mangalaraja, R. V. (2023). Recent advancements in the use of nanomaterials in electrochemical sensors for the detection of pesticides and organic pollutants. In *Materials Technology for the Energy and Environmental Nexus* (Vol. 2, pp. 10-1–10-31). IOP Publishing. https://doi.org/10.1088/978-0-7503-5729-6ch10

Bakirhan, N. K., Uslu, B., & Ozkan, S. A. (2018). The detection of pesticide in foods using electrochemical sensors. *Food Safety and Preservation*, 91–141. https://doi.org/10.1016/B978-0-12-814956-0.00005-6

Balasubramanian, K., Karuppiah, C., Alagarsamy, S., Mohandoss, S., Arunachalam, P., Govindasamy, C., Velmurugan, M., Yang, C. C., Lee, H. J., & Ramaraj, S. K. (2024). Highly sensitive detection of environmental toxic fenitrothion in fruits and water using a porous graphene oxide nanosheets based disposable sensor. *Environmental Research, 259*, 119500. https://doi.org/10.1016/J.ENVRES.2024.119500

Barut, B. B., Erkmen, C., İpek, S., Yıldırım, S., Üstündağ, A., & Uslu, B. (2023). Analytical studies on some pesticides with antifungal effects: Simultaneous determination by HPLC, investigation of interactions with DNA and DNA damages. *Journal of Chromatography B, 1229*, 123862. https://doi.org/10.1016/J.JCHROMB.2023.123862

Bhambri, A., Singh, P., Karn, S. K., & Dey, A. (2023). Agriculture pesticide and their remediation. In *Development in Wastewater Treatment Research and Processes*. Advances in Industrial Wastewater Treatment Technologies: Removal of Contaminants and Recovery of Resources, pp. 145–162. https://doi.org/10.1016/B978-0-323-95684-0.00010-5

Bhatt, P., Chandra Suyal, D., Bhandari, G., Mohan Pathak, V., Verma, V. K., Singh Rawat, B., Kaur, B., Babu, N., Sharma, A., Dewali, S., Yadav, M., Kumari, R., Singh, S., Mohapatra, A., Pandey, V., Rana, N., & Maria Cunill, J. (2022). Current status of pesticide effects on environment, human health and it's eco-friendly management as bioremediation: A comprehensive review. *Frontiers in Microbiology*. https://doi.org/10.3389/fmicb.2022.962619

Bolat, G., Yaman, Y. T., Abacı, S., & Seyyar, S. (2023). Poly-arginine/graphene oxide functionalized disposable sensor for monitoring fenitrothion pesticide residues in water and cucumber samples. *Materials Today Chemistry, 30*, 101517. https://doi.org/10.1016/J.MTCHEM.2023.101517

Buledi, J. A., Solangi, A. R., Mallah, A., Ameen, S., & López-Maldonado, E. A. (2024). Development of sensitive electrochemical sensor Ni@CuO/rGO/PtE for the determination of 4-Nitrophenol: An electroanalytical approach for pollution control. *Diamond and Related Materials, 142*, 110801. https://doi.org/10.1016/J.DIAMOND.2024.110801

Celik, M., Bozal-Palabiyik, B., Kanbes Dindar, C., & Uslu, B. (2024). Voltammetric determination of metrafenone using an iridium oxide nanoparticle–ionic liquid nanocomposite modified glassy carbon electrode. *Analytical Letters*. https://doi.org/10.1080/00032719.2024.2318653

Celik, M., Kanbes-Dindar, C., Karayel, A., Bozal-Palabiyik, B., & Uslu, B. (2023). Voltammetric analysis of thiram with bimetallic nanosensor and investigation of adsorption mechanism by DFT-D3 method. *Journal of the Electrochemical Society*, *170*(12), 127510. https://doi.org/10.1149/1945-7111/ad129a

Chen, Y., He, K., Sun, F., Wei, D., & Li, H. (2021). Efficient electrochemical sensor based on multi-walled carbon nanotubes/gold nanocomposite for detection of dichlorvos pesticide in agricultural food. *International Journal of Electrochemical Science*, *16*(3), 210321. https://doi.org/10.20964/2021.03.53

Chuasontia, I., Sirisom, W., Nakpathomkun, N., Toommee, S., Pechyen, C., Tangnorawich, B., & Parcharoen, Y. (2023). Development and characterization of nano-ink from silicon carbide/multi-walled carbon nanotubes/synthesized silver nanoparticles for non-enzymatic paraoxon residuals detection. *Micromachines*, *14*(8), 1613. https://doi.org/10.3390/mi14081613

Clesceri, L. S., & Clesceri, E. J. (2021). *Environmental Control of Pests and Vectors* (pp. 387–410). Springer. https://doi.org/10.1007/978-3-030-54626-7_10

Disouza, F. P. D., Ruspika, S., Chen, S. M., Balaji, R., Meena Devi, J., Peng, J. Y., & Irudaya Jothi, A. (2023). Synergetic effect of ZnS: SnS_2/reduced graphene oxide heterostructures for electrochemical detection of carcinogenic pollutant Maleic hydrazide. *Microchemical Journal, 193*, 109087. https://doi.org/10.1016/J.MICROC.2023.109087

Elugoke, S. E., Fayemi, O. E., Adekunle, A. S., Nkambule, T. T. I., Mamba, B. B., & Ebenso, E. E. (2021). *Conductive Nanodiamond-Based Detection of Neurotransmitters: One Decade, Few Sensors*. https://doi.org/10.1021/acsomega.1c01534

Erkmen, C., Unal, D. N., Kurbanoglu, S., & Uslu, B. (2022). Basics of Electrochemical Sensors. In *Engineering Materials* (pp. 81–99). Springer Science and Business Media Deutschland GmbH. https://doi.org/10.1007/978-3-030-98021-4_5

Facure, M. H. M., Schneider, R., Lima, J. B. S., Mercante, L. A., & Correa, D. S. (2021). Graphene quantum dots-based nanocomposites applied in electrochemical sensors: A recent survey. *Electrochem*, *2*(3), 490–519. https://doi.org/10.3390/electrochem2030032

Freitas, M., Carvalho, A., Nouws, H. P. A., & Delerue-Matos, C. (2022). *Tracking Arachis hypogaea Allergen in Pre-Packaged Foodstuff: A Nanodiamond-Based Electrochemical Biosensing Approach*. https://doi.org/10.3390/bios12060429

Gao, B., Poma, G., Malarvannan, G., Dumitrascu, C., Bastiaensen, M., Wang, M., & Covaci, A. (2022). Development of an analytical method based on solid-phase extraction and LC-MS/MS for the monitoring of current-use pesticides and their metabolites in human urine. *Journal of Environmental Sciences, 111*, 153–163. https://doi.org/10.1016/J.JES.2021.03.029

Gonçalves, D. A., Alencar, L. M., Anjos, J. P. B., Orth, E. S., & Souza, V. H. R. (2023). Ultrasensitive electrochemical detection of glyphosate using crumpled graphene/copper oxide nanocomposite. *Materials Today Communications, 36*, 106746. https://doi.org/10.1016/J.MTCOMM.2023.106746

Gopal, G., Roy, N., & Mukherjee, A. (2023). *Recent Developments in the Applications of GO/rGO-Based Biosensing Platforms for Pesticide Detection*. https://doi.org/10.3390/bios13040488

Guo, M., Han, J., Mishchenko, Y., Butenko, A., Kovalenko, V., & Zhao, H. (2023). Fabrication of methyl parathion electrochemical sensor based on β-cyclodextrin decorated single-wall carbon nanotubes. *Materials Research Innovations*, *28*(3), 154–160. https://doi.org/10.1080/14328917.2023.2244742

Han, J., Guo, X., Li, F., Zhao, M., Liu, Y., Guo, M., Wang, Z., Ran, Q., & Zhao, H. (2024). Facile fabrication of single-walled carbon nanotubes@cerium oxide composite sensing platform for ultrasensitive electrochemical analysis of methyl parathion. *Journal of Food Composition and Analysis, 131*, 106240. https://doi.org/10.1016/J.JFCA.2024.106240

Hareesha, N., & Manjunatha, J. G. (2021). Electro-oxidation of formoterol fumarate on the surface of novel poly(thiazole yellow-G) layered multi-walled carbon nanotube paste electrode. *Scientific Reports*, *11*, 12797. https://doi.org/10.1038/s41598-021-92099-x

Hossain, M. I., & Hasnat, M. A. (2023). Recent advancements in non-enzymatic electrochemical sensor development for the detection of organophosphorus pesticides in food and environment. *Heliyon*, *9*(9). https://doi.org/10.1016/j.heliyon.2023.e19299

Hu, X., Dang, W., & Jiang, Y. (2023). Application of nanomaterial-modified electrochemical sensors in pesticide residue detection. In *Highlights in Science, Engineering and Technology NSCBM* (Vol. 2023).

Huang, P., Wu, W., Li, M., Li, Z., Pan, L., Ahamad, T., Alshehri, S. M., Bando, Y., Yamauchi, Y., & Xu, X. (2024). Metal-organic framework-based nanoarchitectonics: A promising material platform for electrochemical detection of organophosphorus pesticides. *Coordination Chemistry Reviews, 501*, 215534. https://doi.org/10.1016/J.CCR.2023.215534

Itsoponpan, T., Thanachayanont, C., & Hasin, P. (2021). Sponge-like CuInS2 microspheres on reduced graphene oxide as an electrocatalyst to construct an immobilized acetylcholinesterase electrochemical biosensor for chlorpyrifos detection in vegetables. *Sensors and Actuators B: Chemical, 337*, 129775. https://doi.org/10.1016/J.SNB.2021.129775

Júnior, M., Junior, S., Design, S., Franciné Maia Júnior, F., Sales Junior, R., Ferreira Barbosa, G., Hussain, S., Jara-Cornejo, E., & Khan, S. (2024). Design and fabrication of 0061 biomimetic smart material for electrochemical detection of carbendazim pesticides in real samples with enhanced selectivity. *Biosensors, 14*(6), 304. https://doi.org/10.3390/bios14060304

Kadiru, S., Patil, S., & D'Souza, R. (2022). Effect of pesticide toxicity in aquatic environments: A recent review. *International Journal of Fisheries and Aquatic Studies, 10*(3), 113–118. https://doi.org/10.22271/fish.2022.v10.i3b.2679

Kaur, R., Mavi, G. K., Raghav, S., & Khan, I. (2019). Pesticides classification and its impact on environment. *International Journal of Current Microbiology and Applied Sciences, 8*(03), 1889–1897. https://doi.org/10.20546/ijcmas.2019.803.224

Kennedy, C. M., Lonsdorf, E., Neel, M. C., Williams, N. M., Ricketts, T. H., Winfree, R., Bommarco, R., Brittain, C., Burley, A. L., Cariveau, D., ısa Carvalheiro, L. G., Chacoff, N. P., Cunningham, S. A., Danforth, B. N., Dudenh, J.-H., Elle, E., Gaines, H. R., Garibaldi, L. A., Gratton, C., ... Kremen, C. (2013). A global quantitative synthesis of local and landscape effects on wild bee pollinators in agroecosystems. *Ecology Letters, 16*, 584–599. https://doi.org/10.1111/ele.12082

Khan, B. A., Nadeem, M. A., Nawaz, H., Amin, M. M., Abbasi, G. H., Nadeem, M., Ali, M., Ameen, M., Javaid, M. M., Maqbool, R., Ikram, M., & Ayub, M. A. (2023). *Pesticides: Impacts on Agriculture Productivity, Environment, and Management Strategies* (pp. 109–134). https://doi.org/10.1007/978-3-031-22269-6_5

Khosropour, H., Saboohi, M., Keramat, M., Rezaei, B., & Ensafi, A. A. (2023). Electrochemical molecularly imprinted polymer sensor for ultrasensitive indoxacarb detection by tin disulfide quantum dots/carbon nitride/multiwalled carbon nanotubes as a nanocomposite. *Sensors and Actuators B: Chemical, 385*, 133652. https://doi.org/10.1016/J.SNB.2023.133652

Köhler, H. R., & Triebskorn, R. (2013). Wildlife ecotoxicology of pesticides: Can we track effects to the population level and beyond? *Science, 341*, 759–765.

Kowalska, G., Pankiewicz, U., & Kowalski, R. (2020). Estimation of pesticide residues in selected products of plant origin from Poland with the use of the HPLC-MS/MS technique. *Agriculture (Switzerland), 10*(6), 1–20. https://doi.org/10.3390/agriculture10060192

Lee, S.-K., Song, M.-J., Kim, J.-H., Kan, T.-S., Lim, Y.-K., Ahn, J.-P., & Lim, D.-S. (2014). 3D-networked carbon nanotube/diamond core-shell nanowires for enhanced electrochemical performance. *NPG Asia Materials, 6*, 115. https://doi.org/10.1038/am.2014.50

Li, M., Pan, Q., Wang, J., Wang, Z., & Peng, C. (2024). Machine learning-assisted fluorescence sensor array for qualitative and quantitative analysis of pyrethroid pesticides. *Food Chemistry, 433*, 137368. https://doi.org/10.1016/J.FOODCHEM.2023.137368

Liu, J., Zhong, X., Gong, X., Deng, L., Tan, G., Zhang, Q. e., Xiao, Z., Yao, Q., Liu, S., Gao, Y., Wang, L., & Lu, L. (2024). Highly sensitive turn-on electrochemical sensing of organophosphorus pesticides by integration of homogeneous reaction and heterogeneous catalytic signal amplification. *Food Chemistry, 458*, 140275. https://doi.org/10.1016/J.FOODCHEM.2024.140275

Liu, Y., Xiao, Y., Zhang, Y., Gao, X., Wang, H., Niu, B., & Li, W. (2024). ZnO-rGO-based electrochemical biosensor for the detection of organophosphorus pesticides. *Bioelectrochemistry, 156*, 108599. https://doi.org/10.1016/J.BIOELECHEM.2023.108599

Lynch, S. M., Rusiecki, J. A., Blair, A., Dosemeci, M., Lubin, J., Sandler, D., Hoppin, J. A., Lynch, C. F., & Alavanja, M. C. R. (2006). Cancer incidence among pesticide applicators exposed to cyanazine in the Agricultural Health Study. *Environmental Health Perspectives, 114*(8), 1248–1252. https://doi.org/10.1289/ehp.8997

Mahmood, I., Imadi, S. R., Shazadi, K., Gul, A., & Hakeem, K. R. (2016). Effects of pesticides on environment. In *Plant, Soil and Microbes: Volume 1: Implications in Crop Science* (pp. 253–269). Springer International Publishing. https://doi.org/10.1007/978-3-319-27455-3_13

Malhotra, A., Rawat, A., Prakash, O., Kumar, R., Srivastava, R. M., & Kumar, S. (2023). Chemical composition and pesticide activity of essential oils from *Artemisia annua* L. harvested in the rainy and winter seasons. *Biochemical Systematics and Ecology, 107*, 104601. https://doi.org/10.1016/J.BSE.2023.104601

Meskher, H., Ragdi, T., Thakur, A. K., Ha, S., Khelfaoui, I., Sathyamurthy, R., Sharshir, S. W., Pandey, A. K., Saidur, R., Singh, P., & Lynch, I. (2023). Critical reviews in analytical chemistry a review on CNTs-based electrochemical sensors and biosensors: Unique properties and potential applications. *Critical Reviews in Analytical Chemistry*, *54*(7), 2398–2421. https://doi.org/10.1080/10408347.2023.2171277

Moldovan, R., Iacob, B. C., Farcău, C., Bodoki, E., & Oprean, R. (2021). Strategies for SERS detection of organochlorine pesticides. *Nanomaterials*, *11*(2), 1–21. MDPI AG. https://doi.org/10.3390/nano11020304

Murathan, Z., İnam, R., & Demir, E. (2024). A new strategy for the determination of dinobuton fungicide by square wave stripping voltammetry on the multi-walled carbon nanotube electrode. *Materials Research Express*, *11*(1). https://doi.org/10.1088/2053-1591/ad1bfc

Nunn, N., Torelli, M., McGuire, G., & Shenderova, O. (2017). Nanodiamond: A high impact nanomaterial. *Current Opinion in Solid State and Materials Science*, *21*(1), 1–9. https://doi.org/10.1016/J.COSSMS.2016.06.008

Parra-Arroyo, L., González-González, R. B., Castillo-Zacarías, C., Melchor Martínez, E. M., Sosa-Hernández, J. E., Bilal, M., Iqbal, H. M. N., Barceló, D., & Parra-Saldívar, R. (2022). Highly hazardous pesticides and related pollutants: Toxicological, regulatory, and analytical aspects. *Science of The Total Environment, 807*, 151879. https://doi.org/10.1016/J.SCITOTENV.2021.151879

Pengsomjit, U., Alabdo, F., Karuwan, C., Kraiya, C., Alahmad, W., & Ozkan, S. A. (2024). Innovative graphene-based nanocomposites for improvement of electrochemical sensors: Synthesis, characterization, and applications. *Critical Reviews in Analytical Chemistry*. https://doi.org/10.1080/10408347.2024.2343854

Pimalai, D., Putnin, T., & Bamrungsap, S. (2025). A highly sensitive electrochemical sensor based on poly(3-aminobenzoic acid)/graphene oxide-gold nanoparticles modified screen printed carbon electrode for paraquat detection. *Journal of Environmental Sciences, 148*, 139–150. https://doi.org/10.1016/J.JES.2023.10.006

Popp, J., Pető, K., & Nagy, J. (2013). Pesticide productivity and food security. A review. *Agronomy for Sustainable Development*, *33*(1), 243–255. https://doi.org/10.1007/s13593-012-0105-x

Pratap, A., Raman, S., Thakur, G., Pandey, G., Kumari, K., & Singh, P. (2024). An updated review on functionalized graphene as sensitive materials in sensing of pesticides. *Chemistry Biodiversity, 21*, 202302080–202302081. https://doi.org/10.1002/cbdv.202302080

Rajapaksha, R. D. A. A., Hashim, U., Gopinath, S. C. B., Parmin, N. A., & Fernando, C. A. N. (2021). Nanoparticles in electrochemical bioanalytical analysis. *Nanoparticles in Analytical and Medical Devices*, 83–112. https://doi.org/10.1016/B978-0-12-821163-2.00006-6

Rajkumar, C., & Kim, H. (2023). 4-Cyanophenol herbicide sensor using multi-walled carbon nanotube embedded dual-microporous polypyrrole nanoparticles as metal-free and environmentally friendly hybrid electrode. *Microchimica Acta, 190*, 197. https://doi.org/10.1007/s00604-023-05773-4

Ramirez, C. A., Thiruppathi, A. R., Ozoemena, O., & Chen, A. (2024). Recent advances in graphene oxide-based electrochemical sensors. *Canadian Journal of Chemistry*. https://doi.org/10.1139/cjc-2024-0010

Rani, L., Thapa, K., Kanojia, N., Sharma, N., Singh, S., Grewal, A. S., Srivastav, A. L., & Kaushal, J. (2021). An extensive review on the consequences of chemical pesticides on human health and environment. *Journal of Cleaner Production, 283*, 124657. https://doi.org/10.1016/J.JCLEPRO.2020.124657

Rasheed, T. (2023). Carbon dots as robust class of sustainable and environment friendlier nano/optical sensors for pesticide recognition from wastewater. *TrAC Trends in Analytical Chemistry, 160*, 116957. https://doi.org/10.1016/J.TRAC.2023.116957

Sadasivan Pilla, K. (2023). Classifying chemicals into toxicity categories based on LC values - Pros and con. *Journal of Environmental Biology*, *44*(5), I–III.

Şener, D., Erden, P. E., & Kaçar Selvi, C. (2024). Disposable biosensor based on nanodiamond particles, ionic liquid and poly-l-lysine for determination of phenolic compounds. *Analytical Biochemistry, 688*, 115464. https://doi.org/10.1016/J.AB.2024.115464

Shahrokhian, S., & Khafaji, M. (2010). Application of pyrolytic graphite modified with nano-diamond/graphite film for simultaneous voltammetric determination of epinephrine and uric acid in the presence of ascorbic acid. *Electrochimica Acta*, *55*(28), 9090–9096. https://doi.org/10.1016/J.ELECTACTA.2010.08.043

Silah, H., Demir, E., Yıldırım, S., & Uslu, B. (2022). Carbon nanomaterial-based sensors for the development of sensitive sensor platform. *Carbon Nanomaterials-Based Sensors: Emerging Research Trends in Devices and Applications*, 191–246. https://doi.org/10.1016/B978-0-323-91174-0.00009-3

Silva, L. R. G., Carvalho, J. H. S., Stefano, J. S., Oliveira, G. G., Prakash, J., & Janegitz, B. C. (2023). Electrochemical sensors and biosensors based on nanodiamonds: A review. *Materials Today Communications, 35*, 106142. https://doi.org/10.1016/J.MTCOMM.2023.106142

Songa, E. A., & Okonkwo, J. O. (2016). Recent approaches to improving selectivity and sensitivity of enzyme-based biosensors for organophosphorus pesticides: A review. *Talanta, 155*, 289–304. https://doi.org/10.1016/J.TALANTA.2016.04.046

Sood, P. (2024). Pesticides usage and its toxic effects–a review. *Indian Journal of Entomology*, *86*(1), 339–347. https://doi.org/10.55446/IJE.2023.505

Srivastava, A. K., & Kesavachandran, C. (2019). *Health Effects of Pesticides*. CRC Press.

Su, S., Han, L., Gao, L., Ren, Q., Nie, M., Zhu, Z., Wan, H., Liu, H., & Wu, N. (2020). Spark plasma sintering compaction of hybrid nanodiamond/carbon nanotubes/metal electrode and its application. *Diamond and Related Materials, 104*, 107746. https://doi.org/10.1016/J.DIAMOND.2020.107746

Sun, B., Yu, S., Ma, Q., Shi, H., Dang, Q., Liu, Y., Hu, J., Bao, L., Yang, L., & Shi, X. (2024). A rapid and easy procedure of enzyme biosensor based on nitrogen-doped graphene for detection of methyl parathion in CHM. *Journal of the Electrochemical Society, 171*(3), 037522. https://doi.org/10.1149/1945-7111/ad31f0

Tallapragada, S., & L. R. (2022). Effect of pesticides on crop, soil microbial flora and determination of pesticide residue in agricultural produce: A review. *International Journal of Environment and Climate Change, 12*, 36–56.

Tanwar, S., Sharma, A., & Mathur, D. (2023). A graphene quantum dots–glassy carbon electrode-based electrochemical sensor for monitoring malathion. *Beilstein Journal of Nanotechnology, 14*, 701–710. https://doi.org/10.3762/BJNANO.14.56

Tian, J., Qin, L., Li, D., Qin, S., Gao, W., & Jia, Y. (2022). Carbofuran-imprinted sensor based on a modified electrode and prepared via combined multiple technologies: Preparation process, performance evaluation, and application. *Electrochimica Acta, 404*, 139600. https://doi.org/10.1016/J.ELECTACTA.2021.139600

Tigari, G., & Manjunatha, · J G. (2020). Poly(glutamine) film-coated carbon nanotube paste electrode for the determination of curcumin with vanillin: an electroanalytical approach. *Monatsh Chem, 151*, 1681–1688. https://doi.org/10.1007/s00706-020-02700-8

Tudi, M., Ruan, H. D., Wang, L., Lyu, J., Sadler, R., Connell, D., Chu, C., & Phung, D. T. (2021). Agriculture development, pesticide application and its impact on the environment. *International Journal of Environmental Research and Public Health, 18*(3), 1–24. https://doi.org/10.3390/ijerph18031112

Tümay, S. O., Şenocak, A., Sarı, E., Şanko, V., Durmuş, M., & Demirbas, E. (2021). A new perspective for electrochemical determination of parathion and chlorantraniliprole pesticides via carbon nanotube-based thiophene-ferrocene appended hybrid nanosensor. *Sensors and Actuators B: Chemical, 345*, 130344. https://doi.org/10.1016/J.SNB.2021.130344

Uslu, B. (2021). Electroanalytical applications of quantum dot-based biosensors: A volume in micro and nano technologies. In *Electroanalytical Applications of Quantum Dot-Based Biosensors*. https://doi.org/10.1016/B978-0-12-821670-5.09994-2

Wahyudi, S., Bahtiar, A., Panatarani, C., Anas, & Risdiana. (2023). Recent advanced carbon dots derived natural products and aptasensor-based carbon dots for detection of pesticides. *Sensing and Bio-Sensing Research, 41*, 100576. https://doi.org/10.1016/J.SBSR.2023.100576

Wang, D., Saleh, N. B., Byro, A., Zepp, R., Sahle-Demessie, E., Luxton, T. P., Ho, K. T., Burgess, R. M., Flury, M., White, J. C., & Su, C. (2022). Nano-enabled pesticides for sustainable agriculture and global food security. *Nature Nanotechnology*, *17*(4), 347–360. https://doi.org/10.1038/s41565-022-01082-8

Wang, L., Li, M., Li, B., Wang, M., Zhao, H., & Zhao, F. (2023). *Electrochemical sensor based on laser-induced graphene for carbendazim detection in water. Foods*, *12*(12), 2277 https://doi.org/10.3390/foods12122277

Wang, X., He, L., Xu, L., Liu, Z., Xiong, Y., Zhou, W., Yao, H., Wen, Y., Geng, X., & Wu, R. (2023). Intelligent analysis of carbendazim in agricultural products based on a ZSHPC/MWCNT/SPE portable nanosensor combined with machine learning methods. *Analytical Methods*. https://doi.org/10.1039/d2ay01779b

Wang, Z., Wu, H., He, Y., Yan, Y., Zhou, W., Zhang, G., Liu, D., Ye, Z., & Qiu, F. (2022). *An Electrochemical Sensor Based on Molecularly-Imprinted-Polymer-Modified Carbon Quantum Dots@hexagonal Boron Nitride Nanosheets Nanocomposites for Triclosan Determination.* https://doi.org/10.1002/slct.202201141

Weheabby, S., You, S., Pašti, I. A., Al-Hamry, A., & Kanoun, O. (2024). Direct electrochemical detection of pirimicarb using graphene oxide and ionic liquid composite modified by gold nanoparticles. *Measurement, 236*, 115132. https://doi.org/10.1016/J.MEASUREMENT.2024.115132

WHO. (2019). *The WHO Recommended Classiication of Pesticides by Hazard and Guidelines to Classification 2019.*

Zambianco, N. A., Silva, T. A., Zanin, H., Fatibello-Filho, O., & Janegitz, B. C. (2019). Novel electrochemical sensor based on nanodiamonds and manioc starch for detection of diquat in environmental samples. *Diamond and Related Materials, 98*, 107512. https://doi.org/10.1016/J.DIAMOND.2019.107512

Zhang, H., Jin, J., Wang, X., Han, W., Qin, L., Chen, L., Mao, X., Liu, Z., & Xiang, X. (2024). Bioenzyme–nanoenzyme–chromogen all-in-one test strip for convenient and sensitive detection of malathion in water environment. *Science of the Total Environment, 924*, 171512. https://doi.org/10.1016/J.SCITOTENV.2024.171512

Zhang, Y., Rhee, K. Y., Hui, D., & Park, S. J. (2018). A critical review of nanodiamond based nanocomposites: Synthesis, properties and applications. *Composites Part B: Engineering, 143*, 19–27. https://doi.org/10.1016/J.COMPOSITESB.2018.01.028

Zhenzhen Yang, L. C. (2024). Electrochemical sensor based on silver nanoparticles/ carboxylated multi-walled carbon nanotubes modified glassy carbon electrode for detection of carbamate pesticide in food. *Journal of Food Measurement and Characterization*, 1–12. https://doi.org/10.1007/s11694-024-02642-3

Zhu, J., He, Y., Luo, L., Li, L., & You, T. (2023). Electrochemical determination of hazardous herbicide diuron using MWCNTs-CS@NGQDs composite-modified glassy carbon electrodes. *Biosensors, 13*(8), 808. https://doi.org/10.3390/bios13080808

16 Carbon Sensors and Significance

Nida Aydogdu Ozdogan, Seyda Nur Samanci, and Ersin Demir

16.1 INTRODUCTION

Carbon, the fundamental building block of organic life, boasts a captivating history spanning billions of years and encompassing numerous scientific disciplines (Figure 16.1). Carbon's existence extends beyond its crucial role in biological processes to encompass the development of various nanomaterials, demonstrating its versatility and significance in modern technology and materials science. Carbon nanotubes (CNTs), in particular, have emerged as a significant area of research due to their unique properties and potential applications in energy storage systems, representing a remarkable evolution in our understanding of this element's capabilities in the context of contemporary innovation [1].

A sensor is a device that detects the presence of a particular condition or quality attribute. Sensor and biosensor research is a highly active and interdisciplinary field that has significant socio-economic relevance. Sensors play a crucial role in various fields, including healthcare, industry, and daily life. There are many different types of sensors available, and many more are being developed. Sensors are available for various environmental characteristics, including temperature, light, humidity, and the presence of specific chemicals such as hydrogen, carbon monoxide, and ammonia [2]. Among them, and especially in the electrochemical field, carbon sensors have gained significant importance. Carbon sensors are a type of sensor that is categorized based on the material being sensed. These sensors have physical properties that enable low-level detection in a variety of applications. Carbon sensors belong to the large family of chemical sensors and biosensors. They include devices whose surface consists mainly of carbon in different chemical configurations (graphite, graphene, CNTs s, and carbon particles), as well as those on whose surface carbon compounds are immobilized. Carbon sensors are used for electrochemical detection of various species, in gas sensors, biosensors for various analytes, and many other applications [3,4].

Carbon-based nanomaterials have gained significant attention in sensor development due to their unique properties, including high conductivity, biocompatibility, large surface area, and versatile modification potential [5]. These materials, such as fullerenes, CNTs, and graphene, offer improved sensitivity, selectivity, and lower detection limits for various analytes [5,6]. Carbon composite sensors have shown particular promise in electrochemical applications, providing enhanced accuracy and portability through miniaturization [7]. They have been successfully applied in diverse fields, including environmental monitoring, pharmaceutical analysis, and food safety [7,8]. For instance, carbon-based sensors have been developed for the detection of carbamate pesticides in environmental, food, and biological matrices [8]. The versatility and performance of carbon-based sensors make them valuable tools for both routine applications and fundamental research in electrochemistry and sensor development [6,7].

DOI: 10.1201/9781003500322-16

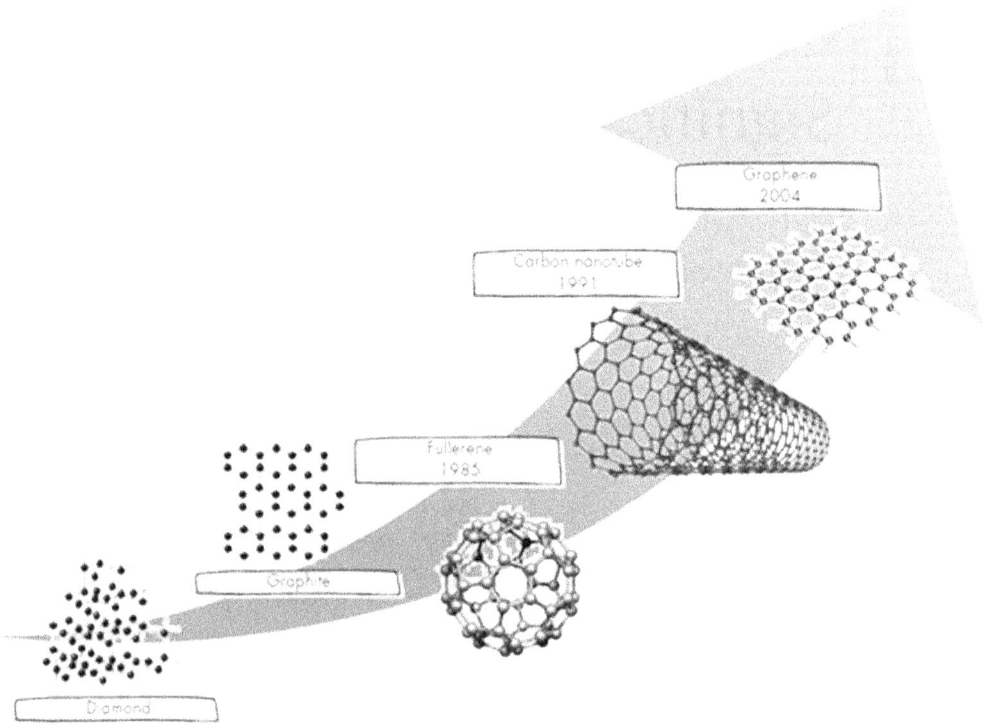

FIGURE 16.1 The discovery of carbon nanostructures.

The utilization of nanomaterials in sensors represents a significant breakthrough for the detection of analytes and target molecules. Carbon-based electrochemical sensors are noteworthy due to their biocompatibility, lower detection limits, high sensitivity, and exceptional selectivity for a broad range of chemical and biological molecules [2]. This chapter provides a comprehensive overview of the most recent advancements in carbon-based electrochemical sensor technology. The chapter delves into the characteristic properties of various nanomaterials, such as fullerenes, CNTs, graphene, carbon quantum dots, carbon nanohorns (CNHs), carbon nanodiamonds, carbon nanofibers, and carbon black, among others. The chapter also focuses on small molecule electroanalytical detection, with an emphasis on the application areas of carbon sensors in the determination of various food additives, pesticides, heavy metals, drugs, and other types of analytes. Additionally, the recent application of these nanomaterials as electrochemical biosensors for the detection of various biomolecules has been highlighted, and the future prospects and possibilities in this field have been outlined.

16.1.1 Types of Carbon Sensors

Carbon, as one of the most prevalent elements on earth, is a versatile element with a tetravalent value, which allows it to form various types of bonds, including single, double, and triple bonds, as well as long chains and undergo polymerization. Additionally, carbon can exhibit sp, sp^2, and sp^3 hybridizations. The two widely recognized allotropic forms of carbon are diamond, which has an sp^3 hybridization, and graphite, which has an sp^2 hybridization. Both allotropes possess distinct structural features that contribute to a wide range of physical and chemical properties, making carbon an essential element in both natural and industrial settings [1].

The categorization of carbon nanomaterials relies on their geometric configuration. These particles come in a variety of shapes, such as tubes, horns, spheres, or ellipsoids. CNTs or CNHs are tube or

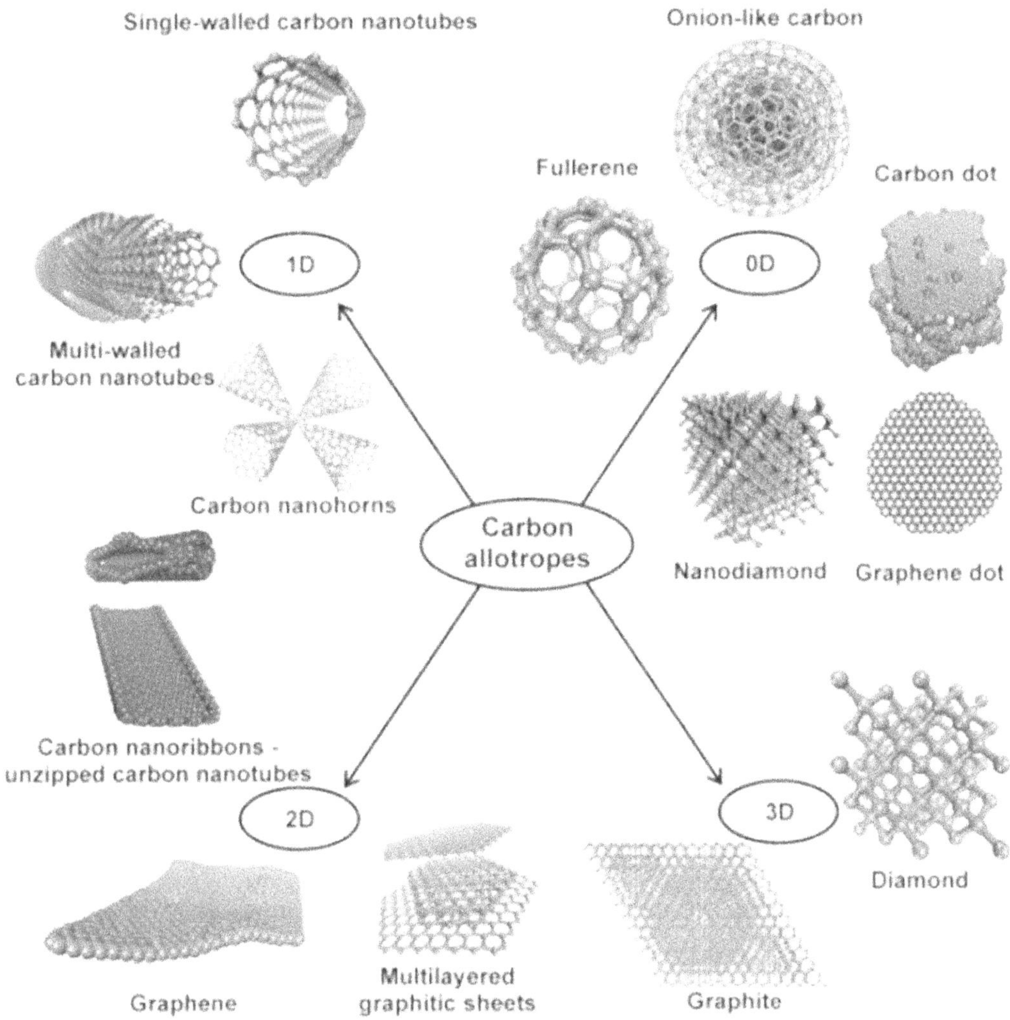

FIGURE 16.2 Structures of different carbon nanomaterials. (Reproduced under the terms of the CC-BY Creative Commons Attribution 4.0 International license (https://creativecommons.org/licenses/by/4.0) [11] Copyright 2021, The Authors, published by Wiley-VCH.)

horn-shaped particles, while fullerenes contain spherical or ellipsoidal nanoparticles (Figure 16.2). The carbon nanomaterials with zero, one, two, and three-dimensional structures (e.g., fullerenes, CNTs, graphene, carbon quantum dots, CNHs, nanodiamonds, carbon nanofibers, and carbon black) possess inherent traits that can be easily harnessed for the development of cutting-edge technology for sensing applications. These materials, including carbon quantum dots, nanotubes, graphene, and nanofibers, exhibit exceptional electronic, optical, and mechanical properties. The discovery of carbon nanostructures has significantly advanced the field of nanotechnology, providing insights into the atomic arrangements and interactions that underlie the remarkable properties of these materials, which have far-reaching implications. Over the past three decades, extensive scientific research has been conducted worldwide in the field of carbon nanomaterials due to their significant electronic, optical, mechanical, chemical, and thermal properties [9]. Electrochemical biosensors based on carbon nanomaterials have demonstrated improved sensitivity, selectivity, and biocompatibility for detecting various biomolecules. Additionally, hybrid carbon materials and composites have emerged as potential candidates for applications in superconductivity and high-current density devices. The versatility of carbon nanostructures continues to drive research in various fields, including biosensing, energy storage, and environmental monitoring [10].

Fullerenes are a type of carbon molecule with distinct shapes, such as spheres and cylinders, which were first discovered in 1985 [12]. These molecules have unique chemical and physical properties that make them appealing for a wide range of applications. Due to their unique characteristics, fullerenes have emerged as promising materials for electrochemical applications. For instance, a composite film composed of fullerenes-C60 and bimetalic nanoparticles has been developed as a high-performance electrochemical sensor for detecting vitamin D3 in cancer cell samples, demonstrating high sensitivity and a wide dynamic range [13]. Additionally, functionalized fullerenes, such as C60@UiO-66-NH_2 nanocomposites, have been used to create rapid and sensitive aptasensors for detecting tobramycin [14]. In electrochemical sensors and biosensors, fullerenes are often used as modifying materials to improve sensitivity and analysis speed for monitoring drugs and metabolites in pharmaceutical and biological samples. These carbon-based nanomaterials offer advantages such as low cost, ease of manufacturing, and rapid analysis, making them valuable for various sensing applications in the pharmaceutical and medical fields [15].

CNTs were initially discovered by Iijima in 1991, leading to a significant advancement within the realm of nanotechnology [4]. These minuscule tubular structures are composed of graphite sheets and typically have diameters smaller than 100 nm, although their lengths often surpass their diameters [9]. CNTs are categorized as single-walled or multi-walled tubes as cylindrical structures formed by rolled graphene layers [16]. The exceptional properties of CNTs, such as their high electrical and thermal conductivities, mechanical strength, and optical properties, make them highly valuable for a wide range of applications, including drug delivery, sensors, and various technological fields [17,18]. Owing to their unique characteristics, CNTs are promising materials for electrochemical applications. They serve as excellent catalyst carriers and can enhance the electrochemical signals in sensors used for drug and food analysis [19,20]. Furthermore, their ability to confine guest molecules enables tuning of the redox properties and facilitates enhanced electron transfer [21].

- *Single-walled carbon nanotubes (SWCNTs)* are materials with a wide range of applications due to their unique properties, including high electrical and thermal conductivity, mechanical strength, and flexibility [22]. These properties are influenced by both geometrical and chemical factors, and different treatments can affect the performance of inner sphere redox probes [23]. SWCNTs have gained significant attention in electrochemical analysis due to their unique properties. SWCNTs have been combined with zinc oxide nanoparticles to create sensitive electrodes for chloramphenicol detection [24]. Additionally, a glucose biosensor was developed using SWCNT electrodes modified with glucose oxidase, ferrocenecarboxylic acid, and polyethylene glycol, achieving high sensitivity and stability [25]. Another study investigated hybrid nanocomposites of cellulose nanocrystals and SWCNTs, which showed superior electroanalytical performance compared to SWCNT-based and screen-printed carbon electrodes [26]. Notably, cellulose nanocrystal type II/SWCNT hybrids demonstrated significantly higher sensitivity for glycoprotein detection. These studies showcase the potential of SWCNTs in advancing electrochemical sensing technologies.
- *Multi-walled carbon nanotubes (MWCNTs)* are highly advanced materials that possess exceptional properties, such as exceptional strength, low density, and impressive thermal and electrical conductivity. These materials have become versatile for electrochemical sensing applications. MWCNT-modified electrodes have exhibited improved catalytic properties, low detection limits, and potential for real-world applications in environmental and clinical analyses. By collaborating with gold nanoparticles, MWCNTs exhibit heightened sensitivity in detecting herbicides like paraquat, thereby achieving a detection limit of 32 nM [27]. Additionally, MWCNTs combined with zinc oxide nanoparticles have been used to detect

the herbicide linuron in soil and water samples [28]. A molecularly imprinted sensor using MWCNTs has been developed for the detection of tetrahydrocannabinol in blood plasma [29]. MWCNTs decorated with cobalt phthalocyanine on glassy carbon electrodes have shown high sensitivity in detecting heavy metals like cadmium and lead in water samples [30]. These studies illustrate the versatility of MWCNT-based electrochemical sensors in detecting various substances, including herbicides, drugs, and heavy metals, with high sensitivity and selectivity.

Graphene is a two-dimensional carbon allotrope comprising a single layer of atoms arranged in a honeycomb lattice [31]. This remarkable nanomaterial has garnered significant attention due to its unique properties, including high mechanical stiffness, thermal stability, electrical conductivity, and biocompatibility, as well as its versatile applications in industrial, environmental, and biomedical fields. Graphene-based electrochemical sensors offer enhanced sensitivity, stability, and rapid detection, although performance may vary depending on factors such as film morphology, composition, and graphene loading. Recent studies have explored the use of doped graphene, reduced graphene oxide, and their nanocomposites to further enhance analytical performance. The high surface area, excellent conductivity, and ease of functionalization of graphene have rendered it significant in nanocomposite-based electrochemical sensors, particularly in its potential to detect environmental pollutants, drugs, and pesticides [32,33]. Although these sensors show promise in detecting a wide range of analytes, from heavy metals and antibiotics to pesticides, further research is needed to address the challenges in real sample detection and to better understand the fundamental aspects of electron transfer reactions through graphene electrodes [34–36]. While the field is experiencing rapid growth, challenges remain in developing stable, sensitive, and practical sensors for the real world [37,38].

Graphite is a crystalline form of carbon with a layered structure of hexagonal carbon atoms arranged in graphene sheets [39]. Recent research has focused on developing graphite-based electrochemical sensors for various applications. Wang et al. (2021) fabricated a nanometer-sized graphite/boron-doped diamond electrode for acetaminophen sensing, demonstrating high sensitivity and a wide linear range [40]. Tasić et al. (2022) repurposed graphite rods from waste batteries to create a sensor for L-tryptophan determination in food samples, showcasing good analytical performance and environmental benefits [41]. These studies highlight the versatility and effectiveness of graphite-based electrochemical sensors for diverse analytes, offering cost-effective and sensitive detection methods for environmental and pharmaceutical applications.

16.2 CARBON SENSOR DEVELOPMENT: SIGNIFICANCE FOR VARIOUS ANALYTE DETERMINATIONS

Food additives are substances deliberately added to foods to improve various qualities, such as taste, appearance, and shelf life [42]. Today, it is known that more than 25,000 food additives are used to improve and preserve nutrients [43,44]. A study of the French market found that 53.8% of food products contained at least one additive, with artificially sweetened drinks, ice creams, and industrial sandwiches among the categories most likely to contain additives. The study also identified six clusters of additives that are often found together in food products [45]. Given the widespread use of additives and possible health concerns, it is crucial to balance their benefits against possible risks and enable consumers to make informed choices. Although they play an important role in the modern food industry, concerns are growing about their potential health effects, including cancer, allergies, and behavioral disorders in children [46]. For this reason, the determination of food additives is of great importance. Recent advances in carbon-based nanomaterial sensors have shown promise for detecting food additives. These sensors demonstrated high sensitivity, broad linear ranges, and practical applicability in various food samples. The use of carbon-based materials in conjunction with metal oxides or different nanomaterials has shown promise in enhancing the electrocatalytic activity and overall performance of these sensors for food additive detection (Table 16.1).

Wang and co-workers [62], successfully synthesized Ni_3Se_4/MWCNTs-OH nanocomposites for the electrochemical determination of maltol. MWCNTs-OH/GCE was obtained by dropping MWCNTs-OH (5.0 μL) dispersion onto a GCE surface and drying at room temperature. Then, Ni_3Se_4 particles were electroplated onto the MWCNTs-OH/GCE surface at –0.95 V in 60 mL solution containing 0.10 M NH_4Cl, 0.05 M H_2SeO_3 and 0.05 M $NiSO_4$, and Ni_3Se_4/MWCNTs-OH/GCE was prepared (Figure 16.3). The chemical composition and morphology properties of Ni_3Se_4/MWCNTs-OH composites were characterized by X-ray diffraction (XRD), X-ray photoelectron spectroscopy (XPS), scanning electron microscopy (SEM), and energy dispersive X-Ray (EDX). Cyclic voltammetry (CV) and differential pulse voltammetry (DPV) were used to investigate the electrochemical behavior of maltol in Ni_3Se_4/MWCNTs-OH/GCE. The maltol oxidation peak current on carbon sensor increased linearly in the range of 0.1–2000 μM, and the limit of detection (LOD) was calculated as 17.5 nM. Additionally, the fabricated electrocheemical sensor has been proven to exhibit stability, reliable repeatability, and excellent selectivity and can be successfully used to detect maltol with accuracy in real samples. As a result, Ni_3Se_4/MWCNTs-OH/GCE prepared to detect maltol has been proven to exhibit good electrocatalyst performance due to the synergistic effect of Ni_3Se_4 and MWCNTs-OH [62].

Pesticides are substances used to control organisms that damage crops, transmit diseases, or cause other unwanted effects [73]. These substances can be classified as insecticides, herbicides, and fungicides, designed to secure crops and prevent post-harvest losses by ensuring food safety [74]. While pesticides are indispensable in agriculture for better crop production, they can be dangerous to humans and other species if not handled carefully. Environmental pollution occurs when pesticides contaminate water and soil, potentially accumulating in the food chain and damaging ecosystems [75]. Due to these concerns, new agricultural approaches that will significantly reduce the use of chemical pesticides and new methods that can detect pesticides even at the lowest concentration are needed. Recent advances in electrochemical detection of pesticides have shown promise as alternatives to traditional chromatographic methods for on-site monitoring. These techniques offer advantages such as ease of use, high sensitivity, and rapid results [74,76,77]. Electrode surface modifications using carbon-based materials, metal nanoparticles, and composites have improved sensitivity and selectivity for pesticide detection [78]. Table 16.2 includes studies showing that electrochemical carbon sensors are sensitive, selective, and applicable to different samples in pesticide determination.

Han and friends [87] prepared a multifunctional carbon sensor with ultrasonic-assisted strategy for ultrasensitive analysis of methyl parathion (MP). 40 mg of nanostructured cerium oxide nanoparticles (CeO_2) was uniformly dispersed in 10 mL of DMF solution with the help of ultrasonic technology Then, 20 mg of SWCNTs were added to the above suspension containing CeO_2 nanoparticles, followed by ultrasonic-assisted dispersing treatment for 2 h. Then, 5 μL of the obtained SWCNT@CeO_2 suspension was used to decorate the polished GCE. Finally, the SWCNT@CeO_2/GCE sensor was successfully prepared by drying the modified sensing electrode with the help of an oven lamp (Figure 16.4). The microstructures of SWCNT, CeO_2, and SWCNT@CeO_2 samples were explored by SEM, EDX, XPS, and XRD. The electrochemical features were examined by CV and DPV measurements. It was observed that SWCNT and CeO_2 nanoparticles had a synergistic effect in increasing the MP oxidation signal. The produced SWCNT@CeO_2/GCE sensor was calculated as LOD 9.12 nM in MP electrochemical analysis. This value proves the sensitivity of the developed carbon sensor. Regarding the electrochemical detection of MP in real food samples (peach juice and cabbage), both recovery rates of 87.68–107.31% and low relative standard deviation (RSD) of 1.09–2.65% were achieved. This study demonstrates that the produced sensor has satisfactory repeatable, reproducible, and anti-interference capabilities in the field of pesticide residue analysis [87].

TABLE 16.1
Electro-analysis of food additives using carbon sensors

Food addivite	Electrode	Method	Lineer range	LOD	LOQ	Samples	Ref
Ascorbic acid	rGO-AuNPs/ITO	LSV	20–150 µM	3.1 µM	9.4 µM	Milk, juice, urine	[47]
Butylated hydroxyanisole	Gr-AuNP-Peach-ext/GCE	DPV	0.2–9.8 µM	0.07 µM	-	Mayonnaise	[48]
Caffein	MoS_2/PANI@g-C_3N_4/GCE	DPV	9.8–140.0 µM	0.062 µM	-	Ice cream, Red bull (energy drink) and theophylline tablets	[49]
Caffeine	Poly(PRO)-MWCNT/GCE	DPV	10.04–93.26 µM	0.11 µM	-	Tea, coffe, infusion of yerba mate, Coke zero, energy drink (speed) and human breast milk	[50]
Curcumin	MWCT/GCE	AdSSWV	0.011–5.0 µM	0.005 µM	-	Turmeric powder	[51]
Curcumin	Azobenzene-rGO@CNTs/GCE	SWV	0.008–2.0 µM 2.0–10.0 µM	0.003 µM	-	Blood serum and urine	[52]
Curcumin	f-CNTs/GCE	SWV	2.0–14.0 µM	0.06 µM	-	Commercial turmeric powder	[53]
Curcumin	Av-GO	CV	0.0001–0.1 µM	0.0039 µM	-	-	[54]
Curcumin	Av-rGO	CV	$1.0x10^{-6}$–0.1 µM	0.0049 µM	-	-	[54]
Curcumin	CNTs-CMC/AuE	CV	1.0–48.0 µM	0.084	-	Turmeric powder	[55]
Curcumin	rGO/CPE	DPV	10.0–6000.0 µM	3.2 µM	-	Human blood serum	[56]
Curcumin	β-CD/rGO/GCE	DPV	0.05–10.0 µM	0.0003 µM	-	-	[57]
Curcumin	MnO_2-c-MWCNT/GCE	SDLSV	0.01–1.0 and 1.0–80.0 µM	0.006 µM	-	Turmeric powder; curry; mustard; instant noodle seasoning; ginger powder	[58]
Cyanidin	f-CNTs/GCE	DPV	10–560 ug/L	18 ug/L	-	Fruit and food	[59]
Kojic acid	(Hpy)(PF6)/N-CQDs/Fe_3O_4/CPE	DPV	0.03–300.0 µM	0.008 µM	0.0266 µM	Vinegar, Tomato sauce	[60]
L-arginine	PANI/MWCNTs/Fe_3O_4/GCE	CV, DPV, Amperometry	10.0–1000.0 µM	20.0 µM	-	Normal and leukemic blood sample	[61]
Maltol	Ni_3Se_4/MWCNTs-OH/GCE	DPV	0.1 µM–2000 µM	17.5 nM		Juice, beer	[62]
Neohesperidin dihydrochalcone (NHDC)	ZIF-8/ErGO/GCE	LSV	0.08–80 µM	0.035 µM	-	Milk	[63]
Nicotine	HBN/Gr/ GCE	Amperometry	1.0–1000.0 µM	0.42 µM		Tobacco	[64]
Propyl gallate	AQ/MWCNT/CS/GCE	DPV	1.0–30.0 µM	0.67 µM	0.76 µM	Sesame oil, blend oil, peanut oil	[65]

(continued)

TABLE 16.1 (Continued)
Electro-analysis of food additives using carbon sensors

Food addivite	Electrode	Method	Lineer range	LOD	LOQ	Samples	Ref
Sodium Metabisülfite	Fe(TPFPP)Cl/rGO@Pd^0	SWV	1.0x10^{-4}–1.0 μM	3x10^{-5} μM	1x10^{-4} μM	-	[66]
Sodium Metabisulfite	TPP/rGO@Pd0	SWV	1.0x10^{-5}–0.1 μM	3x10^{-6} μM	1x10^{-5} μM	Cookies, bean flakes, horseradish paste, urine	[66]
Theophylline	MoS_2/PANI@g-C_3N_4/GCE	DPV	6.6–130.0 μM	0.051 μM	-	Ice cream, energy drink, theophylline tablets	[49]
Theophylline	Poly(PRO)-MWCNT/GCE	DPV	9.66–100.5 μM	0.044 μM	-	Tea, coffe, infusion of yerba mate, Coke zero, energy drink (speed) and human breast milk	[50]
Triclosan	AgNPs/C3N4NTs@GNQDs/ILs/GCE	DPV	1x10^{-5} – 0.01 μM	1x10^{-5} μM	2x10^{-6} μM	Wastewater	[67]
Vanillin	f-MWCNTs-FNTs/CPE	CV	0.05–9.0 and 10.0–100.0 μM	0.035 μM	-	Sugar samples	[68]
Vanillin	poly(phenylalanine)/TiO_2/CPE	DPSV	1.6–98.6 μM	0.21 μM	-	Chocolate sample, vanillin powder	[69]
Vitamin A	Gr/Nafion/SPE	AdSSWV	0.06–3.49 μM	0.03 μM	0.0998 μM	Infant milk, yogurt, parsley	[70]
Vitamin B_{12}	N,S-GQD-CS/GCE	AdSSWV	0.001–8.0 μM	0.00032 μM	0.00108 μM	Energy drink	[71]
Vitamin B_2	N,S-GQD-CS/GCE	AdSSWV	0.001–8.0 μM	0.0003 μM	0.00101 μM	Energy drink	[71]
Vitamin B_2	HMS@GO/GCE	SWV	0.001–4.0 μM	0.00026 μM	-	Egg yolk, shrimp, milk powder, honey	[72]
Vitamin B_6	N,S-GQD-CS/GCE	AdSSWV	0.1–18.0 μM	0.0301 μM	0.1001 μM	Energy drink	[71]
Vitamin D	Gr/Nafion/SPE	AdSSWV	0.05–2.59 μM	0.0163 μM	0.0548 μM	Infant milk, yogurt, parsley	[70]

TABLE 16.2
Electro-analysis of pesticides using carbon sensors

Pesticide	Electrode	Method	Lineer range	LOD	LOQ	Samples	Ref
Acetamiprid	3D-CS/ rGO/GCE	SWV	0.1 pM–0.1 μM	71.2 fM	-	Tea samples	[79]
Carbendazim	MoS_2QDs@ MWCNTs /GCE	SWV	0.04–1.0 μM	26 nM	-	Platycodon grandiflorum and pear	[80]
Carbendazim	Fullerene/MWCNT/Nafion®/GCE	SWV	20–350 nM	17 nM	55.7 nM	Ferritic soils	[81]
Carbofuran	Gd2S3/RGO/GCE	DPV	0.001–1381 μM	0.012 μM	-	Potato stock solution and river water	[82]
Chloridazon	MIP-CNT/CPE	DPV	5.0×10^{-7} to 4.0×10^{-4} M	6.2×10^{-8} M	-	Water	[83]
Dichlorvos	Au@MWCNTs/GCE	DPV	1–120 μM	5 nM	-	Agricultural Food	[84]
Dinobuton	MWCNTPE	SWSV	3.74–25.8 μM	0.73 μM	2.43 μM	Water and juice (apple, grape)	[85]
Fluometuron	FePc / MWCNTPE	DPSV	0.4–15.0 mg/L	69.8 μg/L	233.0 μg/L	Tap water Commercial formulation	[86]
Fluometuron	FePc / MWCNTPE	SWSV	0.4–7.5 mg/L	101.0 μg/L	337.0 μg/L	-	[86]
Methyl parathion	SWCNT@CeO_2/GCE	DPV	0.01–10.0 μM	9.12 nM		Peach juice, cabbage	[87]
Methyl parathion	MWCNT/zirconia	DPV	19.9–176.8 μM	9.0 nM	-	Soybean sample	[88]
Methyl parathion	SWCNT-ZnPc/ GCE	DPV	2.45– 40 nM	1.49 nM	-	Orange, peach juice	[89]
Methyl parathion	SWCNT-SubPc-Pc/ GCE	DPV	2.5– 37.5 nM	1.78 nM	-	Orange and cherry juice	[90]
Methyl parathion	ZrO_2/rGO/GCE	GECT	10 pg/mL–10 μg/mL	10 pg/mL	-	Chinese cabbage samples.	[91]
Trifluralin	PGE	DPV	1.0–75.0 μM (ox) and 0.50–100.0 μM (red)	0.39 μM (ox) 0.20 μM (red)	-	Water	[92]
Trifluralin	Au@Ag cores-shell/GQDs/ wearable sensing systems	Amperometry	10 nM–1 mM	-	10 nM	Water Real leaf Apple	[93]

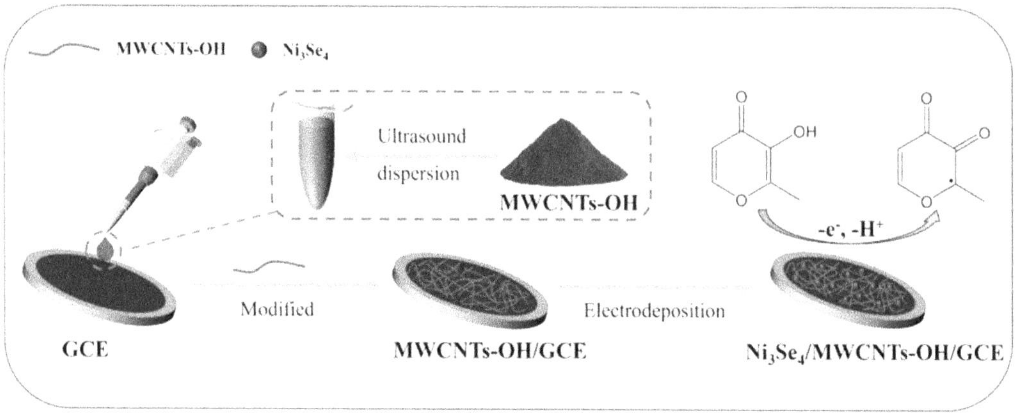

FIGURE 16.3 Illustration of preparation procedure of Ni_3Se_4/MWCNTs/GCE. (Reproduced with permission [62]. Copyright 2024, Elsevier.)

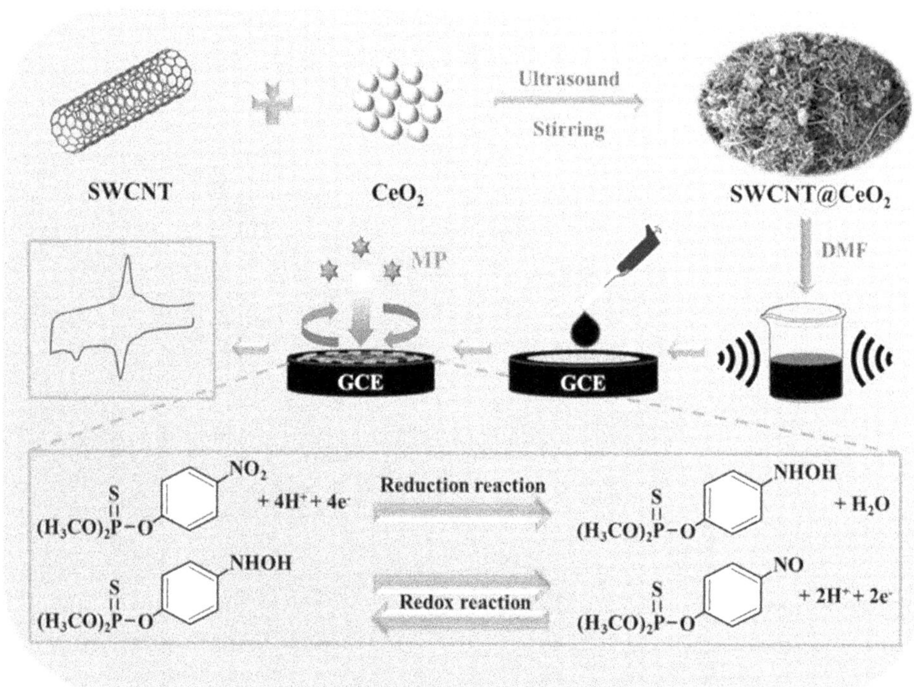

FIGURE 16.4 Design of the SWCNT@CeO_2/GCE sensor for the determination of MP. (Reproduced with permission [87]. Copyright 2024, Elsevier.)

Heavy metals are elements with atomic weights and densities at least five times greater than those of water [94]. Some researchers propose to define heavy metals as naturally occurring elements with atomic number greater than 20 and density exceeding 5 g/cm³ [95]. Arsenic, cadmium, chromium, lead, and mercury are of particular concern owing to their high toxicity. These elements are widely distributed in the environment because of their various applications, raising concerns regarding their effects on human health and ecosystems. The toxicity of heavy metals depends on factors such as dose, route of exposure, and individual characteristics [94]. Although some heavy metals are necessary at permissible levels for biological processes, excessive amounts can accumulate in the body, leading to serious health problems and environmental degradation [96]. Therefore, heavy metal analysis is very important to assess the health risks associated with herbal medicines and supplements [97]. Regular monitoring of heavy metals in soil, water, and air is vital because contamination can affect crop yield, soil quality, and human health along the food chain [98,99]. Carbon nanomaterials have emerged as sensitive, selective, and promising platforms for determining heavy metal concentrations in environmental and human samples, offering advantages over traditional techniques [100]. Table 16.3 summarizes the studies conducted in the last five years for the determination of heavy metals using different carbon materials.

In their study, Chaudhary and colleagues [104] produced boron (B) and nitrogen (N) co-doped laser-formed porous graphene (LIG_{BN}) nanostructured electrodes using a direct laser writing technique on polyimide (PI) sheets (Figure 16.5). They investigated the morphology and microstructure of LIG_{BN} using field emission scanning electron microscopy (FE-SEM) and high-resolution transmission electron microscopy. They studied the binding properties, defects, and chemical states of LIG_{BN} using a Raman spectrometer and XPS. They performed a square wave voltammetry (SWV) technique for the separate and simultaneous electrochemical detection of lead (Pb^{2+}) and cadmium (Cd^{2+}) ions using the produced electrodes. They calculated the sensitivity of the electrode to Pb^{2+} and Cd^{2+} ions as 0.725 μA/μM and 0.661 μA/μM, respectively, due to the synergistic effect of B and N co-doping. Additionally, the carbon sensor provides a low detection limit of 0.21 μM and 0.25 μM for Pb^{2+} and Cd^{2+} ions and wide linear ranges of 8.0 to 80 μM for Pb^{2+} and Cd^{2+} ions. They proved that it provides high linearity ($R^2 = 0.99$) in case of simultaneous detection. They emphasized with this study that this fast and easy method of producing heteroatom-doped porous graphene opens a new path in electrochemical detection studies to detect various hazardous metal ions [104].

A drug is a chemical substance primarily used for therapeutic purposes in the diagnosis, treatment, and prevention of diseases [116]. Drugs play a crucial role in healthcare, with rational drug use being essential for effective treatment [117]. Drug analysis plays a crucial role in ensuring the safety, efficacy, and quality of pharmaceutical products throughout the drug development process [118]. Various analytical techniques are employed for this purpose, particularly useful for analyzing drugs in biological samples. These analytical methods are essential for identifying and characterizing compounds, elucidating molecular structures, and maintaining quality control in pharmaceutical manufacturing [118]. The importance of drug analysis extends to various fields, including the pharmaceutical industry, control laboratories, and biomedicine, contributing significantly to public health and safety [119]. Electrochemical analysis using carbon-based sensors has emerged as a powerful tool for drug detection, offering high sensitivity, low cost, and rapid results [120,121]. These sensors have been applied to various drug classes, including anti-inflammatory, anti-depressant, and anticancer drugs [122,123]. Recent developments include the creation of hybrid platforms that combine different carbon nanomaterials or integrate them with conductive polymers and ionic liquids to further improve sensor performance [124]. Additionally, metal oxide nanoparticles are often combined with carbon-based materials to create a synergistic effect, increasing catalytic activity and lowering detection limits to nanomolar or even picomolar levels [125–128]. The growing significance of electrochemical sensors in drug analysis is evident in their application to biological, pharmaceutical, and environmental samples, where they have become increasingly valuable (Table 16.4).

Xu et al. [139] developed a new ultrasensitive carbon-based sensor based on copper metal-organic framework (CuMOF) and SWCNTs@AuNPs to detect the antileukemic drug imatinib. Poly(diallyldimethylammonium chloride) (PDDA) was used as dispersant for the development of CuMOF-SWCNTs@AuNP. The internal van der Waals interactions of carboxylic single-walled carbon nanotubes (SWCNT-COOH) were reduced. Thus, the AuNP loading was increased (Figure 16.6). The morphology, structure, composition, and electrochemical properties of the proposed CuMOF-SWCNTs@AuNP composite film were characterized using SEM, TEM, FT-IR, UV-visible spectroscopy, XRD, XPS, CV, and EIS. The proposed sensor exhibited a broad linear

TABLE 16.3
Electro-analysis of heavy metals using carbon sensors

Heavy Metals	Electrode	Method	Linear range	LOD	LOQ	Samples	Ref
Cd (II)	MWCNTs- COOH/UiO-66-NH_2/ MWCNTs-COOH/GCE	DPASV	1–511 ppb	0.09 ppb	0.27 ppb	Seawater, rainwater, and air conditioning	[101]
Pb (II)	MWCNTs- COOH/UiO-66-NH_2/ MWCNTs-COOH/GCE	DPASV	1–271 ppb	0.071 ppb	0.213 ppb	seawater, rainwater, and air conditioning	[101]
Cd (II)	Sb_2O_3/MWCNTs/CPE	SWASV	10–100 ppb	1.7 ppb	-	Tap water	[102]
Pb (II)	Sb_2O_3/MWCNTs/CPE	SWASV	10–100 ppb	1.2 ppb	-	Tap water	[102]
Cd (II)	MWCNTs/Nafion/HgE	DPASV	0.2 100 µg/L	0.02 µg/L		Tap and lake waters	[103]
Pb (II)	MWCNTs/Nafion/Hg	DPASV	0.2–100 µg/L	0.17 µg/L		Tap and lake waters	[103]
Cd (II)	LIG_{BN}	SWV	8–80 µM	0.25 µM	-	-	[104]
Pb (II)	LIG_{BN}	SWV	8–80 µM	0.21 µM	-	-	[104]
Cr (III)	MWCNTs–COOH/CPE	Potentiometry	0.1 nM–10 mM	30 pM	-	EDTA titration	[105]
Cr (VI)	PANI@ZrO_2-SO_4^{2-}@MWCNT/GCE	DPV	0.55–39.5 µM	64.3 nM	-	Waste water	[106]
Cs (I)	MWCNTs@Cs-IIP/GCE	Potentiometry	0.1–100 µM	0.04 µM	-	River; industrial waste water and brine	[107]
Cu (II)	MWCNT/CPE	Potentiometry	10 µM – 1 M	3.3 µM	-	Real water samples and urine and serum	[108]
Cu (II)	3DOM/Chitosan/PB/SWCNTs/Au electrode	Amperometry	10^{-18} – 10^{-5} M	10^{-19} M	-	River water	[109]
Cu (II)	NH_2-MIL-101(Fe)/GCE	DPV	1.0–10 µM	8.53 nM	-	-	[110]
Hg (II)	NH_2-MIL-101(Fe)/GCE	DPV	1.0–12 µM	18.0 nM	-	-	[110]
Pb (II)	NH_2-MIL-101(Fe)/GCE	DPV	0.1–10 µM	12.0 nM	-	-	[110]
La(III)	Chitosan/MWCNTs/CPE	Potentiometry	1 µM – 10 mM	1 µM	-	Zoshk river and tap water	[111]
Mercury	PANI/MWCNTs/AuNPs/ITO	DPV	0.01–10.00 ppm	0.08 ppm	-	Cosmetic products	[112]
Pb (II)	TCPP-MWCNTs@Fe_3O_4/MGCE	FSV	0.2–2.0 ng/L	0.067 ng/L	-	Seawater	[113]
Pb (II)	Co-Nx-C@/MWCNTs/GCE	DPASV	2–60 µg/L	0.7 µg/L	-	Tap and river water	[114]
Pb (II)	Bi/Ag@CC electrode	DPV	20 to 400 ppb	0.15 ppb	-	Tap water and lake water	[115]

TABLE 16.4
Electro-analysis of drugs using carbon sensors

Drugs	Electrode	Method	Linear range	LOD	LOQ	Samples	Ref
Pazopanib	4-ABA/PZB@MIP/GCE	DPV	0.1–1.0 pM	10.4 fM	34.8 fM	Human serum, tablets	[123]
Cefditoren	MWCNT/ChitosanNCs/Fe_2O_3/GCE	AdSDPV	0.2–10 µM	1.65 nM	5.50 nM	Pharmaceutical dosage form and human serum	[126]
Linezolid	TiO_2/MWCNT-COOH/GCE	AdSDPV	1.0–20.0 µM	0.0365 µM	0.1215 µM	Tap water, natural spring water, and soil	[127]
Mepivacaine	AgNP/MWCNTPE	DPV	0.1–8.0 mg/L	31.0 µg/L	-	Pharmaceutical dosage form and synthetic human serum	[128]
5-fluorouracil	poly(GLY)/PGE	DPV LSV CV	1.0–13.0 µM	0.012 µM	-	Urine sample	[129]
5-fluorouracil	CeO_2–CuO/CPE	CV DPV	0.1–40 µM	0.01 µM	-	Human urine samples	[130]
Amitriptyline	C60-RGO-H20-NH_2/CoNPs	Immunosensor	1nM– 100 µM	0.5 nM	-	Urine	[131]
Capecitabine	DPB/Cu-MCM-41/CPE	DPV CV LSV	2.0–80.0 µM	0.144 µM	-	Blood serum sample	[132]
Ciprofloxacin	Nafion and Fullerenes (N- F/CPE)	DPV	1.0–18 µM	1.0 µM	3.0 µM	Beef	[133]
Dasatinib	ZnO@f-MWCNT	DPV	0.03–82.65 µM	0.052 µM	-	Pharmaceutical and biological samples	[134]
Dasatinib	ZnO@f-MWCNT	SWV	0.01–122.45 µM	0.48 µM	-	Pharmaceutical and biological samples	[134]
Diclofenac	DCF/Aptamer/EDC-NHS/f-MWCNTs/GCE	EIS	0.25–1 pM; 1 pM–500 nM	1.62×10^{-4} nM	-	-	[135]
Doxorubicin	N-CNOs/GCE	DPV	0.2 nM –10.0 µM	60.0 pM	-	Human blood serum	[136]
Flutamide	DNPs/SPCE	CV DPV	0.025– 606.65 µM	0.023 µM	-	River samples	[137]
Idarubicin	dsDNA/PP/La_2O_3NP@SF-L Cu_2S NS/PGE	DPV	0.01– 500.0 µM	1.3 nM		Human blood serum	[138]

(continued)

TABLE 16.4 (Continued)
Electro-analysis of drugs using carbon sensors

Drugs	Electrode	Method	Linear range	LOD	LOQ	Samples	Ref
Imatinib	CuMOF-SWCNTs@AuNPs/GCE	CV EIS	0.05 to 20.0 µM	5.2 nM	-	Human serum	[139]
Isoniazid	ds-DNA/Carbon/La^{3+}/CuO/CPE	DPV	1.0 to 165.0 µM	35 nM	-	Pharmaceutical dosage form, Urine	[140]
Levodopa	C60–CNT/GCE	DPV	0.5–2000 µM	0.035 µM	-	Synthetic solutions	[141]
Paracetamol	CuNPs/fullerene-C60/MWCNT/CPE	SWV	4.0–400.0 nM	73 pM	2.4 nM	Serum, plasma and urine	[142]
Primaquine	Fullerene C60-MIP/PGE	DPASV	2.7–848.5 nM	0.80 nM	-	Human blood plasma, urine, and Pharmaceutics	[143]
Rifampicin	IL-f-TiO_2NPs/ MWCNTs/GCE	DPV	0.015–2.8 µM	21.8 nM	312.0 nM	-	[144]
Rifaximin	MnO_2 NPs/fullerene-C60/CPE	DPV	0.8–31.5 mg/L	0.76 mg/L	2.31 mg/L	Pharmaceutic	[145]
Salbutamol	Ponceau 4R-oxidized polypyrrole/ SWCNT/GCE	LSV	0.02–10.0 µM	6.0 nM	-	Pharmaceutical and clinical samples	[146]
Triclosan	NA-DNA/IrO_2 NPs/MWCNTs-Gr/GCE	EIS	0.01–10.0 µM	1.2 nM	3.7 nM	Toothpaste, handwashing	[147]
Valacyclovir	AuNPs/Poly-AHP/CPE	DPV	0.0050.08 µM	1.89 nM	-	Urine, serum, and pharmaceutical form	[148]
Vismodegib	Co-MOF/MIP/GCE	DPV	0.1–1.0 pM	14.1 fM	47.3 fM	Synthetic human serum	[149]

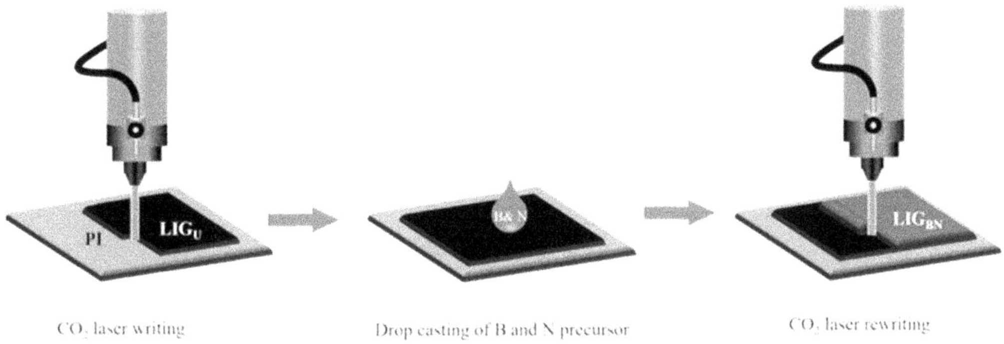

FIGURE 16.5 Schematic illustration on the synthesis of LIG_{BN}. (Reproduced under the terms of the open access Creative Common CC BY license. Copyright 2024, MPDI. [104])

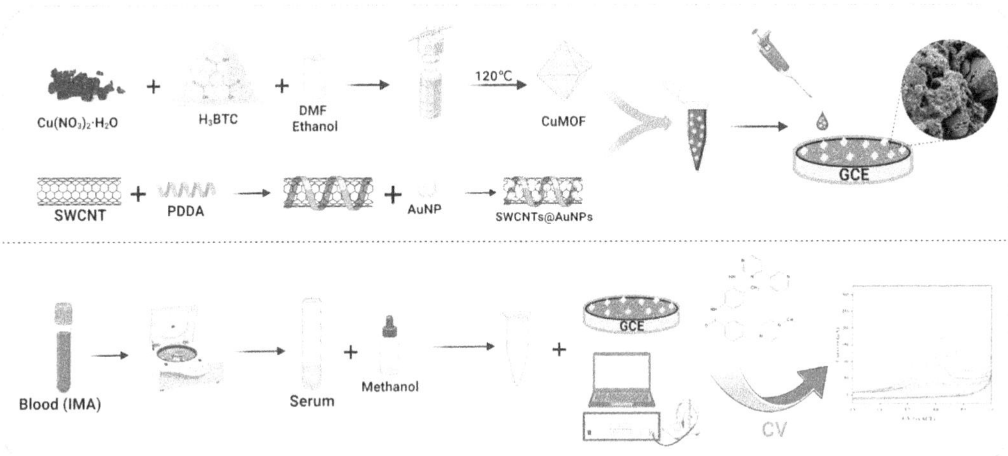

FIGURE 16.6 Engineering of CuMOF-SWCNTs@AuNPs-based carbon sensors for determination of imatinib. (Reproduced under the terms of the CC-BY-NC-ND 4.0 Creative Commons Attribution 4.0 International license (https://creativecommons.org/licenses/by-nc-nd/4.0/) [139]. Copyright 2024, The Authors, published by ACS Publications.)

response in the range of 0.05 to 20.0 μM and a low detection limit of 5.2 nM under optimal conditions. The developed carbon sensor was used for the imatinib analysis of blood serum samples. Therefore, the CuMOF-SWCNTs@AuNP-based electrochemical sensor was proven to have distinct sensing responses for imatinib, which can be used as a prospective approach in clinical applications [139].

Carbon-based electrochemical sensors have gained significant attention for the analysis of various analytes due to their unique properties, including non-toxicity, low cost, and good electrical conductivity [150]. These sensors have been successfully applied to food, cosmetics, and biological samples. Furthermore, major molecules found in plants are detected with carbon sensor [151,152]. These sensors offer advantages such as high sensitivity, low cost, and ease of use [153,154]. Table 16.5 shows the carbon sensor developed for the determination of different analytes and its applicability in different samples as a result of the linearity studies of these substances.

A study conducted by Priya et al. [166], presented a zerovalent iron in mesoporous carbon (Fe@C) material prepared from an metal-organic framework (MOF) via a simple hydrothermal and pyrolysis technique for the determination of metol. The MOF-derived zerovalent iron in mesoporous carbon

TABLE 16.5
Electro-analysis of different analytes using carbon-based sensors

Analyte	Electrode	Method	Lineer range	LOD	LOQ	Samples	Ref
Chlorogenic acids	ZIF-67/PEDOT/CP	DPV	0.1 – 40 μM	0.003 μM		*Honeysuckle*	[151]
Quercetin	TrpMA@QUE/MIP-GCE	DPV	1.0–25 pM	0.235 pM	-	*Rubus sanctus* and *Fragaria vesca*	[152]
Hexanal	AuNPs/PB-ZIF-8@SWCNT/GCE	DPV	10^{-16}–10^{-9} M	10^{-16} M	-	-	[153]
Heat shock protein 70	Fullerene C60/GCE	EIS	0.8–12.8 ng/L	0.273 ng/L	0.909 ng/L	Human blood serum	[154]
Caffeine	Pt/CNT/GCE	LSV	5.0 to 25 μM	0.54 μM	1.80 μM	real samples of pharmaceutical formulation	[155]
Catechol	Pt/C60/PGE	DPV	50.0 – 1500.0 μM	2.97 μM	-	local river sanitary waste water	[156]
Hydroquinone	Pt/C60/PGE	DPV	50.0 -1100.0 μM	2.19 μM	-	local river sanitary waste water	[156]
CD146 (Melanoma cell adhesion molecule)	Cu-UIO-66@SWCNT/PtNPs/GCE	SWV	10^{-9}–10^{-4} mg/mL	12 fg/mL		Lung and liver cancer patients' serum samples	[157]
Dopamine	CE-fullerene-Py- Py-3-COOH	EIS	25–250 μg/L	8.77 μg/L	26.6 μg/L	Urine	[158]
Dopamine	C70-COOH/Au Electrode	DPV	1.0–13.0 μM	10 nM	-	Dopamine hydrochloride injection	[159]
Dopamine	Ga_2O_3·ZnO@SWCNT/GCE	Amperometry	1.0–2056 μM	0.052 μM	0.158 μM	Human blood serum and commercial dopamine injection	[160]
E. coli	PEI-SWCNT-modified 1.6 cfu/mL	EIS	101–108 cfu/mL	1.6 cfu/mL	-	-	[161]
H_2O_2	$CuFeS_2$-CB/GCE	Amperometry	0.1 to 100 mM	0.04 mM	-	-	[162]
H_2O_2	$CuFeS_2$-CB/GCE	CV	2.5 to 60 mM	1.0 mM	-	Drinking water	[162]
Homovanilic acid	Fullerene–graphene oxide (C60–GO)/nitrophenyl /GCE	SWV	0.1–7.2 μM	0.03 μM	-	Urine	[163]
L-cysteine	Co/CNT/GCE	EIS	10—500 μM	0.2 μM	0.6 μM	Acetylcysteine drug sample	[164]
Mangiferin	C60/MWCNTs-IL/GCE	Amperometry	2–40 nM	0.61 nM	-	human plasma	[165]
Metol (p-methylamino phenol sulfate)	Fe@C/CB/SPCE	DPV	0.01–362.0 μM	0.0013 μM		Tap, pond and river water	[166]
Prostate-specific antigen (PSA)	PdNP@PANI-C60/GCE	Immunosensor	0.16 ng/L–38 μg/L	19.5 pg/L	-	Serum and urine	[167]
Zearalenone	CNH/GNP/GPE	Immunosensor	10 to 1000 μg/Kg	4.40 μg/Kg	14.90 μg/Kg	Corn	[168]

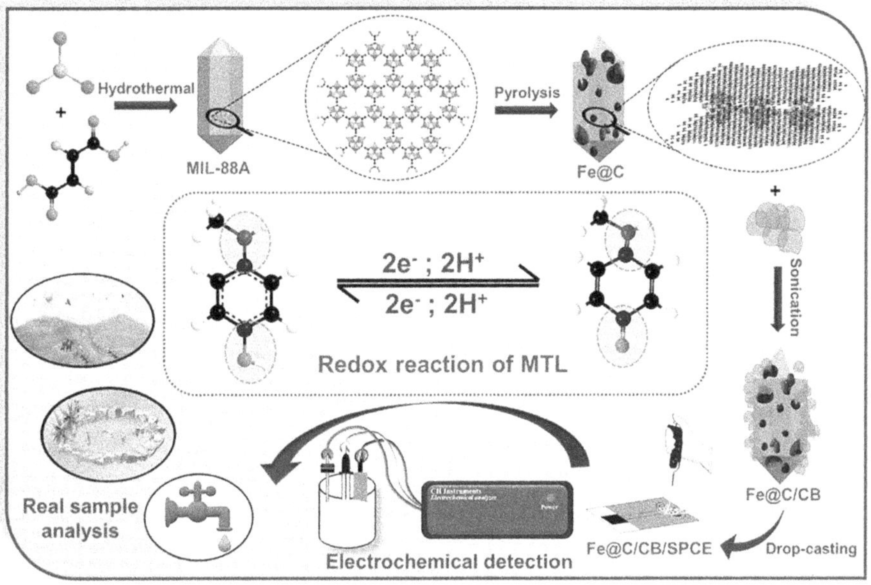

FIGURE 16.7 Preparation of the Fe@C/CB/SPCE sensor and metol detection. (Reproduced with permission. [166] Copyright 2024, Elsevier.)

(Fe@C) with a carbon black (CB) (Fe@C/CB) composite was created through ultrasonication and used to modify the SPCE for metol detection (Figure 16.7). The material's structural and elemental characteristics were analyzed using powder X-ray diffraction (PXRD), fourier transform infrared spectroscopy (FT-IR), Raman, X-ray photoelectron spectroscopy (XPS), field emission scanning electron microscopy (FE-SEM), transmission electron microscopy (TEM) and Brunauer-Emmett-Teller (BET) methods. The electrochemical activity of the sensing platform was assessed using EIS and CV measurements, and the sensing parameters were investigated via multiple CV experiments. The electroanalytical performance of the electrode was evaluated using a highly sensitive DPV technique. The carbon sensor exhibited low LOD (0.003 μM), high sensitivity (12.948 $\mu A\ \mu M^{-1} cm^{-2}$), and impressive stability for the metol determination. The Fe@C/CB/SPCE sensor was employed to monitor the metol in diverse water samples, demonstrating that it yielded remarkable recovery values (98.2–99.8%). As a result, it can be concluded that our Fe@C/CB-based electrochemical sensor is promising for metol detection [166].

16.4 FUTURE TRENDS AND INNOVATIONS IN CARBON SENSOR TECHNOLOGY

Carbon, the most important element of the earth, has been an indispensable component of the sensor world since the years it first entered the electroanalytical world due to its extraordinary conductivity, stability, robustness, and large surface area properties. The unique properties of carbon, one of the most abundant elements in the world, are opening up new areas of application every day with the discovery of its unique properties. Carbon-based materials, including graphene, CNTs, and carbon quantum dots, have become important nanomaterials for energy storage, CO_2 capture, and sensors. Due to their advantageous properties such as high conductivity, chemical stability, and biocompatibility, they have found a place in countless application areas. The most important of these can be said to be the sensor world. In addition to factors such as conductivity, sensitivity, stability, and low cost of carbon and its derivatives, chemical modification of their surfaces increases the variety of sensors. These sensors, referred to as hybrid or composite, exhibit incredible selectivity, accuracy, and sensitivity. Even miniaturized sensors allow for field (on-site) analysis. In addition, they have become attractive equipment with sensor kits as they become portable. Thus, it has given rise to new ideas in sensor development in the scientific world. In addition, expanding interdisciplinary research in electrochemical sensors based on carbon materials is directing advances in production techniques and has found a place in a variety of multidisciplinary fields of study. This chapter focuses on carbon and carbon-based nanomaterials with special emphasis on food additives, pesticides, heavy metals, pharmaceuticals, and various other analytes and their applications over the last five years.

Carbon nanomaterials have become possible to produce sensors with the desired dimensions and hardness, 2D or 3D features with various technological tools. This is a promising development regarding the construction of carbon sensors that can be used in health industry and defense industry, and become commercial. The production of sensor kits in microchip sizes and integrated into the human body will benefit from the unique properties of carbon indicators. Moreover, carbon materials have an important place in the development of sensors that can be controlled remotely and provide instant data. Carbon sensors will contribute significantly to the development of sensors that can control the dose levels of pesticides that need to be controlled not only in terms of human health but also in agriculture. It will lead to the determination of pesticides on-site, simply, selectively, quickly, and sensitively without any pre-treatment. Thus, carbon sensors will provide serious gains in terms of time and budget. In addition, with the functional feature provided by carbon material, it opens up horizons for the development of more effective and catalytic hybrid sensors by combining different polymer, organic, and inorganic nanomaterials. The importance of carbon sensors in the production and commercialization of wearable sensing systems, which are of great importance in the defense industry, is increasing day by day. There will always be a need for sensors to be improved in terms of lifespan, repeatability, and sensitivity. In addition, wearable sensors need to be flexible and, in this case, the need for carbon paper sensors and flexible solid sensors is increasing day by day. Although flexible, they should show excellent functionality even under pressure, temperature, and biological factors. Although biosensing systems are relatively resistant to these conditions due to the stability and robustness of carbon nanomaterials, they need to be improved. Moreover, the rapid development of technology causes the sensor world to look differently. It brings a difference to the examination of sensor–analyte interfaces with robotic sensing sensors. These analysis kits can be controlled remotely and cause trace-level analytes to be determined qualitatively and quantitatively within seconds. It also leads to the trend of artificial intelligence in the production of robotic (mechanical) analysis systems for very sensitive detection. Considering all these situations, the development of detection kits has a significant impact not only on scientific but also on the social contribution process.

16.5 CONCLUSION

Carbon nanomaterial is an extremely popular substance that began with its discovery in the middle of the 19th century and is still being researched today. Especially with the endless innovations that contribute to the sensor world, many studies are being carried out today for the improvement, development, and production of nanomaterials. In particular, the number of more sensitive, selective, portable, and commercial carbon sensors is increasing day by day. The discovery of carbon and its derivatives, nanotubes, has made a positive contribution to this development. In addition, the fact that carbon nanomaterials can be treated with polymers, inorganics, clay, and nanoparticles has contributed to the development of new generation composite sensors. Thus, more sensitive, long-lasting, and highly selective hybrid sensors have been developed. In addition, carbon nanosensors have an important place in the production of online analysis kits that are portable and provide field analysis. Thus, on-site analytes can be determined easily in a very short time. Not only the quantitative analyses of countless substances with carbon sensors but also their qualitative analyses have been made possible using electrochemical techniques with extraordinary selectivity. In addition, carbon nanosensors compete very easily with other traditional analytical methods due to their wide working range and low detection limits. In fact, carbon sensors used in electrochemical techniques are the first indicator that comes to mind due to their selectivity, reproducibility, portability, short analysis time, and advantages such as not requiring any pretreatment. Consequently, carbon nanomaterials with unique properties have become promising and indispensable biomarkers in many fields of study, especially in sensors.

ACKNOWLEDGEMENTS

Seyda Nur Samanci thanks the financial support from the Scientific and Technological Research Council of Türkiye (TUBITAK) under the BIDEB/2211-A Ph.D. and ARDEB/1004 Ph.D. Scholarship Programs.

ABBREVIATIONS

µM	Micromolar
AdSSWV	Absorptive stripping square wave voltammetry
AuE	Gold electrode
CB	Carbon black
CC	Carbon cloth
CD	Carbon dot
CMC	Carboxymethylcellulose
CNH	Carbon nanohorns
CNT	Carbon nanotube
CP	Carbon paper electrode
CPE	Carbon paste electrode
CV	Cyclic voltammetry
DPV	Differential pulse voltammetry
DPSV	Differential pulse stripping voltammetry
EIS	Electrochemical impedance spectroscopy
ERGO	Electrochemical reduced graphene oxide
fM	Fentomolar
GCE	Glassy carbon electrode
GECT	Graphene electrochemical transistor
GNP	Nanoporous Gold
GO	Graphene oxide
GPE	Graphitized paper electrode
Gr	Graphene
GQD	Graphene quantum dot
HgE	Hg electrode
ITO	Indium tin oxide
LIG_{BN}	Boron (B) and nitrogen (N) co-doped laser-induced porous graphene
LOD	Limit of detection
LOQ	Limit of quantification
LSV	Linear sweep voltammetry
MIP	Molecularly imprinted polymer
MWCNT	Multi-walled carbon nanotubes
nM	Nanomolar
OMPD	Origami microfluidic paper-based

PGE	Pencil graphite electrode
pM	Picomolar
rGO	Reduced graphene oxide
SDLSV	Second-order derivative linear sweep voltammetry
SPCE	Screen-printed carbon electrode
SPE	Screen-printed electrode
SWCNT	Single-walled carbon nanotubes
SWV	Square wave voltammetry
SWSV	Square wave stripping voltammetry

REFERENCES

[1] R. Amin, P. Ramesh Kumar and I. Belharouak, *Carbon Nanotubes: Applications to Energy Storage Devices, in Carbon Nanotubes – Redefining the World of Electronics*, IntechOpen, 2021.

[2] M. Javaid, A. Haleem, S. Rab, R. Pratap Singh and R. Suman, *Sensors for daily life: A review*, *Sensors Int.* 2 (2021), pp. 100121.

[3] X. Wang, J. Wang, S. Biswas, H. Kim and I. Nam, *Mechanical, electrical, and piezoresistive sensing characteristics of epoxy-based composites incorporating hybridized networks of carbon nanotubes, graphene, carbon nanofibers, or graphite nanoplatelets*, *Sensors* 20 (2020), pp. 2094.

[4] M.N. Norizan, M.H. Moklis, S.Z. Ngah Demon, N.A. Halim, A. Samsuri, I.S. Mohamad et al., *Carbon nanotubes: Functionalisation and their application in chemical sensors*, *RSC Adv.* 10 (2020), pp. 43704–43732.

[5] G. Speranza, *Carbon nanomaterials: Synthesis, functionalization and sensing applications*, *Nanomaterials* 11 (2021), pp. 967.

[6] S. Morais, *Special issue: Carbon-based nanomaterials for (bio)sensors development*, *Nanomaterials* 11 (2021), pp. 2430.

[7] J.G. Manjunatha, *Electrochemical Sensors Based on Carbon Composite Materials*, IOP Publishing, 2022.

[8] T.M.B.F. Oliveira, F.W.P. Ribeiro, C.P. Sousa, G.R. Salazar-Banda, P. de Lima-Neto, A.N. Correia et al., *Current overview and perspectives on carbon-based (bio)sensors for carbamate pesticides electroanalysis*, *TrAC Trends Anal. Chem.* 124 (2020), pp. 115779.

[9] R. Kour, S. Arya, S.-J. Young, V. Gupta, P. Bandhoria and A. Khosla, *Review—Recent advances in carbon nanomaterials as electrochemical biosensors*, *J. Electrochem. Soc.* 167 (2020), pp. 037555.

[10] N. Slepičková Kasálková, P. Slepička and V. Švorčík, *Carbon nanostructures, nanolayers, and their composites*, *Nanomaterials* 11 (2021), pp. 2368.

[11] C. Verma, M.A. Quraishi, E.E. Ebenso and C.M. Hussain, *Recent advancements in corrosion inhibitor systems through carbon allotropes: Past, present, and future*, *Nano Sel.* 2 (2021), pp. 2237–2255.

[12] J.H.Z.Z.W.Z. Zijian Xu, *Fullerene*, in *Definitions*, Qeios, 2020.

[13] T. Anusha, K.S. Bhavani, J.V.S. Kumar and P.K. Brahman, *Designing and fabrication of electrochemical nanosensor employing fullerene-C60 and bimetallic nanoparticles composite film for the detection of vitamin D3 in blood samples*, *Diam. Relat. Mater.* 104 (2020), pp. 107761.

[14] H.-W. Zhang, H.-K. Li, Z.-Y. Han, R. Yuan and H. He, *Incorporating fullerenes in nanoscale metal–organic matrixes: An ultrasensitive platform for impedimetric aptasensing of tobramycin*, *ACS Appl. Mater. Interfaces* 14 (2022), pp. 7350–7357.

[15] C. Laghlimi, A. Moutcine, A. Chtaini, J. Isaad, A. Soufi, Y. Ziat et al., *Recent advances in electrochemical sensors and biosensors for monitoring drugs and metabolites in pharmaceutical and biological samples*, ADMET DMPK (2023).

[16] A.E.-M.A. Mohamed and M.A. Mohamed, *Carbon nanotubes: Synthesis, characterization, and applications*, in *Carbon Nanomaterials for Agri-Food and Environmental Applications*, Elsevier, 2020, pp. 21–32.

[17] S. Morais, *Advances and applications of carbon nanotubes*, *Nanomaterials* 13 (2023), pp. 2674.

[18] M.N. Karim, M.A.S. Patwary, S.M.A. Abedin, M.R. Hossen and M.S. Rahman, *A review on carbon nanotubes: Preparation, properties and applications*, *J. Res. Updat. Polym. Sci.* 9 (2020), pp. 96–103.

[19] F. Karimi, N. Zakariae, R. Esmaeili, M. Alizadeh and A.-M. Tamadon, *Carbon nanotubes for amplification of electrochemical signal in drug and food analysis; A mini review*, *Curr. Biochem. Eng.* 6 (2020), pp. 114–119.

[20] Q. He, S. Qiao, Y. Zhou, R. Vajtai, D. Li, P.M. Ajayan et al., *Carbon nanotubes-based electrocatalysts: Structural regulation, support effect, and synchrotron-based characterization*, *Adv. Funct. Mater.* 32 (2022).

[21] J.W. Jordan, W.J.V. Townsend, L.R. Johnson, D.A. Walsh, G.N. Newton and A.N. Khlobystov, *Electrochemistry of redox-active molecules confined within narrow carbon nanotubes*, *Chem. Soc. Rev.* 50 (2021), pp. 10895–10916.

[22] M.V. Kharlamova and C. Kramberger, *Applications of filled single-walled carbon nanotubes: Progress, challenges, and perspectives*, *Nanomaterials* 11 (2021), pp. 2863.

[23] E. Leppänen, E. Gustafsson, N. Wester, I. Varjos, S. Sainio and T. Laurila, *Geometrical and chemical effects on the electrochemistry of single-wall carbon nanotube (SWCNT) network electrodes*, *Electrochim. Acta* 466 (2023), pp. 143059.

[24] W. Wang and D. Sun, *Electrochemical determination of chloramphenicol based on ZnO-NPs/SWCNTS composite modified glassy carbon electrode*, *Int. J. Electrochem. Sci.* 16 (2021), pp. 210216.

[25] J.-H. Han, S. Hwang and J. Hyub Kim, *Electrochemical impedance spectroscopy analysis of plasma-treated, spray-coated single-walled carbon-nanotube film electrodes for chemical and electrochemical devices*, *Microchem. J.* 188 (2023), pp. 108446.

[26] S. Dortez, T. Sierra, M.Á. Álvarez-Sánchez, J.M. González-Domínguez, A.M. Benito, W.K. Maser et al., *Effect of nanocellulose polymorphism on electrochemical analytical performance in hybrid nanocomposites with non-oxidized single-walled carbon nanotubes*, *Microchim. Acta* 189 (2022), pp. 62.

[27] R. Rajaram, T. Gurusamy, K. Ramanujam and L. Neelakantan, *Electrochemical determination of paraquat using gold nanoparticle incorporated multiwalled carbon nanotubes*, *J. Electrochem. Soc.* 169 (2022), pp. 047522.

[28] K. Prabhu, S.J. Malode, N.P. Shetti and R.M. Kulkarni, *Analysis of herbicide and its applications through a sensitive electrochemical technique based on MWCNTs/ZnO/CPE fabricated sensor*, *Chemosphere* 287 (2022), pp. 132086.

[29] Y. Zhao, Y. Moon and R. Savari, *Molecularly imprinted electrochemical sensor for determination of tetrahydrocannabinol in human blood plasma*, *Int. J. Electrochem. Sci.* 17 (2022), pp. 221185.

[30] Mounesh and K.R.V. Reddy, *Decorated CoPc with appliance of MWCNTs on GCE: Sensitive and reliable electrochemical investigation of heavy metals*, *Microchem. J.* 160 (2021), pp. 105610.

[31] D. Sahu, H. Sutar, P. Senapati, R. Murmu and D. Roy, *Graphene, Graphene-Derivatives and composites: Fundamentals, synthesis approaches to applications*, *J. Compos. Sci.* 5 (2021), pp. 181.

[32] H.C.A. Murthy, S. Ghotekar, B. Vinay Kumar and A. Roy, *Retracted: Graphene: A multifunctional nanomaterial with versatile applications*, *Adv. Mater. Sci. Eng.* 2021 (2021).

[33] S. Singh, M.R. Hasan, P. Sharma and J. Narang, *Graphene nanomaterials: The wondering material from synthesis to applications*, *Sensors Int.* 3 (2022), pp. 100190.

[34] P.M.V. Fernandes, C.M. Pereira, J.M. Campiña and A. Fernando Silva, *Graphene-modified glassy carbon electrodes: Correlations between electrochemical performance, film morphology and composition*, *Int. J. Electrochem. Sci.* 17 (2022), pp. 220347.

[35] R. Suresh, S. Rajendran, P.S. Kumar, T.K.A. Hoang, M. Soto-Moscoso and A.A. Jalil, *Recent developments on graphene and its derivatives based electrochemical sensors for determinations of food contaminants*, *Food Chem. Toxicol.* 165 (2022), pp. 113169.

[36] W. Jiang, L. Huang, D. Zhang, Y. Wang and G. Pan, *Graphene-based composites for electrochemical sensor fabrication and their application for drug detection*, *Int. J. Electrochem. Sci.* 16 (2021), pp. 21037.

[37] Y.V.M. Reddy, J.H. Shin, V.N. Palakollu, B. Sravani, C.-H. Choi, K. Park et al., *Strategies, advances, and challenges associated with the use of graphene-based nanocomposites for electrochemical biosensors*, *Adv. Colloid Interface Sci.* 304 (2022), pp. 102664.

[38] Y. Zheng, S. Mao, J. Zhu, L. Fu and M. Moghadam, *A scientometric study on application of electrochemical sensors for detection of pesticide using graphene-based electrode modifiers*, *Chemosphere* 307 (2022), pp. 136069.

[39] L. Abbas, L. Bih, K. Yamni, A. Elyahyaouy, A. El Attaoui and Z. Ramzi, *Graphite carbone structure, Cryst. Struct. Theory Appl.* 12 (2024), pp. 1–10.
[40] P. Wang, X. Yuan, Z. Cui, C. Xu, Z. Sun, J. Li et al., *A nanometer-sized graphite/boron-doped diamond electrochemical sensor for sensitive detection of acetaminophen, ACS Omega* 6 (2021), pp. 6326–6334.
[41] Ž.Z. Tasić, M.B.P. Mihajlović, M.B. Radovanović, A.T. Simonović, D.V. Medić and M.M. Antonijević, *Electrochemical determination of L-tryptophan in food samples on graphite electrode prepared from waste batteries, Sci. Rep.* 12 (2022), pp. 5469.
[42] E.S.R. Rizwana, C.L.G. Aparajita and K. Prateek, *Food additives and hypersensitivity: A review, Int. J. Curr. Microbiol. Appl. Sci.* 10 (2021), pp. 1697–1717.
[43] M. Carocho, M.F. Barreiro, P. Morales and I.C.F.R. Ferreira, *Adding molecules to food, pros and cons: A review on synthetic and natural food additives, Compr. Rev. Food Sci. Food Saf.* 13 (2014), pp. 377–399.
[44] L. Wu, C. Zhang, Y. Long, Q. Chen, W. Zhang and G. Liu, *Food additives: From functions to analytical methods, Crit. Rev. Food Sci. Nutr.* 62 (2022), pp. 8497–8517.
[45] E. Chazelas, M. Deschasaux, B. Srour, E. Kesse-Guyot, C. Julia, B. Alles et al., *Food additives: distribution and co-occurrence in 126,000 food products of the French market, Sci. Rep.* 10 (2020), pp. 3980.
[46] S.P. Ukwo, I.I. Udo and N. Ndaeyo, *Food additives: Overview of related safety concerns, Food Sci. Nutr. Res.* 5 (2022).
[47] F. Mazzara, B. Patella, G. Aiello, A. O'Riordan, C. Torino, A. Vilasi et al., *Electrochemical detection of uric acid and ascorbic acid using r-GO/NPs based sensors, Electrochim. Acta* 388 (2021), pp. 138652.
[48] E.G. Döll, E.R. Santana, J.P. Winiarski, L.G. Baumgarten and I.C. Vieira, *Green synthesis of gold nanoparticles using peach extract incorporated in graphene for the electrochemical determination of antioxidant butylated hydroxyanisole in food matrices, Biosensors* 13 (2023), pp. 1037.
[49] E. Murugan and A. Dhamodharan, *Separate and simultaneous determination of vanillin, theophylline and caffeine using molybdenum disulfide embedded polyaniline/graphitic carbon nitrite nanocomposite modified glassy carbon electrode, Diam. Relat. Mater.* 120 (2021), pp. 108684.
[50] N.A. Aschemacher, S.A. Gegenschatz, C.M. Teglia, Á.S. Siano, F.A. Gutierrez and H.C. Goicoechea, *Highly sensitive and selective electrochemical sensor for simultaneous determination of gallic acid, theophylline and caffeine using poly(l-proline) decorated carbon nanotubes in biological and food samples, Talanta* 267 (2024), pp. 125246.
[51] M. Cittan, E. Altuntaş and A. Çelik, *Multi-walled carbon nanotube modified glassy carbon electrode as curcumin sensor, Chem. Mon.* 151 (2020), pp. 881–888.
[52] K.Z. Mousaabadi, A.A. Ensafi, H. Hadadzadeh and B. Rezaei, *Reduced graphene oxide and carbon nanotubes composite functionalized by azobenzene, characterization and its potential as a curcumin electrochemical sensor, J. Electroanal. Chem.* 873 (2020), pp. 114418.
[53] S. Sahoo and A.K. Satpati, *Determination of curcumin on functionalized carbon nano tube modified electrode and probing its interaction with DNA and copper ion, J. Anal. Test.* 7 (2023), pp. 136–146.
[54] N. Dey, D.S. Raj and T. Devasena, *A comparative study on the feasibility of graphene based avidin conjugate for potential curcumin detection, J. Phys. Conf. Ser.* 1964 (2021), pp. 032009.
[55] R. Wada, S. Takahashi, H. Muguruma and N. Osakabe, *Electrochemical detection of curcumin in food with a carbon nanotube - carboxymethylcellulose electrode, Anal. Sci.* 36 (2020), pp. 1113–1118.
[56] M. Rahimnejad, R. Zokhtareh, A.A. Moghadamnia and M. Asghary, *An electrochemical sensor based on reduced graphene oxide modified carbon paste electrode for curcumin determination in human blood serum, Port. Electrochim. Acta* 38 (2020), pp. 29–42.
[57] B. Mirzaei, A. Zarrabi, A. Noorbakhsh, A. Amini and P. Makvandi, *A reduced graphene oxide-β-cyclodextrin nanocomposite-based electrode for electrochemical detection of curcumin, RSC Adv.* 11 (2021), pp. 7862–7872.
[58] P. Deng, Y. Wei, W. Li, S. Shi, C. Zhou, J. Li et al., *A novel platform based on MnO2 nanoparticles and carboxylated multi-walled carbon nanotubes composite for accurate and rapid determination of curcumin in commercial food products, J. Food Compos. Anal.* 115 (2023), pp. 104940.

[59] L. Yang, T. Yang, G. Li, J. xi Ma and Y. Yu, *Electrochemical studies of polyphenols, anthocyanins, and flavonoids extracted from blueberry fruit, Int. J. Electrochem. Sci.* 17 (2022), pp. 22038.
[60] S. Cheraghi, F. Shalali and M.A. Taher, *Kojic acid exploring as an essential food additive in real sample by a nanostructure sensor amplified with ionic liquid, J. Food Meas. Charact.* 17 (2023), pp. 1728–1734.
[61] A.K. Singh, R. Sharma, M. Singh and N. Verma, *Electrochemical determination of L-arginine in leukemic blood samples based on a polyaniline-multiwalled carbon nanotube—magnetite nanocomposite film modified glassy carbon electrode, Instrum. Sci. Technol.* 48 (2020), pp. 400–416.
[62] X. Wang, Y. Jiang, S. Wang, L. Zhang, T. Zhang and P. He, *Electrochemical sensor based on nickel selenide/hydroxylated multiwalled carbon nanotubes nanocomposites for sensitive detection of maltol, Microchem. J.* 201 (2024), pp. 110692.
[63] L. Zhang, J. Xu, J. Qian and Y. Wang, *A novel detection method for neohesperidin dihydrochalcone (NHDC) based on ZIF-8/ErGO composite modified electrode in milk, J. Food Compos. Anal.* 123 (2023), pp. 105584.
[64] R. Jerome and A.K. Sundramoorthy, *Preparation of hexagonal boron nitride doped graphene film modified sensor for selective electrochemical detection of nicotine in tobacco sample, Anal. Chim. Acta* 1132 (2020), pp. 110–120.
[65] C. Yin, Y. Wang and Q. Zhuang, *Dual-ratiometric electrochemical sensor for propyl gallate detection, J. Electroanal. Chem.* 880 (2021), pp. 114817.
[66] R.-M. Ilie-Mihai, B.C. Ion and J. (Koos) F. van Staden, *Sodium metabisulfite in food and biological samples: A rapid and ultra-sensitive electrochemical detection method, Micromachines* 13 (2022), pp. 1707.
[67] O. Akyıldırım, *A sensitive voltammetric sensor based on silver nanoparticles/carbon nitride nanotubes@graphene quantum dots/a novel organic liquid: Determination of triclosan in wastewater, Bull. Mater. Sci.* 43 (2020), pp. 195.
[68] L. Taouri, M. Bourouina, S. Bourouina-Bacha and D. Hauchard, *Fullerene-MWCNT nanostructured-based electrochemical sensor for the detection of Vanillin as food additive, J. Food Compos. Anal.* 100 (2021), pp. 103811.
[69] I. Ali, M. Mısır, E. Demir, İ. Dinçer, M. Locatelli and Z.A. ALOthman, *Nano solid phase micro membrane tip and electrochemical methods for vanillin analysis in chocolate samples, Anal. Biochem.* 677 (2023), pp. 115268.
[70] J. Thangphatthanarungruang, A. Yakoh, R. Laocharoensuk, C. Chotsuwan, O. Chailapakul and W. Siangproh, *High-efficient of graphene nanocomposite: Application to rapidly simultaneous identification and quantitation of fat-soluble vitamins in different matric samples, J. Electroanal. Chem.* 873 (2020), pp. 114361.
[71] E.C. Martins, E.R. Santana and A. Spinelli, *Nitrogen and sulfur co-doped graphene quantum dot-modified electrode for monitoring of multivitamins in energy drinks, Talanta* 252 (2023), pp. 123836.
[72] L. Zhang, X. Liu, L. Luo, C. Hu, J. Fu, X. Chang et al., *A high-performance voltammetric methodology for the ultra-sensitive detection of riboflavin in food matrices based on graphene oxide-covered hollow MnO_2 spheres, Food Chem.* 352 (2021), pp. 129368.
[73] F.M. Rubino, S. Mandić-Rajčević and C. Colosio, *Pesticides*, in *Kirk-Othmer Encyclopedia of Chemical Technology*, Wiley, 2020, pp. 1–48.
[74] E. Demir, H. Silah and N. Aydogdu, *Electrochemical Analysis of Pesticide in Food Samples*, in *Medicine & Health*, H. Kaya and H. Kafadar, eds., Efe Akademi, 2021, pp. 213–234.
[75] R. Kaur, D. Choudhary, S. Bali, S.S. Bandral, V. Singh, M.A. Ahmad et al., *Pesticides: An alarming detrimental to health and environment, Sci. Total Environ.* 915 (2024), pp. 170113.
[76] J.S. Noori, J. Mortensen and A. Geto, *Recent development on the electrochemical detection of selected pesticides: A focused review, Sensors* 20 (2020), pp. 2221.
[77] R.D. Kadu, R.S. Keri, N.D.H. and S. Budagumpi, *State-of-the-art electrochemical sensors for quantitative detection of pesticides, Appl. Organomet. Chem.* 37 (2023).
[78] M.Z. Momčilović, J.S. Milićević and M.S. Ranđelović, *Recent advances in electrochemical determination of pesticides, J. Nanosci. Nanotechnol.* 21 (2021), pp. 5795–5811.
[79] J. Yi, Z. Liu, J. Liu, H. Liu, F. Xia, D. Tian et al., *A label-free electrochemical aptasensor based on 3D porous CS/rGO/GCE for acetamiprid residue detection, Biosens. Bioelectron.* 148 (2020), pp. 111827.

[80] X. Zhang, J. Du, D. Wu, X. Long, D. Wang, J. Xiong et al., *Anchoring metallic MoS 2 quantum dots over MWCNTs for highly sensitive detection of postharvest fungicide in traditional Chinese medicines*, *ACS Omega* 6 (2021), pp. 1488–1496.
[81] D.N. Teadoum, S.K. Noumbo, K.T. Arnaud, T.T. Ranil, A.D. Mvondo Zé and I.K. Tonle, *Square wave voltammetric determination of residues of carbendazim using a fullerene/multiwalled carbon nanotubes/Nafion® coated glassy carbon electrode*, *Int. J. Electrochem.* 2016 (2016), pp. 1–9.
[82] V. Mariyappan, M. Keerthi and S.-M. Chen, *Highly selective electrochemical sensor based on gadolinium sulfide rod-embedded RGO for the sensing of carbofuran*, *J. Agric. Food Chem.* 69 (2021), pp. 2679–2688.
[83] A. Ghorbani, M.R. Ganjali, R. Ojani and J. Raoof, *Detection of chloridazon in aqueous matrices using a nano-sized chloridazon-imprinted polymer-based voltammetric sensor*, *Int. J. Electrochem. Sci.* 15 (2020), pp. 2913–2922.
[84] Y. Chen, *Efficient electrochemical sensor based on multi-walled carbon nanotubes/gold nanocomposite for detection of dichlorvos pesticide in agricultural food*, *Int. J. Electrochem. Sci.* 16 (2021), pp. 210321.
[85] Z. Murathan, R. İnam and E. Demir, *A new strategy for the determination of dinobuton fungicide by square wave stripping voltammetry on the multi-walled carbon nanotube electrode*, *Mater. Res. Express* 11 (2024), pp. 015604.
[86] E. Demir, Ö. Göktug, R. İnam and D. Doyduk, *Development and characterization of iron (III) phthalocyanine modified carbon nanotube paste electrodes and application for determination of fluometuron herbicide as an electrochemical sensor*, *J. Electroanal. Chem.* 895 (2021), pp. 115389.
[87] J. Han, X. Guo, F. Li, M. Zhao, Y. Liu, M. Guo et al., *Facile fabrication of single-walled carbon nanotubes@cerium oxide composite sensing platform for ultrasensitive electrochemical analysis of methyl parathion*, *J. Food Compos. Anal.* 131 (2024), pp. 106240.
[88] K. dos S. Caetano, D.S. da Rosa, T.M. Pizzolato, P.A.M. dos Santos, R. Hinrichs, E.V. Benvenutti et al., *MWCNT/zirconia porous composite applied as electrochemical sensor for determination of methyl parathion*, *Microporous Mesoporous Mater.* 309 (2020), pp. 110583.
[89] B. Köksoy, D. Akyüz, A. Şenocak, M. Durmuş and E. Demirbas, *Sensitive, simple and fast voltammetric determination of pesticides in juice samples by novel BODIPY-phthalocyanine-SWCNT hybrid platform*, *Food Chem. Toxicol.* 147 (2021), pp. 111886.
[90] B. Köksoy, D. Akyüz, A. Şenocak, M. Durmuş and E. Demirbaş, *Novel SWCNT-hybrid nanomaterial functionalized with subphthalocyanine substituted asymmetrical zinc (II) phthalocyanine conjugate: Design, synthesis, characterization and sensor properties for pesticides*, *Sensors Actuators B Chem.* 329 (2021), pp. 129198.
[91] T. Tao, Y. Zhou, M. Ma, H. He, N. Gao, Z. Cai et al., *Novel graphene electrochemical transistor with ZrO_2/rGO nanocomposites functionalized gate electrode for ultrasensitive recognition of methyl parathion*, *Sensors Actuators B Chem.* 328 (2021), pp. 128936.
[92] Ü. Gege, S. Karakaya and Y. Dilgin, *Sensitive electrochemical determination of trifluralin at a disposable pencil graphite electrode*, *Electroanalysis* (2021), pp. elan.202060618.
[93] M. Mahmoudpour, A. Saadati, M. Hasanzadeh and H. Kholafazad-kordasht, *A stretchable glove sensor toward rapid monitoring of trifluralin: A new platform for the on-site recognition of herbicides based on wearable flexible sensor technology using lab-on-glove*, *J. Mol. Recognit.* 34 (2021).
[94] P.B. Tchounwou, C.G. Yedjou, A.K. Patlolla and D.J. Sutton, *Heavy Metal Toxicity and the Environment*, 2012, pp. 133–164.
[95] H. Ali and E. Khan, *What are heavy metals? Long-standing controversy over the scientific use of the term 'heavy metals' – proposal of a comprehensive definition*, *Toxicol. Environ. Chem.* 100 (2018), pp. 6–19.
[96] M. Oves and M. SaghirKhan, *Heavy metals: Biological importance and detoxification strategies*, *J. Bioremediation Biodegrad.* 07 (2016).
[97] R. Bojanala and D.P.N. Shivappa, *Review on comprehensive heavy metals analysis: A crucial tool for health assessment when using herbal drugs/supplements*, *Int. J. Res. -Granthaalayah* 11 (2023).
[98] M. Arjomandi and H. Shirkhanloo, *A review: Analytical methods for heavy metals determination in environment and human samples*, *Anal. Methods Environ. Chem. J.* 2 (2019), pp. 97–126.
[99] A. Bayata, *Assessment, accumulation, toxicity and importance of heavy metals in agricultural soil and living system – Review*, *Am. J. Environ. Prot.* 9 (2020), pp. 121.

[100] Q. Zhou, L. Fu and J. Zhu, *Electrochemical sensors go nano: Carbon nanomaterials for ultrasensitive heavy metal analysis, Curr. Nanosci.* 20 (2024).
[101] Z. Lu, W. Zhao, L. Wu, J. He, W. Dai, C. Zhou et al., *Tunable electrochemical of electrosynthesized layer-by-layer multilayer films based on multi-walled carbon nanotubes and metal-organic framework as high-performance electrochemical sensor for simultaneous determination cadmium and lead, Sensors Actuators B Chem.* 326 (2021), pp. 128957.
[102] M. Majidian, J.B. Raoof, S.R. Hosseini, R. Ojani, J. Barek and J. Fischer, *Novel type of carbon nanotube paste electrode modified by Sb_2O_3 for square wave anodic stripping voltammetric determination of Cd^{2+} and Pb^{2+}, Electroanalysis* 32 (2020), pp. 2260–2265.
[103] Z. Lai, F. Lin, Y. Huang, Y. Wang and X. Chen, *Automated determination of Cd^{2+} and Pb^{2+} in natural waters with sequential injection analysis device using differential pulse anodic stripping voltammetry, J. Anal. Test.* 5 (2021), pp. 60–68.
[104] Y. Chaudhary, S. Suman, B. Rakesh, G.P. Ojha, U. Deshpande, B. Pant et al., *Boron and nitrogen co-doped porous graphene nanostructures for the electrochemical detection of poisonous heavy metal ions, Nanomaterials* 14 (2024), pp. 806.
[105] Azam Hosseini Fakhrabad, R.S. Khoshnood, M. Ebrahimi and M.R. Abedi, *Fabrication of carbon paste electrodes modified with multi-walled carbon nanotubes for the potentiometric determination of chromium(III), J. Anal. Chem.* 75 (2020), pp. 951–957.
[106] M.H. Motaghedifard, S.M. Pourmortazavi and S. Mirsadeghi, *Selective and sensitive detection of Cr(VI) pollution in waste water via polyaniline/sulfated zirconium dioxide/multi walled carbon nanotubes nanocomposite based electrochemical sensor, Sensors Actuators B Chem.* 327 (2021), pp. 128882.
[107] Z. Wang, L. Wang, C. Zhou and C. Sun, *Determination of cesium ions in environmental water samples with a magnetic multi-walled carbon nanotube imprinted potentiometric sensor, RSC Adv.* 11 (2021), pp. 10075–10082.
[108] E. Frag, M. Mohamed, A. Ali and G. Mohamed, *Potentiometric sensors selective for Cu(II) determination in real water samples and biological fluids based on graphene and multi-walled carbon nanotubes modified graphite electrodes, Indian J. Chem. -Section A* 59 (2020), pp. 162–173.
[109] R. Tian, X. Chen, D. Liu and C. Yao, *A sensitive biosensor for determination of Cu^{2+} by one-step electrodeposition, Electroanalysis* 28 (2016), pp. 1617–1624.
[110] P. Chen, H. Wang, H. Li, B. Niu, H. Guo and Z. Chen, *A high-activity Fe-based MOFs fabricated through ultrasound strategy for electrochemical sensor of heavy metal ions and dopamine, J. Electroanal. Chem.* 957 (2024), pp. 118129.
[111] N. Rajabi, M. Masrournia and M. Abedi, *Potentiometric determination of La (III) Using chitosan modified carbon paste electrode with an experimental design, Chem. Methodol.* 4(2020), pp. 660–670.
[112] N.A. Bohari, S. Siddiquee, S. Saallah, M. Misson and S.E. Arshad, *Electrochemical behaviour of real-time sensor for determination mercury in cosmetic products based on PANI/MWCNTs/AuNPs/ITO, Cosmetics* 8 (2021), pp. 17.
[113] C. Zhang, T. Hao, H. Lin, Q. Wang, Y. Wu, K. Kang et al., *One-step electrochemical sensor based on an integrated probe toward sub-ppt level Pb^{2+} detection by fast scan voltammetry, Anal. Chim. Acta* 1128 (2020), pp. 174–183.
[114] R. Zhao, X. Wu, Y. Gao, Y. Liu, J. Gao, Y. Chen et al., *A unique bimetallic MOF derived carbon–MWCNTs hybrid structure for selective electrochemical determination of lead ion in aqueous solution, Microchem. J.* 158 (2020), pp. 105271.
[115] Y. Cao, X. Zhou, Z. Wang, Y. Li, M. Yan, Y. Zeng et al., *Synthesis of bismuth film assembly on flexible carbon cloth for the electrochemical detection of heavy metal ions, Chemosensors* 12 (2024), pp. 103.
[116] R. Kumar, *Impact of drug abuse and addiction in daily life: A review, Biomed. J. Sci. Tech. Res.* 46 (2022).
[117] A. Kumari, *Rational drug use, J. Pharm. Res. Reports* 4(2) (2023), pp. 1–2.
[118] S.K. Kosuru, S.Kumar G, S. Rafi, A. Jyothi M, J. D and S. Durga M.M.V.V, *Pharmaceutical analysis in drug discovery and drug development, J. Clin. Pharm. Res.* 3(2) (2023), pp. 24–26.
[119] C. Fanali, S. Della Posta and S. Fanali, *Capillary electrochromatography applied to drug analysis, J. Chromatogr. Open* 1 (2021), pp. 100015.

[120] S.I. Kaya, A. Cetinkaya and S.A. Ozkan, *Carbon nanomaterial-based drug sensing platforms using state-of-the-art electroanalytical techniques*, *Curr. Anal. Chem.* 18 (2022), pp. 79–101.

[121] J.G. Manjunatha, P.A. Pushpanjali and N. Hareesha, *An overview of recent developments of carbon-based sensors for the analysis of drug molecules*, *J. Electrochem. Sci. Eng.* 11(3) (2021), pp. 161–177.

[122] L. Qian, S. Durairaj, S. Prins and A. Chen, *Nanomaterial-based electrochemical sensors and biosensors for the detection of pharmaceutical compounds*, *Biosens. Bioelectron.* 175 (2021), pp. 112836.

[123] S.N. Samanci, A. Cetinkaya, N. Doufene, N. Bounoua, S.I. Kaya, E.B. Atici et al., *Electrochemical determination of anti-cancer drug pazopanib with high selectivity and sensitivity using molecularly imprinted polymer-modified glassy carbon electrode*, *J. Electrochem. Soc.* 171 (2024), pp. 037504.

[124] A. Cetinkaya, S.I. Kaya, G. Ozcelikay, F. Budak and S.A. Ozkan, *Carbon Nanomaterials-based novel hybrid platforms for electrochemical sensor applications in drug analysis*, *Crit. Rev. Anal. Chem.* (2022), pp. 1–16.

[125] B. Bozal-Palabiyik, C. Erkmen, S. Kurbanoglu, S.A. Ozkan and B. Uslu, *Electrochemical analysis for pharmaceuticals by the advantages of metal oxide nanomaterials*, *Curr. Anal. Chem.* 17 (2021), pp. 1322–1339.

[126] N. Aydogdu, G. Ozcelikay and S.A. Ozkan, *Rapid and sensitive electrochemical assay of cefditorena with MWCNT/Chitosan NCs/Fe2O3 as a Nanosensor*, *Micromachines* 13 (2022), pp. 1348.

[127] N. Aydogdu Ozdogan, E. Demır and S.A. Ozkan, *Sensitive and selective electrochemical characterization and analytical determination of linezolid in environmental samples using* TiO_2 *nanoparticles and MWCNT-COOH modified glassy carbon electrode*, *Microchem. J.* 199 (2024), pp. 109920.

[128] E. Demir, H. Silah, C. Erkmen and B. Uslu, *Electrochemical sensor design based on silver nanoparticles modified multiwalled carbon nanotubes paste electrode for the sensitive determination of mepivacaine*, *Comb. Chem. High Throughput Screen.* 26 (2023), pp. 2644–2655.

[129] E. Pradeepa, Y. Arthoba Nayaka and H.R. Sahana, *Electrochemical investigation of an anticancer drug 5-fluorouracil in the presence of Theophylline using low-cost and disposable poly(GLY) modified pencil graphite electrode*, *Anal. Biochem.* 687 (2024), pp. 115451.

[130] M. Vazan, J. Tashkhourian and B. Haghighi, *Facile electrochemical determination of 5-fluorouracil as an important anti-cancer drug using* CeO_2*–CuO nanocomposite modified carbon paste electrode*, *Mater. Chem. Phys.* 315 (2024), pp. 128950.

[131] E.P. Medyantseva, D.V. Brusnitsyn, E.R. Gazizullina, R.M. Varlamova, O.A. Konovalova and H.C. Budnikov, *Hybrid nanocomposites as electrode modifiers in amperometric immunosensors for the determination of amitriptyline*, *J. Anal. Chem.* 75 (2020), pp. 536–543.

[132] M. Mazloum-Ardakani, Z. Mokari, Z. Alizadeh, H. Mohammadian-Sarcheshmeh, M. Abdollahi-Alibiek, B.B.F. Mirjalili et al., *Electrochemical sensor for sensitive detection of an anticancer drug capecitabine by modified carbon paste electrode with tetrahydrodipyrazolo pyridine derivative and Cu-MCM-41 nanoparticles*, *Microchem. J.* 199 (2024), pp. 109887.

[133] P. Hernández, G.Y. Aguilar-Lira, G. Islas and J.A. Rodriguez, *Development of a new voltammetric methodology for the determination of ciprofloxacin in beef samples using a carbon paste electrode modified with nafion and fullerenes*, *Electroanalysis* (2021), pp. 1–9.

[134] K. Korgaonkar, N.P. Agadi and J. Seetharamappa, *Porous zinc oxide and functionalized multiwalled carbon nanotubes composite as electrode material for enhanced electrochemical sensing of an anticancer drug, dasatinib*, *J. Electrochem. Soc.* 171 (2024), pp. 037521.

[135] Y. Zou, S. Griveau, A. Ringuedé, F. Bedioui, C. Richard and C. Slim, *Functionalized multi-walled carbon nanotube–based aptasensors for diclofenac detection*, *Front. Chem.* 9 (2022).

[136] M.H. Ghanbari and Z. Norouzi, *A new nanostructure consisting of nitrogen-doped carbon nanoonions for an electrochemical sensor to the determination of doxorubicin*, *Microchem. J.* 157 (2020), pp. 105098.

[137] N. Baskaran, S.B. Prasanna, Y.-C. Lin, Y.-F. Duann, R.-J. Chung and Y. Wei, *The high stability and selectivity of electrochemical sensor using low-cost diamond nanoparticles for the detection of anti-cancer drug flutamide in environmental samples*, *Sensors* 24 (2024), pp. 985.

[138] M.M. Foroughi and S. Jahani, *Investigation of a high-sensitive electrochemical DNA biosensor for determination of Idarubicin and studies of DNA-binding properties*, *Microchem. J.* 179 (2022), pp. 107546.

[139] X. Xu, W. Li, H. Xin, L. Tang, X. Zhou, T. Zhou et al., *Engineering of CuMOF-SWCNTs@AuNPs-based electrochemical sensors for ultrasensitive and selective monitoring of imatinib in human serum*, *ACS Omega* 9 (2024), pp. 4744–4753.

[140] H. Mahmoudi-Moghaddam and Z. Garkani-Nejad, *Development of the electrochemical biosensor for determination of antibiotic drug isoniazid using ds-DNA/Carbon/La3+/CuO/CPE*, *Measurement* 189 (2022), pp. 110580.

[141] M. Mazloum-Ardakani, S.H. Ahmadi, Z. Safaei Mahmoudabadi and A. Khoshroo, *Nano composite system based on fullerene-functionalized carbon nanotubes for simultaneous determination of levodopa and acetaminophen*, *Measurement* 91 (2016), pp. 162–167.

[142] P.K. Brahman, L. Suresh, V. Lokesh and S. Nizamuddin, *Fabrication of highly sensitive and selective nanocomposite film based on CuNPs/fullerene-C 60 /MWCNTs: An electrochemical nanosensor for trace recognition of paracetamol*, *Anal. Chim. Acta* 917 (2016), pp. 107–116.

[143] B.B. Prasad, A. Kumar and R. Singh, *Molecularly imprinted polymer-based electrochemical sensor using functionalized fullerene as a nanomediator for ultratrace analysis of primaquine*, *Carbon N. Y.* 109 (2016), pp. 196–207.

[144] R. Chokkareddy and G.G. Redhi, *Novel electrochemical sensor for rifampicin based on ionic liquid functionalised TiO_2 nanoparticles*, *Curr. Anal. Chem.* 18 (2022), pp. 475–482.

[145] R. Abdellatef, E. Khaled, H.A. Hendawy and R.Y.A. Hassan, *Manganese dioxide (MnO_2)/fullerene-C60-modified electrodes for the voltammetric determination of rifaximin*, *J. Anal. Test.* 5 (2021), pp. 341–349.

[146] S. Panahi, S. Shahrokhian and M. Ghalkhani, *Single-walled carbon nanotube coated with Ponceau4R-doped over-oxidized polypyrrole as a sensitive sensor for voltammetric monitoring of salbutamol*, *Diam. Relat. Mater.* 141 (2024), pp. 110705.

[147] A.R. Jalalvand, *A novel impedimetric biosensor for monitoring the DNA damage induced by photoexcited triclosan and investigation of protective effects of the grape seed oil on the DNA damage: Application to impedimetric biosensing of the triclosan*, *Microchem. J.* 179 (2022), pp. 107512.

[148] S. Kummari, V. Sunil Kumar, K. Yugender Goud and K. Vengatajalabathy Gobi, *Nano-Au particle decorated poly-(3-amino-5-hydroxypyrazole) coated carbon paste electrode for in-vitro detection of valacyclovir*, *J. Electroanal. Chem.* 904 (2022), pp. 115859.

[149] B. Hosseinzadeh, S.I. Kaya, A. Çetinkaya, E. Bellur Atici and S.A. Ozkan, *Development of a molecularly imprinted polymer-based electrochemical sensor with metal-organic frameworks for monitoring the antineoplastic drug vismodegib*, *Talanta* 278 (2024), pp. 126510.

[150] S. Michalkiewicz, A. Skorupa and M. Jakubczyk, *Carbon materials in electroanalysis of preservatives*: A Review, *Materials (Basel).* 14 (2021), pp. 7630.

[151] X. Ma, Y. Gao, W. Pang, X. Chang, Z. Hu and T. Hu, *ZIF-67/PEDOT modified carbon paper electrode for sensitive electrochemical determination of chlorogenic acids*, *J. Electroanal. Chem.* 961 (2024), pp. 118235.

[152] M.M. Hurkul, A. Cetinkaya, S. Yayla, S.I. Kaya, F. Budak, K.C. Tok et al., *Highly selective and sensitive molecularly imprinted sensors for the electrochemical assay of quercetin in methanol extracts of Rubus sanctus and Fragaria vesca*, *Talanta* 273 (2024), pp. 125883.

[153] J. Liu, Y. Ping Chen, P. He, Z. Ding, Y. Guo, S. Cui et al., *A novel olfactory biosensor based on ZIF-8@SWCNT integrated with nanosome-AuNPs/Prussian blue for sensitive detection of hexanal*, *Food Chem.* 442 (2024), pp. 138349.

[154] B. Demirbakan and M.K. Sezgintürk, *A novel immunosensor based on fullerene C60 for electrochemical analysis of heat shock protein 70*, *J. Electroanal. Chem.* 783 (2016), pp. 201–207.

[155] A.H. Kalinke, L.H. Marcolino-Junior, M.F. Bergamini and A.J.G. ZARBIN, *Platinum nanoparticles supported by carbon nanotubes: Improvement in electrochemical sensor performance for caffeine determination*, *An. Acad. Bras. Cienc.* 96 (2024).

[156] Y. Zhu, S. Huai, J. Jiao, Q. Xu, H. Wu and H. Zhang, *Fullerene and platinum composite-based electrochemical sensor for the selective determination of catechol and hydroquinone*, *J. Electroanal. Chem.* 878 (2020), pp. 114726.

[157] D. Chen, Y. He, Y. Wang, Z. Zhang, Y. Pei, Y. Lei et al., *An immune sandwich electrochemical biosensor based on triple-modified zirconium derivatives for detection of CD146 in serum*, *Colloids Surfaces B Biointerfaces* 239 (2024), pp. 113902.

[158] H.D. Ertuğrul Uygun and M.N. Demir, *A novel fullerene-pyrrole-pyrrole-3-carboxylic acid nanocomposite modified molecularly imprinted impedimetric sensor for dopamine determination in urine*, *Electroanalysis* 32 (2020), pp. 1971–1976.
[159] D.-W. Han, X.-L. Zhang, B. Yu, H.-L. Cong, C.-B. Mu and H. Yuan, *Electrochemical behavior of dopamine on La@C 82 -COOH/C 60 -COOH/C 70 -COOH modified electrodes*, *Integr. Ferroelectr.* 171 (2016), pp. 131–139.
[160] J. Ahmed, M. Faisal, J.S. Algethami, M.M. Rahman and F.A. Harraz, *A novel* Ga_2O_3*-doped ZnO decorated SWCNT nanocomposite based amperometric sensor for efficient detection of dopamine in real samples*, *J. Sci. Adv. Mater. Devices* 9 (2024), pp. 100668.
[161] S. Joksović, I. Kundačina, I. Milošević, J. Stanojev, V. Radonić and B. Bajac, *Single-walled carbon nanotube-modified gold leaf immunosensor for* Escherichia coli *detection*, *ACS Omega* 9 (2024), pp. 22277–22284.
[162] T. Ma, Y. Wang, Y. Gu and Y. Hasebe, *A carbon black–doped chalcopyrite–based electrochemical sensor for determination of hydrogen peroxide*, *Ionics (Kiel).* 30 (2024), pp. 5651–5661.
[163] J.A. Rather, E.A. Khudaish, A. Munam, A. Qurashi and P. Kannan, *Electrochemically reduced fullerene–graphene oxide interface for swift detection of Parkinsons disease biomarkers*, *Sensors Actuators B Chem.* 237 (2016), pp. 672–684.
[164] H. Kivrak, K. Selcuk, A. Caglar and N. Aktas, *Selective electrochemical determination of L-cysteine by a cobalt carbon nanotube (CNT)-modified glassy carbon electrode (GCE)*, *Anal. Lett.* (2024), pp. 1–15.
[165] A.R. Jalalvand, *Fabrication of a novel amperometric sensing platform for determination of mangiferin*, *Sens. Bio-Sensing Res.* 29 (2020), pp. 100352.
[166] T. Sakthi Priya, T.-W. Chen, S.-M. Chen, T. Kokulnathan, B.-S. Lou, T. saad Algarni et al., *MIL-88A derived zerovalent iron embedded mesoporous carbon with carbon black composite based electrochemical sensor for the detection of metol*, *Carbon N. Y.* 223 (2024), pp. 119026.
[167] L. Suresh, J.S. Bondili and P.K. Brahman, *Fabrication of immunosensor based on polyaniline, fullerene-C 60 and palladium nanoparticles nanocomposite: An electrochemical detection tool for prostate cancer*, *Electroanalysis* 32 (2020), pp. 1439–1448.
[168] A. Laza, S.V. Pereira, G.A. Messina, M.A. Fernández-Baldo, J. Raba, M.D. Regiart et al., *Origami paper-based electrochemical immunosensor with carbon nanohorns-decorated nanoporous gold for zearalenone detection*, *Chemosensors* 12 (2024), pp. 10.

Index

G

H

Q

R

S

T

U

V

W

X

Y

Z

For Product Safety Concerns and Information please contact our EU
representative GPSR@taylorandfrancis.com
Taylor & Francis Verlag GmbH, Kaufingerstraße 24, 80331 München, Germany

www.ingramcontent.com/pod-product-compliance
Ingram Content Group UK Ltd.
Pitfield, Milton Keynes, MK11 3LW, UK
UKHW050057200926
475702UK00019B/2888